Mathematics for Electrical Engineering and Computing

Mathematics for Electrical Engineering and Computing

Mary Attenborough

Newnes

AMSTERDAM BOSTON LONDON HEIDELBERG NEW YORK
OXFORD PARIS SAN DIEGO SAN FRANCISCO
SINGAPORE SYDNEY TOKYO

Newnes
An imprint of Elsevier
Linacre House, Jordan Hill, Oxford OX2 8DP
200 Wheeler Road, Burlington MA 01803

First published 2003

British Library Cataloguing in Publication Data
A catalogue record for this book is available from the British Library

Library of Congress Cataloguing in Publication Data
A catalogue record for this book is available from the Library of Congress

ISBN 0 7506 5855 X

For information on all Newnes publications
visit our website at www.newnespress.com

Typeset by Newgen Imaging Systems (P) Ltd, Chennai, India
Printed and bound in Great Britain

Contents

Part 3   Functions of more than one variable

Part 4   Graph and language theory

Part 5   Probability and statistics

Preface

This book is based on my notes from lectures to students of electrical, electronic, and computer engineering at South Bank University. It presents a first year degree/diploma course in engineering mathematics with an emphasis on important concepts, such as algebraic structure, symmetries, linearity, and inverse problems, clearly presented in an accessible style. It encompasses the requirements, not only of students with a good maths grounding, but also of those who, with enthusiasm and motivation, can make up the necessary knowledge. Engineering applications are integrated at each opportunity. Situations where a computer should be used to perform calculations are indicated and 'hand' calculations are encouraged only in order to illustrate methods and important special cases. Algorithmic procedures are discussed with reference to their efficiency and convergence, with a presentation appropriate to someone new to computational methods.

Developments in the fields of engineering, particularly the extensive use of computers and microprocessors, have changed the necessary subject emphasis within mathematics. This has meant incorporating areas such as Boolean algebra, graph and language theory, and logic into the content. A particular area of interest is digital signal processing, with applications as diverse as medical, control and structural engineering, non-destructive testing, and geophysics. An important consideration when writing this book was to give more prominence to the treatment of discrete functions (sequences), solutions of difference equations and z transforms, and also to contextualize the mathematics within a systems approach to engineering problems.

Acknowledgements

I should like to thank my former colleagues in the School of Electrical, Electronic and Computer Engineering at South Bank University who supported and encouraged me with my attempts to re-think approaches to the teaching of engineering mathematics.

I should like to thank all the reviewers for their comments and the editorial and production staff at Elsevier Science.

Many friends have helped out along the way, by discussing ideas or reading chapters. Above all Gabrielle Sinnadurai who checked the original manuscript of *Engineering Mathematics Exposed*, wrote the major part of the solutions manual and came to the rescue again by reading some of the new material in this publication. My partner Michael has given unstinting support throughout and without him I would never have found the energy.

Part 1   Sets, functions, and calculus

1 Sets and functions

1.1 Introduction

Finding relationships between quantities is of central importance in engineering. For instance, we know that given a simple circuit with a $1000\,\Omega$ resistance then the relationship between current and voltage is given by Ohm's law, $I = V/1000$. For any value of the voltage V we can give an associated value of I. This relationship means that I is a function of V. From this simple idea there are many other questions that need clarifying, some of which are:

1. Are all values of V permitted? For instance, a very high value of the voltage could change the nature of the material in the resistor and the expression would no longer hold.
2. Supposing the voltage V is the equivalent voltage found from considering a larger network. Then V is itself a function of other voltage values in the network (see Figure 1.1). How can we combine the functions to get the relationship between this current we are interested in and the actual voltages in the network?
3. Supposing we know the voltage in the circuit and would like to know the associated current. Given the function that defines how current depends on the voltage can we find a function that defines how the voltage depends on the current? In the case where $I = V/1000$, it is clear that $V = 1000I$. This is called the inverse function.

Another reason exists for better understanding of the nature of functions. In Chapters 5 and 6, we shall study differentiation and integration. This looks at the way that functions change. A good understanding of functions and how to combine them will help considerably in those chapters.

The values that are permitted as inputs to a function are grouped together. A collection of objects is called a set. The idea of a set is very simple, but studying sets can help not only in understanding functions but also help to understand the properties of logic circuits, as discussed in Chapter 10.

Figure 1.1 *The voltage V is an equivalent voltage found by considering the combined effect of circuit elements in the rest of the network.*

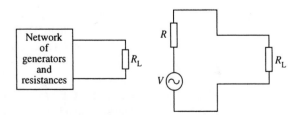

1.2 Sets

A *set* is a collection of objects, called *elements*, in which the order is not important and an object cannot appear twice in the same set.

Example 1.1 Explicit definitions of sets, that is, where each element is listed, are:

A = {a, b, c}

B = {3, 4, 6, 7, 8, 9}

C = {Linda, Raka, Sue, Joe, Nigel, Mary}

a $\in$ A means 'a is an element of A' or 'a belongs to A'; therefore in the above examples:

3 $\in$ B

Linda $\in$ C

The *universal set* is the set of all objects we are interested in and will depend on the problem under consideration. It is represented by $\mathcal{E}$.

The *empty set* (or *null set*) is the set with no elements. It is represented by $\emptyset$ or { }.

Sets can be represented diagrammatically – generally as circular shapes. The universal set is represented as a rectangle. These are called *Venn diagrams*.

Example 1.2

$\mathcal{E}$ = {a, b, c, d, e, f, g}, A = {a, b, c}, B = {d, e}

This can be shown as in Figure 1.2.

We shall mainly be concerned with sets of numbers as these are more often used as inputs to functions.

Some important sets of numbers are (where '...' means continue in the same manner):

The set of *natural numbers* $\mathbb{N}$ = {1, 2, 3, 4, 5, ...}
The set of *integers* $\mathbb{Z}$ = {... −3, −2, −1, 0, 1, 2, 3 ...}
The set of *rationals* (which includes fractional numbers) $\mathbb{Q}$
The set of *reals* (all the numbers necessary to represent points on a line) $\mathbb{R}$

Sets can also be defined using some rule, instead of explicitly.

Example 1.3 Define the set A explicitly where $\mathcal{E}$ = $\mathbb{N}$ and A = {$x \mid x < 3$}.

Solution The A = {$x \mid x < 3$} is read as 'A is the set of elements x, such that x is less than 3'. Therefore, as the universal set is the set of natural numbers, A = {1, 2}

Example 1.4 $\mathcal{E}$ = days of the week and A = {$x \mid x$ is after Thursday and before Sunday}. Then A = {Friday, Saturday}.

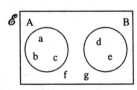

Figure 1.2 *A Venn diagram of the sets $\mathcal{E}$ =* {a, b, c, d, e, f, g}, A = {a, b, c}, *and* B = {d, e}.

Subsets

We may wish to refer to only a part of some set. This is said to be a subset of the original set.

$A \subseteq B$ is read as 'A is a subset of B' and it means that every element of A is an element of B.

Example 1.5

$\mathcal{E} = \mathbb{N}$

$A = \{1, 2, 3\}, \quad B = \{1, 2, 3, 4, 5\}$

Then $A \subseteq B$

Note the following points:

All sets must be subsets of the universal set, that is, $A \subseteq \mathcal{E}$ and $B \subseteq \mathcal{E}$
A set is a subset of itself, that is, $A \subseteq A$
If $A \subseteq B$ and $B \subseteq A$, then $A = B$

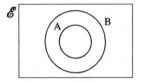

Figure 1.3 *A Venn diagram of a proper subset of* B*:* $A \subset B$.

Proper subsets

$A \subset B$ is read as 'A is a proper subset of B' and means that A is a subset of B but A is not equal to B. Hence, $A \subset B$ and simultaneously $B \subset A$ are impossible.

A proper subset can be shown on a Venn diagram as in Figure 1.3.

1.3 Operations on sets

In Chapter 1 of the background Mathematics notes available on the companion website for this book, we study the rules obeyed by numbers when using operations like negation, multiplication, and addition. Sets can be combined in various ways using set operations. Sets and their operations form a Boolean Algebra which we look at in greater detail in Chapter 4, particularly its application to digital design. The most important set operations are as given in this section.

Complement

$\bar{A}$ or A' represents the complement of the set A. The complement of A is the set of everything in the universal set which is not in A, this is pictured in Figure 1.4.

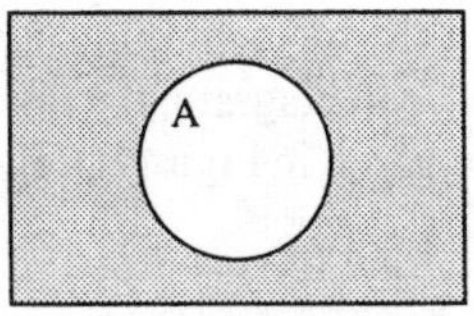

Example 1.6

$\mathcal{E} = \mathbb{N}$

$A = \{x \mid x > 5\}$

then $A' = \{1, 2, 3, 4, 5\}$

Figure 1.4 *The shaded area is the complement,* A'*, of the set* A.

Figure 1.5   A = {x|x < 5} and A' = {x|x ⩾ 5}.

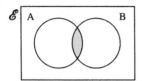

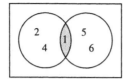

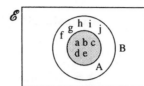

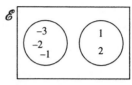

Figure 1.6   *The shaded area represents the intersection of* A *and* B.

Figure 1.7   *The intersection of two sets* {1, 2, 4} ∩ {1, 5, 6} = {1}.

Figure 1.8   *The intersection of two sets:* {a, b, c, d, e} ∩ {a, b, c, d, e, f, g, h, i, j} = {a, b, c, d, e}.

Figure 1.9   *The intersection of the two sets:* {−3, −2, −1} ∩ {1, 2} = Ø, *the empty set, as they have no elements in common.*

Example 1.7   The universal set is the set of real numbers represented by a real number line.

If A is the set of numbers less than 5, A = {x | x < 5} then A' is the set of numbers greater than or equal to 5. A' = {x | x ⩾ 5}. These sets are shown in Figure 1.5.

Intersection

$A \cap B$ represents the intersection of the sets A and B. The intersection contains those elements that are in A and also in B, this can be represented as in Figure 1.6 and examples are given in Figures 1.7–1.10.

Note the following important points:

Figure 1.10   *Disjoint sets* A *and* B.

If $A \subseteq B$ then $A \cap B = A$. This is the situation in the example given in Figure 1.8.

If A and B have no elements in common then $A \cap B = \emptyset$ and they are called *disjoint*. This is the situation given in the example in Figure 1.9. Two sets which are known to be disjoint can be shown on the Venn diagram as in Figure 1.10.

Union

$A \cup B$ represents the union of A and B, that is, the set containing elements which are in A or B or in both A and B. On a Venn diagram, the union can be shown as in Figure 1.11 and examples are given in Figures 1.12–1.15.

Note the following important points:

If $A \subseteq B$, then $A \cup B = B$. This is the situation in the example given in Figure 1.13.

The union of any set with its complement gives the universal set, that is, $A \cup A' = \mathcal{E}$, the universal set. This is pictured in Figure 1.15.

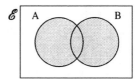

Figure 1.11 *The shaded area represents to union of sets* A *and* B.

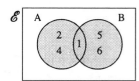

Figure 1.12 *The union of two sets:* $\{1, 2, 4\} \cup \{1, 5, 6\} = \{1, 2, 4, 5, 6\}$.

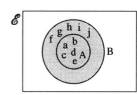

Figure 1.13 *The union of two sets:* $\{a, b, c, d, e\} \cup \{a, b, c, d, e, f, g, h, i, j\} = \{a, b, c, d, e, f, g, h, i, j\}$.

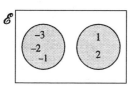

Figure 1.14 *The union of the two sets:* $\{-3, -2, -1\} \cup \{1, 2\} = \{-3, -2, -1, 1, 2\}$.

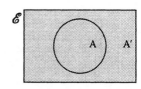

Figure 1.15 *The shaded area represents the union of a set with its complement giving the universal set.*

Cardinality of a finite set

The number of elements in a set is called the cardinality of the set and is written as $n(A)$ or $|A|$.

Example 1.8

$$n(\emptyset) = 0, \quad n(\{2\}) = 1, \quad n(\{a, b\}) = 2$$

For finite sets, the cardinality must be a natural number.

Example 1.9 In a survey, 100 people were students and 720 owned a video recorder; 794 people owned a video recorder or were students. How many students owned a video recorder?

$\mathcal{E} = \{x \mid x$ is a person included in the survey$\}$

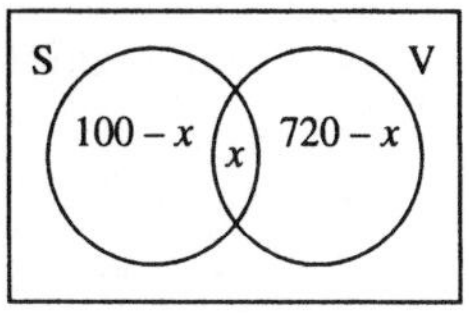

Setting $S = \{x \mid x$ is a student$\}$ and $V = \{x \mid x$ owns a video recorder$\}$, we can solve this problem using a Venn diagram as in Figure 1.16.

x is the number of students who own a video recorder. From the diagram we get

$$100 - x + x + 720 - x = 794$$
$$\Leftrightarrow \quad 820 - x = 794$$
$$\Leftrightarrow \quad x = 26$$

Therefore, 26 students own a video recorder.

Figure 1.16 S *is the set of students in a survey and* V *is the set of people who own a video. The numbers in the sets give the cardinality of the sets,* $n(S) = 100, n(S \cup V) = 794, n(V) = 720, n(S \cap V) = x$.

1.4 Relations and functions

Relations

A relation is a way of pairing up members of two sets. This is just like the idea of family relations. For instance, a child can be paired with its mother, brothers can be paired with sisters, etc. A relation is such that it may not always be possible to find a suitable partner for each element in the first set whereas sometimes there will be more than one. For instance, if we try to pair every boy with his sister there will be some boys who have no sisters and some boys who have several. This is pictured in Figure 1.17.

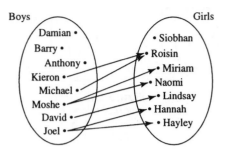

Figure 1.17 *The relation boy → sister. Some boys have more than one sister and some have none at all.*

Functions

Functions are relations where the pairing is always possible. Functions are like mathematical machines. For each input value there is always exactly one output value.

Calculators output function values. For instance, input 2 into a calculator, press $1/x$ and the calculator will display the number 0.5. The output value is called the *image* of the input value. The set of input values is called the domain and the set containing all the images is called the codomain.

The function $y = 1/x$ is displayed in Figure 1.18 using arrows to link input values with output values.

Functions can be represented by letters. If the function of the above example is given the letter f to represent it then we can write

$$f : x \mapsto \frac{1}{x}$$

This can be read as 'f is the function which when input a value for x gives the output value $1/x$'. Another way of giving the same information is:

$$f(x) = \frac{1}{x}$$

$f(x)$ represents the image of x under the function f and is read as 'f of x'. It does not mean the same as f times x.

$f(x) = 1/x$ means 'the image of x under the function f is given by $1/x$' but is usually read as 'f of x equals $1/x$'.

Even more simply, we usually use the letter y to represent the output value, the image, and x to represent the input value. The function is therefore summed up by $y = 1/x$.

x is a variable because it can take any value from the set of values in the domain. y is also a variable but its value is fixed once x is known. So x is called the *independent variable* and y is called the *dependent variable*.

The letters used to define a function are not important. $y = 1/x$ is the same as $z = 1/t$ is the same as $p = 1/q$ provided that the same input values (for x, t, or q) are allowed in each case.

More examples of functions are given in arrow diagrams in Figures 1.19(a) and 1.20(a). Functions are more usually drawn using a graph, rather than by using an arrow diagram. To get the graph the codomain is moved to be at right angles to the domain and input and output values are marked by a point at the position (x, y). Graphs are given in Figures 1.19(b) and 1.20(b).

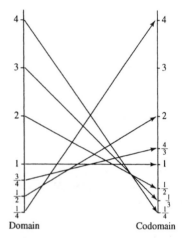

Figure 1.18 *An arrow diagram of the function $y = 1/x$.*

Continuous functions and discrete functions applied to signals

Functions of particular interest to engineers are either functions of a real number or functions of an integer. The function given in Figure 1.19 is an example of a real function and the function given in Figure 1.20 is an example of a function of an integer, also called a discrete function.

Often, we are concerned with functions of time. A variable voltage source can be described by giving the voltage as it depends on time, as also can the current. Other examples are: the position of a moving robot arm, the extension or compression of car shock absorbers and the heat emission of a thermostatically controlled heating system. A voltage or current varying with time can be used to control instrumentation or to convey information. For this reason it is called a signal. Telecommunication signals may be radio waves or voltages along a transmission line or light signals along an optical fibre.

Time, t, can be represented by a real number, usually non-negative. Time is usually taken to be positive because it is measured from some reference instant, for example, when a circuit switch is closed. If time is used to describe relative events then it can make sense to refer to negative time. If lightning is seen 1 s before a thunderclap is heard then this can be described by saying the lightning happened at -1 s or alternatively that the thunderclap was heard at 1 s. In the two cases, the time origin has been chosen differently. If time is taken to be continuous and represented by a real variable then functions of time will be continuous or piecewise continuous. Examples of graphs of such functions are given in Figure 1.21.

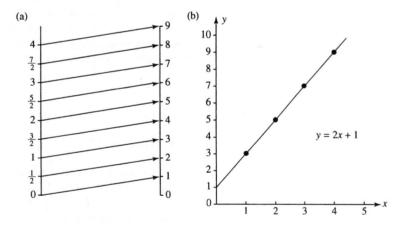

Figure 1.19   *The function $y = 2x + 1$ where x can take any real value (any number on the number line). (a) is the arrow diagram and (b) is the graph.*

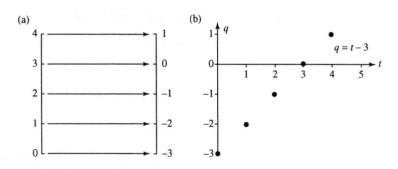

Figure 1.20   *The function $q = t - 3$ where t can take any integer value (a) is the arrow diagram and (b) is the graph.*

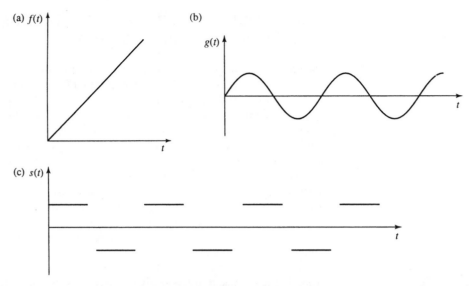

Figure 1.21 *Continuous and piecewise functions where time is represented by a real number > 0. (a) A ramp function; (b) a wave (c) a square wave. (a) and (b) are continuous, while (c) is piecewise continuous.*

A continuous function is one whose graph can be drawn without taking your pen off the paper. A piecewise continuous function has continuous bits with a limited number of jumps. In Figure 1.21, (a) and (b) are continuous functions and (c) is a piecewise continuous function. If we have a digital signal, then its values are only known at discrete moments of time. Digital signals can be obtained by using an analog to digital (A/D) convertor on an originally continuous signal. Digital signals are represented by discrete functions as in Figure 1.22(a)–(c)

A digital signal has a sampling interval, T, which is the length of time between successive values. A digital functions is represented by a discrete function. For example, in Figure 1.22(a) the digital ramp can be represented by the numbers

$$0, 1, 2, 3, 4, 5, \ldots$$

If the sample interval T is different from 1 then the values would be

$$0, T, 2T, 3T, 4T, 5T, \ldots$$

This is a discrete function also called a sequence. It can be represented by the expression $f(t) = t$, where $t = 0, 1, 2, 3, 4, 5, 6, \ldots$ or using the sampling interval, T, $g(n) = nT$, where $n = 0, 1, 2, 3, 4, 5, 6, \ldots$

Yet another common way of representing a sequence is by using a subscript on the letter representing the image, giving

$$f_n = n, \text{ where } n = 0, 1, 2, 3, 4, 5, \ldots$$

or, using the letter a for the image values,

$$a_n = n, \text{ where } n = 0, 1, 2, 3, 4, 5, \ldots$$

Substituting some values for n into the above gives

$$a_0 = 0, \quad a_1 = 1, \quad a_2 = 2, \quad a_3 = 3, \ldots$$

As a sequence is a function of the natural numbers and zero (or if negative input values are allowed, the integers) there is no need to specify

(a)

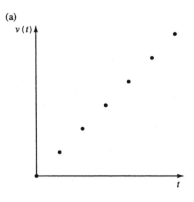

(b)

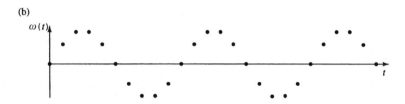

(c)

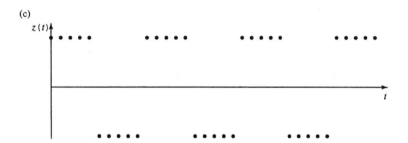

Figure 1.22 *Examples of discrete functions. (a) A digital ramp; (b) a digital wave; (c) a digital square wave.*

the input values and it is possible merely to list the output values in order. Hence the ramp function can be expressed by $0, 1, 2, 3, 4, 5, 6, \ldots$

Time sequences are often referred to as 'series'. This terminology is not usual in mathematics books, however, as the description 'series' is reserved for describing the sum of a sequence. Sequences and series are dealt with in more detail in Chapter 18.

Example 1.10 Plot the following analog signals over the values of t given (t real):

(a) $x = t^3$ $t \geqslant 0$

(b)
$$y = \begin{cases} 0 & t \leqslant 3 \\ t - 3 & 3 < t \leqslant 5 \\ 2 & t > 5 \end{cases}$$

(c) $z = \dfrac{1}{t^2}$ $t > 0$

Solution In each case, choose some values of t and calculate the function values at those points. Plot the points and join them.

(a)

t	0	0.5	1	1.5	2	2.5	3	3.5
$x = t^3$	0	0.125	1	3.375	8	15.625	27	42.875

These values are plotted in Figure 1.23(a).

(b)

t	1	1.5	2	2.5	3	3.5	4	4.5	5	5.5	6	6.5	7
y	0	0	0	0	0	0.5	1	1.5	2	2	2	2	2

$$\underbrace{\qquad\qquad}_{y = 0} \quad \underbrace{\qquad\qquad}_{y = t - 3} \quad \underbrace{\qquad\qquad}_{y = 2}$$

These values are plotted in Figure 1.23(b).

(c)

t	0.0001	0.001	0.01	0.1	1	10	100	1000	10000
z	10^8	10^6	10^4	100	1	0.01	10^{-4}	10^{-6}	10^{-8}

These values are plotted in Figure 1.23(c).

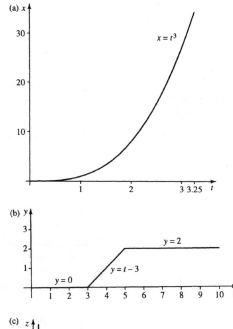

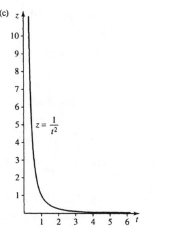

Figure 1.23 *The analog signals described in Example 1.10.*
(a) $x = t^3$ $t \geqslant 0$

$$(b)\ y = \begin{cases} 0 & t \leqslant 3 \\ t - 3 & 3 < t \leqslant 5 \\ 2 & t > 5 \end{cases}$$

(c) $z = 1/t^2$ $t > 0$

Example 1.11 Plot the following discrete signals over the values of t given (t an integer):

(a) $x = \dfrac{1}{t-1}$ $\quad t > 2$

(b)
$$y = \begin{cases} 0 & t \leqslant 4 \\ 1/t - 0.25 & 4 < t < 10 \\ -0.15 & t \geqslant 10 \end{cases}$$

(c) $z = 4t - 2$ $\quad t > 0$

Solution In each case, choose successive values of t and calculate the function values at those points. Mark the points with a dot.

(a)

t	2	3	4	5	6	7	8	9	10
x	1	0.5	0.33	0.25	0.2	0.17	0.14	0.13	0.11

These values are plotted in Figure 1.24(a).

(b)

t	3	4	5	6	7	8	9	10	11	12
y	0	0	−0.05	−0.08	−0.11	−0.12	−0.14	−0.15	−0.15	−0.15

$$y = 0 \qquad\qquad y = \frac{1}{t} - 0.25 \qquad\qquad y = -0.15$$

These values are plotted in Figure 1.24(b).

(c)

t	1	2	3	4	5	6	7	8
z	2	6	10	14	18	22	26	30

These values are plotted in Figure 1.24(c).

Undefined function values

Some functions have 'undefined values', that is, numbers that cannot be input into them successfully. For instance input 0 on a calculator and try getting the value of $1/x$. The calculator complains (usually displaying '-E-') indicating that an error has occurred. The reason that this is an error is that we are trying to find the value of 1/0 that is 1 divided by 0. Look at Chapter 1 of the Background Mathematics Notes, given on the accompanying website for this book, for a discussion about why division by 0 is not defined. The number 0 cannot be included in the domain of the function $f(x) = 1/x$. This can be expressed by saying

$$f(x) = 1/x, \quad \text{where } x \in \mathbb{R} \text{ and } x \neq 0$$

which is read as 'f of x equals $1/x$, where x is a real number not equal to 0'.

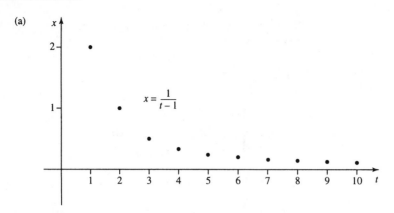

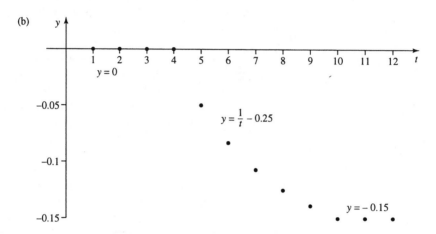

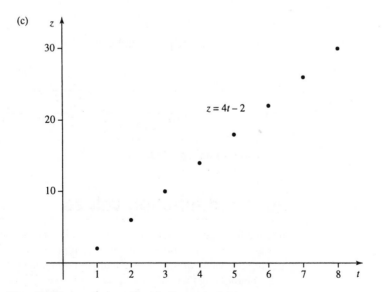

Figure 1.24 *The digital signals described in Example 1.11.*

(a) $x = \dfrac{1}{t-1}$ $t > 2$ *(b)* $y = \begin{cases} 0 & t \leqslant 4 \\ 1/t - 0.25 & 4 < t < 10 \\ -0.15 & t \geqslant 10 \end{cases}$ *(c)* $z = 4t - 2$ $t > 0$

Often, we assume that we are considering functions of a real variable and only need to indicate the values that are not allowed as inputs for the function. So we may write

$$f(x) = 1/x \quad \text{where } x \neq 0$$

Things to look out for as values that are not allowed as function inputs are :

1. Numbers that would lead to an attempt to divide by zero
2. Numbers that would lead to negative square roots
3. Numbers that would lead to negative inputs to a logarithm.

Examples 1.12(a) and (b) require solutions to inequalities which we shall discuss in greater detail in Chapter 2. Here, we shall only look at simple examples and use the same rules as used for solving equations. We can find equivalent inequalities by doing the same thing to both sides, with the extra rule that, for the moment, we avoid multiplication or division by a negative number.

Example 1.12   Find the values that cannot be input to the following functions, where the independent variable (x or r) is real:

(a) $y = 3\sqrt{x - 2} + 5$

(b) $y = 3\log_{10}(2 - 4x)$

(c) $R = \dfrac{r + 1000}{1000(r - 2)}$

Solution

(a) $y = 3\sqrt{x - 2} + 5$

Here $x - 2$ cannot be negative as we need to take the square root of it.

$$x - 2 \geqslant 0 \Leftrightarrow x \geqslant 2$$

therefore, the function is

$$y = 3\sqrt{x - 2} + 5 \quad \text{where } x \geqslant 2$$

(b) $y = 3\log_{10}(2 - 4x)$.

Here $2 - 4x$ cannot be 0 or negative else we could not take the logarithm.

$$2 - 4x > 0 \quad \Leftrightarrow \quad 2 > 4x \quad \Leftrightarrow \quad 2/4 > x$$

or equivalently, $x < \frac{1}{2}$. So the function is

$$y = 3\log_{10}(2 - 4x) \quad \text{where } x < 0.5$$

(c) $R = \dfrac{r + 1000}{1000(r - 2)}$

Here $1000(r - 2)$ cannot be 0, else we would be trying to divide by 0. Solve the equation for the values that r cannot take

$$1000(r - 2) = 0$$
$$r - 2 = 0$$
$$r = 2$$

The function is

$$R = \frac{r + 1000}{1000(r - 2)} \quad \text{where } r \neq 2$$

Example 1.13   Find the values that can be input to the following discrete functions where the independent variable is an integer

(a) $y = \dfrac{1}{k - 4}$ where k $\in \mathbb{Z}$

(b) $f(k) = \dfrac{1}{(k - 3)(k - 2.2)}$ where k $\in \mathbb{Z}$

(c) $a_n = n^2$ where n $\in \mathbb{Z}$

Solution

(a) $y = \dfrac{1}{k - 4}$

Here $k - 4$ cannot be 0 else there would be an attempt to divide by 0. We get $k - 4 = 0$ when $k = 4$ so the function is:

$$y = \frac{1}{k - 4} \quad \text{where } k \neq 4 \text{ and } k \in \mathbb{Z}$$

(b) $f(k) = \dfrac{1}{(k - 3)(k - 2.2)}$ where $k \in \mathbb{Z}$

Solve for $(k - 3)(k - 2.2) = 0$ giving $k = 3$ or $k = 2.2$. As 2.2 is not an integer then there is not need to specifically exclude it from the function input values, so the function is

$$f(k) = \frac{1}{(k - 3)(k - 2.2)} \quad \text{where } k \neq 3 \text{ and } k \in \mathbb{Z}$$

(c) $a_n = n^2, \quad n \in \mathbb{Z}$

Here there are no problems with the function as any integer can be squared. There are no excluded values from the input of the function.

Using a recurrence relation to define a discrete function

Values in a discrete function can also be described in terms of its values for preceeding integers.

Example 1.14 Find a table of values for the function defined by the recurrence relation:

$$f(n) = f(n-1) + 2 \qquad\qquad (1.1)$$

where $f(0) = 0$.

Solution Assuming that the function is defined for $n = 0, 1, 2, \ldots$ then we can take successive values of n and find the values taken by the function. $n = 0$ gives $f(0) = 0$ as given.

Substituting $n = 1$ into Equation (1.1) gives

$$f(1) = f(1-1) + 2$$
$$\Leftrightarrow \quad f(1) = f(0) + 2 = 0 + 2 = 2 \text{ (using } f(0) = 0)$$

hence, $f(1) = 2$.

Substituting $n = 2$ into Equation (1.1) gives

$$f(2) = f(2-1) + 2$$
$$\Leftrightarrow f(2) = f(1) + 2$$
$$\Leftrightarrow f(2) = f(1) + 2 = 2 + 2 = 4 \text{ (using } f(1) = 2)$$

hence, $f(2) = 4$.

Substituting $n = 3$ into Equation (1.1) gives

$$f(3) = f(3-1) + 2$$
$$\Leftrightarrow f(3) = f(2) + 2 = 4 + 2 \text{ (using } f(2) = 4)$$

hence, $f(3) = 6$.

Continuing in the same manner gives the following table:

n	0	1	2	3	4	5	6	7	8	9	10	$\cdots$	n	$\cdots$
f	0	2	4	6	8	10	12	14	16	18	20	$\ldots$	$2n$	$\cdots$

Notice we have filled in the general term $f(n) = 2n$. This was found in this case by simple guess work.

1.5 Combining functions

The sum, difference, product, and quotient of two functions, *f* and *g*

Two functions with $\mathbb{R}$ as their domain and codomain can be combined using arithmetic operations. We can define the sum of f and g by

$$(f + g) : x \mapsto f(x) + g(x)$$

The other operations are defined as follows:

$$(f - g) : x \mapsto f(x) - g(x) \quad \text{difference,}$$
$$(f \times g) : x \mapsto f(x) \times g(x) \quad \text{product,}$$
$$(f/g) : x \mapsto \frac{f(x)}{g(x)} \quad \text{quotient.}$$

Example 1.15   Find the sum, difference, product, and quotient of the functions:

$$f : x \mapsto x^2 \text{ and } g : x \mapsto x^6$$

Solution

$$(f + g) : x \mapsto x^2 + x^6$$
$$(f - g) : x \mapsto x^2 - x^6$$
$$(f \times g) : x \mapsto x^2 \times x^6 = x^8$$
$$(f/g) : x \mapsto \frac{x^2}{x^6} = x^{-4}$$

The specification of the domain of the quotient is not straightforward. This is because of the difficulty which occurs when $g(x) = 0$. When $g(x) = 0$ the quotient function is undefined and we must remove such elements from its domain. The domain of f/g is $\mathbb{R}$ with the values where $g(x) = 0$ omitted.

Composition of functions

This method of combining functions is fundamentally different from the arithmetical combinations of the previous section. The composition of two functions is the action of performing one function followed by the other, that is, a function of a function.

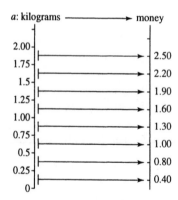

Figure 1.25   *The function a : kilograms → money used in Example 1.16.*

Example 1.16   A post office worker has a scale expressed in kilograms which gives the cost of a parcel depending on its weight. He also has an approximate formula for conversion from pounds (lbs) to kilograms. He wishes to find out the cost of a parcel which weighs 3 lb.
     The two functions involved are:

$$a : \text{kilograms} \to \text{money and } c : \text{lbs} \to \text{kilograms}$$

a is defined by Figure 1.25 and the function c is given by

$$c : x \mapsto x/2.2$$

Solution   The composition '$a \circ c$' will be a function from lbs to money.
     Hence, 3 lb after the function c gives 1.364 and 1.364 after the function a gives €1.90 and therefore

$$(a \circ c)(3) = €1.90.$$

Example 1.17   Supposing $f(x) = 2x + 1$ and $g(x) = x^2$, then we can combine the functions in two ways.

1. A composite function can be formed by performing f first and then g, that is, $g \circ f$. To describe this function, we want to find what happens

to x under the function $g \circ f$. Another way of saying that is we need to find $g(f(x))$. To do this call $f(x)$ a new letter, say y.

$$y = f(x) = 2x + 1$$

Rewrite g as a function of y

$$g(y) = y^2$$

Now substitute $y = 2x + 1$ giving

$$g(2x + 1) = (2x + 1)^2$$

Hence,

$$g(f(x)) = (2x + 1)^2$$
$$(g \circ f)(x) = (2x + 1)^2.$$

2. A composite function can be formed by performing g first and then f, that is, $f \circ g$. To describe this function, we want to find what happens to x under the function $f \circ g$. Another way of saying that is we need to find $f(g(x))$. To do this call $g(x)$ a new letter, say y.

$$y = g(x) = x^2$$

Rewrite f as a function of y

$$f(y) = 2y + 1$$

Now substitute $y = x^2$ giving

$$f(x^2) = 2x^2 + 1$$

Hence,

$$f(g(x)) = 2x^2 + 1$$
$$(f \circ g)(x) = 2x^2 + 1.$$

Example 1.18   Supposing $u(t) = 1/(t - 2)$ and $v(t) = 3 - t$ then, again, we can combine the functions in two ways.

1. A composite function can be formed by performing u first and then v, that is, $v \circ u$. To describe this function, we want to find what happens to t under the function $v \circ u$. Another way of saying that is we need to find $v(u(t))$. To do this call $u(t)$ a new letter, say y.

$$y = u(t) = \frac{1}{t - 2}$$

Rewrite v as a function of y

$$v(y) = 3 - y$$

Now substitute $y = 1/(t - 2)$ giving

$$v\left(\frac{1}{t-2}\right) = 3 - \frac{1}{t-2}$$
$$= \frac{3(t-2) - 1}{t-2}$$

(rewriting the expression over a common denominator)

$$= \frac{3t - 6 - 1}{t-2} = \frac{3t - 7}{t-2}$$

Hence,

$$v(u(t)) = \frac{3t - 7}{t-2}$$
$$(v \circ u)(t) = \frac{3t - 7}{t-2}$$

2. A composite function can be formed by performing v first and then u, that is $u \circ v$. To describe this function, we want to find what happens to t under the function $u \circ v$. Another way of saying that is we need to find $u(v(t))$. To find this call $v(t)$ a new letter, say y.

$$y = v(t) = 3 - t$$

Rewrite u as a function of y

$$u(y) = \frac{1}{y - 2}$$

Now substitute $y = 3 - t$ giving

$$v(3 - t) = \frac{1}{(3 - t) - 2} = \frac{1}{1 - t}$$

Hence,

$$u(v(t)) = \frac{1}{1 - t}$$
$$(u \circ v)(t) = \frac{1}{1 - t}$$

Decomposing functions

In order to calculate the value of a function, either by hand or using a calculator, we need to understand how it decomposes. That is we need to understand to order of the operations in the function expression

Example 1.19   Calculate $y = (2x + 1)^3$ when $x = 2$

Solution   Remember the order of operations discussed in Chapter 1 of the Background Mathematics booklet available on the companion website. The operations are performed in the following order:

Start with $x = 2$ then

$2x = 4$

$2x + 1 = 5$

$(2x + 1)^3 = 125$

So, there are three operations involved

1.   multiply by 2,
2.   add on 1,
3.   take the cube.

This way of breaking down functions can be pictured using boxes to represent each operation that makes up the function, as was used to represent equations in Chapter 3 of the Background Mathematics booklet available on the companion website. The whole function can be thought of as a machine, represented by a box. For each value x, from the domain of the function that enters the machine, there is a resulting image, y, which comes out of it. This is pictured in Figure 1.26.

Inside of the box, we can write the name of the functions or the expression which gives the function rule. A composite function box can be broken into different stages, each represented by its own box. The function $y = (2x + 1)^3$ breaks down as in Figure 1.27.

$y = (3x - 4)^4$ can be broken down as in Figure 1.28.

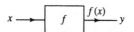

Figure 1.26   *A function pictured as a machine represented by a box.*
x represents the input value, any value of the domain, y represents the output, the image of x under the function.

Figure 1.27   *The function* $y = (2x + 1)^3$ *decomposed into its composite operations.*

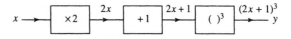

Figure 1.28   *The function* $y = (3x - 4)^4$ *decomposed into its composite operations.*

The inverse of a function

The inverse of a function is a function which will take the image under the function back to its original value. If $f^{-1}(x)$ is the inverse of $f(x)$ then

$f^{-1}(f(x)) = x$

$(f^{-1} \circ f) : x \mapsto x$

Example 1.20

$$f(x) = 2x + 1$$

$$f^{-1}(x) = \frac{x-1}{2}$$

To show this is true, look at the combined function $f^{-1}(f(x)) = (2x + 1 - 1)/2 = x$.

Finding the inverse of a linear function

One simple way of finding the inverse of a linear function is to:

1. Decompose the operations of the function.
2. Combine the inverse operations (performed in the reverse order) to give the inverse function.

This is a method similar to that used to solve linear equations in Chapter 3 of the Background Mathematics Notes available on the companion website for this book.

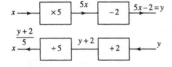

Figure 1.29 *The top line represents the function $f(x) = 5x - 2$ (read from left to right) and the bottom line the inverse function.*

Example 1.21 Find the inverse of the function $f(x) = 5x - 2$.

The method of solution is given in Figure 1.29.

The inverse operations give that $x = (y + 2)/5$. Here y is the input value into the inverse function and x is the output value. To use x and y in the more usual way, where x is the input and y the output, swap the letters giving the inverse function as

$$y = \frac{x+2}{5}$$

This result can be achieved more quickly by rearranging the expression so that x is the subject of the formula and then swap x and y.

Example 1.22 Find the inverse of $f(x) = 5x - 2$.

$$y = 5x - 2 \quad \Leftrightarrow \quad y + 2 = 5x$$
$$\Leftrightarrow \quad \frac{y+2}{5} = x$$
$$\Leftrightarrow \quad x = \frac{y+2}{5}$$

Now swap x and y to give $y = (x + 2)/5$. Therefore,

$$f^{-1}(x) = (x + 2)/5.$$

Example 1.23 Find the inverse of

$$g(x) = \frac{1}{2-x} \quad \text{where } x \neq 2$$

Set

$$y = \frac{1}{2-x} \Leftrightarrow y(2-x) = 1$$

$$\Leftrightarrow 2y - xy = 1$$

$$\Leftrightarrow 2y = 1 + xy$$

$$\Leftrightarrow 2y - 1 = xy$$

$$\Leftrightarrow xy = 2y - 1$$

$$\Leftrightarrow x = \frac{2y-1}{y} \quad \text{where } y \neq 0$$

$$\Leftrightarrow x = 2 - \frac{1}{y}$$

Swap x and y to give $y = 2 - (1/x)$
So

$$g^{-1}(x) = 2 - \frac{1}{x} \quad x \neq 0$$

To check, try a couple of values of x.
Try $x = 4$,

$$g(x) = \frac{1}{2-x} = \frac{1}{2-4} = -\frac{1}{2}$$

Perform g^{-1} on the output value $-(1/2)$.
Substitute $g(4) = -(1/2)$ into $g^{-1}(x)$:

$$g^{-1}\left(-\frac{1}{2}\right) = 2 - \frac{1}{-(1/2)} = 2 + 2 = 4.$$

The function followed by its inverse has given us the original value of x.

The range of a function

When combining functions, for example, $f(g(x))$, we have to ensure that $g(x)$ will only output values that are allowed to be input to f. The set of images of $g(x)$ becomes an important consideration. The set of images of a function is called its range. The range of a function is a subset of its codomain.

1.6 Summary

1. Functions are used to express relationships between physical quantities.
2. The allowed inputs to a function are grouped into a set, called the domain of the function. The set including all the outputs is called the codomain.
3. A set is a collection of objects called elements.

4. $\mathcal{E}$ is the universal set, the set of all objects we are interested in.
5. $\emptyset$ is the empty set, the set with no elements.
6. The three most important operations on sets are:
    (a) intersection: $A \cap B$ is the set containing every element in both A and B;
    (b) union: $A \cup B$ is the set of elements in A or in B or both;
    (c) complement: A' is the set of everything, in the universal set, not in A.
7. A relation is a way of pairing members of two sets.
8. Functions are a special type of relation which can be thought of as mathematical machines. For each input value there is exactly one output value.
9. Many functions of interest are functions of time, used to represent signals. Analogue signals can be represented by functions of a real variable and digital signals by functions of an integer (discrete functions). Functions of an integer are also called sequences and can be defined using a recurrence relation.
10. To find the domain of a real or discrete function exclude values that could lead to a division by zero, negative square roots, or negative logarithms or other undefined values.
11. Functions can be combined in various ways including sum, difference, product, and quotient. A special operation of functions is composition. A composite function is found by performing a second function on the result of the first.
12. The inverse of a function is a function which will take the image under the function back to its original value.

1.7 Exercises

1.1. Given $\mathcal{E} = \{a, b, c, d, e, f, g\}$, $A = \{a, b, e\}$, $B = \{b, c, d, f\}, C = \{c, d, e\}$.

Write down the following sets:
(a) $A \cap B$
(b) $A \cup B$
(c) $A \cap C'$
(d) $(A \cup B) \cap C$
(e) $(A \cap C) \cup (B \cap C)$
(f) $(A \cap B) \cup C$
(g) $(A \cup C) \cap (B \cup C)$
(h) $(A \cap C)'$
(i) $A' \cup C'$.

1.2. Use Venn diagrams to show that:
(a) $(A \cap B) \cap C = A \cap (B \cap C)$
(b) $(A \cup B) \cup C = A \cup (B \cup C)$
(c) $(A \cap B) \cup C = (A \cup C) \cap (B \cup C)$
(d) $(A \cup B) \cap C = (A \cap C) \cup (B \cap C)$
(e) $(A \cap B)' = A' \cup B'$
(f) $(A \cup B)' = A' \cap B'$.

1.3. Let $\mathcal{E} = \{0, 1, 2, 3, 4, 5, 6, 7, 8, 9\}$ and given $P = \{x | x < 5\}, Q = \{x | x \geqslant 3\}$ find explicitly:
(a) P
(b) Q
(c) $P \cup Q$
(d) P'
(e) $P' \cap Q$.

1.4. Below are various assertions for any sets A and B. Write true or false for each statement and give a counter-example if you think the statement is false.
(a) $(A \cap B)' = A' \cap B'$
(b) $(A \cap B)' \subseteq A$
(c) $A \cap B = B \cap A$
(d) $A \cap B' = B \cap A'$.

1.5. Using a Venn diagram simplify the following:
(a) $A \cap (A \cup B)$
(b) $A \cup (B \cap A')$
(c) $A \cap (B \cup A')$.

1.6. A computer screen has 80 columns and 25 rows:
(a) Define the set of positions on the screen.
(b) Taking the origin as the top left hand corner define:
    (i) the set of positions in the lower half of the screen as shown in Figure 1.30(a);
    (ii) the set of positions lying on or below the diagonal as shown in Figure 1.30(b).

1.7. A certain computer system breaks down in two main ways: faults on the network and power supply faults. Of the last 50 breakdowns, 42 involved network faults and 20 power failures. In 13 cases, both the power supply and the network were faulty. How many breakdowns were attributable to other kinds of failure?

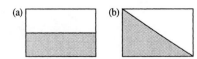

Figure 1.30 *(a) Points lying in shaded area represent the set of positions on the lower half of the computer screen as in Exercise 1.6(a). (b) Points on the diagonal line and lying in the shaded area represent the set of positions for Exercise 1.6(b).*

1.8. Draw arrow diagrams and graphs of the following functions:

(a) $f(t) = (t-1)^2 \quad t \in \{0, 1, 2, 3, 4\}$

(b) $g(z) = 1/z \quad z \in \{-1, -0.5, 0.5, 1, 1.5, 2\}$

(c) $y = \begin{cases} x & x \in \{-2, -1\} \\ 2x & x \in \{0, 1, 2, 3\} \end{cases}$

(d) $h : t \mapsto 3 - t \quad t \in \{5, 6, 7, 8, 9, 10\}$

1.9. Given that $f : x \mapsto 2x - 1$, $\quad g : x \mapsto (1/3)x^2$, $h : x \mapsto 3/x$

(a) Find the following:

(i) $f(2)$ (ii) $g(3)$ (iii) $h(5)$

(iv) $h(2) + g(2)$ (v) $h/g(5)$ (vi) $(h \times g)(2)$

(vii) $h(g(2))$ (viii) $h(h(3))$

(b) Find the following functions:

(i) $f \circ g$ (ii) $g \circ f$ (iii) $h \circ g$ (iv) f^{-1}

(v) h^{-1}

(c) Confirm the following:

(i) $(f^{-1} \circ f) : x \mapsto x$ (ii) $(h^{-1} \circ h) : x \mapsto x$

(iii) $(f \circ f^{-1}) : x \mapsto x$

(d) Using the results from sections (b) and (c), find the following:

(i) $(h^{-1} \circ h)(1)$ (ii) $h(g(5))$ (iii) $g(f(4))$

1.10. An analog signal is sampled using an A/D convertor and represented using only integer values. The original signal is represented by $g(t)$ and the digital signal by $h(t)$ sampled at $t \in \{2, 3, 4, 5, 6, 7, 8, 9, 10\}$. The definitions of g and h are as below

$$g(t) = \begin{cases} t - 2.25 & t < 5 \\ 6.8 - t & t \geqslant 5 \end{cases}$$

$h : 2 \mapsto 0$ $h : 3 \mapsto 1$ $h : 4 \mapsto 2$

$h : 5 \mapsto 2$ $h : 6 \mapsto 1$ $h : 7 \mapsto 0$

$h : 8 \mapsto -1$ $h : 9 \mapsto -2$ $h : 10 \mapsto -3$

If $e(t)$ is the error function (called quantization error), defined at the sample points, find $e(t)$ and represent it on a graph.

2 Functions and their graphs

2.1 Introduction

The ability to produce a picture of a problem is an important step towards solving it. From the graph of a function, $y = f(x)$, we are able to predict such things as the number of solutions to the equation $f(x) = 0$, regions over which it is increasing or decreasing, and the points where it is not defined.

Recognizing the shape of functions is an important and useful skill. Oscilloscopes give a graphical representation of voltage against time, from which we may be able to predict an expression for the voltage. The increasing use of signal processing means that many problems involve analysing how functions of time are effected by passing through some mechanical or electrical system.

In order to draw graphs of a large number of functions, we need only remember a few key graphs and appreciate simple ideas about transformations. A sketch of a graph is one which is not necessarily drawn strictly to scale but shows its important features. We shall start by looking at special properties of the straight line (linear function) and the quadratic. Then we look at the graphs of $y = x$, $y = x^2$, $y = 1/x$, $y = a^x$ and how to transform these graphs to get graphs of functions like $y = 4x - 2$, $y = (x - 2)^2$, $y = 3/x$, and $y = a^{-x}$.

2.2 The straight line: $y = mx + c$

$y = mx + c$ is called a linear function because its graph is a straight line. Notice that there are only two terms in the function; the x term, mx, where m is called the coefficient of x and c which is the constant term. m and c have special significance. m is the gradient, or the slope, of the line and c is the value of y when $x = 0$, that is, when the graph crosses the y-axis. This graph is shown in Figure 2.1(a) and two particular examples shown in Figure 2.1(b) and (c).

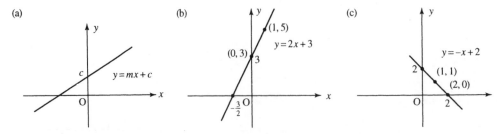

Figure 2.1   *(a) The graph of the function $y = mx + c$. m is the slope of the line, if m is positive then travelling from left to right along the line of the function is an uphill climb, if m is negative then the journey is downhill. The constant c is where the graph crosses the y-axis. (b) $m = 2$ and $c = 3$ (c) $m = -1$ and $c = 2$.*

The gradient of a straight line

The gradient gives an idea of how steep the climb is as we travel along the line of the graph. If the gradient is positive then we are travelling uphill as we move from left to right and if the gradient is negative then we are travelling downhill. If the gradient is zero then we are on flat ground. The gradient gives the amount that y increases when x increases by 1 unit. A straight line always has the same slope at whatever point it is measured. To show that in the expression $y = mx + c$, m is the gradient, we begin with a couple of examples as in Figure 2.1(b) and (c)

In Figure 2.1(b), we have the graph of $y = 2x + 3$. Take any two values of x which differ by 1 unit, for example, $x = 0$ and $x = 1$. When $x = 0, y = 2 \times 0 + 3 = 3$ and when $x = 1, y = 2 \times 1 + 3 = 5$. The increase in y is $5 - 3 = 2$, and this is the same as the coefficient of x in the function expression.

In Figure 2.1(c), we see the graph of $y = -x + 2$. Take any two values of x which differ by 1 unit, for example, $x = 1$ and $x = 2$. When $x = 1, y = -(1) + 2 = 1$ and when $x = 2, y = -(2) + 2 = 0$. The increase in y is $0 - 1 = -1$ and this is the same as the coefficient of x in the function expression.

In the general case, $y = mx + c$, take any two values of x which differ by 1 unit, for example, $x = x_0$ and $x = x_0 + 1$. When $x = x_0, y = mx_0 + c$ and when $x = x_0 + 1, y = m(x + 1) + c = mx + m + c$. The increase in y is $mx + m + c - (mx + c) = m$.

We know that every time x increases by 1 unit y increases by m. However, we do not need to always consider an increase of exactly 1 unit in x. The gradient gives the ratio of the increase in y to the increase in x. Therefore, if we only have a graph and we need to find the gradient then we can use any two points that lie on the line.

To find the gradient of the line take any two points on the line (x_1, y_1) and (x_2, y_2).

$$\text{The gradient} = \frac{\text{change in } y}{\text{change in } x} = \frac{y_2 - y_1}{x_2 - x_1}$$

Example 2.1 Find the gradient of the lines given in Figure 2.2(a)–(c) and the equation for the line in each case.

Solution

(a) We are given the coordinates of two points that lie on the straight line in Figure 2.2(a) as (0,3) and (2,5),

$$\text{gradient} = \frac{\text{change in } y}{\text{change in } x} = \frac{5 - 3}{2 - 0} = \frac{2}{2} = 1.$$

To find the constant term in the expression $y = mx + c$, we find the value of y when the line crosses the y-axis. From the graph this is 3, so the equation is $y = mx + c$ where $m = 1$ and $c = 3$, giving

$$y = x + 3$$

(b) Two points that lie on the line in Figure 2.2(b) are $(-1, -3)$ and $(-2, -6)$. These are found by measuring the x and y values for some points on the line.

$$\text{gradient} = \frac{\text{change in } y}{\text{change in } x} = \frac{-6 - (-3)}{-2 - (-1)} = \frac{-3}{-1} = 3.$$

To find the constant term in the expression $y = mx + c$, we find the value of y when the line crosses the y-axis. From the graph this

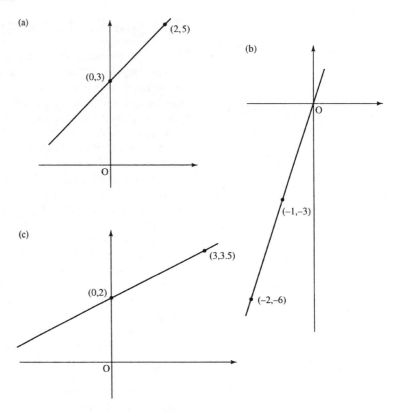

Figure 2.2 *Graphs for Example 2.1.*

is 0, so the equation is $y = mx + c$ where $m = 3$ and $c = 0$ giving

$$y = 3x$$

(c)   Two points that lie on the line in Figure 2.2(c) are $(0,2)$ and $(3,3.5)$.

$$\text{gradient} = \frac{\text{change in } y}{\text{change in } x} = \frac{3.5 - 2}{3 - 0} = \frac{1.5}{3} = 0.5$$

To find the constant term in the expression $y = mx + c$, we find the value of y when the line crosses the y-axis. From the graph this is 2, so the equation is $y = mx + c$ where $m = 0.5$ and $c = 2$ giving

$$y = 0.5x + 2$$

Finding the gradient from the equation for the line

To find the gradient from the equation of the line we look for the value of m, the number multiplying x in the equation. The constant term gives the value of y when the graph crosses the y-axis, that is, when $x = 0$.

Example 2.2   Find the gradient and the value of y when $x = 0$ for the following lines:

(a) $y = 2x + 3$,   (b) $3x - 4y = 2$,

(c) $x - 2y = 4$,   (d) $\dfrac{x - 1}{2} = 1 - \dfrac{y}{3}$.

Solution

(a) In the equation $y = 2x + 3$, the value of m, the gradient, is 2 as this is the coefficient of x. $c = 3$ which is the value of y when the graph crosses the y-axis, that is, when $x = 0$.

(b) In the equation $3x - 4y = 2$, we rewrite the equation with y as the subject of the formula in order to find the value of m and c.

$$3x - 4y = 2 \quad \Leftrightarrow \quad 3x = 2 + 4y$$
$$\Leftrightarrow \quad 3x - 2 = 4y$$
$$\Leftrightarrow \quad \frac{3x}{4} - \frac{2}{4} = y$$
$$\Leftrightarrow \quad y = \frac{3x}{4} - \frac{1}{2}$$

We can see, by comparing the expression with $y = mx + c$, that m, the gradient, is $3/4$ and $c = -1/2$.

(c) Write y as the subject of the formula:

$$x - 2y = 4 \quad \Leftrightarrow \quad x = 4 + 2y$$
$$\Leftrightarrow \quad x - 4 = 2y$$
$$\Leftrightarrow \quad 2y = x - 4$$
$$\Leftrightarrow \quad y = \frac{x}{2} - 2$$

We can see, by comparing the expression with $y = mx + c$, that m, the gradient, is $1/2$ and $c = -2$.

(d) Write y as the subject of the formula

$$\frac{x - 1}{2} = 1 - \frac{y}{3}$$
$$\Leftrightarrow \quad \frac{x}{2} - \frac{1}{2} = 1 - \frac{y}{3}$$
$$\Leftrightarrow \quad \frac{3x}{2} - \frac{3}{2} = 3 - y$$
$$\Leftrightarrow \quad y = 3 - \left(\frac{3x}{2} - \frac{3}{2} \right)$$
$$\Leftrightarrow \quad y = -\frac{3x}{2} + \frac{9}{2}$$

We can see, by comparing the expression with $y = mx + c$, that m, the gradient, is $-3/2$ and $c = 9/2$.

Finding the equation of a line which goes through two points

Supposing we have been given two points, (x_1, y_1) and (x_2, y_2), which lie on a line and we want to find the equation of that line. We already found that the gradient of the line is given by:

$$\text{The gradient} = \frac{\text{change in } y}{\text{change in } x} = \frac{y_2 - y_1}{x_2 - x_1}$$

We know that the equation of a line is of the form $y = mx + c$, but we would like to express the equation just in terms of the two variables, x

and y, and the known points which the line passes through, (x_1, y_1) and (x_2, y_2). To do this we use the fact that a line is of constant gradient. That means that for any point (x, y), the gradient between (x, y) and the point (x_1, y_1) must be the same as the gradient between (x_1, y_1) and the point (x_2, y_2). This gives:

$$\frac{y - y_1}{x - x_1} = \frac{y_2 - y_1}{x_2 - x_1}$$

providing that $y_2 \neq y_1$, we can rearrange this equation as:

$$\frac{y - y_1}{y_2 - y_1} = \frac{x - x_1}{x_2 - x_1}$$

Note that if $y_2 = y_1$ and we have chosen two different points on the line, that is, $x_2 \neq x_1$ then y is a constant and the equation of the line will be $y = y_1$.

Example 2.3

(a) Find the equation of a line through the points (2,4) and (0,6).
(b) Find the equation of a line through the points $(1, -6)$ and $(5, -6)$.

Solution

(a) To find the equation of a line through the points (2,4) and (0,6) we use

$$\frac{y - y_1}{y_2 - y_1} = \frac{x - x_1}{x_2 - x_1} \text{ and } (x_1, y_1) = (2, 4) \text{ and } (x_2, y_2) = (0, 6).$$

This gives

$$\frac{y - 4}{6 - 4} = \frac{x - 2}{0 - 2}$$

$$\Leftrightarrow \quad \frac{y - 4}{2} = -\frac{x - 2}{2}$$

$$\Leftrightarrow \quad y - 4 = 2 - x$$

$$\Leftrightarrow \quad y = -x + 6$$

Check: Check that this is the correct equation by substituting the points (2,4) and (0,6) into the equation $y = -x + 6$. (2,4) gives $4 = -2 + 6 \Leftrightarrow 4 = 4$, which is true; (0,6) gives $6 = -0 + 6$, which is true.

This shows that $y = -x + 6$ is the equation of a straight line which goes through the points (2,4) and (0,6).

(b) To find the equation of a line through the points $(1, -6)$ and $(5, -6)$ we try to use $(y - y_1)/(y_2 - y_1) = (x - x_1)/(x_2 - x_1)$ with $(x_1, y_1) = (1, -6)$ and $(x_2, y_2) = (5, -6)$ but find that, because $y_2 = y_1$, the substitution would result in a division by 0, which is undefined. We note that y has constant value and therefore the equation of the line is:

$$y = -6.$$

Sketching a straight line graph

The quickest way to sketch a straight line graph is to find the points where it crosses the axes. However, any two points on the graph can be used and sometimes it is more convenient to find other points.

Example 2.4

(a) Sketch the graph of $y = 4x - 2$.

To find where the graph crosses the y-axis, substitute $x = 0$ into the equation of the line:

$$y = 4(0) - 2 = -2.$$

This means that the graph passes through the point $(0, -2)$.

To find where the graph crosses the x-axis, substitute $y = 0$, that is,

$$4x - 2 = 0$$

$$\Leftrightarrow \quad 4x = 2$$

$$\Leftrightarrow \quad x = \frac{2}{4} = 0.5.$$

Therefore, the graph passes through $(0.5, 0)$.

Mark the points $(0,2)$ and $(0.5,0)$, on the x- and y-axes and join the two points. This is done in Figure 2.3(a).

(b) Sketch the graph of $y = -4x$ When $x = 0$ we get $y = 0$, that is the graph goes through the point $(0,0)$. In this case, as the graph passes through the origin, we need to choose a different value for x for the second point. Taking $x = 2$ gives $y = -8$, so another point is $(2, -8)$. These points on marked on the graph and joined to give the graph as in Figure 2.3(b).

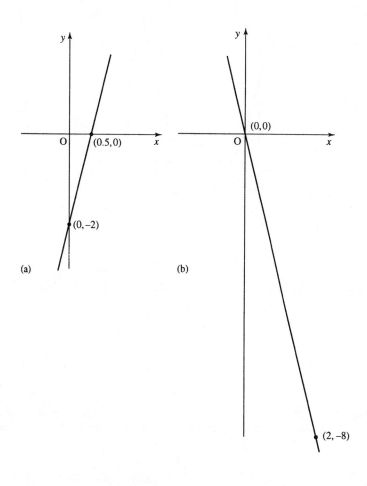

Figure 2.3 *(a) The graph of* $y = 4x - 2$. *(b) The graph of* $y = -4x$.

2.3 The quadratic function: $y = ax^2 + bx + c$

$y = ax^2 + bx + c$ is a general way of writing a function in which the highest power of x is a squared term. This is called the quadratic function and its graph is called a parabola as shown in Figure 2.4.

All the graphs, in this figure, cross the y-axis at $(0, c)$. To find where they cross the x-axis can be more difficult. These values, where $f(x) = 0$, are called the roots of the equation. There is a quick way to discover whether the function crosses the x-axis, only touches the x-axis, or does not cross or touch it. In the latter case there are no solutions to the equation $f(x) = 0$. The three possibilities are given in Figure 2.4.

Crossing the *x*-axis

The function $y = ax^2 + bx + c$ crosses the x-axis when $y = 0$, that is, when $ax^2 + bx + c = 0$. The solutions to $ax^2 + bx + c = 0$ are examined in the Background Mathematics Notes available on the companion website for this book and are given by the formula

$$x = \frac{-b \pm \sqrt{b^2 - 4ac}}{2a}$$

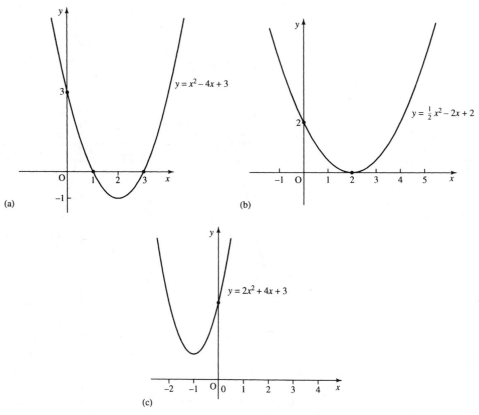

Figure 2.4 (a) The function $y = ax^2 + bx + c$. (a) Case 1 where there are 2 solutions to $f(x) = 0$. (b) Case 2 where there is only one solution to $f(x) = 0$. (c) Case 3, where there are no real solutions to $f(x) = 0$.

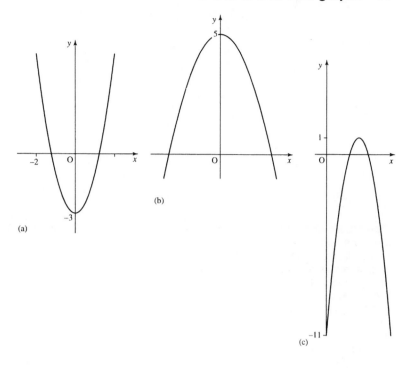

Figure 2.5 *Three quadratic functions with two roots to the equation $f(x) = 0$. Each satisfies $b^2 - 4ac > 0$.(a) $y = 2x^2 - 3, a = 2, b = 0, c = -3, b^2 - 4ac = 0 - 4(2)(-3) = 24$. (b) $y = -x^2 + 5, a = -1, b = 0, c = 5, b^2 - 4ac = 0 - 4(-1)(5) = 20$.(c) $y = -3(x - 2)^2 + 1 \Leftrightarrow y = -3x^2 + 12x - 11, a = -3, b = 12, c = -11, b^2 - 4ac = (12)^2 - 4(-3)(-11) = 144 - 132 = 12$.*

From the graph, we can see there are three possibilities:

1. In Figure 2.4(a) where there are two solutions, that is, the graph crosses the x-axis for two values of x. For this to happen, the square root part of the formula above must be greater than zero:

$$b^2 - 4ac > 0$$

Examples are given in Figure 2.5.
2. Only one unique solution, as in Figure 2.4(b). The graph touches the x-axis in one place only. For this to happen, the square root part of the formula must be exactly 0. Examples of this are given in Figure 2.6.
3. No real solutions, that is, the graph does not cross the x-axis. Examples of these are given in Figure 2.7.

2.4 The function $y = 1/x$

The function $y = 1/x$ has the graph as in Figure 2.8. This is called a hyperbola. Notice that the domain of $f(x) = 1/x$ does not include $x = 0$. The graph does not cross the x-axis so there are no solutions to $1/x = 0$.

2.5 The functions $y = a^x$

Graphs of exponential functions, $y = a^x$, are shown in Figure 2.9. The functions have the same shape for all $a > 1$. Notice that the function is always positive and the graph does not cross the x-axis so there are no solutions to the equation $a^x = 0$.

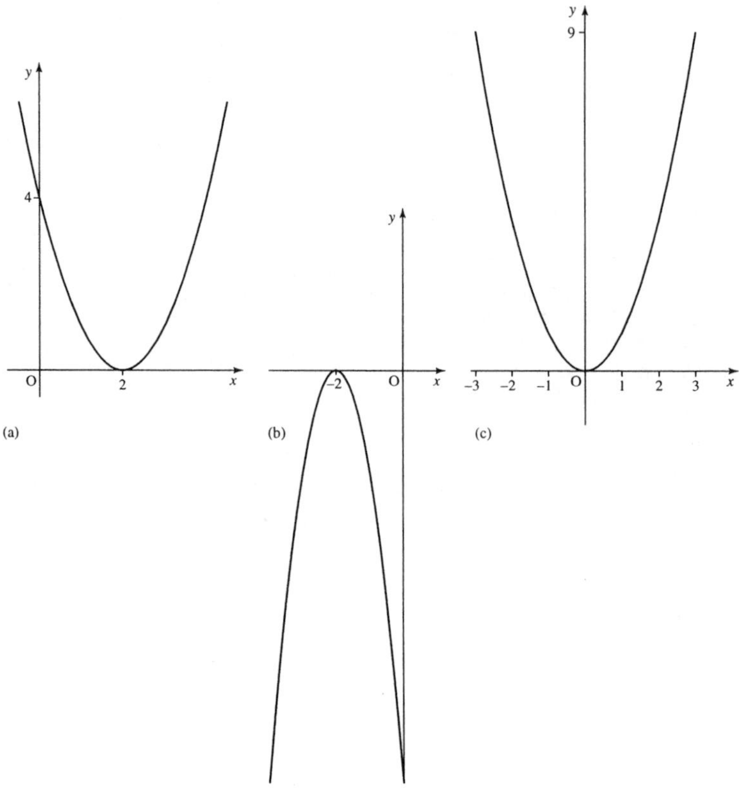

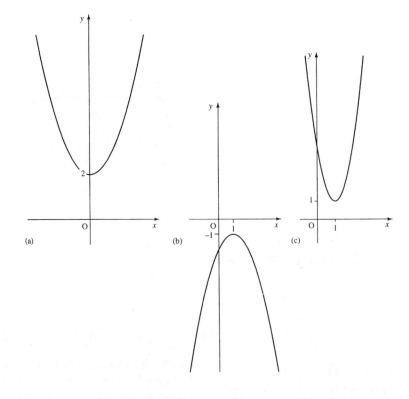

Figure 2.6 *Quadratic functions with only one unique root of the equation $f(x) = 0$. Each satisfies $b^2 - 4ac = 0$.*
(a) $y = x^2 - 4x + 4$, $a = 1$, $b = -4$, $c = 4$, $b^2 - 4ac = (-4)^2 - 4(1)(4) = 16 - 16 = 0$.
(b) $y = -3x^2 - 12x - 12$, $a = -3$, $b = -12$, $c = -12$, $b^2 - 4ac = (-12)^2 - 4(-3)(-12) = 144 - 144 = 0$.
(c) $y = x^2$, $a = 1$, $b = 0$, $c = 0$, $b^2 - 4ac = (0)^2 - 4(1)(0) = 0 - 0 = 0$.

Figure 2.7 *Quadratic functions with no real roots to the equation $f(x) = 0$. In each case $b^2 - 4ac < 0$.*
(a) $y = x^2 + 2$, $a = 1$, $b = 0$, $c = 2$, $b^2 - 4ac = (0)^2 - 4(1)(2) = 0 - 8 = -8$.
(b) $y = -x^2 + 2x - 2$, $a = -1$, $b = 2$, $c = -2$, $b^2 - 4ac = (2)^2 - 4(-1)(-2) = 4 - 8 = -4$.
(c) $y = 3x^2 - 6x + 4$, $a = 3$, $b = -6$, $c = 4$, $b^2 - 4ac = (-6)^2 - 4(3)(4) = 36 - 48 = -12$.

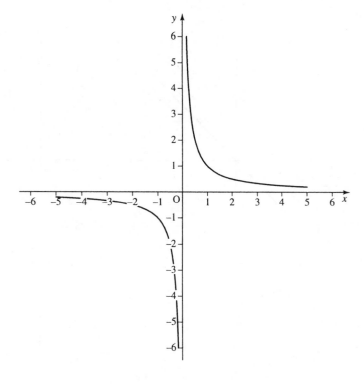

Figure 2.8   *Graph of the hyperbolic function $y = 1/x$.*

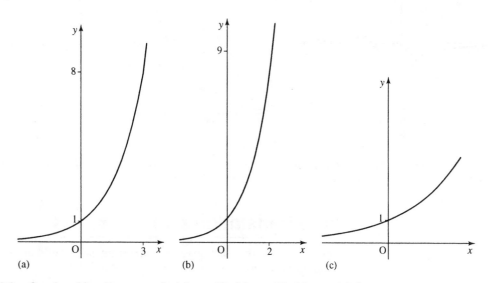

(a)   (b)   (c)

Figure 2.9   *Graphs of functions $y = a^x$: (a) $y = 2^x$; (b) $y = 3^x$; (c) $y = (1.5)^x$.*

2.6 Graph sketching using simple transformations

One way of sketching graphs is to remember the graphs of simple functions and to translate, reflect or scale those graphs to get graphs of other functions. We begin with the graphs below as given in Figure 2.10.

The translation $x \mapsto x + a$

If we have the graph of $y = f(x)$, then the graph of $y = f(x + a)$ is found by translating the graph of $y = f(x)$ a units to the left. Examples are given in Figure 2.11.

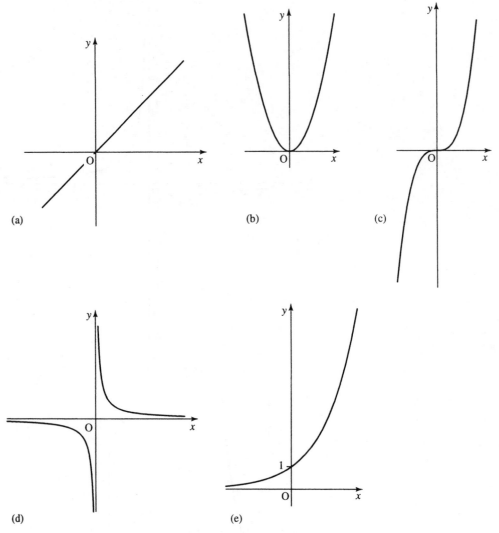

Figure 2.10 *To sketch graphs using transformations we begin with known graphs. In the rest of this section we use: (a) $y = x$; (b) $y = x^2$; (c) $y = x^3$; (d) $y = 1/x$; (e) $y = a^x$.*

The translation $f(x) \mapsto (x) + A$

Adding A on to the function value leads to a translation of A units upwards. Examples are given in Figure 2.12.

Reflection about the y-axis, $x \mapsto -x$

Replacing x by $-x$ in the function has the effect of reflecting the graph in the y-axis – that is, as though a mirror has been placed along the axis and only the reflection can be seen. Examples are given in Figure 2.13.

Reflection about the x axis, $f(x) \mapsto -f(x)$

To find the graph of $y = -f(x)$, reflect the graph of $y = f(x)$ about the x-axis. Examples are given in Figure 2.14.

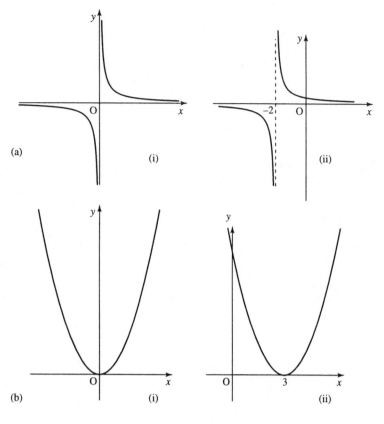

Figure 2.11   *Translations*
$x \mapsto x + a$. *(a) (i) $y = 1/x$;*
(ii) $y = 1/(x + 2)$. Here x has
been replaced by $x + 2$
translating the graph 2 units to
the left. (b) (i) $y = x^2$;
(ii) $y = (x - 3)^2$, x has been
replaced by $x - 3$ translating
the graph 3 units to the right.

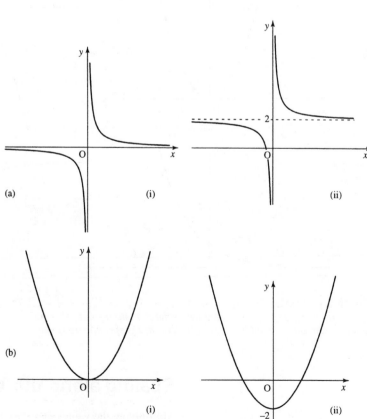

Figure 2.12   *Translations*
$f(x) \mapsto f(x) + A$. *(a) $y = 1/x$;*
(ii) $y = 1/x + 2$. Here the
function value has been
increased by 2 translating the
graph 2 units upwards. (b) (i)
$y = x^2$; (ii) $y = x^2 - 2$. The
function value has had 2
subtracted from it, translating
the graph 2 units downwards.

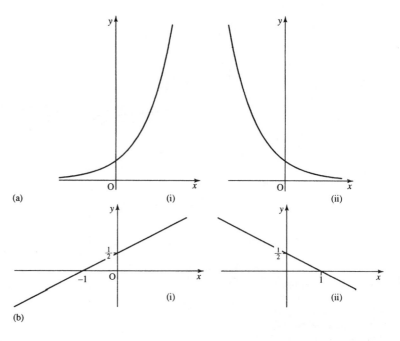

Figure 2.13 *Reflections*
$x \mapsto -x$. (a) (i) $y = a^x$, $a > 1$;
(ii) $y = a^{-x}$, x has been
replaced by −x to get
the second function. This has
the effect of reflecting the
graph in the y-axis. (b) (i)
$y = (x/2) + (1/2)$;
(ii) $y = -(x/2) + (1/2)$, x has
been replaced by −x,
reflecting the graph in the
y-axis.

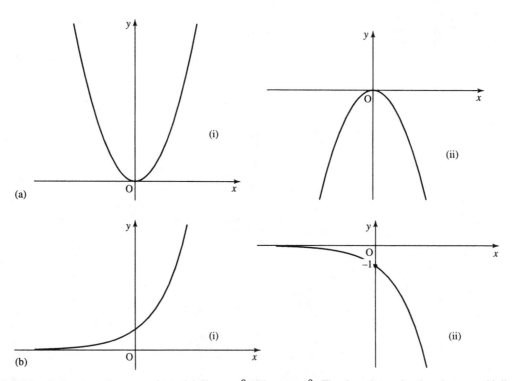

Figure 2.14 *Reflections $f(x) \mapsto -f(x)$. (a) (i) $y = x^2$; (ii) $y = -x^2$. The function value has been multiplied by −1 turning the graph upside down (reflection in the x-axis). (b) (i) $y = 2^x$; (ii) $y = -2^x$. The second function has been multiplied by −1 turning the graph upside down.*

Scaling along the *x*-axis, *x* ↦ *ax*

Multiplying the values of x by a number, a, has the effect of: squashing the graph horizontally if $a > 1$ or stretching the graph horizontally if $0 < a < 1$. Examples are given in Figure 2.15.

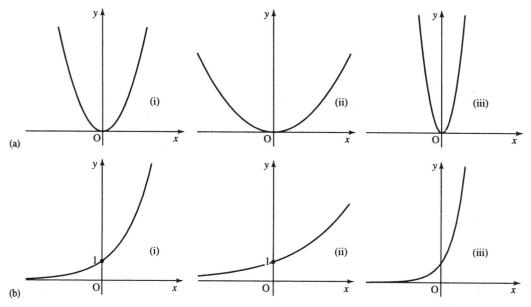

Figure 2.15   *Scalings $x \mapsto ax$. (a) (i) $y = x^2$; (ii) $y = [(1/2)x]^2$; (iii) $y = (2x)^2$. The second function has x replaced by $(1/2)x$ which has stretched the graph horizontally (the multiplication factor is between 0 and 1). The third function has replaced x by 2x, which has squashed the graph horizontally (the multiplication factor is greater than 1). (b) (i) $y = 2^x$; (ii) $y = 2^{(1/2x)}$; (iii) $y = 2^{2x}$. The second function has replaced x by $(1/2)x$ which has stretched the graph horizontally. The third function has x replaced by 2x which has squashed the graph horizontally.*

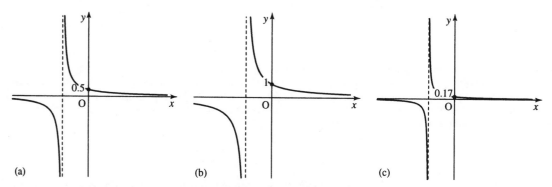

Figure 2.16   *Scalings $f(x) \mapsto Af(x)$. (a) $y = 1/(x + 2)$; (b) $y = 2/(x + 2)$ (c) $y = 1/[3(x + 2)]$. The second graph has the function values multiplied by 2 stretching the graph vertically. The third graph has function values multiplied by 1/3 squashing the graph vertically.*

Scaling along the y-axis, $f(x) \mapsto Af(x)$

Multiplying the function value by a number A has the effect of stretching the graph vertically if $A > 1$, or squashing the graph vertically if $0 < A < 1$. Examples are given in Figure 2.16.

Reflecting in the line $y = x$

If the graph of a function $y = f(x)$ is reflected in the line $y = x$, then it will give the graph of the inverse relation. Examples are given in Figure 2.17.

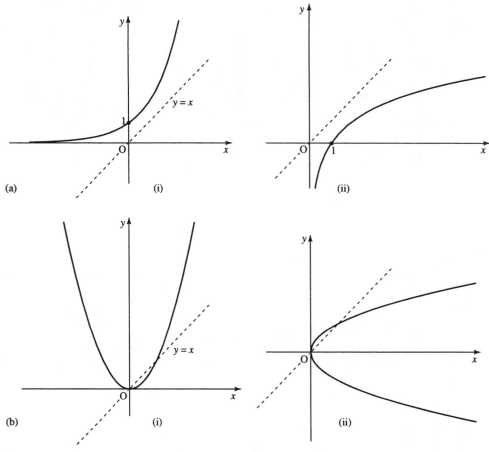

Figure 2.17 *Reflections in the line y = x produce the inverse relation. (a) (i) y = 2^x; (ii) y = log₂(x). The second graph is obtained from the first by reflecting in the dotted line y = x. The inverse is a function as there is only one value of y for each value of x. (b) (i) y = x²; (ii) y = ±√x. The second graph is found by reflecting the first graph in the line y = x. Notice that y = ±√x is not a function as there is more that one possible value of y for each value of x > 0.*

In Chapter 1, we defined the inverse function as taking any image back to its original value. Check this with the graph of $y = 2^x$ in Figure 2.17(a): $x = 1$ gives $y = 2$. In the inverse function, $y = \log_2(x)$, substitute 2, which gives the result of 1, which is back to the original value.

However, the inverse of $y = x^2$, $y \pm \sqrt{x}$, shown in Figure 2.17(b), is not a function as there is more than one y value for a single value of x.

To understand this problem more fully, perform the following experiment. On a calculator enter -2 and square it (x^2) giving 4. Now take the square root. This gives the answer 2, which is not the number we first started with, and hence we can see that the square root is not a true inverse of squaring. However, we get away with calling it the inverse because it works if only positive values of x are considered. To test if the inverse of any function exists, draw a line along any value of $y = $ constant. If, wherever the line is drawn, there is ever more than one x value which gives the same value of y then the function has no inverse function. In this situation, the function is called a 'many-to-one' function. Only 'one-to-one' functions have inverses. Figure 2.18 has examples of functions with an explanation of whether they are 'one-to-one' or 'many-to-one'.

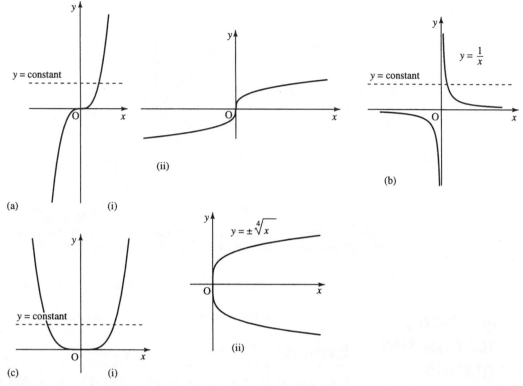

Figure 2.18 *(a) (i) y = x³. This function has only one x value for each value of y as any line y = constant only cuts the graph once. In this case, the function is one-to-one and it has an inverse function (ii) y = ∛x is the inverse function of y = x³. (b) y = 1/x, x ≠ 0, has only one x value for each value of y as any line y = constant only cuts the graph once. It therefore is one-to-one and has an inverse function – in fact, it is its own inverse! (to see this reflect it in the line y = x and we get the same graph after the reflection). (c) (i) y = x⁴. This function has two values of x for each value of y when y is positive (e.g. the line y = 16 cuts the graph twice at x = 2 and at x = −2). This shows that there is no inverse function as the function is many-to-one. (ii) The inverse relation y = ±⁴√x.*

2.7 The modulus function, $y = |x|$ or $y = abs(x)$

The modulus function $y = |x|$, often written as $y = abs(x)$ (short for the absolute value of x) is defined by

$$y = \begin{cases} x & x \geqslant 0 \ (x \text{ positive or zero}) \\ -x & x < 0 \ (x \text{ negative}) \end{cases}$$

The output from the modulus function is always a positive number or zero.

Example 2.5 Find $|-3|$.
Here $x = -3$, which is negative, therefore

$$y = -x = -(-3) = +3.$$

An alternative way of thinking of it is to remember that the modulus is always positive, or zero, so simply replacing any negative sign by a positive one will give a number's modulus or absolute value.

$$|-5| = 5, \quad |-4| = 4, \quad |5| = 5, \quad |4| = 4.$$

The graph of the modulus function can be found from the graph of $y = x$ by reflecting the negative x part of the graph to make the function values positive. This is shown in Figure 2.19.

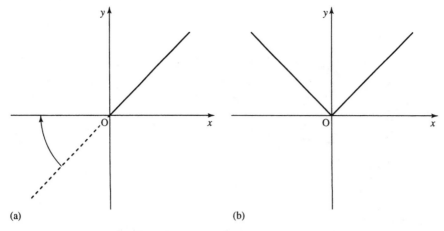

Figure 2.19   *The graph of the modulus function y = |x| obtained from the graph of y = x. (a) The graph of y = x with the negative part of the graph displayed as a dotted line. This is reflected about the x-axis to give y = −x for x < 0. (b) The graph of y = |x|.*

2.8 Symmetry of functions and their graphs

Functions can be classified as even, odd, or neither of these.

Even functions

Even functions are those that can be reflected in the y-axis and then result in the same graph. Examples of even functions are (see Figure 2.20):

$$y = x^2, \quad y = |x|, \quad y = x^4.$$

As previously discussed, reflecting in the y-axis results from replacing x by $-x$ in the function expression and hence the condition for a function to be even is that substituting $-x$ for x does not change the function expression, that is, $f(x) = f(-x)$.

Example 2.6   Show that $3x^2 - x^4$ is an even function. Substitute $-x$ for x in the expression $f(x) = 3x^2 - x^4$ and we get

$$f(-x) = 3(-x)^2 - (-x)^4 = 3(-1)^2(x)^2 - (-1)^4(x)^4 = 3x^2 - x^4.$$

So, we have found that $f(-x) = f(x)$ and therefore the function is even.

Odd functions

Odd functions are those that when reflected in the y-axis result in an upside down version of the same graph. Examples of odd functions are (see Figure 2.21):

$$y = x, \quad y = x^3, \quad y = \frac{1}{x}$$

Reflecting in the y-axis results from replacing x by $-x$ in the function expression and the upside down version of the function $f(x)$ is found by multiplying the function by -1. Hence, the condition for a function to be odd is that substituting $-x$ for x gives $-f(x)$, that is,

$$f(-x) = -f(x).$$

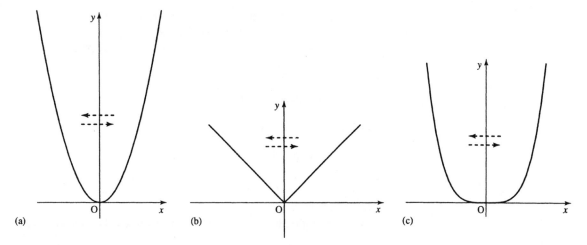

Figure 2.20 $y = x^2, y = |x|$, and $y = x^4$ are even functions.

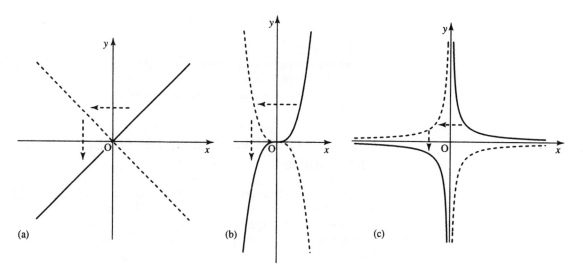

Figure 2.21 (a) $y = x$, (b) $y = x^3$, (c) $y = 1/x$ are odd functions. If they are reflected in the y-axis they result in an upside down version of the original graph.

Example 2.7 Show that $4x - (1/x)$ is an odd function. Substitute x for $-x$ in the expression $f(x) = 4x - (1/x)$ and we get

$$f(-x) = 4(-x) - \frac{1}{-x} = -4x + \frac{1}{x} = -\left(4x - \frac{1}{x}\right).$$

We have found that $f(-x) = -f(x)$, so the function is odd.

2.9 Solving inequalities

For linear and quadratic functions, $y = f(x)$, we have discussed how to find the values where the graph of the functions crosses the x-axis, that is how to solve the equation $f(x) = 0$. It is often of interest to find ranges of values of x where $f(x)$ is negative or where $f(x)$ is positive. This means solving inequalities like $f(x) < 0$ or $f(x) > 0$, respectively.

Like equations, inequalities can be solved by looking for equivalent inequalities. One way of finding these is by doing the same thing to both sides of the expression. There is an important exception for inequalities that if both sides are multiplied or divided by a negative number then the direction of the inequality must be reversed.

To demonstrate these equivalences begin with a true proposition

$3 < 5$ or   '3 is less than 5'.

Add 2 on to both sides and it is still true

$3 + 2 < 5 + 2$,   i.e. $5 < 7$.

Subtract 10 from both sides and we get

$5 - 10 < 7 - 10$,   i.e. $-5 < -3$

which is also true.

Multiply both sides by -1 and if we do not reverse the inequality we get

$(-1)(-5) < (-1)(-3)$,   i.e. $5 < 3$

which is false. However, if we use the correct rule that when multiplying by a negative number we must reverse the inequality sign then we get:

$(-1)(-5) > (-1)(-3)$,   i.e. $5 > 3$

which is true. This process is pictured in Figure 2.22.

Note that inequalities can be read from right to left as well as from left to right: $3 < 5$ can be read as '3 is less than 5' or as '5 is greater than 3' and so it can also be written the other way round as $5 > 3$.

Using a number line to represent inequalities

An inequality can be expressed using a number line as in Figure 2.23. In Figure 2.23(a), the open circle indicates that 3 is not included in the set of values, $t < 3$. In Figure 2.23(b), the closed circle indicates that -2 is included in the set of values, $x \geqslant -2$. In Figure 2.23(c), the closed circle indicates that the value 4.5 is included in the set $y \leqslant 4.5$.

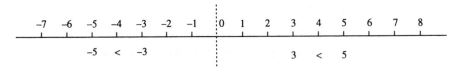

Figure 2.22 *On the number line, numbers to the left are less than numbers to their right: $-5 < -3$. If the inequality is multiplied by -1 we need to reverse the sign to get $5 > 3$.*

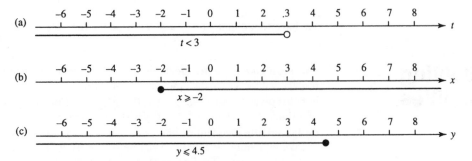

Figure 2.23 *Representing inequalities on a number line.*

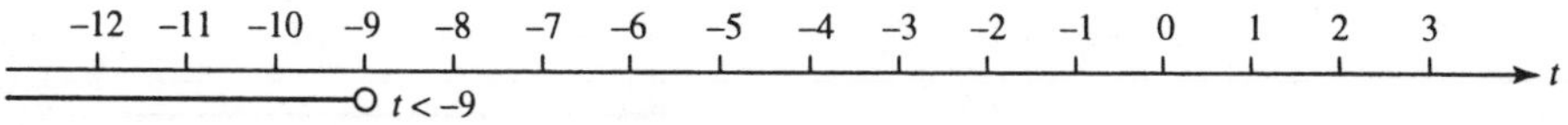

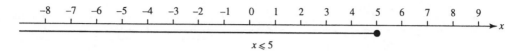

Figure 2.24   *The solution to $2t + 3 < t - 6$ is given by $t < -9$.*

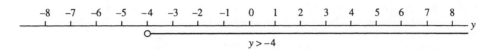

Figure 2.25   *The solution to $x + 5 \geqslant 4x - 10$ is found to be $x \leqslant 5$, here represented on a number line.*

Figure 2.26   *The solution to $16 - y > -5y$ is found to be $y > -4$, here pictured on a number line.*

Example 2.8   Find a range of values for t, x, and y such that the following inequalities hold

(a) $2t + 3 < t - 6$
(b) $x + 5 \geqslant 4x - 10$
(c) $16 - y > -5y$

Solution

(a) $2t + 3 < t - 6 \quad \Leftrightarrow \quad 2t - t + 3 < -6$ (subtract t from both sides)

$\Leftrightarrow \quad t < -6 - 3$ (subtract 3 from both sides)

$\Leftrightarrow \quad t < -9.$

This solution can be represented on a number line as in Figure 2.24.

(b) $x + 5 \geqslant 4x - 10$

$\Leftrightarrow \quad +5 \geqslant 3x - 10$ (subtract x from both sides)

$\Leftrightarrow \quad 15 \geqslant 3x$ (add 10 to both sides)

$\Leftrightarrow \quad 5 \geqslant x$ (divide both sides by 3)

$\Leftrightarrow \quad x \leqslant 5.$

This solution in represented in Figure 2.25.

(c) $16 - y > -5y$

$\Leftrightarrow \quad 16 > -4y$ (add y to both sides)

$\Leftrightarrow \quad -4 < y$ (divide by -4 and reverse the sign)

$\Leftrightarrow \quad y > -4.$

This solution is represented in Figure 2.26.

Representing compound inequalities on a number line

We sometimes need a picture of the range of values given if two inequalities hold simultaneously, for instance $x \geqslant 3$ and $x < 5$. This is analysed

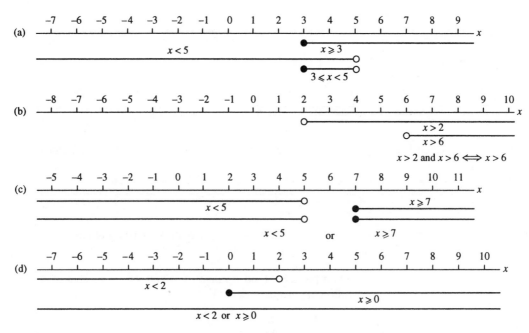

Figure 2.27   *(a) x ⩾ 3 and x < 5. (b) x > 6 and x > 2 combines to give x > 6. (c) x < 5 or x ⩾ 7. (d) x < 2 or x ⩾ 0.*

in Figure 2.27(a) and we can see that for both inequalities to hold simultaneously x must lie in the overlapping region where $3 \leqslant x < 5$. $3 \leqslant x < 5$ is a way of expressing that x lies between 3 and 5 or is equal to 3. In the example in Figure 2.27(b), $x > 6$ and $x > 2$, and for them both to hold then $x > 6$.

Another possible way of combining inequalities is to say that one or another inequality holds. Examples of this are given in Figure 2.27(c) where $x < 5$ or $x \geqslant 7$ and this gives the set of values less than 5 or greater than or equal to 7. Figure 2.27(d) gives the example where $x < 2$ or $x \geqslant 0$ and in this case it results in all numbers lying on the number line, that is, $x \in \mathbb{R}$.

Example 2.9   Find solutions to the following combinations of inequalities and represent them on a number line.

(a) $x + 3 > 4$ and $x - 1 < 5$,
(b) $1 - u < 3u + 2$ or $u + 2 \geqslant 6$,
(c) $t + 5 > 12$ and $-t > 24$.

Solution

(a) $x + 3 > 4$ and $x - 1 < 5$ We solve both inequalities separately and then combine their solution sets

$x + 3 > 4 \quad \Leftrightarrow \quad x > 1 \quad$ (subtracting 3 from both sides)

$x - 1 < 5 \quad \Leftrightarrow \quad x < 6 \quad$ (adding 1 to both sides)

So the combined inequality giving the solution is $x > 1$ and $x < 6$, which from Figure 2.28(a) we can see is the same as $1 < x < 6$.

(b) $1 - u < 3u + 2$ or $u + 2 \geqslant 6$

We solve both inequalities separately and then combine their

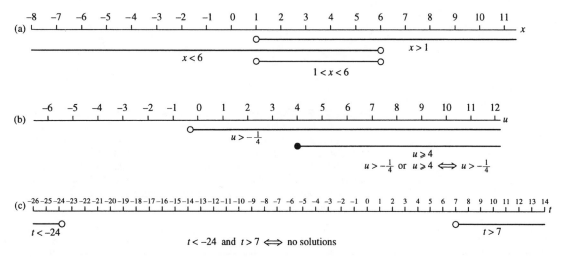

Figure 2.28   *Solutions to compound inequalities as given in Example 2.8 represented on a number line.*

solution sets.

$$1 - u < 3u + 2$$

$$\Leftrightarrow \quad -1 < 4u \quad \text{(subtracting 2 from both sides)}$$

$$\Leftrightarrow \quad -\frac{1}{4} < u \quad \text{(dividing both sides by 4)}$$

$$\Leftrightarrow \quad u > -\frac{1}{4}$$

$$u + 2 \geqslant 6$$

$$\Leftrightarrow \quad u \geqslant 4 \quad \text{(subtracting 2 from both sides)}$$

Combining the two solutions gives $u > -1/4$ or $u \geqslant 4$ and this is represented on the number line in Figure 2.28(b) where we can see that it is the same as $u > -1/4$.

(c) $t + 5 > 12$ and $-t > 24$

$$t + 5 > 12$$

$$\Leftrightarrow \quad t > 7 \quad \text{(subtracting 5 from both sides)}$$

$$-t > 24$$

$$\Leftrightarrow \quad t < -24 \quad \text{(multiply both sides by } -1 \text{ and reverse the inequality sign)}$$

Combining the two solutions sets gives $t > 7$ and $t < -24$ and we can see from Figure 2.28(c) that this is impossible and hence there are no solutions.

Solving more difficult inequalities

To solve more difficult inequalities, our ideas about equivalence are not enough on their own, we also use our knowledge about continuous functions. In the previous chapter, we defined a continuous function as one that could be drawn without taking the pen off the paper. If we wish to solve the inequality $f(x) > 0$ and we know that $f(x)$ is continuous then we can picture the problem graphically as in Figure 2.29. From the graph

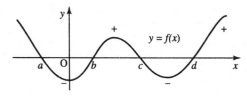

Figure 2.29 *The graph of a continuous function. To solve for*
$f(x) > 0$, *we first find the values where* $f(x) = 0$. *On the graph these*
are marked as a, b, c, and d. If the function is above the x-axis then
the function values are positive, if the function lies below the x-axis
then the function values are negative. The solution to $f(x) > 0$ *is*
given by those values of x for which the function lies above the x-axis,
that is, y positive. For the function represented in the graph the
solution to $f(x) > 0$ *is* $x < a$ *or* $b < x < c$ *or* $x > d$

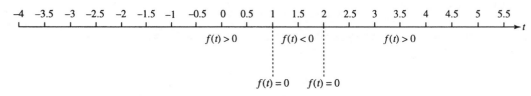

Figure 2.30 *Solving* $t^2 - 3t + 2 < 0$ *(Example 2.10).*

we can see that to solve the inequality we need only find the values where
$f(x) = 0$ (the roots of $f(x) = 0$) and determine whether $f(x)$ is positive
or negative between the values of x where $f(x) = 0$. To do this, we can
use any value of x between the roots. We are using the fact that as $f(x)$
is continuous then it can only change from positive to negative by going
through zero.

Example 2.10 Find the values of t such that $t^2 - 3t < -2$.
   Write the inequality with 0 on one side of the inequality sign

$$t^2 - 3t < -2 \Leftrightarrow t^2 - 3t + 2 < 0 \quad \text{(adding 2 to both sides)}$$

Find the solutions to $f(t) = t^2 - 3t + 2 = 0$ and mark them on a number
line as in Figure 2.30.
   Using the formula

$$t = \frac{-b \pm \sqrt{b^2 - 4ac}}{2a}$$

where $a = 1, b = -3$, and $c = 2$ gives

$$t = \frac{-3 \pm \sqrt{9 - 8}}{2}$$

$$\Leftrightarrow t = \frac{3 \pm 1}{2}$$

$$\Leftrightarrow t = \frac{3 + 1}{2} \quad \text{or} \quad t = \frac{3 - 1}{2}$$

$$\Leftrightarrow t = 2 \quad \text{or} \quad t = 1$$

Substitute values for t which lie on either side of the roots of $f(t)$ in
order to find the sign of the function between the roots. Here we choose
0, 1.5, and 3.

When $t = 0$

$$t^2 - 3t + 2 = (0)^2 - 3(0) + 2 = 0 + 0 + 2 = 2$$

which is positive, giving $f(t) > 0$.
When $t = 1.5$

$$t^2 - 3t + 2 = (1.5)^2 - 3(1.5) + 2 = 2.25 - 4.5 + 2 = -0.25$$

which is negative giving $f(t) < 0$.
When $t = 3$

$$t^2 - 3t + 2 = (3)^2 - 3(3) + 2 = 9 - 9 + 2 = 2$$

which is positive, giving $f(t) > 0$.

By marking the regions on the number line, given in Figure 2.30, with $f(t) > 0$, $f(t) < 0$, or $f(t) = 0$ as appropriate we can now find the solution to our inequality $f(t) < 0$ which is given by the region where $1 < t < 2$.

Example 2.11   Find the values of x such that $(x^2 - 4)(x + 1) > 0$.

Solution   The inequality already has 0 on one side of the inequality sign so we begin by finding the roots to $f(x) = 0$, that is,

$$(x^2 - 4)(x + 1) > 0$$

Factorization gives

$$(x^2 - 4)(x + 1) > 0 \Leftrightarrow (x - 2)(x + 2)(x + 1) = 0$$

$\Leftrightarrow x = 2, x = -2$, or $x = -1$. So the roots are $-2, -1$, and 2. These roots are pictured on the number line in Figure 2.31.

Substitute values for x which lie on either side of the roots of $f(x)$ in order to find the sign of the function between the roots. Here we choose $-3, -1.5, 0$, and 2.5.
When $x = -3$

$$(x - 2)(x + 2)(x + 1) = 0 \quad \text{gives} \quad (-3 - 2)(-3 + 2)(-3 + 1)$$
$$= (-5)(-1)(-2)$$
$$= -10, \quad \text{giving } f(x) < 0.$$

When $x = -1.5$

$$(x - 2)(x + 2)(x + 1) \quad \text{gives} \quad (-1.5 - 2)(-1.5 + 2)(-1.5 + 1)$$
$$= (-3.5)(0.5)(-0.5) = 0.875, \quad \text{giving } f(x) > 0.$$

When $x = 0$

$$(x - 2)(x + 2)(x + 1) \quad \text{gives} \quad (0 - 2)(0 + 2)(0 + 1)$$
$$= (-2)(2)(1) = -4, \quad \text{giving } f(x) < 0.$$

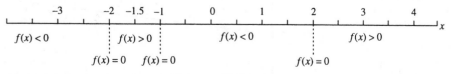

Figure 2.31   *Solving $(x^2 - 4)(x + 1) > 0$ (Example 2.11).*

When $x = 2.5$

$(x - 2)(x + 2)(x + 1)$ gives $(2.5 - 2)(2.5 + 2)(2.5 + 1)$

$\quad = (0.5)(4.5)(3.5) = 7.875$, giving $f(x) > 0$.

These regions are marked on the number line as in Figure 2.31 and the solution is given by those regions where $f(x) > 0$. Looking for the regions where $f(x) > 0$ gives the solution as $-2 < x < -1$ or $x > 2$.

2.10 Using graphs to find an expression for the function from experimental data

Linear relationships

Linear relationships are the easiest ones to determine from experimental data. The points are plotted on a graph and if they appear to follow a straight line then a line can be drawn by hand and the equation can be found using the method given in Section 2.2.

Example 2.12 A spring is stretched by hanging various weights on it and in each case the length of the spring is measured.

Mass (kg)	0.125	0.25	0.5	1	2	3
Length (m)	0.4	0.41	0.435	0.5	0.62	0.74

Approximate the length of the spring when no weight is hung from it and find the expression for the length in terms of the mass.

Solution First, draw a graph of the given experimental data. This is done in Figure 2.32.

A line is fitted by eye to the data. The data does not lie exactly on a line due to experimental error and due to slight distortion of the spring with heavier weights. From the line we have drawn we can find the gradient by choosing any two points on the line and calculating

$$\frac{\text{change in } y}{\text{change in } x}.$$

Taking the two points as $(0, 0.28)$ and $(2, 0.58)$, we get the gradient as

$$\frac{0.58 - 0.38}{2 - 0} = \frac{0.2}{2} = 0.1.$$

The point where it crosses the y-axis, that is, where the mass hung on the spring is 0 can be found by extending the line until it crosses the y-axis. This gives 0.38 m.

Finally, the expression for the length in terms of the mass of the attached weight is given by $y = mx + c$, where y is the length and x is the mass,

Figure 2.32 *The data for length of spring against mass of the weight as given in the Example 2.11. The line is a fitted by eye to the experimental data and the equation of the line can be found using the method of Section 2.2.*

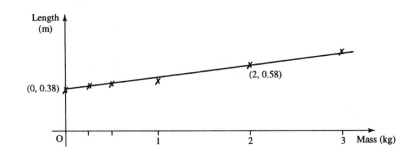

m is the gradient and c is the value of y when $x = 0$, that is, where there is no weight on the string. This gives

$$\text{length} = 0.1 \times \text{mass} + 0.38$$

The initial length of the spring is 0.38 m.

Exponential relationships

Many practical relationships behave exponentially particularly those involving growth or decay. Here it is slightly less easy to find the relationship from the experimental data, however, it is simplified by using a log–linear plot. Instead of plotting the values of the dependent variable, y, we plot the values of $\log_{10}(y)$. If the relationship between y and time, t, is exponential as we suspected then the $\log_{10}(y)$ against t plot will be a straight line.

The reason this works can be explained as follows. Supposing $y = y_0 10^{kt}$ where y_0 is the value of y when $t = 0$ and k is some constant; then, taking the log base 10 of both sides, we get

$$\log_{10}(y) = \log_{10}(y_0 10^{kt})$$
$$= \log_{10}(y_0) + \log_{10}(10^{kt}).$$

As the logarithm base 10 and raising to the power of 10 are inverse operations, we get

$$\log_{10}(y) = \log_{10}(y_0) + kt$$

As y_0 is a constant, the initial value of y, and k is a constant then we can see that this expression shows that we shall get a straight line if $\log_{10}(y)$ is plotted against t. The constant k is given by the gradient of the line and $\log_{10}(y_0)$ is the value of $\log_{10} y$ where it crosses the vertical axis. Setting $Y = \log_{10}(y), c = \log_{10}(y_0)$

$$Y = c + kt$$

which is the equation of the straight line.

Example 2.13   A room was tested for its acoustical absorption properties by playing a single note on a trombone. Once the sound had reached its maximum intensity, the player stopped and the sound intensity was measured for the next 0.2 s at regular intervals of 0.02 s. The initial maximum intensity at time 0 is 1.0. The readings were as follows:

Time(s)	0	0.02	0.04	0.06	0.08	0.1	0.12	0.14	0.16	0.18	0.2
Intensity	1.0	0.63	0.35	0.22	0.13	0.08	0.05	0.03	0.02	0.01	0.005

Draw a graph of intensity against time and log(intensity) against time and use the latter plot to approximate the relationship between the intensity and time.

Solution   The graphs are plotted in Figure 2.33 where, for the second graph (b), we take the $\log_{10}$ (intensity) and use the table below:

Time	0	0.02	0.04	0.06	0.08	0.1	0.12	0.14	0.16	0.18	0.2
$\log_{10}$ (intensity)	0	−0.22	−0.46	−0.66	−0.89	−1.1	−1.3	−1.5	−1.7	−2	−2.3

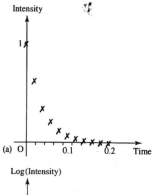

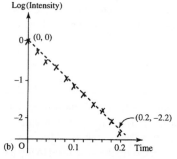

Figure 2.33 *(a) Graph of sound intensity against time as given in Example 2.13. (b) Graph of $\log_{10}$ (intensity) against time and a line fitted by eye to the data. The line goes through the points (0,0) and (0.2, −2.2).*

We can see that the second graph is approximately a straight line and therefore we can assume that the relationship between the intensity and time is exponential and could be expressed as $I = I_0 10^{kt}$. The $\log_{10}$ of this gives

$$\log_{10}(I) = \log_{10}(I_0) + kt.$$

From the graph in Figure 2.33(b), we can measure the gradient, k. To do this we calculate

$$\frac{\text{change in } \log_{10}(\text{intensity})}{\text{change in time}}$$

giving

$$\frac{-2.2 - 0}{0.2 - 0} = -11 = k.$$

The point at which it crosses the vertical axis gives

$$\log_{10}(I_0) = 0 \Leftrightarrow I_0 = 10^0 = 1.$$

Therefore, the expression $I = I_0 10^{kt}$ becomes

$$I = 10^{-11t}.$$

Power relationships

Another common type of relationship between quantities is when there is a power of the independent variable involved. In this case, if $y = ax^n$ where n could be positive or negative then the value of a and n can be found by drawing a log–log plot.

This is because taking $\log_{10}$ of both sides of $y = ax^n$ gives

$$\log_{10}(y) = \log_{10}(ax^n) = \log_{10}(a) + n\log_{10}(x)$$

Replacing $Y = \log_{10}(y)$ and $X = \log_{10}(x)$ we get:

$$Y = \log_{10}(a) + nX,$$

showing that the log–log plot will give a straight line, where the slope of the line will give the power of x and the position where the line crosses the vertical axis will give the $\log_{10}(a)$. Having found a and n, they can be substituted back into the expression

$$y = ax^n.$$

Example 2.14   The power received from a beacon antenna is though to depend on the inverse square of the distance from the antenna and the receiver. Various measurements, given below, were taken of the power received against distance r from the antenna. Could these be used to justify the inverse square law? If so, what is the constant, A, in the expression:

$$p = \frac{A}{r^2}$$

Power received (W)	0.39	0.1	0.05	0.025	0.015	0.01
Distance from antenna (km)	1	2	3	4	5	6

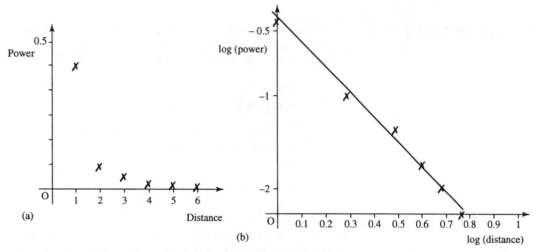

Figure 2.34 *(a) Plot of power against distance (Example 2.14). (b) $\log_{10}$(power) against $\log_{10}$(distance). In (b), the line is fitted by eye to the data, from which the slope of the graph indicates n in the relationship $P = Ar^n$. Two points lying on the line are $(0, -0.38)$ and $(0.7, -1.8)$.*

Solution To test whether the relationship is indeed a power relationship, we draw a log–log plot. The table of values is found below:

$\log_{10}$(power)	-0.41	-1	-1.3	-1.6	-1.8	-2
$\log_{10}$(distance)	0	0.3	0.5	0.6	0.7	0.78

Graphs of power against distance and $\log_{10}$(power) against $\log_{10}$(distance) are given in Figure 2.34(a) and (b).

As the second graph is a straight line, we can assume that the relationship is of the form $P = Ar^n$ where P is the power and r is the distance. In which case, the log–log graph is

$$\log_{10}(P) = \log_{10}(A) + n\log_{10}(r).$$

We can measure the slope by calculating

$$\frac{\text{change in } \log_{10}(P)}{\text{change in } \log_{10}(r)}$$

and, using the two points that have been found to lie on the line, this gives

$$\frac{-1.8 - (-0.38)}{0.7 - 0} = -2.03.$$

As this is very near to -2, the inverse square law would appear to be justified.

The value of $\log_{10}(A)$ is given from where the graph crosses the vertical axis and this gives

$$\log_{10}(A) = -0.38 \quad \Leftrightarrow \quad A = 10^{-0.38} \quad \Leftrightarrow \quad A = 0.42.$$

So the relationship between power received and distance is approximately

$$P = 0.42r^{-2} = \frac{0.42}{r^2}.$$

2.11 Summary

1. The linear function $y = mx + c$ has gradient (slope) m and crosses the y-axis at $y = c$.

2. The gradient, m, of a straight line $y = mx + c$ is given by:

$$m = \frac{\text{change in } y}{\text{change in } x}$$

and this is the same along the length of the line.

3. The equation of a line which goes through two points, (x_1, y_1) and (x_2, y_2) is:

$$\frac{y - y_1}{y_2 - y_1} = \frac{x - x_1}{x_2 - x_1} \quad \text{where } y_2 \neq y_1.$$

4. The graph of the quadratic function $y = ax^2 + bx + c$ is called a parabola. The graph crosses the y-axis (when $x = 0$) at $y = c$.

5. There are three possibilities for the roots of the quadratic equation $a\,x^2 + bx + c = 0$

   Case I: two real roots when $b^2 - 4ac > 0$,

   Case II: only one unique root when $b^2 - 4ac = 0$,

   Case III: no real roots when $b^2 - 4ac < 0$.

6. By considering the graphs of known functions $y = f(x)$, for instance, those given in Figure 2.10, and the following transformations, many other graphs can be drawn.

   (a) Replacing x by $x + a$ in the function $y = f(x)$ results in shifting the graph a units to the left.

   (b) Replacing $f(x)$ by $f(x) + A$ results in shifting the graph A units upwards.

   (c) Replacing x by $-x$ reflects the graph in the y-axis.

   (d) Replacing $f(x)$ by $-f(x)$ reflects the graph in the x-axis (turning it upside down).

   (e) Replacing x by ax squashes the graph horizontally if $a > 1$ or stretches it horizontally if $0 < a < 1$.

   (f) Replacing $f(x)$ by $Af(x)$ stretches the graph vertically if $A > 1$ or squashes it vertically if $0 < A < 1$.

   (g) Reflecting the graph of $y = f(x)$ in the line $y = x$ results in the graph of the inverse relation.

7. A function may be even, or odd, or neither of these.

   (a) An even function is one whose graph remains the same if reflected in the y-axis, that is, when $x \mapsto -x$. This can also be expressed by the condition

$$f(-x) = f(x)$$

      Examples of even functions are $y = x^2$, $y = |x|$, and $y = x^4$.

   (b) An odd function is one which when reflected in the y-axis, that is, when $x \mapsto -x$, gives an upside down version of the original graph (i.e. $-f(x)$). This can also be expressed as the condition:

$$f(-x) = -f(x)$$

      Examples of odd functions are $y = x$ and $y = x^3$.

8. Not all functions have true inverses. Only one-to-one functions have inverse functions. A function is one-to-one if any line $y = \text{constant}$ drawn on the graph $y = f(x)$ crosses the function only once. This means there is exactly one value of x that gives each value of y.

9. Simple inequalities can be solved by finding equivalent inequalities. Inequalities remain equivalent if both sides of the inequality have the same expression added or subtracted. They may also be multiplied or divided by a positive number but if they are multiplied or divided by a negative number then the direction of the inequality sign must be reversed.

10. To solve the inequalities $f(x) > 0$, $f(x) < 0$, $f(x) \leqslant 0$, or $f(x) \geqslant 0$, where $f(x)$ is a continuous function, solve $f(x) = 0$ and choose any value for x around the roots to find the sign of $f(x)$ for each region of values for x.

11. Graphs can be used to find relationships in experimental data. First, plot the data then:

    (a) If the data lies on an approximate straight line then draw a straight line through the data and find the equation of the line.

    (b) If it looks exponential, then take the log of the values of the dependent variable and draw a log–linear graph. If this looks approximately like a straight line then assume there is an exponential relationship $y = y_0 10^{kt}$, where k is given by the gradient of the line and $\log_{10}(y_0)$ is the value where the graph crosses the vertical axis.

    (c) If the relationship looks something like a power relationship, $y = Ax^n$, then take the log of both sets of data and draw a log–log graph. If this is approximately like a straight line, then assume there is a power relationship and n is given by the gradient of the line and $\log_{10} A$ is the value where the graph crosses the vertical axis.

2.12 Exercises

2.1. Sketch the graphs of the following:

(a) $y = 3x - 1$,   (b) $y = 2x + 1$,
(c) $y = -5x$,   (d) $y = \frac{1}{2}x - 3$.

In each case state the gradient of the line.

2.2. A straight line passes through the pair of points given. Find the gradient of the line in each case.

(a) $(0, 1), (1, 4)$ (b) $(1, 1), (2, -4)$
(c) $(-1, -1), (6, 3)$ (d) $(1, 4), (3, 4)$

2.3. A straight line graph has gradient -5 and passes through $(1,6)$. Find the equation of the line.

2.4. In Figure 2.35 are various graphs drawn to the scale 1 unit $= 1$ cm. By finding the gradients of the lines and where they cross the y-axis, find the equation of the line.

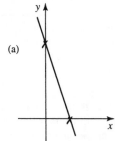

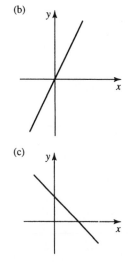

Figure 2.35 *Straight line graphs for Exercise 2.4.*

2.5. A straight line passes through the pair of points given. Find the equation of the line in each case.

(a) $(0, 1), (-1, 4)$ (b) $(1, 1), (-2, -4)$
(c) $(1, 1), (6, 3)$ (d) $(-1, -4), (-3, -4)$

2.6. Find the values of x such that $f(x) = 0$ for the following functions
  (a) $f(x) = x^2 - 4$,
  (b) $f(x) = (2x - 1)(x + 1)$,
  (c) $f(x) = (x - 3)^2$,
  (d) $f(x) = (x - 4)(x + 4)$,
  (e) $f(x) = x^2 + x - 6$,
  (f) $f(x) = x^2 + 7x + 12$,
  (g) $f(x) = 12x^2 - 12x - 144$.
  Using the fact that the peak or trough in the parabola, $y = f(x)$, occurs at a value of x half-way between the values where $f(x) = 0$ then sketch graphs of the above quadratic functions.

2.7. By considering transformations of simple functions sketch graphs of the following:

  (a) $y = \frac{1}{x - (1/2)}$,   (b) $y = 3.2^{-x}$,
  (c) $y = \frac{1}{2}x^3$,   (d) $y = -3^{(1/2)x}$,
  (e) $y = (2x - 1)^2$ (f) $y = (2x - 1)^2 - 2$,
  (g) $y = \log_2(x + 2)$,   (h) $y = 6 - 2^x$,
  (i) $y = 4x - x^2$.

2.8. Consider reflections of the graphs given in Figure 2.36 to determine whether they are even, odd, or neither of these.

(a)

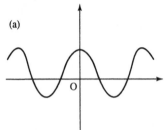

(b)

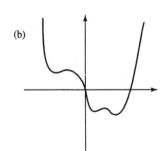

(c)

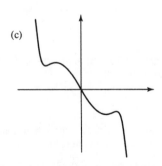

Figure 2.36 *Graphs of functions for Exercise 2.7.*

2.9. By substituting $x \mapsto -x$ in the following functions determine whether they are odd, even, or neither of these:
  (a) $y = -x^2 + \frac{1}{x^2}$ where $x \neq 0$,
  (b) $y = |x^3| - x^2$,
  (c) $y = \frac{-1}{x} + \log_2(x)$ where $x > 0$,
  (d) $y = \frac{-1}{x} + x + x^5$,
  (e) $y = 6 + x^2$,
  (f) $y = 1 - |x|$.

2.10. Draw graphs of the following functions and draw the graph of the inverse relation in each case. Is the inverse a function?
  (a) $f(t) = -t + 2$,   (b) $g(x) = (x - 2)^2$,
  (c) $h(w) = \dfrac{4}{w + 2}$.

2.11. Find the range of values for which the following inequalities hold and represent them on a number line.
  (a) $10t - 2 \leqslant 31$,   (b) $10x - 3x > -2$,
  (c) $3 - 4y \geqslant 11 + y$,   (d) $t + 15 < 6 - 2t$.

2.12. Find the range of values for which the following hold and represent them on a number line:
  (a) $x - 2 > 4$ or $1 - x < 12$,
  (b) $4t + 2 \geqslant 10$ and $3 - 2t < 1$,
  (c) $3u + 10 > 16$ or $3 - 2u > 13$.

2.13. Solve the following inequalities and represent the solutions on a number line:
  (a) $x^2 - 4 < 5$,   (b) $(2x - 3)(x + 1)(x - 5) > 0$,
  (c) $t^2 + 4t \leqslant 21$,   (d) $4w^2 + 4w - 35 \geqslant 0$.

2.14. For the following sets of data, y is thought to depend exponentially on t. Draw log–linear graphs in each case and find constants A and k such that $y = A10^{kt}$.
  (a)

y	75	48	30	19	12	7
t	1	2	3	4	5	6

  (b)

y	2	4.2	8.5	18	35	73
t	0.1	0.2	0.3	0.4	0.5	0.6

2.15 An experiment measuring the change in volume of a gas as the pressure is decreased gave the following measurements:

$P(10^5\,\mathrm{N\,m^{-2}})$	1.5	1.4	1.3	1.2	1.1	1
$V(\mathrm{m^3})$		0.95	1	1.05	1.1	1.16	1.24

If the gas is assumed to be ideal and the expansion is adiabatic then the relationship between pressure and volume should be:

$$pV^\gamma = C$$

where γ and C are constants and p is the pressure and V is the volume. Find reasonable values of γ and C to fit the data and from this expression find the predicted volume at atmospheric pressure, $p = 1.013 \times 10^5\,\mathrm{N\,m^{-2}}$.

3 Problem solving and the art of the convincing argument

3.1 Introduction

Mathematics is used by engineers to solve problems. This usually involves developing a mathematical model. Just as when building a working model aeroplane we would hope to include all the important features, the same thing applies when building a mathematical model. We would also like to indicate the things we have had to leave out because they were too fiddly to deal with, and also those details that we think are irrelevant to the model. In the case of a mathematical model the things that have been left out are listed under assumptions of the model. To build a mathematical model, we usually need to use scientific rules about the way things in the world behave (e.g. Newton's laws of motion, conservation of momentum and energy, Ohm's law, Kirchoff's laws for circuits, etc.) and use numbers, variables, equations, and inequalities to express the problem in a mathematical language.

Some problems are very easy to describe mathematically. For instance: 'Three people sitting in a room were joined by two others, how many people are there in the room in total?' This can be described by the sum $3 + 2 = ?$ and can be solved easily as $3 + 2 = 5$.

The final stage of solving the problem is to translate it back into the original setting – the answer is: 'there are 5 people in the room in total'.

Assumptions were used to solve this problem. We assumed that no one else came in or left the room in the meantime and we made general assumptions about the stability of the room, for example, the building containing it did not fall down. However, these assumptions are so obvious that they do not need to be listed. In more complex problems it is necessary to list important assumptions as they may have relevance as to the validity of the solution.

Another example is as follows: 'There are three resistors in series in a circuit, two of the resistors are known to have resistance of 3 and 4 ohm, respectively. The voltage source is a battery of 12 V and the current is measured as 1 A. What is the resistance of the third resistor?'

To help express the problem in a mathematical form we may draw a circuit diagram as in Figure 3.1.

The problem can be expressed mathematically by using Ohm's law and the fact that an equivalent resistance to resistances in series is given by the sum of the individual resistances. If x is the unknown value of the

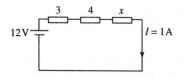

Figure 3.1 *A simple circuit.*

third resistance and $V = RI$ where $R = R_1 + R_2 + R_3$, we obtain:

$$12 = (3 + 4 + x)1$$

The expression of the mathematical problem has taken the form of an equation where we now need to find x, the value of the third resistance.

The main assumptions that have been used to build this mathematical model are:

(1) There are such things as pure resistors that have no capacitance or inductance.
(2) Resistances remain constant and are not affected by any possible temperature changes or other environmental effects.
(3) The battery gives a constant voltage that does not deteriorate with time.
(4) The battery introduces no resistance to the circuit.

These assumptions are simplifications that are acceptable because although the real world cannot behave with the simplicity of the mathematical model, the amount of error introduced by making these assumptions is small.

Once we have the solution of a mathematical model then it should be tested against a real-life situation to see whether the model behaves reasonably closely to reality. Once the model has been accepted then it can be used to predict the behaviour of the system for input values other than those that it has been tested for.

The stages in solving a problem are as follows:

(1) Take to real problem and express it as a mathematical one using any necessary scientific rules and assumptions about the behaviour of the system and using letters to represent any unknown quantities. Include an account of any important assumptions and simplifications made.
(2) Solve the mathematical problem using your knowledge of mathematics.
(3) Translate the mathematical solution back into the setting of your original problem.
(4) Test the model solutions for some values to check that it behaves like the real-life problem.

Most mathematical problems are expressed by using equations, or inequalities, differential or difference equations, or by expressing a problem geometrically or a combination of all of these. We might need to incorporate a random element which results in the need to use a probabilistic model. In many of the following chapters we will look at the modelling process in more detail as we come across new mathematical tools and the situations in which they are used. To perform the entire modelling cycle properly, we need to be able to test our results in a real-life situation in order to reconsider assumptions used in the model. This would require access to engineering situations and tools. For this reason, engineering mathematics books tend to concentrate on those models that are commonly used by engineers. Many of the applied problems presented in the following chapters however do present an opportunity to move from an English language description of a problem to a mathematical language description of a problem, which is an important step in the modelling process.

In this chapter, we will look at translating a problem into mathematical language and, for the main part of the chapter, we concentrate on solving a mathematical problem and the reasoning that is involved in so doing.

To solve the problem using your knowledge of mathematics, we need to use the ideas of mathematical statements and how to decide whether, and express the fact that, one statement leads logically on to the next. We shall mainly use examples of solving equations and inequalities although the same ideas apply to the solving of all problems.

3.2 Describing a problem in mathematical language

The stages in expressing a problem in mathematical language can be summarized as:

(1) Assign letters to represent the unknown quantities.
(2) Write down the known facts using equations and inequalities, and using drawings and diagrams where necessary.
(3) Express the problem to be solved mathematically.

This is not a simple process because it involves a great deal of interpretation of the original problem. It is useful to try to limit the number of unknowns used as much as possible, or the problem may appear more difficult than necessary.

Example 3.1 Express the following problem mathematically: A web development company employs a freelance web designer and a freelance graphic designer to put up listings for new businesses on to their virtual business park website. Business customers are charged €200 per year for a listing. The fixed costs of the web development company amount to €2000 per week over 52 weeks in the year. The web designer charges €80 per listing and the graphic designer €100 per listing and both can prepare these at the rate of 2.4 listings in a day. The freelancers work for up to 200 days per year. How many listings does the company need in the first year to break even?

Solution The mathematical problem can be expressed by firstly assigning letters to some of the unknown quantities and then write down all the known facts as equations or inequalities.

First assign letters: Total number of listings of businesses on the park in the first year is L. L_W is the number of listings prepared by the web designer and L_G is the number of listings prepared by the graphic designer. The costs are K per year and the profit is P. The known facts can be expressed as follows:

$$L = L_W + L_G$$

This expresses the fact that the total number of listings L is made up of those prepared by the web designer and those prepared by the graphic designer. As there are up to 200 working days in a year and they both do a maximum of 2.4 listings per day.

$$0 \leqslant L_W \leqslant 480, \quad 0 \leqslant L_G \leqslant 480$$

The costs, K, are; fixed costs of 2000×52, plus the cost of the freelance web designer at $80L_W$, plus the cost of the freelance graphic designer at

$100L_G$. This can be expressed as:

$$K = 104\,000 + 80L_W + 100L_G.$$

We need to relate the profit to the other variables. As the profit is 200 multiplied by the number of jobs minus the total costs, we get:

$$P = 200L - K$$

Finally, we must express the mathematical problem that we would like to solve. For the web development business to make a profit in the first year then the profit must be positive, hence we get the problem expressed as: Find the minimum L such that $P > 0$.

Example 3.2   Express the following in mathematical language: A car brake pedal, as represented in Figure 3.2(a) is pivoted at point A. What is the force on the brake cable if a constant force of 900 N is applied by the driver's foot and the pedal is stationary.

Solution   First, we assign letters to the unknowns. Let $F =$ the force on the brake cable.

In order to write down the known facts we need to consider what scientific laws can be used. As the force applied on the pedal initially provides a turning motion then we know to use the ideas of moments. The moment of a force about an axis is the product of the force F and its perpendicular distance, x, to the line of action of the force. Furthermore, as the pedal is now stationary, then the moments must be balanced so the clockwise moment must equal the anti-clockwise moment.

To use this fact, we need to use two further measurements, currently unknown, the perpendicular distance from the line of action of the force provided by the driver to the axis, A. This is marked as x_1 m on the diagram in Figure 3.2(b). The other distance is the perpendicular distance from the line of action of the force on the cable to the axis A. This is marked as x_2 m in Figure 3.2(b).

We can now write down the known facts, involving the unknowns x_1, x_2, and F. From the right angle triangle containing x_1, we have

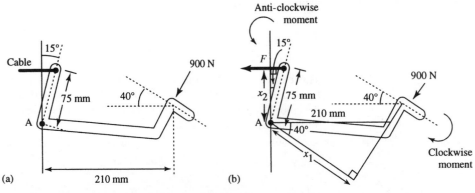

(a)

(b)

Figure 3.2   *(a) A representation of a car brake pedal. (b) The same diagram as (a) with some unknown quantities marked and triangles used to formulate the problem.*

(converting $210\,\text{mm} = 0.21\,\text{m}$),

$$\cos(40°) = \frac{x_1}{0.21}$$

From the right angle triangle containing x_2, we have (converting $75\,\text{mm} = 0.075\,\text{m}$),

$$\cos(15°) = \frac{x_2}{0.075}$$

The moments of the forces can now be calculated and equated. The clockwise moment is $800x_1$ and the anti-clockwise moment is given by Fx_2 and hence we have:

$$800x_1 = Fx_2$$

Finally, we need to express the problem we are trying to solve. In this case it is simply 'what is F?'.

Note that in both Examples 3.1 and 3.2, certain modelling assumptions had been used in order to formulating the 'natural language' description of the problem that we were given. For instance, it is probable that the business park listings for the business park in Example 3.1 are not all identical and therefore average figures for times and costings have been used. Similarly, in Example 3.2 no mention has been made of friction would provide an extra force to consider. Here we have only considered the transition from natural language and accompanying diagrams to the mathematical problem. We have implicitly assumed that the modelling process can be performed in two stages. From real-life problem to a natural language description which incorporates some simplifying assumptions, and then from there to a mathematical description. In reality modelling a system is much more involved. We would probably repeat stages in this process if we decided that the mathematical description was too complex and return to the real-life situation in order to make new assumptions.

We are now in a position to discuss mathematical statements and how to move from the statement of the problem to finding the desired solution.

3.3
Propositions and predicates

When we first set up a problem to be solved, we write down mathematical expressions like:

$$2 + 3 = ? \tag{3.1}$$

and

$$12 = (3 + 4 + x) \cdot 1 \tag{3.2}$$

These are mathematical statements with an unknown value. Statements containing unknowns (or variables) are called predicates. A predicate can be either true or false depending on the value(s) substituted into it. When values are substituted into a predicate it becomes a simple proposition. If in Equation (3.1) we substitute 5 for the question mark we get:

$$2 + 3 = 5 \quad \Leftrightarrow \quad 5 = 5 \quad \text{which is true.}$$

If, however, we substitute 6 we get:

$$2 + 3 = 6 \quad \Leftrightarrow \quad 5 = 6 \quad \text{which is false.}$$

$2 + 3 = 5$ and $2 + 3 = 6$ are examples of propositions. These are simple statements that can be assigned as either true or false. They contain no unknown quantities. Notice that if we simply rewrite a proposition or predicate we use '$\equiv$' or '$\Leftrightarrow$' to mean 'is equivalent to' or 'is the same as'.

In Equation (3.2) if we substitute 4 for x we get:

$12 = 11$ which is false

but, if we substitute 5 for x we get

$12 = 12$ which is true.

Example 3.3   Assign true or false to the following:

(a) $(3x - 2)(x + 5) = 10$ where $x = 1$
(b) $5x^2 - 2x + 1 = 25$ where $x = -2$
(c) $y > 5t + 3$ where $y = 2$ and $t = -3$

Solution

(a)   Substitute $x = 1$ in the expression and we get:

$$(3 \cdot 1 - 2)(1 + 5) = 10$$
$$\Leftrightarrow \quad 1(6) = 10$$
$$\Leftrightarrow \quad 6 = 10 \quad \text{which is false.}$$

(b)   Substituting $x = -2$ into $5x^2 - 2x + 1 = 25$ gives

$$5(-2)^2 - 2(-2) + 1 = 25$$
$$\Leftrightarrow \quad 20 + 4 + 1 = 25$$
$$\Leftrightarrow \quad 25 = 25 \quad \text{which is true.}$$

(c)   Substituting $y = 2$ and $t = -3$ into $y > 5t + 3$ gives

$$2 > 5(-3) + 3$$
$$\Leftrightarrow \quad 2 > -15 + 3$$
$$\Leftrightarrow \quad 2 > -12 \quad \text{which is true.}$$

Like functions, predicates have a domain which is the set of all allowed inputs to the predicate. For instance, the predicate $1/(x - 1) = 1$, where $x \in \mathbb{R}$, has the restriction that $x \neq 1$, as letting x equal 1 would lead to an attempt to divide by 0, which is not defined.

$\sqrt{x - 2} = 25$ where $x \in \mathbb{R}$ has the restriction that $x \geqslant 2$, as values of x less than 2 would lead to an attempt to take the square root of a negative number, which is not defined.

3.4 Operations on propositions and predicates

Consider the problem given in Example 3.1. Notice that the conditions that we discovered when writing down the known facts must all be true in any solution that we come up with. If any one of these conditions is not true then we cannot accept the solution. The first condition must be true *and* the second *and* the third, etc.

Here we have an example of an operation on predicates. In Chapter 1, we defined an operation on numbers is a way of combining two numbers to give a single number. 'And', written as $\wedge$ is an operation on two predicates or propositions which results a single predicate or proposition.

Table 3.1 *Truth table for the operation 'and'. This table can also be expressed by*
$T \wedge T \Leftrightarrow T$,
$T \wedge F \Leftrightarrow F$,
$F \wedge T \Leftrightarrow F$,
$F \wedge F \Leftrightarrow F$. *T stands for 'true' and F stands for 'false'*

p	q	$p \wedge q$
T	T	T
T	F	F
F	T	F
F	F	F

Table 3.2 *Truth table for 'or', This table can also be expressed by*
$T \vee T \Leftrightarrow T$,
$T \vee F \Leftrightarrow T$,
$F \vee T \Leftrightarrow T$,
$F \vee F \Leftrightarrow F$

p	q	$p \vee q$
T	T	T
T	F	T
F	T	T
F	F	F

Table 3.3 *The truth table for 'not', $\neg$. This table can also be expressed by*
$\neg T \Leftrightarrow F$, $\neg F \Leftrightarrow T$

p	$\neg p$
T	F
F	T

Therefore, to express the fact that both $L \geqslant 0$ and $L = L_W + L_G$ we can write

$$L \geqslant 0 \wedge L = L_W + L_G$$

and the compound statement is true if each part is also true.

As propositions can only be either true (T) or false (F), all possible outcomes of the operation can easily be listed in a small table called a truth table. The truth table for the operation of 'and' is given in Table 3.1. p and q represent any two propositions, for instance, for some given values of L, L_W, and L_G, p and q could be defined by:

p: '$L \geqslant 0$'

q: '$L = L_W + L_G$'

Another important operation is that of 'or'. One example of the use of this operation comes about by solving a quadratic equation. One way of solving quadratic equations is to factorize an expression which is equal to 0.

To solve $x^2 - x - 6 = 0$, the left-hand side of the equation can be factorized to give $(x - 3)(x + 2) = 0$.

Now we use the fact that for two numbers multiplied together to equal 0 then one of them, at least, must be 0, to give:

$$(x - 3)(x + 2) = 0 \quad \Leftrightarrow \quad (x - 3) = 0 \quad \text{or} \quad (x + 2) = 0$$

'or' can be written using the symbol $\vee$. The compound statement is true if either $x - 3 = 0$ is true or if $x + 2 = 0$ is true. Therefore, to express the statement that either $x - 3 = 0$ or $x + 2 = 0$ we can write:

$$(x - 3) = 0 \vee (x + 2) = 0$$

$\vee$ is also called 'non-exclusive or' because it is also true if both parts of the compound statement are true. This usage is unlike the frequent use of 'or' in the English language, where it is often used to mean a choice, for example, 'you may have either an apple or a banana' implies either one or the other but not both. This everyday usage of the word 'or' is called 'exclusive or'.

The truth table for 'or' is given in Table 3.2.

A further operation is that of 'not' which is represented by the symbol $\neg$. For instance, we could express the sentence 'x is not bigger than 4' as $\neg(x > 4)$.

The truth table for 'not' is given in Table 3.3.

Example 3.4 Assign truth values to the following:

(a) $x - 2 = 3 \wedge x^2 = 4$ when $x = 2$
(b) $x - 2 = 3 \vee x^2 = 4$ when $x = 2$
(c) $\neg(x - 4 = 0)$ when $x = 4$
(d) $\neg((a - b) = 4) \vee (a + b) = 2$ when $a = 5, b = 3$

Solution

(a) $x - 2 = 3 \wedge x^2 = 4$ when $x = 2$

Substitute $x = 2$ into the predicate, $x - 2 = 3 \wedge x^2 = 4$ and we get $2 - 3 = 3 \wedge 2^2 = 4$.

The first part of the compound statement is false and the second part is true. Overall, as $F \wedge T \Leftrightarrow F$, the proposition is false.

(b) $x - 2 = 3 \vee x^2 = 4$ when $x = 2$

Substitute $x = 2$ into the predicate and we get

$2 - 2 = 3 \vee 2^2 = 4$

The first part of the compound statement is false and second part is true. As $F \vee T \Leftrightarrow T$, the proposition is true.

(c) $\neg(x - 4 = 0)$ when $x = 4$

When $x = 4$ the expression becomes:

$$\neg(4 - 4 = 0) \quad \Leftrightarrow \quad \neg T \quad \Leftrightarrow \quad F$$

(d) $\neg((a - b) = 4) \vee (a + b) = 2$ when $a = 5, b = 3$

$\neg((a - b) = 4)$ when $a = 5, \ b = 3$ gives

$$\neg(5 - 3 = 4) \quad \Leftrightarrow \quad \neg F \quad \Leftrightarrow \quad T$$

$(a + b) = 2$ when $a = 5, \ b = 3$ gives

$$(5 + 3) = 2 \quad \Leftrightarrow \quad 8 = 2 \quad \Leftrightarrow \quad F$$

Overall, $T \vee F \Leftrightarrow T$ so $\neg((a - b) = 4) \vee (a + b) = 2$ when $a = 5$, $b = 3$ is true.

Example 3.5   Represent the following inequalities on a number line:

(a) $x > 2 \wedge x \leqslant 4$
(b) $x < 2 \vee x \geqslant 4$
(c) $\neg(x < 2)$

Solution

(a) $x > 2 \wedge x \leqslant 4$. To represent the operation of 'and', find where the two regions overlap (Figure 3.3a). $x > 2 \wedge x \leqslant 4$ can also be represented by $2 < x \leqslant 4$.
(b) $x < 2 \vee x \geqslant 4$. To represent the operation of $\vee$, 'or', take all points on the first highlighted region as well as all points in the second highlighted region and any end points (Figure 3.3b).
(c) $\neg(x < 2)$. To represent the operation of 'not' take all the points on the number line not in the original region (Figure 3.3c). This can also be expressed by $x \geqslant 2$.

3.5 Equivalence

We can now express an initial problem in terms of a predicate, probably an equation, a number of equations, or a number of inequalities. However, to solve the problem we need to be able to move from the original expression of the problem toward the solution. In Chapter 3 of the Background Mathematics Notes, available on the companion website for this book, we discussed how to solve various types of equations and introduced the idea of equivalent equations. In Chapter 2 we also looked at equivalent inequalities. In both cases we used the idea that in moving from one expression to an equivalent expression the set of solutions remained the

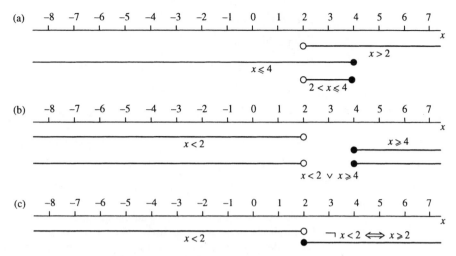

Figure 3.3 *The operations (a)∧ (and); (b) ∨ (or); and (c) ¬ (not).*

same. In general, two predicates are equivalent if they are true for exactly the same set of values. We use our knowledge of mathematics to determine what operations can be performed that will maintain that equivalence. The rule that can be used to move from one equation to another was given as: 'Equations remain equivalent if the same operation is performed to both sides of the equation'. In the case of quadratic equations we can also use a formula for the solution or use a factorization and the fact that:

$$ab = 0 \quad \Leftrightarrow \quad a = 0 \quad \text{or} \quad b = 0.$$

In passing from one equation to an equivalent equation we should use the equivalence symbol. This then makes a mathematical sentence:

$$x + 5 = 3$$
$$\Leftrightarrow \quad x = 3 - 5$$

can be read as 'The equation $x + 5 = 3$ is equivalent to $x = 3 - 5$'.

In all but the most obvious cases, it is a good practice to list a short justification for the equivalence by the side of the expression.

$$x + 5 = 3$$
$$\Leftrightarrow \quad x = 3 - 5 \quad \text{(subtracting 5 from both sides)}$$
$$\Leftrightarrow \quad x = -2.$$

Because of the possibility of making a mistake, the solution(s) should be checked by substituting the values into the original expression of the problem. To check, substitute $x = -2$ into the original equation giving $-2 + 5 = 3$ which is true, indicating that the solution is correct.

Example 3.6 Solve the following equation:

$$x - 3 = 5 - 2x.$$

Solution

$$x - 3 = 5 - 2x$$

$$\Leftrightarrow \quad x + 2x - 3 = 5 \quad \text{(adding } 2x \text{ to both sides)}$$

$$\Leftrightarrow \quad 3x = 8 \qquad\qquad \text{(adding 3 to both sides)}$$

$$\Leftrightarrow \quad x = \tfrac{8}{3} \qquad\qquad \text{(dividing both sides by 3).}$$

Check by substituting $x = 8/3$ into the original equation:

$$\tfrac{8}{3} - 3 = 5 - 2 \times \tfrac{8}{3}$$

$$\Leftrightarrow \quad -\tfrac{1}{3} = -\tfrac{1}{3}, \quad \text{which is true.}$$

We looked at methods of solving inequalities in Chapter 2. The rules for finding equivalent inequalities were: 'Perform the same operation to both sides'; but in the case of a negative number when multiplying or dividing the direction of the inequality sign must be reversed. To solve more complex inequalities, such as $f(x) > 0$, $f(x) < 0$, where $f(x)$ is a continuous but non-linear function, then we solve $f(x) = 0$ and then use a number line to mark regions where $f(x)$ is positive, negative or zero. The important thing in the process is to present a short justification of the equivalence. Finally, when the set of solutions has been found, some of the solutions can be substituted into the original expression of the problem in order to check that no mistakes have been made.

Example 3.7   Solve the following inequalities:

(a) $3x - 1 < 6x + 2$
(b) $x^2 - 5x > -6$

Solution

(a) $3x - 1 < 6x + 2$

$$\Leftrightarrow \quad -1 < 6x + 2 - 3x \quad \text{(subtracting } 3x \text{ from both sides)}$$

$$\Leftrightarrow \quad -1 - 2 < 3x \qquad \text{(subtracting 2 from both sides)}$$

$$\Leftrightarrow \quad -\frac{3}{3} < \frac{3x}{3} \qquad\qquad \text{(dividing both sides by 3)}$$

$$\Leftrightarrow \quad -1 < x$$

$$\Leftrightarrow \quad x > -1$$

Check: Test a few values from the set $x > -1$ and substitute into

$$3x - 1 < 6x + 2$$

Try $x = 0$: this gives $-1 < 2 \Leftrightarrow T$
Try $x = 2$: this gives $3(2) - 1 < 6(2) + 2 \Leftrightarrow 5 < 14 \Leftrightarrow T$

(b) $x^2 - 5x > -6$

Write the inequality with 0 on one side of the inequality sign

$$x^2 - 5x > -6 \quad \Leftrightarrow \quad x^2 - 5x + 6 > 0 \quad \text{(adding 6 to both sides)}$$

Find the solutions to $f(x) = 0$ where $f(x) = x^2 - 5x + 6$ and mark them on a number line as in Figure 3.4.

$$x^2 - 5x + 6 = 0 \quad \Leftrightarrow \quad x = \frac{5 \pm \sqrt{25 - 24}}{2}$$

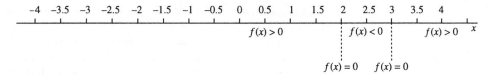

Figure 3.4 *Solving $x^2 - 5x + 6 > 0$.*

(using the formula for solution of quadratic equations)

$$\Leftrightarrow \quad x = \frac{5+1}{2} \lor x = \frac{5-1}{2}$$

$$\Leftrightarrow \quad x = 3 \lor x = 2$$

Using the fact that the function is continuous, we can substitute values for x which lie on either side of the roots of $f(x) = 0$ in order to find the sign of the function in that region. Here, we choose 0, 2.5, and 4 and find that

when $x = 0$: $\quad f(x) = x^2 - 5x + 6 = 6, \ \text{so } f(x) > 0$

when $x = 2.5$: $\quad f(x) = x^2 - 5x + 6 = 6.25 - 12.5 + 6$

$$= -0.25, \ \text{so } f(x) < 0$$

when $x = 4$: $\quad f(x) = x^2 - 5x + 6 = 16 - 20 + 6 = 2,$

$$\text{so } f(x) > 0$$

These regions are marked on the number line as in Figure 3.4 and this gives the solution to $f(x) > 0$ as $x < 2 \lor x > 3$.

Check: A check is to substitute some values from the solution set $x < 2 \lor x > 3$ into the original predicate $x^2 - 5x > -6$

Substitute $x = 1$, this gives $1 - 5 > -6 \Leftrightarrow -4 > -6 \Leftrightarrow T$
Substitute $x = 5$, this gives $25 - 25 > -6 \Leftrightarrow 0 < -6 \Leftrightarrow T$

It therefore appears that this solution is correct.

3.6 Implication

We previously described one method of finding equivalent equations as that of 'doing the same thing to both sides'. This was rather simplistic but a useful way of seeing it at the time. There are only certain things that can be 'done to both sides' like adding, subtracting, multiplying by a non-zero expression, or dividing by a non-zero expression that always maintain equivalence. There are also many operations that can be performed to both sides of an equation which do not give an equivalent equation but give an equation with the same solutions and yet more besides. In this situation we say that the first equation *implies* the second equation. The symbol for implies is $\Rightarrow$.

An example of implication is given by squaring both sides of the equation

$$x - 2 = 2 \Rightarrow (x - 2)^2 = 4$$

The first predicate $x - 2 = 2$ has only one solution, $x = 4$, the second predicate has two solutions $x = 4$ and $x = 0$. By squaring the equation we have found a new equation which includes all the solutions of the first equation, and has one more beside. Implication is expressed in English by using phrases like 'If . . . then . . .'.

An expression involving an implication cannot always be turned the other way around in the same way as those involving equivalence can. An example of this is given by the following statement. It is true that: 'If I am going to work then I take the car' which can be written using the implication symbol as:

'I am going to work' $\Rightarrow$ 'I take the car'

However, it is not true that:

'If I take the car then I am going to work'

This is because there are more occasions when I take the car than simply going to work.

More examples are:

'I only clean the windows if it is sunny'

'I am cleaning the windows' $\Rightarrow$ 'it is sunny'

This does not mean that 'If it is sunny then I clean the windows', as there are some sunny days when I have to go to work or just laze in the garden, or I am on holiday.

An implication sign can be written, and read, from left to right

'It is sunny' $\Leftarrow$ 'I am cleaning the windows'

which I can still read as 'I am cleaning the windows therefore it is sunny' or I could try rearranging the sentence as 'Only if it is sunny will I clean the windows'.

The various ways of expressing these sentiments can get quite involved. The important point to remember is that $p \Rightarrow q$ means that q must be true for all the occasions that p is true, but q could be true on more occasions besides. Going back to equations or inequalities:

$$p \Rightarrow q$$

means that the solution set, P, of p is a subset of the solution set, Q, of q. This is pictured in Figure 3.5.

We can now see that for two equations or inequalities to be equivalent then $p \Rightarrow q$ and $q \Rightarrow p$. This means that their solution sets are exactly the same (Figure 3.6).

Figure 3.5 *P is the solution set of p, Q is the solution set of q. p $\Rightarrow$ q means that P $\subseteq$ Q. D is the domain of p and q.*

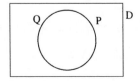

Figure 3.6 *P is the solution set of p and Q is the solution set of q. Then p $\Leftrightarrow$ q means that P = Q.*

Example 3.8   Fill in the correct symbol in each case either $\Rightarrow$, $\Leftarrow$ or $\Leftrightarrow$

(a) $x^2 - 9 = 0 \ldots x = -3$

(b) $x = -\frac{1}{2} \ldots (1/(2x - 5)(x + 1)) = -\frac{1}{3}$ where $x \in \mathbb{R}$, $x \neq 5$, $x \neq -1$

(c) $(x - 3)(x - 1) > 0 \cdots x > 3 \vee x < 1$

Solution

(a) $x^2 - 9 = 0 \cdots x = -3$

Solving $x^2 - 9 = 0$ gives

$$x^2 - 9 = 0 \quad \Leftrightarrow \quad (x - 3)(x + 3) = 0$$
$$\Leftrightarrow \quad x = 3 \lor x = -3.$$

Hence, -3 is only one of the solutions of the first equation so the correct expression is

$$x^2 - 9 = 0 \Leftarrow x = -3.$$

(b) $x = -\frac{1}{2} \ldots (1/(2x - 5)(x + 1)) = -\frac{1}{3}$ where $x \in \mathbb{R}$, $x \neq 5$, $x \neq -1$

Solving

$$\frac{1}{(2x - 5)(x + 1)} = -\frac{1}{3}$$

gives

$$\frac{1}{(2x - 5)(x + 1)} = -\frac{1}{3} \quad \Leftrightarrow \quad -3 = (2x - 5)(x + 1)$$

(multiplying both sides by $(2x-5)(x + 1)$ and as $x \neq 5, x \neq 1$)

$$\Leftrightarrow \quad -3 = 2x^2 - 3x - 5 \quad \text{(multiplying out the brackets)}$$

$$\Leftrightarrow \quad 2x^2 - 3x - 2 = 0$$

(adding 3 on to both sides of the equation)

$$\Leftrightarrow \quad x = \frac{3 \pm \sqrt{9 + 16}}{4}$$

(using the quadratic formula to solve the equation)

$$\Leftrightarrow \quad x = \frac{3 \pm 5}{4}$$

$$\Leftrightarrow \quad x = 2 \lor x = -\frac{1}{2}$$

The second predicate

$$\frac{1}{(2x - 5)(x + 1)} = -\frac{1}{3}$$

has more solutions than the first predicate $x = -\frac{1}{2}$. Thus, the correct expression is:

$$x = -\frac{1}{2} \Rightarrow \frac{1}{(2x - 5)(x + 1)} = -\frac{1}{3}$$

where $x \in \mathbb{R}$, $x \neq 5$, $x \neq -1$.

(c) $(x - 3)(x - 1) > 0 \cdots x > 3 \lor x < 1$

Solve the inequality on the left by firstly solving $f(x) = 0$:

$$(x - 3)(x - 1) = 0 \quad \Leftrightarrow \quad x = 3 \lor x = 1$$

Choosing values on either side of the roots, for example 0,2,4 gives

$$f(0) = (-3)(-1) = 3, \quad \text{i.e. } f(x) > 0$$
$$f(2) = (-1)(1) = -1, \quad \text{i.e. } f(x) < 0$$
$$f(4) = (1)(3), \quad\quad\quad \text{i.e. } f(x) > 0$$

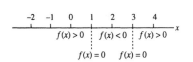

Figure 3.7 *Solving*
$(x - 3)(x - 1) > 0.$

This is then marked on a number line as in Figure 3.7.

As the solution to $(x - 3)(x - 1) > 0$ is $x < 1 \vee x > 3$; we have therefore shown that $(x - 3)(x - 1) > 0 \Leftrightarrow x > 3 \vee x < 1$.

3.7 Making sweeping statements

In Chapter 1 of the Background Mathematics Notes, available on the companion website for this book, we made some statements about numbers which we stated were true for all real numbers. Some of these were the commutative laws:

$$a + b = b + a \quad \text{and} \quad ab = ba$$

and the distributive law:

$$a(b + c) = ab + bc.$$

There is a symbol which stands for 'for all' or 'for every' which allows these laws to be expressed in a mathematical shorthand

$$\forall a, b \in \mathbb{R} \quad a + b = b + a$$
$$\forall a, b \in \mathbb{R} \quad ab = ba$$
$$\forall a, b, c \in \mathbb{R} \quad a(b + c) = ab + bc.$$

Rules, such as the commutative law, are axioms for numbers and need not be proved true. However, more involved expressions, such as

$$\forall a, b \in \mathbb{R} \quad (a - b)(a + b) = a^2 - b^2$$

need to be justified.

If the symbol 'for all' is used with a predicate about its free variable then it becomes a simple proposition which is either true or false. To show that an expression is true we use our knowledge of mathematics to write equivalent expressions until we come across an expression which is obviously true (like $a = a$). To prove it is false is much easier. As we have made a sweeping statement about the expression and said it is true *for all* a, b then we only need to come across one example of numbers which make the expression false.

Example 3.9   Are the following true or false? Justify your answer.

(a) $\forall a, b \in \mathbb{R}, \ a^3 - b^3 = (a - b)(a^2 + ab + b^2)$

(b) $\forall t \in \mathbb{R}$, where $t \neq 1, \ t \neq -1 \quad 1/(t + 1) = (t - 1)/(t^2 - 1)$

(c) $\forall x \in \mathbb{R}$, where $x \neq 0 \quad (x^2 - 1)/x = x - 1$

Solution

(a) $\forall a, b \in \mathbb{R} \quad a^3 - b^3 = (a - b)(a^2 + ab + b^2)$

Looking at the right-hand side of the equality we have

$(a - b)(a^2 + ab + b^2)$

$\quad = a(a^2 + ab + b^2) - b(a^2 + ab + b^2)$ (taking out the brackets)

$\quad = a^3 + a^2 b + ab^2 - ba^2 - ab^2 - b^3$

$\quad\quad$ (taking out the remaining brackets)

$\quad = a^3 - b^3$ (simplifying)

We have shown that the right-hand side is equal to the left-hand side

$\forall a, b \in \mathbb{R} \quad a^3 - b^3 = (a - b)(a^2 + ab + b^2) \Leftrightarrow a^3 - b^3 = a^3 - b^3$

which is true. Therefore

$\forall a, b \in \mathbb{R} \quad a^3 - b^3 = (a - b)(a^2 + ab + b^2)$

is true.

(b) $\forall t \in \mathbb{R}$ where $t \neq 1, t \neq -1 \quad 1/(t + 1) = (t - 1)/(t^2 - 1)$

Take the right-hand side of the equality

$\dfrac{t - 1}{t^2 - 1} = \dfrac{t - 1}{(t - 1)(t + 1)}$ (factorizing the bottom line)

$\quad\quad\quad = \dfrac{1}{(t + 1)}$ (dividing the top and bottom line by $t - 1$ which is allowed as $t \neq 1$)

Hence

$\dfrac{1}{t + 1} = \dfrac{t - 1}{t^2 - 1} \quad \Leftrightarrow \quad \dfrac{1}{t + 1} = \dfrac{1}{t + 1}$

which is true. Thus,

$\forall t \in \mathbb{R}$ where $t \neq 1, t \neq -1 \quad \dfrac{1}{t + 1} = \dfrac{t - 1}{t^2 - 1}$

is true.

(c) $\forall x \in \mathbb{R}$ where $x \neq 0 \quad \dfrac{x^2 - 1}{x} = x - 1$

To show this is false, substitute a value for x, for example, $x = 2$. When $x = 2$

$\dfrac{x^2 - 1}{x} = x - 1$

becomes

$\dfrac{4 - 1}{2} = 2 - 1 \Leftrightarrow \dfrac{3}{2} = 1 \Leftrightarrow \text{F}.$

As the predicate fails for one value of x then

$\forall x \in \mathbb{R}$ where $x \neq 0 \quad (x^2 - 1)/x = x - 1$

is false.

Another useful symbol is ∃, which means, 'there exists'. This can be used to express the fact that every real number has an inverse under addition. Hence, we get

$$\forall a \in \mathbb{R}, \exists b, \quad a + b = 0.$$

If the symbol ∃ is used with a predicate about its free variable, it becomes a simple proposition which is either true or false. In the case of the example given concerning the inverse, this is an axiom of the real numbers and we can just state it is true. Other statements involving existence will need some justification. Proving existence is simpler than disproving it. If I were to state 'There exists a blue moon in the universe', to prove this to be true I only need to find one blue moon but to disprove it I must find all the moons in the universe and show that not one of them is blue.

In other words, to show that some value exists which makes a certain predicate into a true proposition then we only need to find that value and demonstrate that the resulting proposition is true. To show that no value exists, however, is more difficult and if the domain of interest is a set of numbers we need to present an argument about any member of the set.

Example 3.10   Are the following true or false? Justify your answer.

(a) $\exists x \in \mathbb{R}, (x + 2)(x - 1) = 0$
(b) $\exists x \in \mathbb{R}, x^2 + 4 < 0$

Solution

(a) $\exists x \in \mathbb{R}, (x + 2)(x - 1) = 0$

To show this is true, we only need find one value of x which makes the equality correct. For instance, take $x = -2$: when $x = -2$, $(x + 2)(x - 1) = 0$ becomes $(-2 + 2)(-2 - 1) = 0 \Leftrightarrow 0 = 0$, which is true.
   Therefore, $\exists x \in \mathbb{R}, (x + 2)(x - 1) = 0$ is true.

(b) $\exists x \in \mathbb{R}, x^2 + 4 < 0$

Trying a few values for x (e.g. -1, 0, 20, -2) we might suspect that this statement is false. We need to present a general argument in order to convince ourselves of this.
   x^2 is always positive or zero, that is, $x^2 \geqslant 0$ for all x. If we then add on 4, then for all $x, x^2 + 4 \geqslant 4$ and as 4 is bigger than 0.

$x^2 + 4 > 0$ for all x; hence,
$\exists x \in \mathbb{R}, x^2 + 4 < 0$ is false.

3.8 Other applications of predicates

Predicates are often used in software engineering. Some simpler applications are:

(a)   To express the condition under which a program block will be carried out (or a loop will continue execution).
(b)   To express a program specification in terms of its pre- and post-conditions.

Example 3.11   Express the following in pseudo-code: print x and y if y is a multiple of x and x is an integer between 1 and 100 inclusive.

Solution   Pseudo-code is a system of writing algorithms which is similar to some computer languages but not in any particular computer language. We can use any symbols we like as long as the meaning is clear.

y is a multiple of x means that if y is divided by x then the result is an integer. This can be expressed as

$$\frac{y}{x} \in \mathbb{Z}$$

The condition that x must lie between 1 and 100 can be expressed as $x \geqslant 1$ and $x \leqslant 100$. Combining these conditions gives the following interpretation for the algorithm:

if $\left(\dfrac{y}{x} \in \mathbb{Z} \wedge x \geqslant 1 \wedge x \leqslant 100\right)$ then

   print x, y
endif

Example 3.12   A program is designed to take a given whole positive number, x, greater than 1, and find two factors of x, a and b, which multiplied together give x. a and b should be whole positive numbers different from 1, unless x is prime. Express the pre- and post-conditions for the program.

Solution   Pre-condition is $x \in \mathbb{N} \wedge x > 1$.

The post-condition is slightly more difficult to express. Clearly $ab = x$ is a statement of the fact that a and b must multiply together to give x. Also a and b must be elements of $\mathbb{N}$. a and b cannot be 1 unless x is prime, this can be expressed by

$$(a \neq 1 \wedge b \neq 1) \vee (x \text{ is prime}).$$

Finally, we have the post-condition as

$$a \cdot b = x \wedge (a \in \mathbb{N}) \wedge (b \in \mathbb{N}) \wedge ((a \neq 1 \wedge b \neq 1) \vee (x \text{ is prime})).$$

3.9 Summary

(1) The stages in solving a real-life problem using mathematics are:
   (a) Express the problem as a mathematical one, using any necessary scientific rules and assumptions about the behaviour of the system and using letters to represent any unknown quantities. This is called a mathematical model.
   (b) Solve the mathematical problem by moving from one statement to an equivalent statement justifying each stage by using relevant mathematical knowledge.
   (c) Check the mathematical solution(s) by substituting them into the original formulation of the mathematical problem.
   (d) Translate the mathematical solution back into the setting of the original problem.
   (e) Test the model solutions for some realistic values to see how well the model correctly predicts the behaviour of the system. If it is acceptable, then the model can be used to predict more results.

(2) A predicate is a mathematical statement containing a variable. Examples of predicates are equations and inequalities.

(3) If values are substituted into a predicate it becomes a simple proposition which is either true or false.

(4)   The three main operations on predicates and propositions are $\wedge$, $\vee$, $\neg$, and these can be defined using truth tables as in Tables 3.1–3.3.

(5)   Two predicates, p, q, are equivalent ($p \Leftrightarrow q$) if they are true for exactly the same set of values.

(6) $p \Rightarrow q$ means 'p implies q', that is, q is true whenever p is true. If p, q are equations or inequalities and $p \Rightarrow q$ then all solutions of p are also solutions of q and q may have more solutions besides.

(7)   The symbol $\forall$ stands for 'for all' or 'for every' and can be used with a predicate to make it into a simple proposition, for example, $\forall a, b \in \mathbb{R}, a^2 - b^2 = (a + b)(a - b)$, which is true.

(8)   The symbol $\exists$ stands for 'there exists' and can also be used with a predicate to make it into a simple proposition, for example, $\exists x \in \mathbb{R}$, $3x = 45$, which is true.

3.10 Exercises

3.1. Assign T or F to the following
   (a) $2x + 2 = 10$ when $x = 1$
   (b) $2x + 2 = 10$ when $x = 2$
   (c) $3x^2 + 3x - 6 = 0$ when $x = 1$
   (d) $1 - t^2 = -3$ when $t = -2$
   (e) $t - 5 = 6.5 \wedge t + 4 = 2.5$ when $t = 1.5$
   (f) $u + 3 = 6 \wedge 2u - 1 = 4$ when $u = 3$
   (g) $3y + 2 = -2.5 \vee 1 - y = 1$ when $y = -1.5$
   (h) $\neg(x^2 - x + 2 = 0)$ when $x = -1$
   (i) $\neg(t - 2 = 4 \wedge t = 3)$
   (j) $\neg(t - 2 = 4) \wedge (t = 3)$
   (k) $\neg\left(3t - 4 = 6 \vee 1 - t = -2\frac{1}{3}\right)$ when $t = 3\frac{1}{3}$.

3.2. Solve the following, justifying each stage of the solution and checking the result.
   (a) $3 - 2x = -1$,   (b) $1 - 2t^2 = 1 - 10t$
   (c) $50t - 11 = -25t^2$,   (d) $30y - 13 = 8y^2$
   (e) $10t - 4 \leqslant -3$,   (f) $10 - 4x > 12$.

3.3. Find the range of values for which the following hold and represent them on a number line.
   (a) $x + 3 \geqslant 5 \vee 1 - 2x > 3$,   (b) $2 - 4t \leqslant 3 \wedge 2 - t < 1$
   (c) $\neg(2x + 3 \leqslant 9)$.

3.4. Fill the correct sign $\Rightarrow$, $\Leftarrow$ or $\Leftrightarrow$ or indicate none of these. Assume the domain is $\mathbb{R}$ unless indicated otherwise.
   (a) $3x^2 - 1 = 0 \cdots x = \frac{1}{\sqrt{3}}$

   (b) $\sqrt{x - 1} = 5 \cdots x = 26$ (where $x \geqslant 1$)

   (c) $t^2 - 5t = 36 \cdots (t - 4)(t - 9) = 0$

   (d) $(2x - 2)/(x - 3) = 1 \cdots 3x + 4 = -x$ (where $x \neq 3$)

   (e) $3x = 4 \cdots (3x)^2 = (4)^2$

   (f) $t + 1 = 5 \cdots (t + 1)^3 = 5^3$

   (g) $(x+1)(x-3) = (x-3)(x+2) \cdots (x+1) = (x+2)$

   (h) $x - 1 = 25 \cdots \sqrt{x - 1} = 5$ where $x \geqslant 1$

   (i) $w/(w^2 - 1) = 1 \cdots 1/(w - 1) = 1$ where $w \neq 1$ and $w \neq -1$

   (j) $(x - 1)(x - 3) < 0 \cdots (x - 3) < 0 \vee (x - 1) < 0$

   (k) $x > 2 \vee x < -2 \cdots x^2 > 4$.

3.5. Determine whether the following statements are true or false and justify your answer.
   (a) $\forall a, b \in \mathbb{R}, a^4 - b^4 = (a - b)(a^3 - a^2b + ab^2 - b^3)$
   (b) $\forall a, b \in \mathbb{R}, a^3 + b^3 = (a + b)^3$
   (c) $\forall x \in \mathbb{R}, x \neq 0 \ 1/(1/x) = x$
   (d) $\exists t, t \in \mathbb{R}, t^2 - 3 = 4$
   (e) $\exists t, t \in \mathbb{R}, t^2 + 3 = 0$.

3.6. Write the following conditions using mathematical symbols:
   (a) x is not divisible by 3,
   (b) y is a number between 3 and 60 inclusive,
   (c) w is an even number greater than 20,
   (d) t differs from t_{n-1} by less than 0.001.

3.7. Express the following problems mathematically and solve them:
   (a) A set of screwdrivers cost €10 and hammers €6.50. Find the possible combinations of maximum numbers of screwdriver and hammer sets that can be bought for €40.
   (b) An object is thrown vertically upwards from the ground with an initial velocity of $10 \, \text{m s}^{-1}$. The mass of the object is 1 kg. Find the maximum height that the object can reach using
      (i)   Kinetic energy (KE) is given by $\frac{1}{2}mv^2$, where m is its mass and v its velocity.
      (ii) The potential energy (PE) is mgh, where m is the mass, g is the acceleration due to gravity (which can be taken as $10 \, \text{m s}^{-2}$), and h is the height.
      (iii) Assuming that no energy is lost as heat due to friction, then the conservation of energy law gives KE + PE = constant.

3.8. A road has a bend with radius of curvature 100 m. The road is banked at an angle of 10°. At what speed should a car take the bend in order not to experience any side thrust on the tyres? Use the following assumptions:

(a) The sideways force needed on the vehicle in order to maintain it in circular motion (called the centripetal force) $= mv^2/r$ where r is the radius of curvature of the bend, v the velocity, and m the mass of the vehicle.

(b) The only force with a component acting sideways on the vehicle, is the reactive force of the ground. This acts in a direction normal to the ground (i.e. we assume no frictional force in a sideways direction).

(c) The force due to gravity of the vehicle is mg, where m is the mass of the vehicle and g is the acceleration due to gravity ($\approx 9.8 \, \mathrm{m\,s^{-2}}$). This acts vertically downwards. The forces operating on the vehicle and ground, in a lateral or vertical direction, are pictured in Figure 3.8.

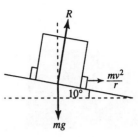

Figure 3.8   *A vehicle rounding a banked bend in the road. R is the reactive force of the ground on the vehicle. The vehicle provides a force of mg, the weight of the vehicle, operating vertically downwards. The vehicle needs a sideways force of mv²/r in order to maintain the locally circular motion.*

4 Boolean algebra

4.1 Introduction

Boolean algebra can be thought of as the study of the set $\{0, 1\}$ with the operations $+$ (or), . (and), and $^-$ (not). It is particularly important because of its use in design of logic circuits. Usually, a high voltage represents TRUE (or 1), and a low voltage represents FALSE (or 0). The operation of OR $(+)$ is then performed on two voltage inputs, using an OR gate, AND(.) using an AND gate and NOT is performed using a NOT gate. This very simple algebra is very powerful as it forms the basis of computer hardware.

You will probably have noticed that the operations of $\wedge$ (AND), $\vee$ (OR), and $\neg$ (NOT) used in Chapter 3 for propositions are very similar to the operations $\cap$ (AND), $\cup$ OR, and $'$ (NOT) (complement) used for sets. This connection is not surprising as membership of a set, A, could be defined using a statements like '3 is a member of A' which is either TRUE or FALSE. In simplifying logic circuits, use is made of the different interpretations that can be put upon the operations and variables. We can use truth tables, borrowed from the theory of propositions, as given in Chapter 3, or we can use Venn diagrams, borrowed from set theory, as given in Chapter 1.

The first thing we shall examine in this chapter is what do we mean by an algebra and why are we able to skip between these various interpretations. Then we look at implementing and minimizing logic circuits.

4.2 Algebra

Before we look at Boolean algebra, we will have a look at some ideas about algebra:

(a) What is an algebra?
(b) What is an operation?
(c) What do we mean by properties (or laws or axioms) of an algebra?

An algebra is a set with operations defined on it. In Chapter 1 of the Background Mathematics Notes, available on the companion website for this book, we looked at the algebra of real numbers and defined an operation is a way of combining two numbers to give a single number. We could therefore define an operation as a way of combining two elements of the set to result in another element of the set.

Example 4.1   The set of real numbers, $\mathbb{R}$, has the operations $+$ and ., for example,

$$3 + 5 = 8 \quad \text{and} \quad 3 \cdot 4 = 12$$

and we could combine any two numbers in this way and we would always get another real number.

Example 4.2 Consider the set of sets in some universal set $\mathcal{E}$, for example,

$\mathcal{E} = \{a, b, c, d, e\}$

$A = \{a, d\}, \quad B = \{a, b, c\}$

then

$A \cap B = \{a\} \quad \text{and} \quad A \cup B = \{a, b, c, d\}.$

The operations of $\cap$ and $\cup$ also result in another set contained in $\mathcal{E}$.

In both of these examples, the operations are *binary* operations because they use *two* inputs to give one output.

There is another sort of operation which is important, called a *unary* operation, because it only has *one* input to give one output. Consider Example 4.2: $A' = \{b, c, e\}$ gives the complement of A. This is a unary operation as only one input, A, was needed to define the output A'.

If we can find a rule which is always true for an algebra then that is called a property, (law or axiom) of that algebra. For example, $(3 + 5) + 4 = 3 + (5 + 4)$ is an application of the associative law of addition which can be expressed in general in the following way for the set of real numbers:

$$\forall a, b, c \in \mathbb{R}, \quad (a + b) + c = a + (b + c)$$

If we can list all the properties of a particular algebra then we can give that algebra a name. For instance, the real numbers with the operations of $+$ and . form a *field*.

4.3 Boolean algebras

Sets as a Boolean algebra

The sets contained in some universal set display a number of properties which can be shown using Venn diagrams.

Example 4.3 Show, using Venn diagrams, that, for any 3 sets A, B, C in some universal set $\mathcal{E}$,

$A \cap (B \cup C) = (A \cap B) \cup (A \cap C).$

Solution This can be shown to be true by drawing a Venn diagram of the left-hand side of the expression and another of the right-hand side of the expression. Operations are performed in the order indicated by the brackets and the result of each operation is given a different shading. This is done in Figure 4.1(a) and (b). The region shaded in Figure 4.1(a) representing $A \cap (B \cup C)$ is the same as that representing $(A \cap B) \cup (A \cap C)$ in Figure 4.1(b), hence, showing that $A \cap (B \cup C) = (A \cap B) \cup (A \cap C)$.

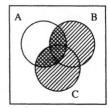

(a) $B \cup C$ ▨
    $A \cap (B \cup C)$ ▧

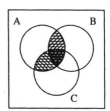

(b) $A \cap B$ ▨
    $A \cap C$ ▨
    $(A \cap B) \cup (A \cap C)$ ≡

Figure 4.1 *(a) A Venn diagram of $A \cap (B \cup C)$. (b) A Venn diagram of $(A \cap B) \cup (A \cap C)$.*

Table 4.1 *The truth tables defining the logical operators*

p	q	$p \wedge q$	p	q	$p \vee q$	p	$\neg p$
T	T	T	T	T	T	T	F
T	F	F	T	F	T	F	T
F	T	F	F	T	T		
F	F	F	F	F	F		

In the same way, other properties can be shown to be true. A full list of the properties gives:

For every A, B, C $\subseteq$ $\mathcal{E}$

(B1) $A \cup A = A$	$A \cap A = A$	Idempotent
(B2) $A \cup (B \cup C)$	$A \cap (B \cap C)$	Associative
$= (A \cup B) \cup C$	$= (A \cap B) \cap C$	
(B3) $A \cup B = B \cup A$	$A \cap B = B \cap A$	Commutative
(B4) $A \cup (A \cap B) = A$	$A \cap (A \cup B) = A$	Absorption
(B5) $A \cup (B \cap C)$	$A \cap (B \cup C)$	Distributive
$= (A \cup B) \cap (A \cup C)$	$= (A \cap B) \cup (A \cap C)$	laws
(B6) $A \cup \mathcal{E} = \mathcal{E}$	$A \cap \emptyset = \emptyset$	Bound laws
(B7) $A \cup \emptyset = A$	$A \cap \mathcal{E} = A$	Identity law
(B8) $A \cup A' = \mathcal{E}$	$A \cap A' = \emptyset$	Complement laws
(B9) $\emptyset' = \mathcal{E}$	$\mathcal{E}' = \emptyset$	0 and 1 laws
(B10) $(A \cup B)' = A' \cap B'$	$(A \cap B)' = A' \cup B'$	De Morgan's laws

Notice that all the laws come in pairs (called duals). A dual of a rule is given by replacing $\cup$ by $\cap$ and $\emptyset$ by $\mathcal{E}$ and vice versa.

Propositions

We looked at propositions in Chapter 3. Propositions can either be given a value of TRUE (T) or FALSE (F). Examples of propositions are $3 = 5$ which is false and $2 < 3$ which is true. The logical operators of AND, OR, and NOT are defined using truth tables, which we repeats in Table 4.1.

Properties of propositions and their operations can be shown using truth tables.

Example 4.4 Show, using truth tables, that for any propositions p, q, r

$$(p \wedge q) \wedge r = p \wedge (q \wedge r)$$

Solution The truth tables are given in Table 4.2. Note that there are eight lines in the truth table in order to represent all the possible states (T, F) for the three variables p, q, and r. As each can be either TRUE or FALSE, in total there are $2^3 = 8$ possibilities. To find $(p \wedge q) \wedge r$, $p \wedge q$ is performed first and the result of that is ANDed with r. To find $p \wedge (q \wedge r)$ then $q \wedge r$ is performed first and p is ANDed with the result. As the resulting columns are equal we can conclude that

$$(p \wedge q) \wedge r \quad \Leftrightarrow \quad p \wedge (q \wedge r)$$

Table 4.2 *A truth table to show*
$(p \wedge q) \wedge r \Leftrightarrow p \wedge (q \wedge r)$. *The fifth column gives
the truth values of* $(p \wedge q) \wedge r$ *and the seventh
column gives the truth value of* $p \wedge (q \wedge r)$. *As
the two columns are the same we can conclude
that* $(p \wedge q) \wedge r \Leftrightarrow p \wedge (q \wedge r)$

p	q	r	$p \wedge q$	$(p \wedge q) \wedge r$	$q \wedge r$	$p \wedge (q \wedge r)$
T	T	T	T	T	T	T
T	T	F	T	F	F	F
T	F	T	F	F	F	F
T	F	F	F	F	F	F
F	T	T	F	F	T	F
F	T	F	F	F	F	F
F	F	T	F	F	F	F
F	F	F	F	F	F	F

Table 4.3 *A truth table to show* $\neg(p \wedge q) \Leftrightarrow (\neg p) \vee (\neg q)$.
The fourth column gives the truth values of $\neg(p \wedge q)$ *and
the seventh column gives the truth value of* $(\neg p) \vee (\neg q)$. *As
the two columns are the same we can conclude that*
$\neg(p \wedge q) \Leftrightarrow (\neg p) \vee (\neg q)$

p	q	$p \wedge q$	$\neg(p \wedge q)$	$\neg p$	$\neg q$	$\neg p \vee \neg q$
T	T	T	F	F	F	F
T	F	F	T	F	T	T
F	T	F	T	T	F	T
F	F	F	T	T	T	T

Example 4.5 Show that for any two propositions p, q:

$$\neg(p \wedge q) \quad \Leftrightarrow \quad (\neg p) \vee (\neg q)$$

Solution The truth table is given in Table 4.3.

It turns out that all the properties we listed for sets are also true for propositions. We list them again, for any three propositions p, q, r

(B1) $p \vee p \Leftrightarrow p$	$p \wedge p \Leftrightarrow p$	Idempotent
(B2) $p \vee (q \vee r)$	$p \wedge (q \wedge r)$	Associative
$\Leftrightarrow (p \vee q) \vee r$	$\Leftrightarrow (p \wedge q) \wedge r$	
(B3) $p \wedge q \Leftrightarrow q \vee p$	$p \wedge q \Leftrightarrow q \wedge p$	Commutative
(B4) $p \vee (p \wedge q) \Leftrightarrow p$	$p \wedge (p \vee q) \Leftrightarrow p$	Absorption
(B5) $p \vee (q \wedge r)$	$p \wedge (q \vee r)$	Distributive laws
$\Leftrightarrow (p \vee q) \wedge (p \vee r)$	$\Leftrightarrow (p \wedge q) \vee (p \wedge r)$	
(B6) $p \vee T \Leftrightarrow T$	$p \wedge F \Leftrightarrow F$	Bound laws
(B7) $p \vee F \Leftrightarrow p$	$p \wedge T \Leftrightarrow p$	Identity laws
(B8) $p \vee \neg p \Leftrightarrow T$	$p \wedge \neg p \Leftrightarrow F$	Complement laws
(B9) $\neg F \Leftrightarrow T$	$\neg T \Leftrightarrow F$	0 and 1 laws
(B10) $\neg(p \vee q) \Leftrightarrow \neg p \wedge \neg q$	$\neg(p \wedge q) \Leftrightarrow \neg p \vee \neg q$	De Morgan's laws

Notice again that all the laws are duals of each other. A dual of a rule is given by replacing $\vee$ by $\wedge$ and F by T, and vice versa.

Table 4.4 *The operations of AND (.), OR (+) and NOT (−) defined for any variables a, b taken from the Boolean set {0, 1}*

a	b	a.b	a	b	a + b	a	ā
0	0	0	0	0	0	0	1
0	1	0	0	1	1	1	0
1	0	0	1	0	1		
1	1	1	1	1	1		

The Boolean set {0, 1}

The simplest Boolean algebra is that defined on the set {0,1}. The operations on this set are AND (.), OR (+), and NOT $^{(-)}$. The operations can be defined using truth tables as in Table 4.1, shown again in Table 4.4. This time notice that the first two are usually ordered in order to mimic binary counting, starting with 0 0, then 0 1, then 1 0, then 1 1. This is merely a convention and the rows may be ordered any way you like. a and b are two variables which may take the values 0 or 1.

This now looks far more like arithmetic. However, beware because although the operation AND behaves like multiplication, $0.0 = 0$, $0.1 = 0$, $1.0 = 0$ and $1.1 = 1$ as in 'ordinary' arithmetic, the operation OR behaves differently as $1 + 1 = 1$.

All the laws as given for sets and for propositions hold again and they can be listed as follows:

For any three variables $a, b, c \in \{0, 1\}$

(B1) $a + a = a$	$a.a = a$	Idempotent
(B2) $a + (b + c)$ $= (a + b) + c$	$a.(b.c) = (a.b).c$	Associative
(B3) $a + b = b + a$	$a + b = b + a$	Commutative
(B4) $a + (a.b) = a$	$a.(a + b) = a$	Absorption
(B5) $a + (b.c)$ $= (a + b).(a + c)$	$a.(b + c)$ $= (a.b) + (a.c)$	Distributive laws
(B6) $a + 1 = 1$	$a.0 = 0$	Bound laws
(B7) $a.1 = a$	$a + 0 = a$	Identity laws
(B8) $a + \bar{a} = 1$	$a.\bar{a} = 0$	Complement laws
(B9) $\bar{0} = 1$	$\bar{1} = 0$	0 and 1 laws
(B10) $\overline{(a + b)} = \bar{a}.\bar{b}$	$\overline{(a.b)} = \bar{a} + \bar{b}$	De Morgan's laws

We often leave out the '.' so that 'ab' means '$a.b$'. We also adopt the convention that . takes priority over + hence miss out some of the brackets.

Example 4.6 Evaluate the following where +, ., and − are Boolean operators.

(a) $1.1.0 + \bar{0}.1$
(b) $(1.\bar{1}) + 1$
(c) $(\bar{1} + 1).0 + (1 + 1).0$

Solution

(a)　We use the conventio

$$1.1.0 + \bar{0}.1 = 0 + 1.1$$

(b)　$(1.\bar{1}) + 1 = (1.0) + 1$

(c)　$(\bar{1} + 1).0 + (1 + 1).0$

The algebraic laws can b

Example 4.7　Simplify

$$abc + \bar{a}bc + b\bar{c}$$

Solution

$abc + \bar{a}b + b\bar{c}$
$= (a + \bar{a})bc + b\bar{c}$　(using a distributive law)
$= 1.bc + b\bar{c}$　　　　(using a complement law)
$= bc + b\bar{c}$　　　　(using an identity law)
$= b(c + \bar{c})$　　　　(by one of the distributive laws)
$= b$　　　　　　　(using a complement and identity law)

Although it is possible to simplify in this way, it can be quite difficult to spot the best way to perform the simplification; hence, there are special techniques used in the design of digital circuits which are more efficient.

4.4 Digital circuits

Switching circuits form the basis of computer hardware. Usually, a high voltage represents TRUE (or 1) while a low voltage represents a FALSE (or 0). Digital circuits can be represented using letters for each input.

There are three basic gates which combine inputs and represent the operators NOT$(-)$, AND $(.)$, and OR $(+)$. These are shown in Figure 4.2.

Other gates

Other common gates used in the design of digital circuits are the NAND gate, $\overline{(ab)}$, that is, not(ab), the NOR gate, $\overline{(a + b)}$, that is, not$(a + b)$ and the EXOR gate, $a \oplus b$, (exclusive or) $a \oplus b = a\bar{b} + \bar{a}b$

These gates are shown in Figure 4.3.

Implementing a logic circuit

First, we need to simplify the expression. Each letter represents an input that can be on or off (1 or 0). The operations between inputs are represented by the gates. The output from the circuit represents the entire Boolean expression.

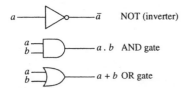

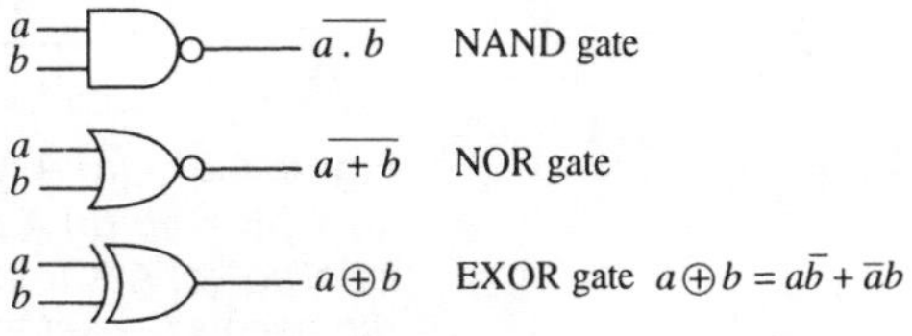

Figure 4.2　*The three basic gates; NOT (−), AND (.), and OR (+).*

Figure 4.3　*Three other common gates; NAND $\overline{(ab)}$, NOR $\overline{(a + b)}$, and EXOR $(a \oplus b)$.*

Example 4.8   Implement $ab\bar{c} + a\bar{b} + a\bar{c}$.

Solution   We can use absorption to write this as $a\bar{b} + a\bar{c}$ and this can be implemented as in Figure 4.4 using AND, OR, and NOT gates. Alternatively, we can use the distributive and De Morgan's laws to write the expression as:

$$a\bar{b} + a\bar{c} = a(\bar{b} + \bar{c}) = a\overline{bc}$$

which can be implemented using an AND and a NAND gate.

Minimization and Karnaugh maps

It is clear that there are numbers of possible implementations of the same logic circuit. However, in order to use less components in building the circuit it is important to be able to minimize the Boolean expression. There are several methods for doing this. A popular method is using a Karnaugh map. Before using a Karnaugh map, the Boolean expression must be written in the form of a 'sum of products'. To do this, we may either use some of the algebraic rules or it may be simpler to produce a truth table and then copy the 0s and 1s into the Karnaugh map. Example 4.9 is initially in the sum of product form and Example 4.10 uses a truth table to find the Karnaugh map.

Example 4.9   Minimize the following using a Karnaugh map:

$$ab + \bar{a}b + a\bar{b}$$

and draw the implementation of the resulting expression as a logic circuit.

Solution   Draw a Karnaugh map as in Figure 4.5(a). If there are two variables in the expression then there are $2^2 = 4$ squares in the Karnaugh map. Figure 4.5(b) shows a Karnaugh map with the squares labelled term

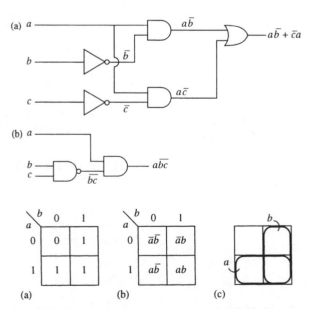

Figure 4.4   *(a) An implementation of $ab\bar{c} + a\bar{b} + a\bar{c} = a\bar{b} + a\bar{c}$. (b) An alternative implementation using $a\bar{b} + a\bar{c} = a\overline{bc}$.*

Figure 4.5   *(a) A two-variable Karnaugh map representing $ab + \bar{a}b + a\bar{b}$ (b) A two-variable Karnaugh map with all the boxes labelled. (c) A Karnaugh map is like a Venn diagram. The second row represents the set a and the second column represents the set b.*

Figure 4.6 *A two-variable Karnaugh map representing $ab + \bar{a}b + a\bar{b}$.*

Figure 4.7 *An implementation of $a + b$.*

by term. Figure 4.5(c) shows that the map is like a Venn diagram of the sets a and b. In Figure 4.5(a) we put a 0 or 1 in the square depending on whether that term is present in our expression. Adjacent 1s indicate that we can simplify the expression. Figure 4.6 indicates how we go about the minimization. We draw a line around any two adjacent 1s and write down the term representing that section of the map. We are able to encircle the second row, representing a, and the second column, representing b. As all the 1s have now been included we know that $a + b$ is a minimization of the expression. Notice that it does not matter if one of the squares with a 1 in it has been included twice but we must not leave any out. The implementation of $a + b$ is drawn in Figure 4.7.

Example 4.10   Minimize $c(b + \overline{(ab)}) + \bar{c}ab$ and draw the implementation of the resulting expression as a logic circuit.

Solution   First, we need to find the expression as a sum of products. This can be done by finding the truth table and then copying the result into the Karnaugh map. The truth table is found in Table 4.5. Notice that we calculate various parts of the expression and build up to the final expression. With practice, the expression can be calculated directly for instance when $a = 0, b = 0$, and $c = 0$ then $c(b + \overline{(ab)}) + \bar{c}ab = 0(0 + \overline{(0.0)}) + \bar{0}.0.0$

$$= 0(0 + 1) + 1.0 = 0.$$

Draw a Karnaugh map as in Figure 4.8(a) and copy in the expression values as found in Table 4.5. There are three variables in the expression, therefore, there are $2^3 = 8$ squares in the Karnaugh map.

Table 4.5   *A truth table to find $ab + \bar{a}b + a\bar{b}$*

a	b	c	ab	$\bar{c}$	$\overline{(ab)}$	$\bar{c}ab$	$b + \overline{(ab)}$	$c(b + \overline{(ab)})$	$c(b + \overline{(ab)}) + \bar{c}ab$
0	0	0	0	1	1	0	1	0	0
0	0	1	0	0	1	0	1	1	1
0	1	0	0	1	1	0	1	0	0
0	1	1	0	0	1	0	1	1	1
1	0	0	0	1	1	0	1	0	0
1	0	1	0	0	1	0	1	1	1
1	1	0	1	1	0	1	1	0	1
1	1	1	1	0	0	0	1	1	1

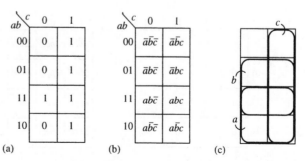

Figure 4.8   *(a) A three-variable Karnaugh map representing $c(b + \overline{(ab)}) + \bar{c}ab$. (b) A three-variable Karnaugh map with all the boxes labelled. (c) A Karnaugh map is like a Venn diagram. The third and fourth rows represent the set a and the second and third rows represent the set b. c is represented by the second column.*

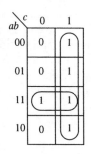

Figure 4.9 *A three-variable Karnaugh map representing* $c(b + \overline{(ab)}) + \bar{c}ab$.

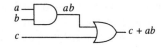

Figure 4.10 *An implementation of* $c + ab$.

Figure 4.8(b) shows a Karnaugh map with the squares labelled term by term. Figure 4.8(c) shows the Venn diagram equivalence with sets a, b, and c. In Figure 4.8(a) we put a 0 or 1 in the square depending on whether that term is present, as given in the truth table in Table 4.5.

Adjacent 1s indicate that we can simplify the expression. Figure 4.9 indicates how we go about the minimization. We draw a line around any four adjacent 1s and write down the term representing that section of the map. The second column represents c and has been encircled. Then we look for any two adjacent 1s. We are able to encircle the third row, representing ab. As all the 1s have now been included we know that $c + ab$ is a minimization of the expression. An implementation of $c + ab$ is drawn in Figure 4.10.

Example 4.11 Minimize $ab\bar{c} + \bar{a}bd + abc\bar{d} + a\bar{b}\bar{c}d + abc$ using a Karnaugh map and draw the implementation of the resulting expression as a logic circuit.

Solution Draw a Karnaugh map as in Figure 4.11(a). There are four variables in the expression therefore there are $2^4 = 16$ squares in the Karnaugh map. Figure 4.11(b) shows a Karnaugh map with the squares labelled term by term. Figure 4.11(c) shows the Venn diagram equivalence with sets $a, b, c,$ and d. In Figure 4.11(a), we put a 0 or 1 in the square depending on whether that term is present in our expression. However, the term $ab\bar{c}$ involves only three out of the four variables. In this case, it must occupy two squares. As d could be either 0 or 1 for '$ab\bar{c}$' to be true, we fill in the squares for $ab\bar{c}d$ and $ab\bar{c}\bar{d}$. The number of squares to be filled with a 1 to represent a certain product is 2^m where m is the number of missing variables in the expression. In this case, $ab\bar{c}$ has no d term in it so the number of squares representing it is 2^1.

Adjacent 1s indicate that we can simplify the expression. Figure 4.12 indicates how we go about the minimisation. We draw a line around any eight adjacent 1s of which there are none. Next we look for any four adjacent 1s and write down the term representing that section of the map. The third row represents ab and has been encircled. The middle four squares represent bd and have been encircled. Then we look for any two adjacent 1s. The bottom two squares of the second column represent $a\bar{c}d$. As all the 1s have now been included we know that $ab+bd+a\bar{c}d$ is a minimization of the expression. This is implemented in Figure 4.13.

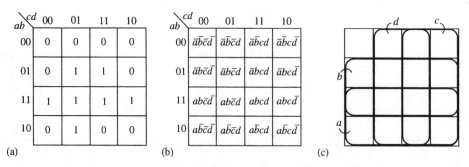

Figure 4.11 *(a) A four variable Karnaugh map representing* $ab\bar{c} + \bar{a}bd + abc\bar{d} + a\bar{b}\,\bar{c}d + abc$. *(b) A four-variable Karnaugh map with all the squares labelled. (c) A Karnaugh map is like a Venn diagram. The third and fourth rows represent the set a and the second and third rows represent the set b. c is represented by the third and fourth columns and d by the second and third columns.*

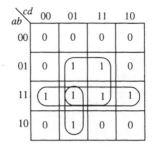

Figure 4.12 *A four-variable Karnaugh map representing $ab\bar{c} + \bar{a}bd + abc\bar{d} + a\bar{b}\,\bar{c}d + abc$.*

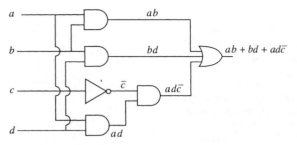

Figure 4.13 *An implementation of $ab + bd + a\bar{c}d$.*

Figure 4.14 *To display the digits 0–9 a seven-segment LED display may be used. For instance, the number 1 requires the segments labelled q and r to light up and the other segments to be off.*

Table 4.6 *A truth table giving the logic control signals for the lamp drivers for the LED segments pictured in Figure 4.14*

Digit displayed	Circuit inputs				Segments						
	a	b	c	d	p	q	r	s	t	u	v
0	0	0	0	0	1	1	1	1	1	1	0
1	0	0	0	1	0	1	1	0	0	0	0
2	0	0	1	0	1	1	0	1	1	0	1
3	0	0	1	1	1	1	1	1	0	0	1
4	0	1	0	0	0	1	1	0	0	1	1
5	0	1	0	1	1	0	1	1	0	1	1
6	0	1	1	0	0	0	1	1	1	1	1
7	0	1	1	1	1	1	1	0	0	0	0
8	1	0	0	0	1	1	1	1	1	1	1
9	1	0	0	1	1	1	1	1	0	1	1
–	1	0	1	0	X	X	X	X	X	X	X
–	1	0	1	1	X	X	X	X	X	X	X
–	1	1	0	0	X	X	X	X	X	X	X
–	1	1	0	1	X	X	X	X	X	X	X
–	1	1	1	0	X	X	X	X	X	X	X
–	1	1	1	1	X	X	X	X	X	X	X

Example 4.12 To display the digits 0–9, a seven-segment light emitting diode (LED) display may be used as shown in Figure 4.14. The various states may be represented using a four-variable digital circuit. The logic control signals for the lamp drivers are given by the truth table given in Table 4.6. The X indicates a 'don't care' condition in the truth table. The column for the segment labelled p can be copied into a Karnaugh map as given in Figure 4.15. Wherever a 1 appears in the truth table representation for p there is a 1 copied to the Karnaugh map. Similarly, the 0s and the 'don't care' crosses are copied. Minimize the Boolean expression for p using the Karnaugh map.

Solution The minimization is represented in Figure 4.15(b). We first look for any eight adjacent squares with a 1 or a X in them. The bottom

Figure 4.15 *(a) A Karnaugh map for the segment labelled p in Figure 4.14. This has been copied from the truth table given in Table 4.5 (b) A minimization of the Karnaugh map. The 'don't care' Xs may be treated as 1s if it is convenient but they can also be treated as 0s.*

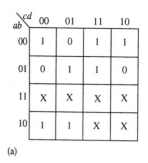

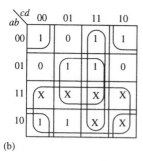

(a)

(b)

two rows are encircled giving the term a. Now we look for groups of four. The central four squares represent bd and the third column represents cd. Finally, we can count the four corner squares as adjacent. This is because two squares may be considered as adjacent if they are located symmetrically with respect to any of the lines which divide the Karnaugh map into equal halves, quarters, or eighths. This means that squares that could be curled round to meet each other, as if the Karnaugh map where drawn on a cylinder, are considered adjacent and also the four corner squares. Here, the four corner squares represent the term $\bar{b}\,\bar{d}$. Hence, the minimization for p gives

$$p = a + cd + bd + \bar{b}\,\bar{d}.$$

4.5 Summary

(1)   An algebra is a set with operations defined on it. A *binary* operation as a way of combining *two* elements of the set to result in another element of the set. A *unary* operation has only *one* input element producing one output.

(2)   A Boolean algebra has the operations of AND, OR, and NOT defined on it and obeys the set of laws given in Section 4.3 as (B1)–(B10). Examples of a Boolean algebra are: the set of sets in some universal set $\mathcal{E}$, with the operations of $\cap$, $\cup$ and $'$; the set of propositions with the operations of $\wedge$, $\vee$, and $\neg$; the set $\{0, 1\}$ with the operations '.' $+$, and $-$.

(3)   Logic circuits can be represented as Boolean expressions. Usually, a high voltage is represented by 1 or TRUE and a low voltage by 0 or FALSE. There are three basic gates to represent the operators AND (.), OR ($+$), and NOT ($-$).

(4)   A Boolean expression may be minimized by first expressing it as a sum of products and then using a Karnaugh map to combine terms.

4.6 Exercises

4.1 Show the following properties of sets using Venn diagrams:
   (a) $A \cup (A \cap B) = A$
   (b) $A \cup (B \cap C) = (A \cup B) \cap (A \cup C)$

4.2 $p = $ 'It rained yesterday'
   $q = $ 'I used an umbrella yesterday'
   (a) Construct English sentences to express, ($\wedge \Leftrightarrow$ 'and', $\vee \Leftrightarrow$ 'or', $\neg \Leftrightarrow$ 'not')

   (i) $p \wedge q$ (ii) $p \vee q$ (iii) $\neg p \vee q$
   (iv) $p \wedge \neg q$ (v) $\neg(p \wedge q)$ (vi) $\neg p \vee \neg q$.
   (b) Given that p is true and q is false, what is the truth value of each part of section (a)?

4.3 Show the following properties of propositions using truth tables
   (a) $p \vee (p \wedge q) \Leftrightarrow p$ (b) $\neg(p \vee q) \Leftrightarrow \neg p \wedge \neg q$.

4.4 Using Venn diagrams or truth tables find simpler expressions for the following:

(a) $ab + a\bar{b}$ (b) $\overline{(ab)}(ac)$

(c) $a + abc$ (d) $\overline{(ab)}a$

4.5 (a) Draw implementations of the following as logic circuits:

(i) $\bar{a}\bar{b} + a\bar{b}$ (ii) $a + bc$

(iii) $\bar{a} + ab$ (iv) $\bar{a}\bar{b}c$

(b) If $a = 1, b = 0$ and $c = 1$, what is the value of each of the expressions in section (a)?

4.6 Minimize the following expressions and draw their logic circuits:

(a) $ab\bar{c} + a\bar{b} + abc$ (b) $\overline{abc}(a + b + c) + a$

(c) $\overline{(a + c)(a + b)} + ab$ (d) $\overline{(a + b)(a + d)} + ab\bar{c} + ab\bar{d} + abcd$

4.7 Obtain a Boolean expression for the logic networks shown in Figure 4.16.

4.8 Consider the LED segment labelled r in Figure 4.14 given in the text. Follow the method given in Example 4.12 to find a minimized expression for r and draw its logic network.

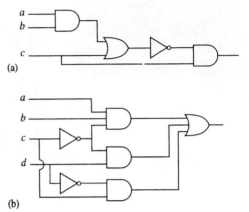

(a)

(b)

Figure 4.16   *(a,b) Logic networks for Exercise 4.7.*

5 Trigonometric functions and waves

5.1 Introduction

Waves occur naturally in a number of situations: the movement of disturbed water, the passage of sound through the air, vibrations of a plucked string. If the movement of a particular particle is plotted against time, then we get the distinctive wave shape, called a sinusoid. The mathematical expression of a wave is found by using the trigonometric functions, sine and cosine. In Chapter 6 of the Background Mathematics Notes on the companion website for this book we looked at right angled triangles and defined the trigonometric ratios. The maximum angle in a right angled triangle is 90° so to find the trigonometric functions, $\sin(t), \cos(t)$, and $\tan(t)$ where t can be extended over the real numbers, we need a new way of defining them. This we do by using a rotating rod. Usually the function will be used to relate, for instance, the height of the rod to time. Therefore, it does not always make sense to think of the input to the cosine and sine functions as being an angle. This problem is overcome by using a new measure for the angle called the radian, which easily relates the angle to the distance travelled by the tip of the rotating rod.

Waves may interfere with each other, as for instance on a plucked string, where the disturbance bounces off the ends producing a standing wave. Amplitude modulation of, for instance, radio waves, works by the superposition of a message on a higher signal frequency. These situations require an understanding of what happens when two, or more cosine or sine functions are added, subtracted, or multiplied and hence we also study trigonometric identities.

5.2 Trigonometric functions and radians

Consider a rotating rod of length 1. Imagine, for instance, that it is a position marked on a bicycle tyre at the tip of one of the spokes, as the bicycle travels along. The distance travelled by the tip of the rod in 1 complete revolution is 2π (the circumference of the circle of radius 1). The height of the rod (measured from the centre of the wheel), y, can be plotted against the distance travelled by the tip as in Figure 5.1.

Similarly, the position to the right or left of the origin, x, can be plotted against the distance travelled by the tip of the rod as in Figure 5.2. Figure 5.1 defines the function $y = \sin(t)$ and Figure 5.2 defines the function $x = \cos(t)$.

This definition of the trigonometric function is very similar to that used for the ratios in the triangle, if the hypotenuse is of length 1 unit. The definitions become the same for angles up to a right angle if radians

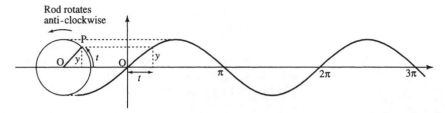

Figure 5.1 *The function y* = sin(*t*), *where t is the distance travelled by the tip of a rotating rod of length 1 unit and y is the height of the rod.*

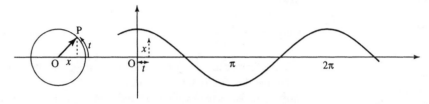

Figure 5.2 *The function x* = cos(*t*), *where t is the distance travelled by the tip of a rotating rod of length 1 unit and x is the position to the right or left of the origin.*

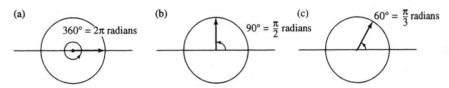

Figure 5.3 *(a)* 360° = 2π *radians. (b)* 90° = π/2 *radians. (c)* 60° = π/3 *radians.*

are used as a measure of the angle in the triangle instead of degrees. Instead of 360° making a complete revolution 2π radians make a complete revolution. Some examples of degree to radian conversion are given in Figure 5.3.

To convert degrees to radians use the fact that 360° is the same as 2π radians or equivalently that 180° is the same as π radians. Hence, to convert degrees to radians multiply by π/180 and to convert radians to degrees multiply by 180/π.

Remember that π is approximately 3.1416 so these conversions can be expressed approximately as: to convert degrees to radians multiply by 0.01745 (i.e. 1° ≈ 0.01745 radians) and to convert radians to degrees multiply by 57.3 (i.e. 1 radian ≈ 57.3°).

Example 5.1

(a) Express 45° in radians.
Multiply 45 by π/180 giving π/4 ≈ 0.785. Hence, 45° ≈ 0.785 radians.

(b) Express, 17° in radians.
Multiply 17 by π/180 giving 17π/180 ≈ 0.297. Hence 17° ≈ 0.297 radians.

(c) Express 120° in radians.
Multiply 120 by π/180 giving 2π/3 ≈ 2.094. Hence, 120° ≈ 2.094.

(d) Express 2 radians in degrees.
Multiply 2 by 180/π giving 114.6. Hence, 2 radians ≈114.6°.

(e) Express (5π/6) radians in degrees.
Multiply 5π/6 by 180/π giving 150. Hence, (5π/6) radians = 150°.

(f) Express 0.5 radians in degrees.
Multiply 0.5 by $180/\pi$ giving 28.6. Hence, 0.5 radians $\approx 28.6°$.

The trigonometric functions can also be defined using a rotating rod of length r as in Figure 5.4.

The function values are given by:

$$\cos(\alpha) = \frac{x}{r}, \quad \sin(\alpha) = \frac{y}{r}, \quad \tan(\alpha) = \frac{y}{x} = \frac{\sin(\alpha)}{\cos(\alpha)}$$

Also

$$\sec(\alpha) = \frac{1}{\cos(\alpha)}, \quad \csc(\alpha) = \frac{1}{\sin(\alpha)}, \quad \cot(\alpha) = \frac{1}{\tan(\alpha)}$$

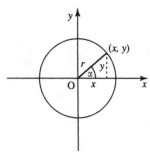

Figure 5.4 *The trigonometric functions defined in terms of a rotating rod of length r.*

where α is measured in radians (one complete revolution is 2π radians).

Notice that the definitions are exactly the same as the trigonometric ratios, where r is the hypotenuse and x and y are the adjacent and opposite sides to the angle, except that x and y can now take both positive and negative values and the angles can be as big as we like or negative (if the rod rotates clockwise):

$$\cos(\alpha) = \frac{\text{adjacent}}{\text{hypotenuse}}$$

$$\sin(\alpha) = \frac{\text{opposite}}{\text{hypotenuse}}$$

$$\tan(\alpha) = \frac{\text{opposite}}{\text{adjacent}}$$

To get the correct function values from the calculator, it should be in 'radian' mode. However, by custom, engineers often use degrees, so we will use the convention that if the 'units' are not specified then radians must be used and for the input to be in degrees that must be expressly marked, for example $\cos(30°)$.

Important relationship between the sine and the cosine

From Pythagoras's theorem, looking at the diagram in Figure 5.4, we have $x^2 + y^2 = r^2$. Dividing both sides by r^2 we get:

$$\frac{x^2}{r^2} + \frac{y^2}{r^2} = 1$$

and using the definitions of

$$\cos(\alpha) = \frac{x}{r} \quad \text{and} \quad \sin(\alpha) = \frac{y}{r}$$

we get

$$(\cos(\alpha))^2 + (\sin(\alpha))^2 = 1$$

and this is written in shorthand as

$$\cos^2(\alpha) + \sin^2(\alpha) = 1$$

where $\cos^2(\alpha)$ means $(\cos(\alpha))^2$.

5.3 Graphs and important properties

We can now draw graphs of the functions for all input values t as in Figures 5.5–5.7.

These are all important examples of periodic functions. To show that the $\cos(t)$ or $\sin(t)$ function is periodic, translate the graph to the left or right by 2π. The resulting graph will fit exactly on top of the original untranslated graph. 2π is called the fundamental period as translating by 4π, 6π, 8π, etc. also results in the graph fitting exactly on top of the original function. The fundamental period is defined as the smallest period that has this property and all other periods are multiples of the fundamental period. This periodic property can be expressed using a letter, n, to represent any integer, giving

$$\sin(t + 2\pi n) = \sin(t)$$
$$\cos(t + 2\pi n) = \cos(t)$$

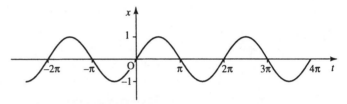

Figure 5.5 *The graph of $y = \sin(t)$, where t can take any value. Notice that the function repeats itself every 2π. This shows that the function is periodic with period 2π. Notice also that the value of $\sin(t)$ is never more than 1 and never less than -1. The function is odd as $\sin(-t) = -\sin(t)$.*

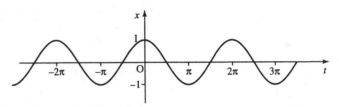

Figure 5.6 *The graph of $x = \cos(t)$, where t can take any value. Notice that the function repeats itself every 2π. This shows that the function is periodic with period 2π. Notice also that the value of $\cos(t)$ is never more than 1 and never less than -1. The function is even as $\cos(-t) = \cos(t)$.*

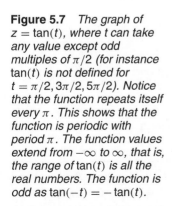

Figure 5.7 *The graph of $z = \tan(t)$, where t can take any value except odd multiples of $\pi/2$ (for instance $\tan(t)$ is not defined for $t = \pi/2, 3\pi/2, 5\pi/2$). Notice that the function repeats itself every π. This shows that the function is periodic with period π. The function values extend from $-\infty$ to ∞, that is, the range of $\tan(t)$ is all the real numbers. The function is odd as $\tan(-t) = -\tan(t)$.*

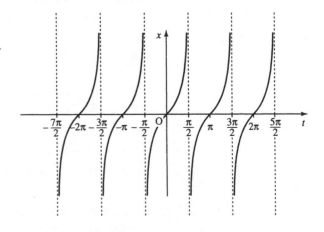

That is, adding or subtracting any multiple of 2π from the value of t gives the same value of the functions $x = \cos(t)$ and $y = \sin(t)$.

The other important thing to remember about $\cos(t)$ and $\sin(t)$ is that although the domain of the functions is all the real numbers, the function values themselves lie between -1 and $+1$

$$-1 \leqslant \cos(t) \leqslant 1$$
$$-1 \leqslant \sin(t) \leqslant 1.$$

We say that the functions are bounded by -1 and $+1$ or, in other words, the range of the cosine and sine functions is $[-1, 1]$.

$z = \tan(t)$ has fundamental period π. If the graph is translated by π to the left or right, then the resulting graph will fit exactly on top of the original graph. This periodic property can be expressed using n to represent any integer, giving

$$\tan(t + \pi n) = \tan(t).$$

The values of $\tan(t)$ are not bounded. We can also say that $-\infty < \tan(t) < \infty$.

Symmetry

Other important properties of these functions are the symmetry of the functions. $\cos(t)$ is even, while $\sin(t)$ and $\tan(t)$ are odd. Unlike the terms odd and even when used to describe numbers, not all functions are either odd or even, most are neither. To show that $\cos(t)$ is even, reflect the graph along the vertical axis. The resulting graph is exactly the same as the original graph. This shows that swapping positive t values for negative ones has no difference on the function values, that is,

$$\cos(-t) = \cos(t).$$

Other examples of even functions were given in Chapter 2 and the general property of even functions was given there as $f(t) = f(-t)$. The functions sine and tangent are odd. If they are reflected along the vertical axis then the resulting graph is an upside down version of the original. This shows that swapping positive t values for negative ones gives the negative of the original function. This property can be expressed by

$$\sin(-t) = -\sin(t)$$
$$\tan(-t) = -\tan(t).$$

For a general function, $y = f(t)$, the property of being odd can be expressed by $f(-t) = -f(t)$.

The relationships between the sine and cosine

Take the graph of $\sin(t)$ and translate it to the left by $90°$ or $\pi/2$ and we get the graph of $\cos(t)$. Equivalently, take the graph of $\cos(t)$ and translate it to the right by $\pi/2$ and we get the graph of $\sin(t)$. Using the ideas of translating functions given in Chapter 2, we get

$$\sin\left(t + \frac{\pi}{2}\right) = \cos(t)$$

$$\cos\left(t - \frac{\pi}{2}\right) = \sin(t).$$

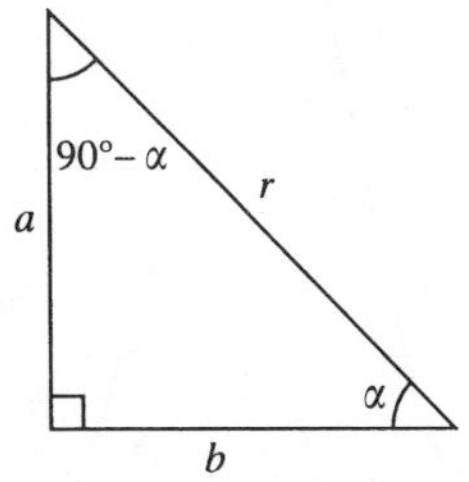

Other relationships can be shown using triangles as in Figure 5.8 giving

$$\cos\left(\alpha - \frac{\pi}{2}\right) = \sin(\alpha) \quad \text{and} \quad \sin\left(\frac{\pi}{2} - \alpha\right) = \cos(\alpha).$$

From Pythagoras theorem we also have that

$$a^2 + b^2 = r^2,$$

dividing both sides by r^2 we get

$$\frac{a^2}{r^2} + \frac{b^2}{r^2} = 1,$$

and using the definitions of $\sin(\alpha) = a/r$ and $\cos(\alpha) = b/r$, we get

$$\cos^2(\alpha) + \sin^2(\alpha) = 1.$$

Rearranging this, we have $\cos^2(\alpha) = 1 - \sin^2(\alpha)$ or $\sin^2(\alpha) = 1 - \cos^2(\alpha)$.

Figure 5.8 *(a)* $\sin(\alpha) = a/r$ *and* $\cos(90° - \alpha) = a/r$. *Then* $\cos(90° - \alpha) = \sin(\alpha)$. *As the cosine is an even function* $\cos(90° - \alpha) = \cos(-(90° - \alpha)) = \cos(\alpha - 90°)$ *which confirms that* $\cos(\alpha - 90°) = \sin(\alpha)$. $\cos(\alpha) = b/r$ *and* $\sin(90° - \alpha) = b/r$, *so* $\cos(\alpha) = \sin(90° - \alpha)$.

Example 5.2 Given $\sin(A) = 0.5$ and $0 \leqslant A \leqslant 90°$, use trigonometric identities to find:

(a) $\cos(A)$
(b) $\sin(90° - A)$
(c) $\cos(90° - A)$.

Solution

(a) Using $\cos^2(A) = 1 - \sin^2(A)$ and $\sin(A) = 0.5$

$$\Rightarrow \quad \cos^2(A) = 1 - (0.5)^2 = 0.75$$

$$\Leftrightarrow \quad \cos(A) \approx \pm 0.866.$$

As A is between $0°$ and $90°$, the cosine must be positive giving $\cos(A) \approx 0.866$.

(b) As $\sin(90° - A) = \cos(A)$, $\sin(90° - A) \approx 0.866$.
(c) As $\cos(90° - A) = \sin(A)$, $\cos(90° - A) = 0.5$.

The functions $A\cos(at + b) + B$ and $A\sin(at + b) + B$

The graph of these functions can be found by using the ideas of Chapter 2 for graph sketching.

Example 5.3 Sketch the graph of y against t, where

$$y = 2\cos\left(2t + \frac{2\pi}{3}\right).$$

The stages in sketching this graph are shown in Figure 5.9.

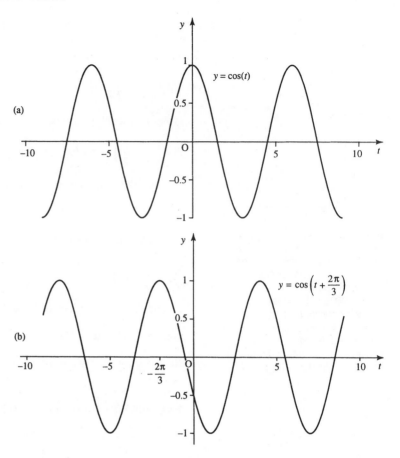

Figure 5.9 *Sketching the graph of $2\cos(2t + 2\pi/3)$: (a) start with $y = \cos(t)$; (b) shift to the left by $2\pi/3$ to give $y = \cos(t + (2\pi/3))$; (c) squash the graph in the t-axis to give $y = \cos(2t + (2\pi/3))$; (d) stretch the graph in the y-axis giving $y = 2\cos(2t + (2\pi/3))$.*

Example 5.4   Sketch the graph of z against q where

$$z = \frac{1}{2}\sin\left(\pi q - \frac{\pi}{4}\right) - \frac{1}{2}.$$

The stages in sketches this graph are given in Figure 5.10.

Amplitude, fundamental period, phase, and cycle rate

In Figure 5.11 are some examples of functions $y = A\cos(ax + b)$ and in each case the amplitude, phase, fundamental period, and cycle rate has been found. In Figure 5.11(a)

$$y = \frac{1}{2}\cos\left(5\pi x + \frac{\pi}{2}\right)$$

is drawn, and has a peak value of 0.5 and a trough value of -0.5. Therefore, the amplitude is half the difference: $(0.5 - (-0.5))/2 = 0.5$. The period, or cycle length, is the minimum amount the graph needs to be shifted to the left or right (excluding no shift) in order to fit over the original graph. In this case the period is 0.4. The phase is found by finding the proportion of the cycle that the graph has been shifted to the left. In this case the proportion of shift is 1/4. Now multiply that by a standard cycle length of 2π to give the phase angle of $\pi/2$. The cycle rate is the number of cycles in unit length given by the reciprocal of the period $= 1/0.4 = 2.5$.

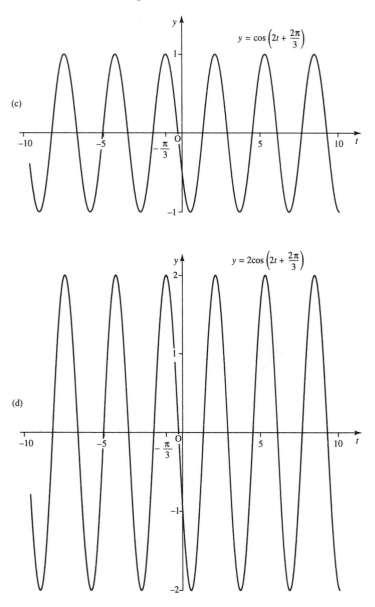

Figure 5.9 *Continued.*

In Figure 5.11(b) $y = 3\cos(2x - 1)$, has a peak value of 3 and a trough value of -3. Therefore, the amplitude is half the difference $= (3 - (-3))/2 = 3$. The period is the minimum amount the graph needs to be shifted to the left or right (excluding no shift) in order to fit over the original graph. In this case the period is π. The phase is found by finding the proportion of the cycle that the graph has been shifted to the left. In this case the proportion of shift is $-0.5/\pi$. Now multiply that by a standard cycle length of 2π to give the phase angle of -1. The cycle rate is the number of cycles in unit length given by the reciprocal of the period $1/\pi \approx 0.32$.

We can generalize from these examples to say that for the function $y = A\cos(ax + b)$, A positive, we have the following.

The *amplitude* is half the difference between the function values at the peak and the trough of the wave and in case where $y = A\cos(ax + b)$ is given by A.

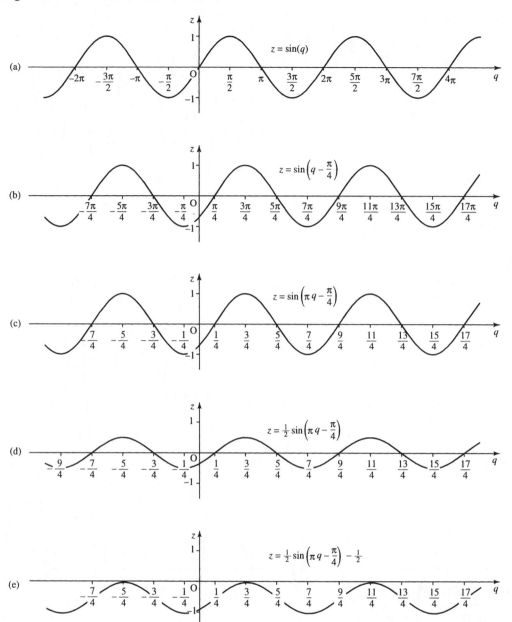

Figure 5.10 *Sketching the graph of $z = \frac{1}{2}\sin(\pi q - (\pi/4)) - \frac{1}{2}$: (a) start with $z = \sin(q)$; (b) shift to the right by $\pi/4$ to give $z = \sin(q - (\pi/4))$; (c) squash the graph in the q-axis to give $z = \sin(\pi q - (\pi/4))$; (d) squash the graph in the z-axis giving $z = \frac{1}{2}\sin(\pi q - (\pi/4))$; (e) translate in the z direction by $\frac{1}{2}$ to get $z = \frac{1}{2}\sin(\pi q - (\pi/4)) - \frac{1}{2}$.*

The *fundamental period, P,* or *cycle length* is the smallest, non-zero, distance that the graph can be shifted to the right or left so that it lies on top of the original graph. This can be found by looking for two consecutive values where the function takes its maximum value, that is when the cosine takes the value 1. Using the fact that $\cos(0) = 1$ and $\cos(2\pi) = 1$,

$\cos(ax + b)$ becomes $\cos(0)$ when $ax + b = 0 \Leftrightarrow x = -b/a$

$\cos(ax + b)$ becomes $\cos(2\pi)$ when $ax + b = 2\pi \Leftrightarrow x = 2\pi/a - b/a$

and the difference between them is $2\pi/a$; giving the fundamental period of the function $\cos(ax + b)$ as $2\pi/a$.

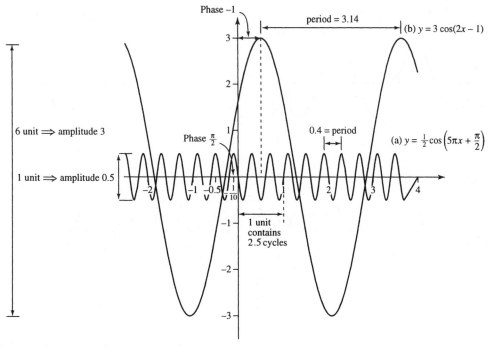

Figure 5.11 *(a) $y = \cos(5\pi x + (\pi/2))$; (b) $y = 3\cos(2x - 1)$.*

The *phase* is given by the number, b, in the expression $A\cos(ax + b)$. The phase is related to the amount the function $A\cos(ax + b)$ is shifted to the left or right with respect to the function $A\cos(ax)$. It expresses the proportion of a standard cycle (maximum 2π) that the graph has been shifted by and therefore a phase can always be expressed between 0 and 2π or more often between $-\pi$ and π. Various phase shifts are given in Figure 5.12.

The cycle rate or frequency is the number of cycles in one unit can be found by relating this to the length of the cycle. The longer the cycle the less cycles there will be in one unit. If the length of one cycle is P (the fundamental period) then there is 1 cycle in P units and $1/P$ cycles in 1 unit.

The cycle rate is the reciprocal of the fundamental period. As for the function $y = A\cos(ax + b)$ the fundamental period is $P = 2\pi/a$, the number of cycles is $1/P$, that is, $a/2\pi$. Examples are given in Figure 5.13.

5.4 Wave functions of time and distance

A wave allows energy to be transferred from one point to another without any particles of the medium moving between the two points. Water waves move along the surface of a pond in response to a child rhythmically splashing a hand in the water. The child's boat floating in the path of the wave merely bobs up and down without moving in the direction of the wave. See Figure 5.14.

If we look at the position of the boat as the wave passes, it moves up and down with the height expressed against time giving a sinusoidal function. This is then a wave function of time and in the expression $y = A\cos(\omega t + \phi)$, the letter A represents the amplitude, ϕ represents the phase, and ω is related to the wave frequency. This is explained in detail in the next section. If we take a snapshot picture of the surface of

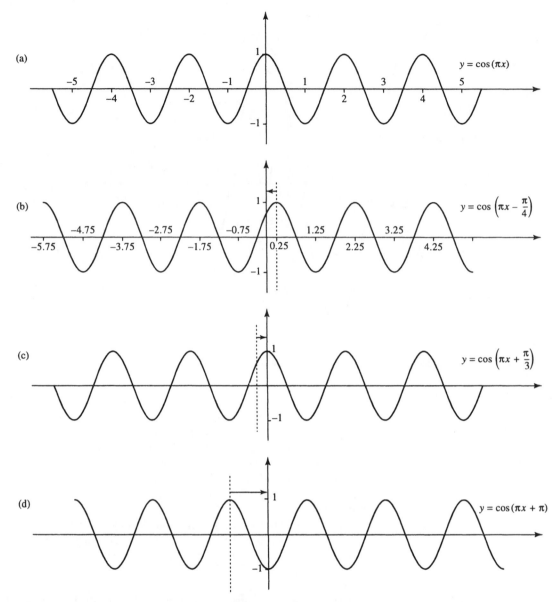

Figure 5.12   *Examples of phase shifting. (a) A graph $y = \cos(\pi x)$. (b) $y = \cos(\pi x - (\pi/4))$ has phase of $-\pi/4$ and is shifted by 1/8 of a cycle (given by the proportion that the phase, $-\pi/4$, represents of a standard cycle of 2π). (c) $y = \cos(\pi x + (\pi/3))$ has phase of $\pi/3$ and is shifted by 1/6 of a cycle (given by the proportion that the phase, $\pi/3$, represents of a standard cycle of 2π). (d) $y = \cos(\pi x + \pi)$ has phase of π and is shifted by 1/2 of a cycle (given by the proportion that the phase, π, represents of a standard cycle of 2π).*

the water at a particular point in time then we will also get a wave shape where we now have an graph of the height of the water expressed against the distance from the waves origin. In this case where $y = A\cos(kx + \phi)$, A still represents the amplitude, ϕ the phase but the coefficient of x, k, is now related to the wavelength. Ideally, we want an expression that can give the height, y, at any position x at any time t. This function is called the progressive wave function and we can combine the two ideas of waves as a function of time and distance to obtain an expression for this function.

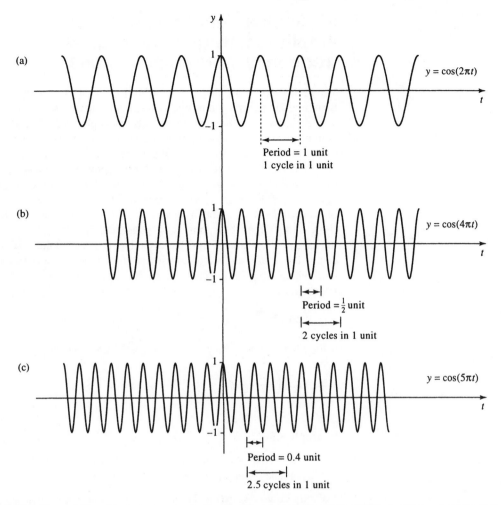

Figure 5.13 *The relationship between cycle length (fundamental period) and the number of cycles in 1 unit. (a) $y = \cos(2\pi t)$ has cycle length $2\pi/2\pi = 1$ and therefore 1 cycle in 1 unit. (b) $y = \cos(4\pi t)$ has cycle length $2\pi/4\pi = 1/2$ and therefore 2 cycles in 1 unit. (c) $y = \cos(5\pi t)$ has cycle length $2\pi/5\pi = 0.4$ and therefore 2.5 cycles in 1 unit.*

Figure 5.14 *A wave created by rhythmically splashing a hand at the edge of a pond. The child's boat bobs up and down without moving in the direction of the wave.*

Sinusoidal functions of time: amplitude, frequency, angular frequency, period, and phase

Waves that represent a displacement from a central fixed position varying with time can be represented by an expression such as y= $A \cos(\omega t + \phi)$ or $y = A \sin(\omega t + \phi)$, where t is in seconds. Examples include an alternating voltage measured across a particular circuit element or the position of the centre of an ear drum as it vibrates in response to a pure sound wave. As we saw in the previous section, A represents the wave amplitude, ω (the Greek letter, omega) is called the angular frequency as it gives the number of cycles in 2π, it is measured in radians per second. The number of cycles in 1 s is called the *frequency*, $f = \omega/2\pi$ and is measured in hertz (Hz). ϕ (the Greek letter, phi) is the phase, the cycle length is $2\pi/\omega$ s. In the case of a function of time the cycle length is called the *periodic time* or just the *period* and we often use the greek letter, τ (tau), to represent this, where $\tau = 2\pi/\omega$. Then we have that $y = A \cos(\omega t + \phi)$ can be rewritten as

$$y = A \cos(2\pi f t + \phi)$$

using the frequency. As $f = 1/\tau$, this can be written as

$$y = A \cos\left(\frac{2\pi}{\tau}t + \phi\right)$$

Example 5.5

(a) $y = 3\cos(t + 1)$; find the amplitude, frequency, period, angular frequency, and phase where t is expressed in seconds.

Compare $y = 3\cos(t + 1)$ with $y = A\cos(\omega t + \phi)$. Then we can see that the angular frequency $\omega = 1$, the phase $\phi = 1$, and the amplitude $A = 3$. As the frequency, $f = \omega/2\pi$, $f = 1/2\pi$, and the period $\tau = 1/f = 1/(1/2\pi) = 2\pi$ s.

(b) $V = 12\cos(314t + 1.6)$; find the amplitude, frequency, period, angular frequency, and phase where t is expressed in seconds.

Compare $V = 12\cos(314t + 1.6)$ with $V = A\cos(\omega t + \phi)$. Then the angular frequency, $\omega = 314$, the phase $\phi = 1.6$, and the amplitude $A = 12$. As $f = \omega/2\pi$, $f = 314/2\pi \approx 50$ Hz, and the period $\tau = 1/f = 1/50 = 0.02$ s.

Sinusoidal functions of distance: amplitude, cycle rate, wavelength, and phase

Waves that give the displacement from a central fixed position of various different points at a fixed moment in time can be represented by an expression such as $y = A\cos(kx + \phi)$ or $y = A\sin(kx + \phi)$, where x is in metres. Examples include the position of a vibrating string at a particular moment or the surface of pond in response to a disturbance. As we saw in the previous section, A represents the wave amplitude; k is called the *wavenumber* and represents the number cycles in 2π. The spatial frequency gives the number of cycles in 1 m $(= k/2\pi)$. The cycle length is called the *wavelength* and we often use the greek letter λ (lambda), to represent this. The phase is ϕ.

The expression for y can be written, using the wavelength, as

$$y = A \cos\left(\frac{2\pi}{\lambda}x + \phi\right).$$

Example 5.6

(a) $y = 4\cos(x + 0.5)$; find the amplitude, wavelength, wavenumber, spatial frequency, and phase where x is expressed in metres.

   Compare $y = 4\cos(x + 0.5)$ with $y = A\cos(kx + \phi)$. Then the wavenumber $k = 1$, the phase $= 0.5$, and the amplitude $A = 4$. As spatial frequency $= k/2\pi$, this gives $1/2\pi$ wavelengths per metre and the wavelength $\lambda = 2\pi/k = 2\pi/1 = 2\pi$ m.

(b) $y = 2\sin(2\pi x)$; find the amplitude, wavelength, wavenumber, spatial frequency, and phase where x is expressed in metres.

   Using $\sin(\theta) = \cos(\theta - (\pi/2))$, we get $2\sin(2\pi x) = \cos(2\pi x - (\pi/2))$. Compare $y = \cos(2\pi x - (\pi/2))$ with $y = A\cos(kx + \phi)$. We can see that the wavenumber $k = 2\pi$, ϕ the phase $= -\pi/2$, and the amplitude $A = 2$. As spatial frequency $= k/2\pi$, this gives 1 wavelength per metre, and the wavelength $\lambda = 2\pi/k = 2\pi/2\pi = 1$ m.

Waves in time and space

The two expressions for a wave function of time and space can be combined as

$$y = A\cos(\omega t - kx)$$

and this is called a progressive wave equation. The $-$ sign is used to give a wavefront travelling from left to right and t should be taken as positive with $\omega t \geqslant kx$.

   Notice that if we look at the movement of a particular point by fixing x, then we replace x by x_0 and this just gives a function of time, $y = A\cos(\omega t + \phi)$ where $\phi = -kx_0$.

   If we look at the wave at a single moment in time, then we fix time and replace t by t_0 and this just gives a function of distance x, $y = A\cos(kx + \phi)$ where $\phi = -\omega t_0$.

   Waves are of two basic types. Mechanical waves need a medium through which to travel, for example, sound waves, water waves, and seismic waves. Electromagnetic waves can travel through a vacuum, for example, light rays, X-rays. In all cases where they can be expressed as a progressive or travelling wave, the frequency, wavelength, etc. can be found from the expression of the wave in the same way.

   Figure 5.15 shows three snapshot pictures of the progressive wave $y = \cos(15t - 3x)$ at $t = 0, t = 2$, and $t = 5$. This wave has angular frequency $\omega = 15$, and wavenumber, $k = 3$ and therefore the frequency f is $2\pi/15$ and the wavelength $\lambda = 2\pi/3$. By considering the amount the wavefront moves in a period of time, we are able to find the wave velocity.

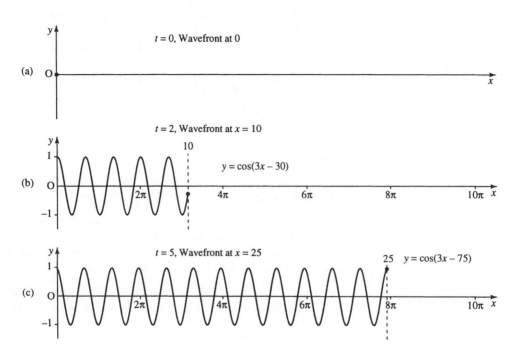

Figure 5.15   *The progressive wave given by the function $y = \cos(15t - 3x)$ where $t > 0$ and $3x < 15t$, $y = 0$ otherwise. (a) No wave at $t = 0$; (b) $t = 2$ gives $y = \cos(30 - 3x) = \cos(3x - 30)$ for $3x < 30$, that is, $x < 10$; (c) $t = 5$ gives $y = \cos(75 - 3x) = \cos(3x - 75)$ for $3x < 75$, that is, $x < 25$. Notice that the wavefront has moved 25 m in 5 s giving a velocity of $25/5 = 5\,\text{m s}^{-1}$.*

Velocity of a progressive wave

The progressive wave $y = A\cos(\omega t - kx)$ vibrates f times per second and the length of each cycle, the wavelength, is λ. In which case the wavefront must move through a distance of λf metres per second and hence the velocity v is given by

$$v = f\lambda$$

where $f = \omega/2\pi$ and $\lambda = 2\pi/k$; hence, $v = \omega/k$.

Example 5.7   A wave is propagated from a central position as in Figure 5.16 and is given by the function $y = 2\cos(6.28t - 1.57r)$ where $t > 0$ and $1.57r \leqslant 6.28t$. Find the frequency, periodic time, spatial frequency, wavenumber, and wavelength.

The wave is pictured for $t = 5$ in Figure 5.16.

Solution   Comparing $y = 2\cos(6.28t - 1.57r)$ with $y = A\cos(\omega t - kr)$ gives $A = 2$, angular frequency $\omega = 6.28$, wavenumber $k = 1.57$. Hence, frequency $f = \omega/2\pi = 6.28/2\pi \approx 1\,\text{Hz}$, periodic time $\tau = 1/f = 1\,\text{s}$, spatial frequency $= k/2\pi = 1.57/2\pi \approx \frac{1}{4}$, wavelength $\lambda = 2\pi/k = 2\pi/1.57 \approx 4\,\text{m}$ and velocity $= f\lambda = 1 \times 4 = 4\,\text{ms}^{-1}$.

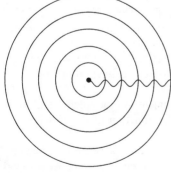

Figure 5.16
$y = 2\cos(6.28t - 1.57r)$ where $1.57r \leqslant 6.28t$ when $t = 5$ giving $y = 2\cos(31.4 - 1.57r)$, $r < 20$. The concentric circles represent the peak amplitudes of the wave. The wavefront has moved to $r = 20$ at $t = 5$ giving a wave velocity of $20/5 = 4\,\text{m s}^{-1}$.

Measuring amplitudes – decibels

In Chapter 2 we looked at sound decay in a room and found that the expression was exponential and could be expressed by using a power of 10. Because of this property of sound decay, and decay of other wave forms, and also because of the need to have a unit which can be used easily to express relatively small quantities, decibels are often used to represent wave amplitudes. In this case the measurement is always in relation to some reference level.

Sound pressure is parallel in electronics to the voltage. The sound pressure level is measured in decibels and is defined as $20 \log_{10}(p/p_0)$ where p is the actual sound pressure and p_0 the reference pressure in $\mathrm{N\,m^{-2}}$. The reference pressure used is approximately the threshold of audibility for sound at $1000\,\mathrm{Hz}$ and is given by

$$p_0 = 2 \times 10^{-5}\,\mathrm{N\,m^{-2}}.$$

Voltage, measured in decibels, is given by $20 \log_{10}(V/V_0)$.

Sound intensity is parallel to power in a circuit. The sound intensity level $= 10 \log_{10}(I/I_0)$ where I is the sound intensity and I_0 is the sound intensity at the threshold of audibility,

$$I_0 = 10^{-12}\,\mathrm{W\,m^{-2}}.$$

Because the reference points used for the measurement of the amplitude of sound are the same whether measuring the sound pressure level or the sound intensity level measurement, of either, will give the same result on the save wave.

Example 5.8   The sound generated by a car has intensity $2 \times 10^{-5}\,\mathrm{W\,m^{-2}}$. Find the sound intensity level and sound pressure level.

Solution   The sound intensity level is

$$10 \log_{10}\left(\frac{2 \times 10^{-5}}{10^{-12}}\right) = 10 \log_{10}(2 \times 10^7)$$
$$= 70 \log_{10}(2) \approx 21.1\,\mathrm{dB}.$$

As this is the same as the sound pressure level, the sound pressure level $= 21.1\,\mathrm{dB}$.

Example 5.9   An amplifier outputs $5\,\mathrm{W}$ when the input power is $0.002\,\mathrm{W}$. Calculate the power gain.

Solution   The power gain is given by

$$10 \log_{10}\left(\frac{5}{0.002}\right) = 10 \log_{10}(2500) \approx 34\,\mathrm{dB}.$$

5.5 Trigonometric identities

Compound angle identities

It can often be useful to write an expression for, for instance, $\cos(A + B)$ in terms of trigonometric ratios for A and B. A common mistake is to assume that $\cos(A + B) = \cos(A) + \cos(B)$ but this can easily be disproved. Take as an example $A = 45°$, $B = 45°$, then

$$\cos(A + B) = \cos(45° + 45°) = \cos(90°) = 0,$$
$$\cos(A) + \cos(B) = \cos(45°) + \cos(45°) \approx 0.707 + 0.707 = 1.414,$$

showing that

$$\cos(A + B) = \cos(A) + \cos(B) \text{ is FALSE.}$$

The correct expression is

$$\cos(A + B) = \cos(A)\cos(B) - \sin(A)\sin(B).$$

The other compound angle identities are as follows:

$$\sin(A + B) = \sin(A)\cos(B) + \cos(A)\sin(B)$$
$$\tan(A + B) = \frac{\tan(A) + \tan(B)}{1 - \tan(A)\tan(B)}.$$

There are various ways of showing these to be true, in Figure 5.17 we show that $\sin(A + B) = \sin(A)\cos(B) + \cos(A)\sin(B)$ by using a geometrical argument. Draw two triangles YZW and YWX so that $\angle A$ and $\angle B$ are adjacent angles and the two triangles are right angled as shown. Draw the lines XX′ and YY′ so that they form right angles to each other, as shown. Notice that $\angle$YXY′ is also $\angle A$. From $\triangle$WX′X, $\sin(A + B) =$ XX′/XW, and as X′Y′YZ is a rectangle then X′Y′ = ZY. So

$$\sin(A + B) = \frac{XY' + ZY}{XW} = \frac{XY'}{XW} + \frac{ZY}{XW} = \frac{ZY}{XW} + \frac{XY'}{XW}$$

As WY/WY = 1 and XY/XY = 1,

$$\sin(A + B) = \frac{ZY}{XW}\frac{WY}{WY} + \frac{XY'}{XW}\frac{XY}{XY}$$
$$= \frac{ZY}{WY}\frac{WY}{XW} + \frac{XY'}{XY}\frac{XY}{XW}.$$

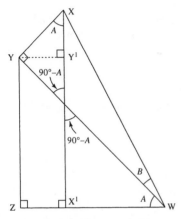

Figure 5.17 $\sin(A + B) = \sin(A)\cos(B) + \cos(A)\sin(B)$.

Looking at the triangles containing these sides we can see that this gives $\sin(A + B) = \sin(A)\cos(B) + \cos(A)\sin(B)$.

A similar argument can be used for $\cos(A + B)$, and $\tan(A + B)$ is usually found by using the expressions for $\sin(A + B)$, $\cos(A + B)$, and the definition of the tangent in terms of A and B.

$$\tan(A + B) = \frac{\sin(A + B)}{\cos(A + B)} = \frac{\sin(A)\cos(B) + \cos(A)\sin(B)}{\cos(A)\cos(B) - \sin(A)\sin(B)}.$$

Divide the top and bottom lines by $\cos(A)\cos(B)$, giving

$$\tan(A + B) = \frac{\dfrac{\sin(A)\cos(B)}{\cos(A)\cos(B)} + \dfrac{\cos(A)\sin(B)}{\cos(A)\cos(B)}}{\dfrac{\cos(A)\cos(B)}{\cos(A)\cos(B)} - \dfrac{\sin(A)\sin(B)}{\cos(A)\cos(B)}}$$

$$\tan(A + B) = \frac{\tan(A) + \tan(B)}{1 - \tan(A)\tan(B)}.$$

From these three identities for $\sin(A + B)$, $\cos(A + B)$, and $\tan(A + B)$ we can obtain many other expressions. A list of important trigonometric identities is given in Table 5.1.

Table 5.1 *Summary of important trigonometric identities*

$$\cos(A \pm B) = \cos(A)\cos(B) \mp \sin(A)\sin(B)$$

$$\sin(A \pm B) = \sin(A)\cos(B) \pm \cos(A)\sin(B)$$

$$\tan(A \pm B) = \frac{\tan(A) \pm \tan(B)}{1 \mp \tan(A)\tan(B)}$$

$$\sin(X) + \sin(Y) = 2\sin\left(\tfrac{1}{2}(X + Y)\right)\cos\left(\tfrac{1}{2}(X - Y)\right)$$

$$\sin(X) - \sin(Y) = 2\cos\left(\tfrac{1}{2}(X + Y)\right)\sin\left(\tfrac{1}{2}(X - Y)\right)$$

$$\cos(X) + \cos(Y) = 2\cos\left(\tfrac{1}{2}(X + Y)\right)\cos\left(\tfrac{1}{2}(X - Y)\right)$$

$$\cos(X) - \cos(Y) = -2\sin\left(\tfrac{1}{2}(X + Y)\right)\sin\left(\tfrac{1}{2}(X - Y)\right)$$

$$\sin(2A) = 2\sin(A)\cos(A)$$

$$\cos(2A) = \cos^2(A) - \sin^2(A)$$

$$\tan(2A) = \frac{2\tan(A)}{1 - \tan(A)}$$

$$\cos(2A) = 2\cos^2(A) - 1$$

$$\cos(2A) = 1 - 2\sin^2(A)$$

$$\cos^2(A) + \sin^2(A) = 1$$

$$\cos^2(A) = \tfrac{1}{2}(\cos(2A) + 1)$$

$$\sin^2(A) = \tfrac{1}{2}(1 - \cos(2A))$$

$$\cos\left(A - \frac{\pi}{2}\right) = \sin(A)$$

$$\sin\left(A + \frac{\pi}{2}\right) = \cos(A)$$

Example 5.10 Using $\cos(2A) = \cos^2(A) - \sin^2(A)$ and $\cos^2(A) + \sin^2(A) = 1$, show that $\cos^2(A) = \tfrac{1}{2}(\cos(2A) + 1)$.

Solution From $\cos^2(A) + \sin^2(A) = 1$, $\sin^2(A) = 1 - \cos^2(A)$ (subtracting $\cos^2(A)$ from both sides).

Substitute this into

$$\cos(2A) = \cos^2(A) - \sin^2(A)$$

$$\cos(2A) = \cos^2(A) - (1 - \cos^2(A))$$

$$\Leftrightarrow \quad \cos(2A) = \cos^2(A) - 1 + \cos^2(A)$$

$$\Leftrightarrow \quad \cos(2A) = 2\cos^2(A) - 1$$

$$\Leftrightarrow \quad \cos(2A) + 1 = 2\cos^2(A) \quad \text{(adding 1 on to both sides)}$$

$$\Leftrightarrow \quad \cos^2(A) = \tfrac{1}{2}(\cos(2A) + 1) \quad \text{(dividing by 2)}$$

Hence

$$\cos^2(A) = \tfrac{1}{2}(\cos(2A) + 1)$$

Example 5.11 From

$$\cos(A \pm B) = \cos(A)\cos(B) \mp \sin(A)\sin(B)$$

$$\sin(A \pm B) = \sin(A)\cos(B) \pm \cos(A)\sin(B),$$

show that

$$\sin(X) + \sin(Y) = 2\sin\left(\tfrac{1}{2}(X + Y)\right)\cos(\tfrac{1}{2}(X - Y)).$$

Solution Use

$$\sin(A + B) = \sin(A)\cos(B) + \cos(A)\sin(B) \qquad (5.1)$$

and

$$\sin(A - B) = \sin(A)\cos(B) - \cos(A)\sin(B), \qquad (5.2)$$

and set

$$X = A + B \qquad (5.3)$$

and

$$Y = A - B. \qquad (5.4)$$

Using Equations (5.3) and (5.4), we can solve for A and B. Add Equations (5.3) and (5.4) giving

$$X + Y = 2A \quad \Leftrightarrow \quad A = \frac{X + Y}{2}.$$

Subtract Equation (5.4) from Equation (5.3) giving

$$X - Y = A + B - (A - B) \quad \Leftrightarrow \quad X - Y = 2B \Leftrightarrow B = \frac{X - Y}{2}.$$

Add Equations (5.1) and (5.2) to give

$$\sin(A + B) + \sin(A - B) = \sin(A)\cos(B) + \cos(A)\sin(B)$$
$$+ \sin(A)\cos(B) - \cos(A)\sin(B)$$
$$\Leftrightarrow \quad \sin(A + B) + \sin(A - B) = 2\sin(A)\cos(B).$$

Substitute for A and B giving

$$\sin(X) + \sin(Y) = 2\sin\left(\tfrac{1}{2}(X + Y)\right)\cos\left(\tfrac{1}{2}(X - Y)\right).$$

Example 5.12 Given that $\cos(60°) = \frac{1}{2}$, find $\cos(30°)$.

Solution Using

$$\cos^2(A) = \tfrac{1}{2}(\cos(2A) + 1)$$

and putting $A = 30°$, we get

$$\cos^2(30°) = \frac{1}{2}(\cos(60°) + 1) = \frac{1}{2}\left(\frac{1}{2} + 1\right) = \frac{1}{2}\left(\frac{3}{2}\right) = \frac{3}{4}$$
$$\Leftrightarrow \quad \cos(30°) = \pm\frac{\sqrt{3}}{2}.$$

From the knowledge of the graph of the cosine we know $\cos(30°) > 0$, so $\cos(30°) = \frac{\sqrt{3}}{2}$.

Example 5.13   Using $\sin(90°) = 1$ and $\cos(90°) = 0$, find $\sin(45°)$

Solution   Using

$$\sin^2(A) = \tfrac{1}{2}(1 - \cos(2A))$$

and putting $A = 45°, 2A = 90°$, we get

$$\sin^2(45°) = \tfrac{1}{2}(1 - \cos(90°)) = \tfrac{1}{2} \quad (\text{as } \cos(90°) = 0)$$

$$\Leftrightarrow \quad \sin(45°) = \pm\sqrt{\tfrac{1}{2}}.$$

From the knowledge of the graph of the sine function we know that $\sin(45°) > 0$, hence,

$$\sin(45°) = \sqrt{\tfrac{1}{2}}.$$

5.6 Superposition

The principle of superposition of waves states that the effect of a number of waves can be found by summing the disturbances that would have been produced by the individual waves separately. This behaviour is quite different from that of travelling particles, which will bump into each other, thereby altering the velocity of both.

The idea of superposition is used to explain the behaviour of:

1. stationary waves formed by two wave trains of the same amplitude and frequency travelling at the same speed in opposite directions.
2. Interference of coherent waves from identical sources.
3. Two wave trains of close frequency travelling at the same speed, causing beats.
4. Diffraction effects.

We look at some examples of these applications.

Standing waves

Suppose that a wave is created by plucking a string of a musical instrument; when the wave reaches the end of the string it is reflected back. The reflected wave will have the same frequency as the initial wave but a different phase and will be travelling in the opposite direction.

The sum of the incident and reflected wave forms a standing wave. An example is shown in Figure 5.18. Figure 5.18(a) shows the incident wave in a string at some instant in time. Its phase is $18°$. Beyond the barrier is shown the hypothetical continuation of the wave as if there were no barrier. The reflected wave is found by turning this section upside down and reflecting it, as shown in Figure 5.18(b). The reflected wave has phase $(180° - \text{phase of the incident wave}) = 180° - 18° = 162°$. In Figure 5.18(c), the sum of the incident and reflected wave produces a standing wave. The maximum and minimum values on this are called antinodes and the zero values are called nodes. As the string is fixed at both ends there must be nodes at the ends.

At different moments in time, the phase of the incident wave will be different. This changes the amplitude of the standing wave but does not change the position of the nodes or antinodes (for a given frequency of wave). Only waves whose wavelengths exactly divide into $2l$ (twice the length of the string) can exist on the string because their amplitude must be 0 at the two end points. Each possible wavelength defines a mode of

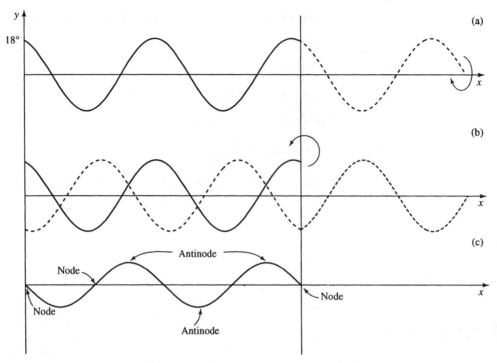

Figure 5.18   *(a) Incident wave in a string at some instant in time. (b) The reflected wave. (c) The sum of the incident and reflected waves.*

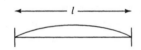

Figure 5.19   *The fundamental mode for a standing wave in a string of length ℓ has wavelength 2ℓ so that half a cycle fits into the length of the string. This is the longest wavelength possible.*

vibration of the string. $\lambda = 2l$ is called the fundamental mode and is shown in Figure 5.19.

The standing wave can be explained using

$$\cos(X) + \cos(Y) = 2\cos\tfrac{1}{2}(X + Y)\cos\tfrac{1}{2}(X - Y).$$

The example given in Figure 5.18 has a wavelength of 4 giving wavenumber $360°/4 = 90°$. The incident wave (phase $18°$) is $y = \cos(90°x + 18°)$ and the reflected wave is $y = \cos(90°x + 162°)$.

Summing these gives

$$\cos(90°x + 18°) + \cos(90°x + 162°)$$
$$= 2\cos\left(\tfrac{1}{2}(90°x + 18° + 90°x + 162°)\right)\cos\left(\tfrac{1}{2}(90°x + 18°\right.$$
$$\left. - (90°x + 162°))\right)$$
$$= 2\cos(90°x + 90°)\cos(72°).$$

As a $90°$ phase-shifted version of a cosine gives a negative sine, this gives $-2\cos(72°)\sin(90°x)$. We see that the result is a sine wave of the same spatial frequency as the incident and reflected wave but with an amplitude of $2\cos 72° \approx 0.62$.

This result can also be found for a general situation – now expressing the phases, etc. in radians. The initial wave is $\cos(kx + \delta)$ and the reflected wave is $\cos(kx + \pi - \delta)$. The standing wave is given by summing these, which gives

$$\cos(kx + \delta) + \cos(kx + \pi - \delta)$$

$$= 2\cos\left(\frac{kx + \delta + kx + \pi - \delta}{2}\right)\cos\left(\frac{kx + \delta - (kx + \pi - \delta)}{2}\right)$$

$$= 2\cos\left(kx + \frac{\pi}{2}\right)\cos\left(\delta - \frac{\pi}{2}\right).$$

The standing wave has the same spatial frequency as the original waves. As $\cos(kx + \pi/2) = -\sin(kx)$ and $\cos(\delta - \pi/2) = \sin(\delta)$, this becomes $-2\sin(kx)\sin(\delta)$. So, the instantaneous amplitude of the standing wave is $2\sin(\delta)$ where δ is the phase of the incident wave.

5.7 Inverse trigonometric functions

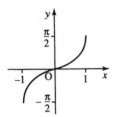

Figure 5.20   *Graph of $y = \sin^{-1}(x)$.*

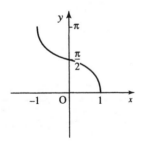

Figure 5.21   *Graph of $y = \cos^{-1}(x)$.*

From the graphs of the trigonometric functions, $y = \sin(x)$, $y = \cos(x)$, and $y = \tan(x)$, we notice that for any one value of y there are several possible values of x. This means that there are no inverse functions if all input values for x are allowed. However, we can see on a calculator that there is a function listed above the sine button and marked as $\sin^{-1}$, so is it in fact the inverse function?

Try the following with the calculator in degree mode. Enter 60 and press sin, then press $\sin^{-1}$. This is shown in Table 5.2(a). The same process is repeated for 120° and for −120°. However, for the latter two cases the inverse function does not work.

If we can restrict the range of values allowed into $\sin(x)$ to the range −90° to +90°, then $\sin^{-1}(x)$ is a true inverse.

The inverse function of $y = \sin(x)$ is defined as $f(x) = \sin^{-1}(x)$ (often written as $\arcsin(x)$ to avoid confusion with $1/\sin(x)$). It is the inverse function only if the domain of the sine function is limited to $-\pi/2 \leqslant x \leqslant \pi/2$. Thus, $\sin^{-1}(\sin(x)) = x$ if x lies within the limits given above and $\sin(\sin^{-1}(x)) = x$ if $-1 \leqslant x \leqslant 1$. The graph of $y = \sin^{-1}(x)$ is given in Figure 5.20.

$f(x) = \cos^{-1}(x)$ is the inverse of $y = \cos(x)$ if the domain of $\cos(x)$ is limited to $0 \leqslant x \leqslant \pi$. $\cos^{-1}(\cos(x)) = x$ if x is limited to the interval above and $\cos(\cos^{-1}(x)) = x$ if $-1 \leqslant x \leqslant 1$. The graph of $y = \cos^{-1}(x)$ is given in Figure 5.21.

$f(x) = \tan^{-1}(x)$ is the inverse of $y = \tan(x)$ if the domain of $\tan(x)$ is limited to $-\pi/2 < x < \pi/2$. Thus, $\tan^{-1}(\tan(x)) = x$ if x is limited as above and $\tan(\tan^{-1}(x)) = x$ for all x. The graph of $y = \tan^{-1}(x)$ is given in Figure 5.22.

Table 5.2   *sin and sin^{-1} on a calculator*

	(a)			(b)			(c)	
$\sin^{-1}$		sin	$\sin^{-1}$		sin	$\sin^{-1}$		sin
60°	→	0.8660	120°	→	0.8660	−120°	→	−0.8660
60°	←	0.8660	60°	←	0.8660	−60°	←	−0.8660

5.8 Solving the trigonometric equations sin *x* = *a*, cos *x* = *a*, tan *x* = *a*

The solutions to the equations $\sin(x) = a$, $\cos(x) = a$, and $\tan(x) = a$ are shown in Figures 5.23–5.25, respectively. Where the lines $y = a$ cross the sine, cosine, or tangent graph gives the solutions to the equations. In Figure 5.23, solutions to $\sin(x) = a$ are given by values of x where the line $y = a$ crosses the graph $y = \sin(x)$. Notice that there are two solutions in every cycle. The first solution is $\sin^{-1}(a)$ and the second is given by $\pi - \sin^{-1}(a)$. Solutions in the other cycles can be found by adding a multiple of 2π to these two solutions.

In Figure 5.24, solutions to $\cos(x) = a$ are given by values of x where the line $y = a$ crosses the graph $y = \cos(x)$. Notice that there are two solutions in every cycle. The two solutions in $[-\pi, \pi]$ are $\cos^{-1}(a)$ and $-\cos^{-1}(a)$. Other solutions can be found by adding a multiple of 2π to these two solutions.

In Figure 5.25, solutions to $\tan(x) = a$ are given by values of x where the line $y = a$ crosses the graph $y = \tan(x)$. Notice there is one solution in every cycle. The solution in $[0, \pi]$ is $\tan^{-1}(a)$. Solutions in the other cycles can be found by adding a multiple of π to this solution.

Example 5.14   Find solutions to $\sin(x) = 0.5$ in the range $[2\pi, 4\pi]$

Solution   From the graph of $y = \sin(x)$ and the line $y = 0.5$ in Figure 5.26, the solutions can be worked out from where the two lines

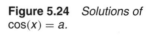

Figure 5.22   *Graph of* $y = \tan^{-1}(x)$.

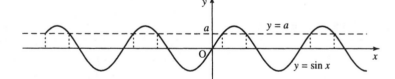

Figure 5.23   *Solutions of* $\sin(x) = a$.

Figure 5.24   *Solutions of* $\cos(x) = a$.

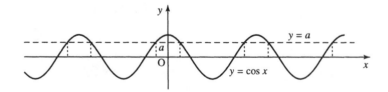

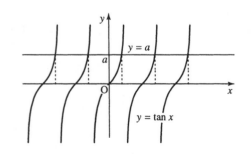

Figure 5.25   *Solutions of* $\tan(x) = a$.

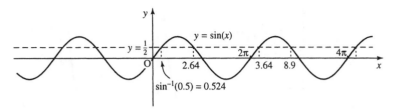

Figure 5.26 *Solutions of*
y = sin(*x*) *and y* = 0.5.

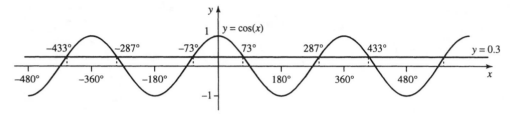

Figure 5.27   *Solutions of y* = cos(*x*) *and y* = 0.3.

cross. The solution nearest $x = 0$ is given by $\sin^{-1}(0.5) \approx 0.524$. The other solution in $[0, 2\pi]$ is given by $\pi - 0.524 \approx 2.62$. Any multiple of 2π added on to these solutions will also give a solution. Therefore, in the range $[2\pi, 4\pi]$ the solutions are 3.64 and 8.9.

Example 5.15 Find solutions to $\cos(x) = 0.3$ in the range $[-480°, 480°]$.

Solution   From the graph of $y = \cos(x)$ and the line $y = 0.3$ in Figure 5.27, the solutions can be worked out from where the two lines cross. The solution nearest $x = 0$ is given by $\cos^{-1}(0.3) \approx 73°$. The other solution in $[0°, 360°]$ is given by $-73°$. Any multiple of $360°$ added on to these solutions will also give a solution. Therefore, in the range $[-480°, 480°]$ the solutions are $-433°, -287°, -73°, 73°, 287°$, and $433°$ (approximately).

Example 5.16   Find solutions to $\tan(x) = 0.1$ in the range $[360°, 540°]$.

Solution   From the graph of $y = \tan(x)$ and the line $y = 0.1$ in Figure 5.28, the solutions can be worked out from where the two lines cross. The solution nearest $x = 0$ is given by $\tan^{-1}(0.1) \approx 6°$. Any multiple of π added on to this solution will also give a solution. Therefore, in the range $[360°, 540°]$ there is only one solution, that is, $366°$.

5.9 Summary

1. Trigonometric functions can be defined using a rotating rod of length 1. The sine is given by plotting the height of the tip of the rod against the distance travelled. The cosine is given by plotting the position that the tip of the rod is to the left or right of the origin against the distance travelled. Hence, if the tip of the rod is at point (x, y) and the tip has travelled a distance of t units, then $\sin(t) = y, \cos(t) = x$, and the tangent is given by

$$\tan(t) = \frac{\sin(t)}{\cos(t)} = \frac{y}{x}.$$

2. If angles are measured in radians, then this definition is the same for angles between 0 and $\pi/2$ as that given by defining the cosine, sine, and tangent from the sides of a triangle (of hypotenuse 1) as

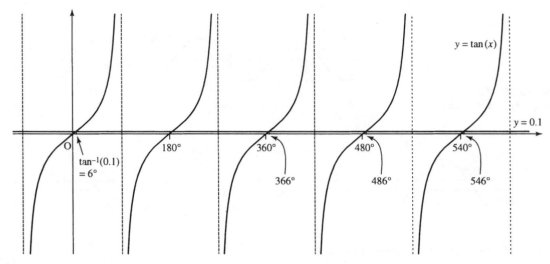

Figure 5.28   *Solutions of y = tan(x) and y = 0.1.*

in Chapter 6 of the Background Mathematics Notes given on the companion website for this book.

There are 2π radians in a complete revolution (360°) and therefore π radians = 180°.

$$1 \text{ radian} = \frac{180°}{\pi}$$

$$1° = \frac{\pi}{180} \text{ radians.}$$

The trigonometric ratios can now be defined using a rotating rod of length r and the angle, α, made by the rod to the x axis. Then, if the tip of the rod is at point (x, y):

$$\cos(\alpha) = \frac{x}{r}, \quad \sin(\alpha) = \frac{y}{r}, \quad \tan(\alpha) = \frac{y}{x} = \frac{\sin(\alpha)}{\cos(\alpha)}$$

α is normally expressed in radians, although engineers often use degrees. If degrees are intended, then they must be explicitly marked.

3. $\sin(t)$ and $\tan(t)$ are odd functions, while $\cos(t)$ is an even function. This can be expressed by

$$\sin(-t) = -\sin(t)$$

$$\cos(-t) = \cos(t)$$

$$\tan(-t) = -\tan(t)$$

$\sin(t)$ and $\cos(t)$ are periodic with period 2π and $\tan(t)$ is periodic with period π. This can be expressed by

$$\sin(t + 2\pi n) = \sin(t)$$

$$\cos(t + 2\pi n) = \cos(t)$$

$$\tan(t + \pi n) = \tan(t)$$

where $n \in \mathbb{Z}$.

4.   For the function $y = A\cos(ax + b)$. A is the amplitude, the cycle rate (number of cycles in 1 unit) = $a/2\pi$, the fundamental period, or cycle length, $P = 2\pi/a$, the phase is b.

For a function of time $y = A\cos(\omega t + \phi)$, ω is the angular frequency and is measured in radians s^{-1}. The number of cycles in

1 s is the frequency, $f = \omega/2\pi$ and is measured in Hz. The cycle length is called the periodic time or period (often represented by τ) $= 2\pi/\omega$ and is measured in seconds. The phase is ϕ.

For a function of distance $y = A\cos(kx + \phi)$, k is the wavenumber. The number of cycles in 1 m is the spatial frequency, $= k/2\pi$ wavelengths per metre, the cycle length is called the wavelength (often represented by λ) $= 2\pi/k$ and is measured in metres, and the phase is ϕ.

The function $y = A\cos(\omega t - kx)$ with $t > 0$ and $\omega t \geqslant kx$ is called the progressive wave equation and has velocity $v = \lambda f$ m s^{-1}.

5. Wave amplitudes are often measured on a logarithmic scale using decibels.

6. There are many trigonometric identities, summarized on Table 5.1. Some of the more fundamental ones are:

$$\cos(A \pm B) = \cos(A)\cos(B) \mp \sin(A)\sin(B)$$

$$\sin(A \pm B) = \sin(A)\cos(B) \pm \cos(A)\sin(B)$$

$$\tan(A) = \frac{\sin(A)}{\cos(A)}$$

$$\cos^2(A) + \sin^2(A) = 1$$

from which others can be derived.

7. The principle of superposition of waves gives that the effect of a number of waves can be found by summing the disturbances that would have been produced by the waves separately. One application of this is the case of stationary waves, which can be explained mathematically using trigonometric identities.

8. $\sin^{-1}(x)$ is the inverse function to $\sin(x)$ if the domain of $\sin(x)$ is limited to $-\pi/2 \leqslant x \leqslant \pi/2$, in which case $\sin^{-1}(\sin(x)) = x$.

$\cos^{-1}(x)$ is the inverse function to $\cos(x)$ if the domain of $\cos(x)$ is limited to $0 \leqslant x \leqslant \pi$, in which case $\cos^{-1}(\cos(x)) = x$.

$\tan^{-1}(x)$ is the inverse function to $\tan(x)$ if the domain of $\tan(x)$ is limited to $-\pi/2 < x < \pi/2$, in which case $\tan^{-1}(\tan(x)) = x$.

9. Inverse trigonometric functions are used in solving trigonometric equations. There are many solutions to trigonometric equations and graphs can be used to help see where the solutions lie.

$\sin(x) = a$ has two solutions: one is $x = \sin^{-1}(a)$ and another is $\pi - \sin^{-1}(a)$. Other solutions can be found by adding or subtracting a multiple of 2π.

$\cos(x) = a$ has two solutions: one is $x = \cos^{-1}(a)$ and another is $-\cos^{-1}(a)$. Other solutions can be found by adding or subtracting a multiple of 2π.

$\tan(x) = a$ has one solution: $x = \tan^{-1}(a)$. Other solutions can be found by adding or subtracting a multiple of π.

5.10 Exercises

5.1. Without using a calculator, express the following angles in degrees (remember π radians $= 180°, \pi \approx 3.142$):

(a) $\frac{2}{3}\pi$ radians   (b) 4π radians   (c) $\frac{3}{5}\pi$ radians
(d) 6.284 radians   (e) 1.571 radians.

Check your answers using a calculator.

5.2. Without using a calculator, express the following angles in radians:

(a) 45°   (b) 135°   (c) 10°   (d) 150°.

Check your answers using a calculator.

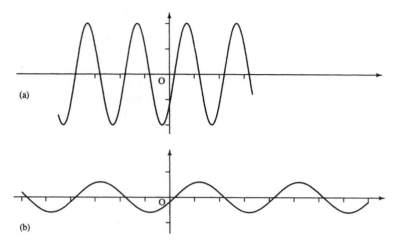

Figure 5.29 *Graphs for Exercise 5.5.*

5.3. Given that $\cos(\pi/3) = 1/2$, without using a calculator, find:

(a) $\sin(\pi/3)$ (b) $\tan(\pi/3)$ (c) $\cos(2\pi/3)$

(d) $\sin(7\pi/3)$ (e) $\tan(4\pi/3)$

Check your answers using a calculator.

5.4. By considering transformations of the graphs of $\sin(x), \cos(x)$, and $\tan(x)$, sketch the graphs of the following:

(a) $y = \sin(x + (\pi/4))$ (b) $y = \tan(x - (\pi/2))$
(c) $y = 3\sin(x)$ (d) $y = \frac{1}{2}\cos(x)$
(e) $y = \sin(\pi x)$ (f) $y = 2\sin(\frac{1}{2}x + (\pi/6))$
(g) $y = \sin(x) + 3$ (h) $y = -\cos(x)$.

5.5. From the graphs in Figure 5.29, find the phase, amplitude, period (cycle length), and number of cycles in one unit.

5.6. For the following functions of time, find the amplitude, period, angular frequency, and phase:

(a) $y = 3\cos(4t + (\pi/2))$ (b) $V = \sin(377t + 0.4)$
(c) $p = 40\cos(3000t - 0.8)$.

5.7. For the following functions of distance, x, find the amplitude, wavelength, spatial frequency, and wavenumber:

(a) $y = 0.5\cos(2x - (\pi/2))$
(b) $y = 2\cos(72x + 0.33)$
(c) $y = 52\sin(80x)$

5.8. Given a progressive wave $t > 0$

$$y = \begin{cases} 3\cos(2t - 5x) & \text{for } 5x \leqslant 2t \\ 0 & \text{otherwise} \end{cases}$$

(a) Sketch the waves for $t = 1, t = 5$, and $t = 10$.
(b) Sketch the wave as a function of time for: (i) $x = 2$ $(t > 5)$; and (ii) $x = 4$ $(t > 10)$.

(c) Find the wave velocity and use your graphs to justify it.

5.9. A pneumatic drill produces a sound pressure of $6\,\mathrm{N\,m^{-2}}$. Given that the reference pressure is $2 \times 10^{-5}\,\mathrm{N\,m^{-2}}$, find the sound pressure level in decibels.

5.10. The reference level on a voltmeter is set as 0.775 V. Calculate the reading in decibels when the voltage reading is 0.4 V.

5.11. Show, using trigonometric identities, that
(a) $\cos(X + \delta) - \cos(X - \delta) = -2\sin(\delta)\sin(X)$
(b) $\sin(X + \delta) + \sin(X - \delta) = 2\sin(X)\cos(\delta)$

5.12. Two wave trains have very close frequencies and can be expressed by the sinusoids $y = 2\sin(6.14t)$ and $y = 2\sin(6.19t)$. Their sum is sketched in Figure 5.30. Use the expression for the summation of two sines to find the beat frequency (the number of times the magnitude of the amplitude envelope reaches a maximum each second).

5.13. A single frequency of 200 Hz (message signal) is amplitude modulated with a carrier frequency of 2 MHz. Express the message signal as $m = a\cos(\omega_1 t)$ and the carrier as $c = b\cos(\omega_2 t)$ and assume that the modulation gives the product $mc = ab\cos(\omega_1 t)\cos(\omega_2 t)$. Use trigonometric identities to show that the modulated signal can be represented by the sum of two frequencies at $2 \times 10^6 \pm 200$ Hz.

5.14. (a) Give the wavelengths of three modes of vibration on a string of length 0.75 m.
(b) The velocity v is approximately given by $v = \sqrt{T/m}$ where T is the tension and m is the mass per unit length of the string. Given that $T = 2200\,\mathrm{N}$ and $m = 0.005\,\mathrm{kg\,m^{-1}}$, find the frequency of the fundamental mode.

5.15. Use $a\cos(\omega t + \delta) = a\cos(\omega t)\sin(\delta) - a\sin(\omega t)\sin(\delta)$ to find c and d in the expression $2\cos(3t + (\pi/3)) = c\cos(3t) + d\sin(3t)$.

5.16. Express as single sines or cosines:
(a) $\sin(43°)\cos(61°) + \cos(43°)\sin(61°)$
(b) $\sin(22°)\cos(18°) - \cos(22°)\cos(18°)$
(c) $\cos(63°)\cos(11°) + \sin(63°)\sin(11°)$

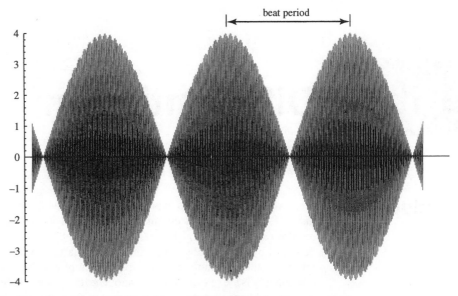

Figure 5.30   *Two sinusoids $y = 2\sin(6.14t)$ and $2\sin(6.19t)$ for Exercise 5.12.*

(d) $\sin(41°)\sin(22°) - \cos(41°)\cos(22°)$

(e) $\sin(2x)\cos(x) + \cos(2x)\sin(x)$

(f) $\cos^2(x) - \sin^2(x)$.

5.17. Express $\cos(x+y+z)$ in terms of the sines and cosines of x, y, and z.

5.18. Find all the solutions to the following equations in the interval $[0, 6\pi]$:

(a) $\sin(x) = -1/2$ (b) $\tan(x) = 1/3$

(c) $\cos(x) = -0.8$ (d) $\sin^2(x) = 0.25$

(e) $\cos(x) + 2\cos^2(x) = 0$ (f) $\sin^2(x) - 2 = 2$.

6 Differentiation

6.1 Introduction

We have used functions to express relationships between variables. For instance, in an electrical circuit the current and the voltage through a resistor can be related by $V = IR$. Here, one physical relationship can be found by simply substituting in the formula the known value of the other. Another common relationship between physical quantities is that one quantity is the rate of change of another: for instance, speed is defined as rate of change of distance with respect to time. It is simple to find the average speed of a moving object if we know how far it has travelled in a certain length of time, which is given by

$$\frac{\text{Distance travelled}}{\text{Time taken}}$$

This does not give an idea of the speed at any particular instant. If I am travelling by coach from London to Birmingham the journey takes about 2.75 h to go about 110 miles. This means that the average speed is about 40 mph. However, I know that the coach by no means travels at a constant speed. Through central London it travels at around 12 mph and on the motorway at around 70 mph. Is there a way that we can estimate the speed at any particular moment as accurately as possible, armed only with a mileometer, to give the distance travelled, and a stop watch? I can measure the distance travelled in a 10 s interval and find that it is 0.2 miles. This means that the average speed is 0.2/10 miles per second $= 0.02$ miles per second $= 0.02 \times 60 \times 60$ mph $= 72$ mph. This gives a pretty good idea of the speed at any moment within that 10 s interval, as the length of time is small enough that the speed probably has not changed too much. The smaller the period of time over which we take the measurement, the more accurately we should be able to estimate the speed at any one instant. The speed is found by looking at the ratio of the distance travelled over the time taken where the time interval is taken as small as possible. This is an approximation to the instantaneous rate of change of distance with respect to time.

The rate of change of one quantity with respect to another is called its derivative. If we know the expression defining the function then we are able to find its derivative. The techniques used for differentiating are described in this chapter.

Many physical quantities are related through differentiation. Some of these are current and charge, acceleration and velocity, force and work done, momentum and force, and power and energy. Applications of differentiation are, therefore, very widespread in all areas of engineering.

6.2 The average rate of change and the gradient of a chord

A ball is thrown from the ground and after t s is at a distance s m from the ground, where

$$s = 20t - 5t^2$$

Find:

(a) the average velocity of the ball between $t = 1$ s and $t = 1.1$ s;
(b) the average velocity between $t = 1$ and $t = 1.01$ s;
(c) the average velocity between $t = 1$ and $t = 1.001$ s;
(d) the average velocity between $t = 1$ and $t = 1.0001$ s;

Guess the velocity at $t = 1$.

Solution The average velocity is given by the distance moved divided by the time taken. That can be represented by

$$\text{Average velocity} = \frac{\text{change in distance}}{\text{change in time}} = \frac{\delta s}{\delta t} = \frac{s(t_2) - s(t_1)}{t_2 - t_1}$$

where t_2 and t_1 are the times between which we are finding the average. δs ('delta s') is used to represent a change in s and δt ('delta t') is used to represent a change in t. The average velocity $\delta s / \delta t = $ 'delta s over delta t'.

So, to solve this problem we can use a table, as given in Table 6.1.

The graph of this function is given in Figure 6.1. If a line that joins two points lying on the graph of the function it is drawn it is called a chord. We found that when $t = 1$ and $s = 15$ the point $(1, 15)$ lies on the graph. When $t = 1.1$ and $s = 15.95$ we can also mark the point

Table 6.1 *Calculation of the average velocities over various intervals*

t_1	t_2	$s(t_1) = 20t_1 - 5t_1^2$	$s(t_2) = 20t_2 - 5t_2^2$	$t_2 - t_1$	$s(t_2) - s(t_1)$	*Average velocity* $= \delta s / \delta t$
1	1.1	15	15.95	0.1	0.95	9.5
1	1.01	15	15.0995	0.01	0.0995	9.95
1	1.001	15	15.009995	0.001	0.009995	9.995
1	1.0001	15	15.00099995	0.0001	9.9995×10^{-4}	9.9995
	$\vdots$					$\vdots$
	Velocity at $t = 1$					10

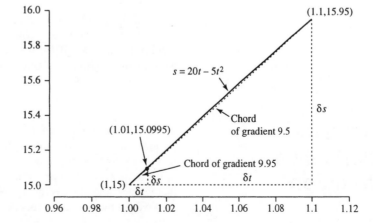

Figure 6.1 *Part of the graph of* $s = 20t - 5t^2$. *The chord joining* $(1, 15)$ *to* $(1.1, 15.95)$ *has gradient* $\delta s / \delta t = 9.5$ *The chord joining* $(1, 15)$ *to* $(1.1, 15.0995)$ *has gradient 9.95.*

(1.1, 15.95). The triangle containing the chord joining these two points has height $\delta s =$ change in $s = 15.95 - 15 = 0.95$ and base length $\delta t =$ change in $t = 1.1 - 1.0 = 0.1$. This means that the gradient $= \delta s/\delta t = 0.95/0.1 = 9.5$. Another chord can be drawn from $t = 1$ to $t = 1.01$. When $t = 1.01$ and $s = 15.0995$ we mark the point $(1.01, 15.0995)$. The triangle containing the chord joining $(1, 15)$ to $(1.01, 15.0995)$ has height $\delta s =$ change in $s = 15.0995 - 15 = 0.0995$ and base length $\delta t =$ change in $t = 1.01 - 1.0 = 0.1$. This means that the gradient $= \delta s/\delta t = 0.0995/0.01 = 9.95$.

As the ends of the chord are put nearer together the gradient of the chord gives a very good approximation to the instantaneous rate of change. Unfortunately, the chord becomes so small that we can hardly see it! To get round this problem we can extend the line at either end. So a chord between two points that are very close together appears to only just touch the function curve. A line that just touches at one point is called a tangent. As the two points on the chord approach each other the line of the chord approaches the tangent. Therefore, the gradient of the chord must also give a good approximation to the gradient of the tangent.

We can guess from Table 6.1 that the instantaneous velocity at $t = 1$ is $10\,\mathrm{m\,s^{-1}}$. Although the length of time over which we take the average gets smaller and smaller, that is, tends towards zero, the average velocity does not get nearer to zero but instead approaches the value of the instantaneous velocity.

The instantaneous velocity is represented by $\mathrm{d}s/\mathrm{d}t$, the derivative of s with respect to t, and can defined using

$$\text{Instantaneous velocity} = \frac{\mathrm{d}s}{\mathrm{d}t} = \lim_{\delta t \to 0} \frac{\delta s}{\delta t}$$

read as '$\mathrm{d}s$ by $\mathrm{d}t$ equals the limit, as delta t tends to zero, of delta s over delta t'.

Note that $\mathrm{d}s/\mathrm{d}t$ is read as '$\mathrm{d}s$ by $\mathrm{d}t$' (not '$\mathrm{d}s$ over $\mathrm{d}t$') because the line between the $\mathrm{d}s$ and the $\mathrm{d}t$ does not mean 'divided by'. However, because it does represent a rate of change it usually 'works' to treat $\mathrm{d}s/\mathrm{d}t$ like a fractional expression. This is because we can always approximate the instantaneous rate of change by the average rate of change (which is a fraction)

$$\frac{\mathrm{d}s}{\mathrm{d}t} \approx \frac{\delta s}{\delta t} \quad \text{for small } \delta t$$

'$\mathrm{d}s$ by $\mathrm{d}t$ is approximately delta s over delta t for small delta t'.

$\delta s/\delta t$ represents the gradient of a chord and $\mathrm{d}s/\mathrm{d}t$ represents the gradient of the tangent. If $\delta s/\delta t$ is used as an approximation to $\mathrm{d}s/\mathrm{d}t$ then we are using the chord to approximate the tangent.

6.3 The derivative function

We saw in Section 6.2 that if the ends of the chord are put closer together the gradient of the chord approaches the gradient of a tangent. The tangent to a curve is a line that only touches the curve at one point. The gradient of the tangent is also more simply referred to as the slope of the curve at that point. If the slope of the curve is found for every point on the curve then we get the derivative function.

The derivative of a function, $y = f(x)$ is defined as

$$\frac{dy}{dx} = \lim_{\delta x \to 0} \frac{\delta y}{\delta x}$$

provided that this limit exists. As δy is the change in y and $y = f(x)$ then, at the points x and $x + \delta x$, y has values $f(x)$ and $f(x + \delta x)$, the increase in y is given by $\delta y = f(x + \delta x) - f(x)$. We have

$$\frac{dy}{dx} = \lim \frac{\delta y}{\delta x} = \lim_{\delta x \to 0} \frac{f(x + \delta x) - f(x)}{\delta x}.$$

The derivative of $y = f(x)$, dy/dx, 'dy by dx' can also be represented as $f'(x)$ (read as 'f dashed of x'). $f'(a)$ is the gradient of the tangent to the curve $f(x)$ at the point $x = a$. This is found by finding the gradient of the chord between two points at $x = a + \delta x$ and $x = a$ and taking the limit as δx tends to zero.

The gradient of a chord gives the average rate of change of a function over an interval and the gradient of the tangent, the derivative, gives the instantaneous rate of change of the function at a point. These definitions are shown in Figure 6.2.

Derivative functions can be found by evaluating the limit shown in Figure 6.2. This is called differentiating from first principles.

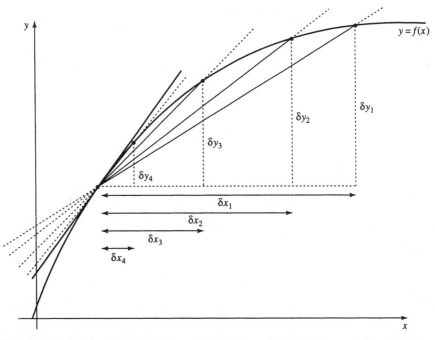

Figure 6.2 *The gradient of the chord $\delta y / \delta x$ approaches the gradient of the tangent (or the slope of the curve) as the ends of the chord get closer together (i.e. δx tends to zero). This is written as*

$$\frac{dy}{dx} = \lim_{\delta x \to 0} \frac{\delta y}{\delta x}$$

read as 'dy by dx equals the limit, as the change in x (delta x) tends to zero, of the change in y (delta y) divided by the change in x (delta x)'.

Example 6.1   If $y = x^2$ find dy/dx using

$$\frac{dy}{dx} = \lim_{\delta x \to 0} \frac{\delta y}{\delta x}$$

Solution   Consider a small change so that x goes from x to $x + \delta x$. As $y = x^2$ we can find the function value at $x + \delta x$ by replacing x by $x + \delta x$ giving $y + \delta y = (x + \delta x)^2$. Therefore, the change in y, δy, is given by

$$(x + \delta x)^2 - x^2 = x^2 + 2x\delta x + (\delta x)^2 - x^2 = 2x\delta x + (\delta x)^2$$

Therefore,

$$\frac{\delta y}{\delta x} = \frac{2x\delta x + (\delta x)^2}{\delta x}$$

As long as δx does not actually equal 0 we can divide the top and bottom line by δx giving

$$\frac{\delta y}{\delta x} = 2x + \delta x$$

and therefore

$$\frac{dy}{dx} = \lim_{\delta x \to 0} 2x + \delta x = 2x$$

This has shown that

$$y = x^2 \Rightarrow \frac{dy}{dx} = 2x.$$

6.4 Some common derivatives

Table 6.2   *The derivative of some simple functions*

$f(x)$	$f'(x)$
C	0
x^n	nx^{n-1}
$\cos(x)$	$-\sin(x)$
$\sin(x)$	$\cos(x)$
$\tan(x)$	$\sec^2(x)$ $= 1/\cos^2(x)$

We begin by listing derivatives of some simple functions (see Table 6.2).

We can also express the lines of Table 6.2 using d/dx as an operator giving, for instance,

$$\frac{d}{dx}(x^n) = nx^{n-1}$$

This can be read as 'the derivative of x^n is nx^{n-1}' or 'd by dx of x^n is nx^{n-1}'.

To see the validity of a couple of entries in Table 6.2, refer to Figure 6.3 for the derivative of a constant, C, and Figure 6.4 for the derivative of $\cos(x)$.

Example 6.2   Differentiate

(a) x (b) x^5 (c) $1/x^3$ (d) $\sqrt{x}$ (e) $1/\sqrt{x^3}$

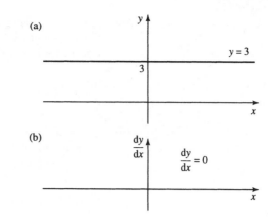

Figure 6.3 *(a) The derivative of a constant, for instance y = 3. The slope at any point is zero (the line has zero gradient everywhere). (b) The graph of the derivative is given as $dy/dx = 0$.*

Figure 6.4 *(a) The graph of $y = \cos(x)$ with a few tangents marked. On travelling from left to right, when we are going uphill the slope (and therefore the derivative) must be positive and when going downhill the derivative must be negative. At the top of the hills and the bottom of the troughs the slope is 0. Joining up the points on the bottom graph gives something like an upside-down sine wave. (b) The graph of the derivative $dy/dx = -\sin(x)$.*

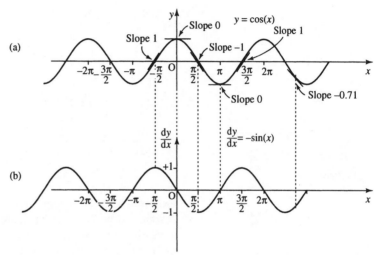

Solution   To differentiate a power of x we must first write the expression in the form x^n where n is some number. We can then use the fact that

$$\frac{d}{dx}(x^n) = nx^{n-1}$$

from Table 6.2.

(a) $x = x^1$; therefore, $n = 1$. Substitute $n = 1$ in

$$\frac{d}{dx}(x^n) = nx^{n-1}$$

to give

$$\frac{d}{dx}(x^1) = 1x^{1-1} = 1x^0 = 1$$

as $x^0 = 1$. Hence,

$$\frac{d}{dx}(x) = 1$$

(b)

$$\frac{d}{dx}(x^5) = 5x^{5-1} = 5x^4$$

(c) $1/x^3 = x^{-3}$ (using properties of negative powers given in Chapter 4 of the Background Mathematic Notes available on the companion website for this book), so $n = -3$

$$\frac{d}{dx}(x^{-3}) = (-3)x^{-3-1} = -3x^{-4} = \frac{-3}{x^4}$$

(d) $\sqrt{x} = x^{1/2}$ (using properties of roots given in Chapter 4 of the Background Mathematic Notes available on the companion website for this book), so $n = 1/2$

$$\frac{d}{dx}(x^{1/2}) = \frac{1}{2}x^{1/2-1} = \frac{1}{2}x^{-1/2} = \frac{1}{2}\frac{1}{x^{1/2}} = \frac{1}{2\sqrt{x}}$$

(e)

$$\frac{1}{\sqrt{x^3}} = \frac{1}{x^{3/2}} = x^{-3/2}$$

so $n = -3/2$. Hence,

$$\frac{d}{dx}(x^{-3/2}) = -\frac{3}{2}x^{-3/2-1} = -\frac{3}{2}x^{-5/2} = -\frac{3}{2}\frac{1}{x^{5/2}} = -\frac{3}{2}\frac{1}{\sqrt{x^5}}$$

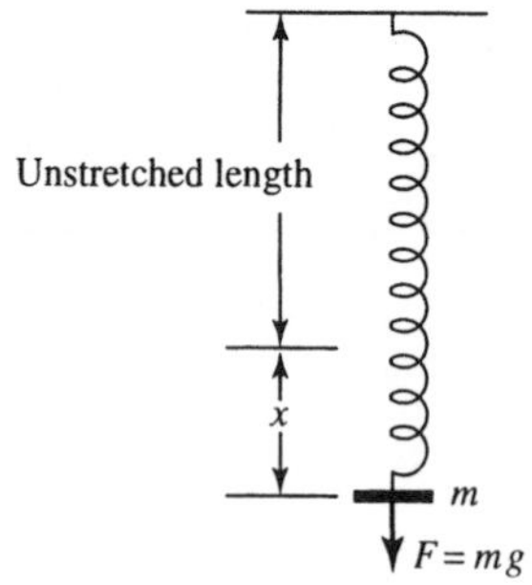

Figure 6.5 *A spring has a weight hanging from it of unknown mass m. The spring extends by an amount x and the energy stored in the spring is known to be E = x². The force due to gravity is F = mg where g is the acceleration due to gravity.*

Example 6.3 The energy stored in a stretched spring of extension x m is found to be $E = x^2$ J (Figure 6.5). The force exerted by hanging a weight on the spring is given by mg where g is the acceleration due to gravity $g \approx 10\,\text{m s}^{-1}$ and m is the mass. Given that the spring is extended by 0.5 m and that $F = dE/dx$ find the mass hanging on the spring.

Solution Find the expression for the force by differentiating $E = x^2$:

$$F = \frac{dE}{dx} = 2x$$

As $x = 0.5$, $F = 2(0.5) = 1\,\text{N}$. Now, $F = mg$, and as $g \approx 10$ then

$$1 = m \times 10 \Leftrightarrow m = \tfrac{1}{10} = 0.1\,\text{kg}$$

Thus, the mass on the spring is 0.1 kg.

6.5 Finding the derivative of combinations of functions

To find the derivative of functions that are the sum, difference quotient, product, or composite of any of the functions given in Table 6.2, we use the entries given in Table 6.2 the rules given in this section.

Derivatives of *af*(*x*) where *a* is a constant

$$\frac{d}{dx}(af(x)) = af'(x)$$

This is only true if a is a constant, not if it is a function of x.

Example 6.4 Differentiate $y = 2x^3$. Notice that this a constant, 2, multiplied by x^3. The derivative of x^3 is found by looking Table 6.2. The derivative of x^n is given by nx^{n-1}. In this case $n = 3$ so

$$\frac{d}{dx}(x^3) = 3x^2$$

and hence

$$\frac{dy}{dx} = 2(3x^2) = 6x^2$$

Derivatives of a sum (or a difference) of functions

If y can be written as the sum of two functions, that is, $y = u + v$ where u and v are functions of x then

$$\frac{dy}{dx} = \frac{du}{dx} + \frac{dv}{dx}$$

Example 6.5 Differentiate $y = \sin(t) + \sqrt{t}$.

Solution

$$y = \sin(t) + \sqrt{t} \quad \Leftrightarrow \quad y = \sin(t) + t^{1/2}$$

To differentiate a sum differentiate each part

$$\frac{dy}{dt} = \cos(t) + \frac{1}{2}t^{-1/2}$$

$$\frac{dy}{dt} = \cos(t) + \frac{1}{2\sqrt{t}}$$

Derivatives of composite functions (function of a function)

If $y = f(x)$ is a composite function, so that we can write $y = h(u)$ where $u = g(x)$, then

$$\frac{dy}{dx} = \frac{dy}{du}\frac{du}{dx}$$

This is called the chain rule.

Example 6.6 Differentiate $y = \sin(2x)$.

Solution   We can substitute $u = 2x$ giving $y = \sin(u)$:

$$\frac{du}{dx} = 2 \quad \text{and} \quad \frac{dy}{du} = \cos(u).$$

Therefore,

$$\frac{dy}{dx} = \frac{dy}{du}\frac{du}{dx} = 2\cos(u).$$

Finally, resubstitute for u giving

$$\frac{dy}{dx} = 2\cos(2x).$$

Note that we always need to make substitutions so that our function is the composite of the simple functions that we know how to differentiate, that is, x^n, $\sin(x)$, $\cos(x)$, $\tan(x)$ (or a constant times these functions, or the sum of these functions). One simple way to guess the required substitution is to look for a bracket.

Example 6.7   Differentiate $y = (5x - 2)^3$.

Solution   Substitute $u = 5x - 2$ (the function in the bracket) so that $y = u^3$. Then

$$\frac{du}{dx} = 5 \quad \text{and} \quad \frac{dy}{du} = 3u^2.$$

Therefore,

$$\frac{dy}{dx} = \frac{dy}{du}\frac{du}{dx}$$

gives

$$\frac{dy}{dx} = 5(3u^2) = 15(5x - 2)^2 \quad \text{(resubstituting } u = 5x - 2\text{)}.$$

Example 6.8   Differentiate $y = \cos(2x^2 + 3)$.

Solution   Substitute $u = 2x^2 + 3$, giving $y = \cos(u)$. Then

$$\frac{du}{dx} = 4x \quad \text{and} \quad \frac{dy}{du} = -\sin(u).$$

Therefore,

$$\frac{dy}{dx} = \frac{dy}{du}\frac{du}{dx} = -4x\sin(u) = -4x\sin(2x^2 + 3).$$

Because of the widespread use of the chain rule it is useful to be able to differentiate a composite function 'in your head'. This is a technique that comes with practice (like mental arithmetic).

Example 6.9   Differentiate $V = 1/(t + 1)$.

Solution   Rewrite $V = 1/(t + 1)$ as $V = (t + 1)^{-1}$ and think of this as $V = (\ )^{-1}$ where $(\ ) = t + 1$.

Now differentiate V with respect to $(\ )$ and multiply by the derivative of $(\ )$ with respect to t. That is,

$$\frac{dV}{dt} = \frac{dV}{d(\ )}\frac{d(\ )}{dt}$$

where $(\ )$ can be any expression. So

$$\frac{dV}{dt} = (-1(t + 1)^{-2})\frac{d}{dt}(t + 1)$$

$$\frac{dV}{dt} = (-1(t + 1)^{-2})1$$

$$\frac{dV}{dt} = \frac{-1}{(t + 1)^2}.$$

Example 6.10 Differentiate

$$y = \frac{1}{(3t^2 + 2t)^2}$$

Solution Rewrite

$$y = \frac{1}{(3t^2 + 2t)^2}$$

as $y = (3t^2 + 2t)^{-2}$, and think of this as $y = (\ )^{-2}$ where $(\ ) = 3t^2 + 2t$.

Now differentiate y with respect to $(\ )$ and multiply by the derivative of $(\ )$ with respect to t. That is,

$$\frac{dy}{dt} = \frac{dy}{d(\ )} \frac{d(\ )}{dt}$$

where $(\ )$ can be any expression. So

$$\frac{dy}{dt} = (-2(3t^2 + 2t)^{-3})\frac{d}{dt}(3t^2 + 2t)$$

$$\Leftrightarrow \quad \frac{dy}{dt} = (-2(3t^2 + 2t)^{-3})(6t + 2)$$

$$\Leftrightarrow \quad \frac{dy}{dt} = -\frac{12t + 4}{(3t^2 + 2t)^3}.$$

Derivative of inverse trigonometric functions

By the definition of the inverse function we know that $y = \sin^{-1}(x) \Leftrightarrow \sin(y) = x$, where $-1 \leq x \leq 1$, and therefore the derivatives are related. We can find the derivative of $\sin^{-1}(x)$ as in Example 6.11.

Example 6.11 Given

$$y = \sin^{-1}(x) \quad \text{and} \quad \frac{d}{dx}(\sin(x)) = \cos(x)$$

find dy/dx.

Solution As $y = \sin^{-1}(x)$ then by the definition of the inverse (assuming x is limited to $[-1, 1]$)

$$\sin(y) = x$$

The left-hand side of this is a function of y, which we can call w; hence, $w = \sin(y)$ and $w = x$. By the chain rule:

$$\frac{dw}{dx} = \frac{dw}{dy} \frac{dy}{dx}$$

Differentiating $w = x$ with respect to x gives $dw/dx = 1$.

Differentiating $w = \sin(y)$ with respect to y gives

$$\frac{dw}{dy} = \cos(y)$$

Hence,

$$\frac{dw}{dx} = \frac{dw}{dy}\frac{dy}{dx}$$

becomes

$$1 = \cos(y)\frac{dy}{dx}$$

Dividing both sides by $\cos(y)$ (if $\cos(y) \neq 0$) gives

$$\frac{dy}{dx} = \frac{1}{\cos(y)}$$

This is an expression for the derivative we want to find but it is a function of y instead of x. Use $\sin(y) = x$ and the trigonometric identity $\cos^2(y) = 1 - \sin^2(y)$, giving $\cos(y) = \sqrt{1 - \sin^2(y)}$ for $(-\pi/2 \leq y \leq \pi/2)$. As $\sin(y) = x$ and $\cos(y) = \sqrt{1 - x^2}$, we get

$$\frac{dy}{dx} = \frac{1}{\sqrt{1 - x^2}}.$$

The same method can be used to find the derivatives of $\cos^{-1}(x)$ and $\tan^{-1}(x)$ and we can now add these functions to the list, giving a new table of standard derivatives, as in Table 6.3.

Table 6.3 *The derivatives of some simple functions*

$f(x)$	$f'(x)$
C	0
x^n	nx^{n-1}
$\cos(x)$	$-\sin(x)$
$\sin(x)$	$\cos(x)$
$\tan(x)$	$\sec^2(x)$
$\sin^{-1}(x)$	$1/\sqrt{1 - x^2}$
$\cos^{-1}(x)$	$-1/\sqrt{1 - x^2}$
$\tan^{-1}(x)$	$1/(1 + x^2)$

Derivatives of a product of two functions

If y can be written as the product of two functions so that

$$y = uv$$

where u and v are functions of x, then

$$\frac{dy}{dx} = u\frac{dv}{dx} + v\frac{du}{dx}$$

Example 6.12   Find the derivative of $y = 5x \sin(x)$.

Solution $y = uv$, where $u = 5x$ and $v = \sin(x)$.

$$\frac{du}{dx} = 5 \quad \text{and} \quad \frac{dv}{dx} = \cos(x)$$

Using the derivative of a product formula:

$$\frac{dy}{dx} = 5x \cos(x) + 5 \sin(x)$$

Example 6.13   Find the derivative of $y = \sin(t) \cos(3t)$

Solution $y = uv$, where $u = \sin(t)$ and $v = \cos(3t)$:

$$\frac{du}{dt} = \cos(t) \quad \text{and} \quad \frac{dv}{dt} = -3\sin(3t)$$

Using the derivative of a product formula:

$$\frac{dy}{dt} = \cos(t)\cos(3t) - 3\sin(t)\sin(3t).$$

Derivatives of a quotient of two functions

If y can be written as the quotient of two functions so that $y = (u/v)$, where u and v are functions of x, then

$$\frac{dy}{dx} = \frac{v(du/dx) - u(dv/dx)}{v^2}.$$

Example 6.14   Find the derivative of

$$y = \frac{\sin(3x)}{x+1}.$$

Solution   We have $u = \sin(3x)$ and $v = x + 1$, so

$$\frac{du}{dx} = 3\cos(3x) \quad \text{and} \quad \frac{dv}{dx} = 1$$

Hence

$$\frac{dy}{dx} = \frac{(x+1)(3\cos(3x)) - (\sin(3x))1}{(x+1)^2}$$

$$= \frac{3(x+1)\cos(3x) - \sin(3x)}{(x+1)^2}$$

Example 6.15   Find the derivative of

$$z = \frac{12t}{1+t^3}$$

Solution   Setting $u = 12t$ and $v = 1 + t^3$, we have

$$\frac{du}{dt} = 12 \quad \text{and} \quad \frac{dv}{dt} = 3t^2$$

Hence

$$\frac{dz}{dt} = \frac{(1+t^3)12 - 3t^2(12t)}{(1+t^3)^2} = \frac{12 + 12t^3 - 36t^3}{(1+t^3)^2} = \frac{12 - 24t^3}{(1+t^3)^2}.$$

6.6
Applications of differentiation

As mentioned in the introduction to this chapter, many physical quantities important in engineering are related by the derivative. Here is a list of just some of these.

Mechanics

$v = \dfrac{dx}{dt}$, where v = velocity, x = distance, t = time.

$a = \dfrac{dv}{dt}$, where a = acceleration, v = velocity, t = time.

$F = \dfrac{dW}{dx}$, where F = force, W = work done (or energy used), x = distance moved in the direction of the force.

$F = \dfrac{dp}{dt}$, where F = force, p = momentum, t = time.

$P = \dfrac{dW}{dt}$, where P = power, W = work done (or energy used), t = time.

$\dfrac{dE}{dv} = p$, where E = kinetic energy, v = velocity, p = momentum.

Gases

$\dfrac{dW}{dV} = p$, where p = pressure, W = work done under isothermal expansion, V = volume.

Circuits

$I = \dfrac{dQ}{dt}$, where I = current, Q = charge, t = time.

$V = \left(L\dfrac{dI}{dt} \right)$, where V is the voltage drop across an inductor, L = inductance, I = current, t = time.

Electrostatics

$E = -\dfrac{dV}{dx}$, where V = potential, E = electric field, x = distance.

Example 6.16   A ball is thrown in the air so that the height of the ball is found to be $s = 3t - 5t^2$. Find

(a)   the ball's initial velocity when first thrown into the air;
(b)   the time when it returns to the ground;
(c)   its final velocity as it hits the ground.

Solution
(a)

$$v = \frac{ds}{dt}, \quad s = 3t - 5t^2 \Rightarrow \frac{ds}{dt} = 3 - 10t$$

The ball is initially thrown into the air when $t = 0$, so

$$\frac{ds}{dt} = 3 - 10(0) = 3 \, \text{m s}^{-1}$$

(b) The ball returns to the ground on the second occasion so that the distance travelled, s, is equal to 0. This time is given by solving the equation for t when $s = 0$.

$$0 = 3t - 5t^2 \iff t(3 - 5t) = 0$$
$$\iff t = 0 \text{ or } 3 - 5t = 0$$
$$\iff t = 0 \text{ or } t = 3/5$$

Therefore, it must return to the ground when $t = 3/5 = 0.6\,\text{s}$.

(c) When $t = 0.6$, using $v = \mathrm{d}s/\mathrm{d}t = 3 - 10t$

$$v = 3 - 10(0.6) = 3 - 6 = -3\,\text{m s}^{-1}$$

Therefore, the velocity as it hits the ground is, $-3\,\text{m s}^{-1}$.

Example 6.17   A rocket is moving with a velocity of $v = 4t^2 + 10\,000\,\text{m s}^{-1}$ over a brief period of time while leaving the Earth's atmosphere. Find its acceleration after 2 s.

Solution   Use $a = \mathrm{d}v/\mathrm{d}t$ as $v = 4t^2 + 10\,000$. Then, $a = 8t$, and at $t = 2$ this gives $a = 16$, so the acceleration after 2 s is $16\,\text{m s}^{-2}$.

Example 6.18   The potential due to a point charge Q at a position r from the charge is given by

$$V = \frac{Q}{4\pi\varepsilon_0 r}$$

where ε_0, the permittivity of free space, $\approx 8.85 \times 10^{-12}\,\text{F m}^{-1}$ and $\pi \approx 3.14$.

Given that $Q = 1\,C$, find the electric field strength at a distance of 5 m using $E = -\mathrm{d}V/\mathrm{d}r$.

Solution

$$V = \frac{Q}{4\pi\varepsilon_0 r}$$

substituting for ε_0 and π and using $Q = 1$, we get

$$V = \frac{1}{4 \times 3.14 \times 8.85 \times 10^{-12}r} \approx \frac{9 \times 10^9}{r} = 9 \times 10^9 r^{-1}$$

Now

$$E = -\frac{\mathrm{d}V}{\mathrm{d}r} = -9 \times 10^9(-r^{-2}) = 9 \times 10^9 r^{-2}$$

When $r = 5\,\text{m}$,

$$E = \frac{9 \times 10^9}{25} = 3.6 \times 10^8\,\text{V m}^{-1}.$$

6.7 Summary

(1) The average rate of change of a function over a certain interval is the same as the gradient of the chord drawn on the graph of the function. This chord gradient, for a function, $y = f(x)$, is given by

$$\frac{\delta y}{\delta x} = \frac{\text{change in } y}{\text{change in } x}$$

(2) If the chord is very short then the gradient of the chord is approximately the gradient of the tangent to the graph at a particular spot, that is, the slope of the graph at that point. The slope of the graph gives the instantaneous rate of change of the function with respect to its independent variable, known as its derivative. This is represented by dy/dx. Then we have the definition

$$\frac{dy}{dx} = \lim_{\delta x \to 0} \frac{\delta y}{\delta x} = \lim_{\delta x \to 0} \frac{f(x + \delta x) - f(x)}{\delta x}$$

This is read as 'dy by dx is the limit, as delta x tends to 0, of delta y over delta x'. The derivative of $y = f(x)$, dy/dx, ('dy by dx'), can also be represented by $f'(x)$ (read as 'f dashed of x').

(3) Derivatives of simple functions are given in Table 6.3. Rules are used to differentiate combinations of these functions. These are:

Product with a constant

$$\frac{d}{dx}(af(x)) = af'(x)$$

Sum

If $y = u + v$ then

$$\frac{dy}{dx} = \frac{du}{dx} + \frac{dv}{dx}$$

Composite function (function of a function) called the chain rule
If $y = f(x)$ where $y = h(u)$ and $u = g(x)$ then

$$\frac{dy}{dx} = \frac{dy}{du}\frac{du}{dx}$$

Product
If $y = uv$ then

$$\frac{dy}{dx} = u\frac{dv}{dx} + v\frac{du}{dx}$$

Quotient
If $y = u/v$ then

$$\frac{dy}{dx} = \frac{v(du/dx) - u(dv/dx)}{v^2}$$

(4) There are many applications of differentiation in all areas of engineering, some of which are listed in Section 6.6.

6.9 Exercises

6.1 A car is travelling such that its distance, s (m), from its starting position after time t (s) is

$$s = \frac{1}{15}t^3 + 2t, \qquad 0 < t < 10$$

$$s = 22(t - 10) + 86.67, \quad t \geq 10$$

(a) What is its average velocity in the first 10 s?
(b) Give the velocity as a function of time.
(c) What is the instantaneous velocity when (i) $t = 5$, (ii) $t = 10$, and (iii) $t = 15$?
(d) What is the average acceleration for the first 10 s?
(e) What is the average acceleration between $t = 10$ and $t = 15$?
(f) Give the acceleration as a function of time.
(g) What is the instantaneous acceleration when (i) $t = 5$, (ii) $t = 10$, and (iii) $t = 15$?

6.2 Differentiate the following:

(1) $3x^2 + 6x - 12$

(2) $x^{1/2} - x^{-1/2}$

(3) $\sqrt{2x^3} - (5/6x^2)$

(4) $\sin(3x^3 + x)$

(5) $2\cos(6x - 2)$

(6) $\tan(x^2)$

(7) $1/(2x - 3)$

(8) $(4x - 5)^6$

(9) $1/\sqrt{x^2 - 1}$

(10) $\sin^{-1}(5 - 2x)$

(11) $\tan(1/x)$

(12) $\sqrt{x^2 + 2}$

(13) $(x + 4)^{-3/2}$

(14) $\sin^2(x)$

(15) $5\cos^3(x)$

(16) $1/\sin^3(x)$

(17) $\cos^2(5x)$

(18) $x^3\sqrt{x + 1}$

(19) $5x\cos(x)$

(20) $6x^2\sin(x)$

(21) $(3x + 1)\tan(5x)$

(22) $x^3\cos^{-1}(x)$

(23) $x^3/\cos(x)$

(24) $1/\sin^2(x)$

(25) $\sin(x)/(2x + 10)$

(26) $x^2/\tan(x)$

(27) $3x^2/\sqrt{x - 1}$

(28) $(5x^2 - 1)/(5x^2 + 1)$

(29) $(x - 1)\cos(x)/(x^2 - 1)$

(30) $\sin^{-1}(x^2)$

(31) $x^2\sqrt{x - 1}\sin(x)$

(32) $\cos^2(x^2)$

(33) $\tan^2(\sqrt{5x - 1})$

12.3 A current i is travelling through a single turn loop of radius 1 m. A four-turn search coil of effective area $0.03\,\text{m}^2$ is placed inside the loop. The magnetic flux linking the search coil is given by

$$\phi = \mu_0\frac{iA}{2r}Wb$$

where r (m) is the radius of the current carrying loop, A (m^2) is the area of the search coil and μ_0 is the permeability of free space $= 4 \times 10^{-7}\,\text{H\,m}^{-1}$. Find the e.m.f. induced in the search coil, given by $\varepsilon = -N(d\phi/dt)$, where N is the number of turns in the search coil and the current is given by $i = 20\sin(20\pi t) + 50\sin(30\pi t)$.

7 Integration

7.1 Introduction

In Chapter 6, we saw that many physical quantities are related by one being the rate of change, the derivative, of the other. It follows that there must be a way of expressing the 'inverse' relationship. This is called integration. Velocity is the rate of change of distance with time, distance is the integral of velocity with respect to time.

Unfortunately, there are two issues that complicate the simple idea that 'integration is the inverse of differentiation'. First, we find that there are many different functions which are the integral of the same function. Luckily, these functions only differ from each other by a constant. To find all the possible integrals of a function we can find any one of them and add on some constant, called the constant of integration. This type of integral is called the indefinite integral. As it is not satisfactory to have an unknown constant left in the solution to a problem we employ some other information to find its value. Once the unknown constant is replaced by some value to fit a certain problem, we have the particular integral.

The second problem with integration is that most functions cannot be integrated exactly, even apparently simple functions like $\sin(x)/x$.

For this reason, numerical methods of integration are particularly important. These methods all depend on understanding the idea of integration as area under the graph. The definite integral of a function, $y = f(x)$, is the integral between two values of x and therefore gives a number (not a function of x) as a result. There is no uncertainty, hence it is called the definite integral.

This chapter is concerned with methods of finding the integral, the formulas that can be used for finding exact integrals and also with numerical integration. We also look more closely at the definitions of definite and indefinite integrals and at applications of integration.

7.2 Integration

Integration is the inverse process to differentiation. Consider the following examples (where C is a constant):

Function	$\rightarrow$	derivative
Integral	$\leftarrow$	function
$x^2 + C$		$2x$
$\sin(x) + C$		$\cos(x)$

The derivative of $x^2 + C$ with respect to x is $2x$, therefore, the integral of $2x$ with respect to x is $x^2 + C$. This can be written as

$$\frac{d}{dx}(x^2 + C) = 2x \quad \Leftrightarrow \quad \int 2x \, dx = x^2 + C$$

$\int 2x\,dx$ is read 'the integral of $2x$ with respect to x'. The $\int$ notation is like an 's' representing a sum, and its origin will be explained more when we look at the definite integral.

The second example gives

$$\frac{d}{dx}(\sin(x) + C) = \cos(x) \quad \Leftrightarrow \quad \int \cos(x)\,dx = \sin(x) + C$$

The derivative of $\sin(x) + C$ with respect to x is $\cos(x)$, which is the same as saying that the integral of $\cos(x)$ with respect to x is $\sin(x) + C$.

Constant of integration

Because the derivative of a constant is zero, it is not possible to determine the exact integral simply through using inverse differentiation. For instance, the derivatives of $x^2 + 1, x^2 - 2$, and $x^2 + 1000$, all give $2x$. Therefore, we express the integral of $2x$ as $x^2 + C$, where C is some constant called the constant of integration.

To find the value that the constant of integration should take in the solution of a particular problem we use some other known information. Supposing we know that a ball has velocity $v = 20 - 10t$. We want to find the distance travelled in time t and we also know that the ball was thrown from the ground which is at distance 0. We can work out the distance travelled by doing 'inverse differentiation' giving the distance $s = 20t - 5t^2 + C$, where C is the constant of integration. As we also know that $s = 0$ when $t = 0$ we can substitute these values to give $0 = C$, hence, the solution is that $s = 20t - 5t^2$.

In solving this problem we used the fact that $v = ds/dt$ and therefore we know that $ds/dt = 20 - 10t$. This is called a differential equation because it is an equation and contains an expression including a derivative. This is one sort of differential equation which can be solved directly by integrating. Some other sorts of differential equations are solved in Chapters 8, 10, and 14.

In this case, the solution $s = 20t - 5t^2 + C$ represents all possible solutions of the differential equation and is therefore called the general solution. If a value of C is found to solve a given problem, then this is the particular solution.

Example 7.1 Find y such that $dy/dx = 3x^2$ given that y is 5 when $x = 0$.

Solution We know that x^3, on differentiation, gives $3x^2$, so

$$\frac{dy}{dx} = 3x^2 \quad \Leftrightarrow \quad y = \int 3x^2\,dx \quad \Leftrightarrow \quad y = x^3 + C$$

where C is some constant. This is the general solution to the differential equation. To find the particular solution for this example use the fact that y is 5 when x is 0. Substitute in $y = x^3 + C$ to give $5 = 0 + C$, so $C = 5$ giving the particular solution as $y = x^3 + 5$.

7.3 Finding integrals

To find the table of standard integrals we take Table 6.3 for differentiation, swap the columns, rewrite a couple of the entries in a more convenient form and add on the constant of integration. This gives Table 7.1.

As integration is 'anti-differentiation' we can spot the integral in the standard cases, that is, those listed in Table 7.1.

Table 7.1   *A table of standard integrals*

$f(x)$	$\int f(x)\,dx$
1	$x + C$
$x^n (n \neq -1)$	$\dfrac{x^{n+1}}{n+1} + C$
$\sin(x)$	$-\cos(x) + C$
$\cos(x)$	$\sin(x) + C$
$\sec^2(x)$	$\tan(x) + C$
$\dfrac{1}{\sqrt{1-x^2}}$	$\sin^{-1}(x) + C$
$\dfrac{-1}{\sqrt{1-x^2}}$	$\cos^{-1}(x) + C$
$\dfrac{1}{1+x^2}$	$\tan^{-1}(x) + C$

Example 7.2   (a) Find $\int x^3\,dx$.
From Table 7.1

$$\int x^n\,dx = \frac{x^{n+1}}{n+1} + C \quad \text{where } n \neq -1$$

Here $n = 3$, so

$$\int x^3\,dx = \frac{x^{3+1}}{3+1} + C = \frac{x^4}{4} + C.$$

Check: Differentiate $(x^4/4) + C$ to give $(4x^3/4) = x^3$ which is the original expression that we integrated, hence showing that we integrated correctly.

(b) Find

$$\int \frac{1}{1+x^2}\,dx.$$

From Table 7.1

$$\int \frac{1}{1+x^2}\,dx = \tan^{-1}(x) + C$$

Check: Differentiate $\tan^{-1}(x) + C$ to give $1/(1+x^2)$.

(c) Find $\int x^{-1/2}\,dx$.
From Table 7.1

$$\int x^n\,dx = \frac{x^{n+1}}{n+1} + C$$

where $n \neq -1$ and in this case $n = -1/2$, so,

$$\int x^{-1/2}\,dx = \frac{x^{-(1/2)+1}}{-(1/2)+1} + C = \frac{x^{1/2}}{1/2} + C = 2x^{1/2} + C.$$

Check: Differentiate $2x^{1/2} + C$ to give $2(1/2)x^{(1/2)-1} = x^{-1/2}$.

(d) Find $\int 1 \, dx$.
From Table 7.1, $\int 1 \, dx = x + C$.

Check: Differentiate $x + C$ to give 1.

We can also find integrals of some combinations of the functions listed in Table 7.1. To do this, we need to use rules similar to those for differentiation. However, because when integrating we are working 'backwards', the rules are not so simple as those used to perform differentiation and furthermore, they will not always give a method that will work in finding the desired integral.

Integration of sums and *af*(*x*)

We can use the fact that

$$\int (f(x) + g(x)) \, dx = \int f(x) \, dx + \int g(x) \, dx$$

and also that

$$\int af(x) \, dx = a \int f(x) \, dx.$$

Example 7.3

(a) $\displaystyle\int (3x^2 + 2x - 1) \, dx = x^3 + x^2 - x + C.$

Check:

$$\frac{d}{dx}(x^3 + x^2 - x + C) = 3x^2 + 2x - 1.$$

(b) $\displaystyle\int 3\sin(x) + \cos(x) dx = -3\cos(x) + \sin(x) + C.$

Check:

$$\frac{d}{dx}(-3\cos(x) + \sin(x) + C) = 3\sin(x) + \cos(x).$$

(c) $\displaystyle\int \frac{1}{\sqrt{1-x^2}} - \frac{2}{1+x^2} \, dx = \sin^{-1}(x) - 2\tan^{-1}(x) + C.$

Check:

$$\frac{d}{dx}(\sin^{-1}(x) - 2\tan^{-1}(x) + C) = \frac{1}{\sqrt{1-x^2}} - \frac{2}{1+x^2}.$$

Changing the variable of integration

In Chapter 6 we looked at differentiating composite functions. If $y = f(x)$ where we can make a substitution in order to express y in terms of u, that is, $y = g(u)$, where $u = h(x)$, then

$$\frac{dy}{dx} = \frac{dy}{du}\frac{du}{dx}.$$

We can use this to integrate in very special cases by making a substitution for a new variable. The idea is to rewrite the integral so that we end up with one of the functions in Table 7.1. To see when this might work as a method of integration, we begin by looking at differentiating a composite function. Consider the derivative of

$$y = (3x + 2)^3.$$

We differentiate this using the chain rule, giving

$$\frac{dy}{dx} = 3(3x + 2)^2 \frac{d}{dx}(3x + 2) = 3(3x + 2)^2 3.$$

As integration is backwards differentiation, therefore

$$\int 3(3x + 2)^2 3 \, dx = (3x + 3)^3 + C.$$

Supposing then we had started with the problem to find the following integral

$$\int 3(3x + 2)^2 3 \, dx.$$

If we could spot that the expression to be integrated comes about from differentiating using the chain rule then we would be able to perform the integration. We can substitute $u = 3x + 2$ to give $du/dx = 3$, and the integral becomes:

$$\int 3u^2 \frac{du}{dx} \, du$$

we then use the 'trick' of replacing $(du/dx) \, dx$ by du giving

$$\int 3u^2 \, du.$$

As the expression to be integrated only involves the variable u, we can perform the integration and we get

$$\int 3u^2 \, du = u^3 + C.$$

Substituting again for $u = 3x + 2$, we get the integral as

$$\int 3(3x + 2)^2 3 \, dx = (3x + 2)^3 + C.$$

We used the trick of replacing $(du/dx) dx$ by du, this can be justified in the following argument. By the definition of the integral as inverse differentiation, if y is differentiated with respect to x and then integrated with respect to x we will get back to y, give or take a constant. This is expressed by

$$\int \frac{dy}{dx} dx = y + C. \tag{7.1}$$

If y is a composite function that can be written in terms of the variable u, then

$$\frac{dy}{dx} = \frac{dy}{du} \frac{du}{dx}.$$

Substituting the chain rule for dy/dx into Equation (7.1) gives

$$\int \frac{dy}{du} \frac{du}{dx} dx = y + C. \tag{7.2}$$

If y is a function of u, then we could just differentiate with respect to u and then integrate again and we will get back to the same expression, give or take a constant, that is

$$\int \frac{dy}{du} du = y + C. \tag{7.3}$$

Considering Equations (7.2) and (7.3) together, we have

$$\int \frac{dy}{du} \frac{du}{dx} dx = \int \frac{dy}{du} du$$

so that we can represent this result symbolically by $(du/dx) dx = du$.

In practice, we make a substitution for u and change the variable of integration by finding du/dx and substituting $dx = du/(du/dx)$.

Example 7.4 Find the integral $\int -(4 - 2x)^3 dx$.

Make the substitution $u = 4 - 2x$. Then $du/dx = -2$, so $du = -2 dx$ and $dx = -du/2$. The integral becomes

$$\int -u^3 \left(\frac{-du}{2} \right) = \int \frac{u^3}{2} du = \frac{u^4}{8} + C$$

Re-substitute for $u = 4 - 2x$, giving

$$\int -(4 - 2x)^3 dx = \frac{1}{8} (4 - 2x)^4 + C.$$

Check: Differentiate the result.

$$\frac{d}{dx} \left(\frac{(4 - 2x)^4}{8} + C \right) = \frac{1}{8} 4(4 - 2x)^4 \frac{d}{dx} (4 - 2x)$$

$$= \frac{1}{8} 4(4 - 2x)^3 (-2) = -(4 - 2x)^3$$

As this is the original expression that we integrated, this has shown that our result was correct.

When using this method, to find a good thing to substitute, look for something in a bracket, or an 'implied' bracket. Such substitutions will not always lead to an expression which it is possible to integrate. However, if the integral is of the form $\int f(u)\,dx$, where u is a linear function of x, or if the integral is of the form

$$\int f(u)\frac{du}{dx}\,dx$$

then a substitution will work providing $f(u)$ is a function with a known integral (i.e. a function listed in Table 7.1).

Integrations of the form $\int f(ax + b)\,dx$

For the integral $\int f(ax + b)\,dx$, make the substitution $u = ax + b$.

Example 7.5   Find $\int \sin(3x + 2)\,dx$.

Solution   Substitute $u = 3x + 2$. Then $du/dx = 3 \Rightarrow du = 3\,dx \Rightarrow dx = du/3$. Then the integral becomes

$$\int \sin(u)\frac{du}{3} = -\frac{\cos(u)}{3} + C$$

Re-substitute $u = 3x + 2$ to give

$$\int \sin(3x + 2)\,dx = -\frac{\cos(3x + 2)}{3} + C.$$

Check:

$$\frac{d}{dx}\left(-\frac{\cos(3x + 2)}{3} + C\right) = \frac{\sin(3x + 2)}{3}\frac{d}{dx}(3x + 2)$$

$$= \frac{3\sin(3x + 2)}{3} = \sin(3x + 2).$$

Example 7.6   Integrate

$$\frac{1}{\sqrt{1 - (3 - x)^2}}$$

with respect to x.

Solution   Notice that this is very similar to the expression which integrates to $\sin^{-1}(x)$ or $\cos^{-1}(x)$. We substitute for the expression in the bracket $u = 3 - x$ giving $du/dx = -1 \Rightarrow dx = -du$. The integral becomes

$$\int \frac{1}{\sqrt{1-(u)^2}}(-\mathrm{d}u) = \int \frac{(-\mathrm{d}u)}{\sqrt{1-(u)^2}}$$

From Table 7.1, this integrates to give

$$\cos^{-1}(u) + C$$

Re-substituting $u = 3 - x$ gives

$$\int \frac{1}{\sqrt{1-(3-x)^2}}\,\mathrm{d}x = \cos^{-1}(3-x) + C.$$

Check:

$$\frac{\mathrm{d}}{\mathrm{d}x}(\cos^{-1}(3-x) + C) = -\frac{1}{\sqrt{1-(3-x)^2}}\frac{\mathrm{d}}{\mathrm{d}x}(3-x)$$

$$= \frac{1}{\sqrt{1-(3-x)^2}}.$$

Integrals of the form $\int f(u)(\mathrm{d}u/\mathrm{d}x)\,\mathrm{d}x$

Make the substitution for $u(x)$. This type of integration will often work when the expression to be integrated is of the form of a product. One of the terms will be a composite function. This term could well involve an expression in brackets, in which case substitute the expression in the bracket for a new variable u. The integral should simplify, provided the other part of the product is of the form $\mathrm{d}u/\mathrm{d}x$.

Example 7.7   Find $\int x \sin(x^2)\,\mathrm{d}x$.

Solution   Substitute $u = x^2 \Rightarrow \mathrm{d}u/\mathrm{d}x = 2x \Rightarrow \mathrm{d}u = 2x\,\mathrm{d}x \Rightarrow \mathrm{d}x = \mathrm{d}u/2x$ to give

$$\int x \sin(x^2)\,\mathrm{d}x = \int x \sin(u)\frac{\mathrm{d}u}{2x} = \int \frac{1}{2}\sin(u)\,\mathrm{d}u$$

$$= -\frac{1}{2}\cos(u) + C.$$

As $u = x^2$, we have

$$\int x \sin(x^2)\,\mathrm{d}x = -\frac{1}{2}\cos(x^2) + C.$$

Check:

$$\frac{\mathrm{d}}{\mathrm{d}x}\left(-\frac{1}{2}\cos(x^2) + C\right) = \frac{1}{2}\sin(x^2)\frac{\mathrm{d}}{\mathrm{d}x}(x^2)$$

$$= \frac{1}{2}\sin(x^2)(2x) = x\sin(x^2).$$

Example 7.8   Find

$$\int \frac{3x}{(x^2+3)^4}\,dx.$$

Solution   Substitute $u = x^2 + 3$. Then $du/dx = 2x \Rightarrow du = 2x\,dx \Rightarrow dx = du/2x$. The integral becomes

$$\int \frac{3x}{(u)^4}\frac{du}{2x} = \int \frac{3}{2}u^{-4}\,du$$

which can be integrated, giving

$$\frac{3}{2}\frac{u^{-4+1}}{(-4+1)} + C = \frac{1}{2}u^{-3} + C.$$

Re-substituting for $u = x^2 + 3$ gives

$$\int \frac{3x}{(x^2+3)^4}\,dx = -\frac{1}{2}(x^2+3)^{-3} + C = -\frac{1}{2(x^2+3)^3} + C.$$

Check:

$$\frac{d}{dx}\left(-\frac{1}{2}(x^2+3)^{-3} + C\right) = -\frac{1}{2}(-3)(x^2+3)^{-4}\frac{d}{dx}(x^2+3).$$

Using the function of a function rule, we get

$$-\frac{1}{2}(-3)(x^2+3)^{-4}(2x) = 3x(x^2+3)^{-4} = \frac{3x}{(x^2+3)^4}.$$

Example 7.9   Find $\int \cos^2(x)\sin(x)\,dx$.

Solution   This can be rewritten as $\int (\cos(x))^2\sin(x)\,dx$. Substitute $u = \cos(x)$, then $du/dx = -\sin(x)$, so $du = -\sin(x)\,dx$, or $dx = -du/\sin(x)$. The integral becomes

$$\int u^2\sin(x)\frac{du}{-\sin(x)} = \int -u^2\,du.$$

Integrating gives

$$-\frac{u^3}{3} + C.$$

Re-substitute for u, giving

$$\int (\cos(x))^2\sin(x)\,dx = -\frac{\cos^3(x)}{3} + C.$$

Check:

$$-\frac{1}{3}\cos^3(x) + C = -\frac{1}{3}(\cos(x))^3 + C.$$

Differentiate

$$\frac{\mathrm{d}}{\mathrm{d}x}\left(-\frac{1}{3}(\cos(x))^3 + C\right) = -3\left(\frac{1}{3}\right)(\cos(x))^2(-\sin(x))$$

$$= \cos^2(x)\sin(x).$$

This method of integration will only work when the integral is of the form

$$\int f(u)\frac{\mathrm{d}u}{\mathrm{d}x}\,\mathrm{d}x$$

that is, there is a function of a function multiplied by the derivative of the substituted variable, or where the substituted variable is a linear function.

Sometimes you may want to try to perform this method of integration and discover that it fails to work, in this case, another method must be used.

Example 7.10 Find

$$\int \frac{x^2}{(x^2+1)^2}\,\mathrm{d}x.$$

Substitute $u = x^2 + 1$, then $\mathrm{d}u/\mathrm{d}x = 2x \Rightarrow \mathrm{d}x = \mathrm{d}u/2x$. The integral becomes

$$\int \frac{x^2}{u^2}\frac{\mathrm{d}u}{2x} = \int \frac{x}{2u^2}\,\mathrm{d}u.$$

This substitution has not worked. We are no nearer being able to perform the integration. There is still a term in x involved in the integral, so we are not able to perform an integration with respect to u only.

In some of these cases, integration by parts may be used.

Integration by parts

This can be useful for integrating some products, for example, $\int x\sin(x)\,\mathrm{d}x$. The formula is derived from the formula for differentiation of a product.

$$\frac{\mathrm{d}}{\mathrm{d}x}(uv) = \frac{\mathrm{d}u}{\mathrm{d}x}v + u\frac{\mathrm{d}v}{\mathrm{d}x}$$

$$\Leftrightarrow \frac{\mathrm{d}}{\mathrm{d}x}(uv) - \frac{\mathrm{d}u}{\mathrm{d}x}v = u\frac{\mathrm{d}v}{\mathrm{d}x}$$

(subtracting $(\mathrm{d}u/\mathrm{d}x)\,v$ from both sides)

$$\Leftrightarrow u\frac{\mathrm{d}v}{\mathrm{d}x} = \frac{\mathrm{d}}{\mathrm{d}x}(uv) - \frac{\mathrm{d}u}{\mathrm{d}x}v$$

$$\Rightarrow \int u\frac{\mathrm{d}v}{\mathrm{d}x}\,\mathrm{d}x = uv - \int v\frac{\mathrm{d}u}{\mathrm{d}x}\,\mathrm{d}x \quad \text{(integrating both sides)}$$

As we found before $(du/dx)\,dx$ can be replaced by du so $(dv/dx)\,dx$ can be replaced by dv, and this gives a compact way of remembering the formula:

$$\int u\,dv = uv - \int v\,du.$$

To use the formula, we need to make a wise choice as to which term is u (which we then need to differentiate to find du) and which term is dv (which we then need to integrate to find v). Note that the second term $\int v\,du$ must be easy to integrate.

Example 7.11   Find $\int x \sin x \, dx$

Solution   Use $u = x$; $dv = \sin(x)\,dx$. Then

$$\frac{du}{dx} = 1 \ \text{ and } \ v = \int \sin x \, dx = -\cos(x).$$

Substitute in $\int u\,dv = uv - \int v\,du$ to give

$$\int x \sin x \, dx = -x\cos(x) - \int -\cos(x)1\,dx$$

$$= -x\cos(x) + \sin(x) + C.$$

Check:

$$\frac{d}{dx}(-x\cos(x) + \sin(x) + C) = -\cos(x) + x\sin(x) + \cos(x)$$

$$= x\sin(x).$$

We can now solve the problem that we tried to solve using a substitution, but had failed.

Example 7.12   Find

$$\int \frac{x^2}{(x^2+1)^2}\,dx.$$

Solution
   We can spot that if we write this as

$$\int x \frac{x}{(x^2+1)^2}\,dx$$

then the second term in the product can be integrated. We set $u = x$ and

$$dv = \frac{x}{(x^2+1)^2}\,dx = x(x^2+1)^{-2}\,dx.$$

Then $du = dx$ and $v = -\frac{1}{2}(x^2+1)^{-1}$. (To find v we have performed the integration $\int x(x^2+1)^{-2}\,dx = -\frac{1}{2}(x^2+1)^{-1}$. Check this result by substituting for $x^2 + 1$). Substitute in $\int u\,dv = uv - \int v\,du$ to give

$$\int x \frac{x}{(x^2+1)^2}\,dx = -\frac{x}{2}(x^2+1)^{-1} - \int \frac{-(x^2+1)^{-1}}{2}\,dx$$

$$= \frac{-x}{2(x^2+1)} + \frac{1}{2}\int \frac{dx}{(x^2+1)}.$$

Note that the remaining integral is a standard integral given in Table 7.1 as $\tan^{-1}(x)$, so the integral becomes

$$\int \frac{x^2}{(x^2+1)^2}\,dx = \frac{-x}{2(x^2+1)} + \frac{1}{2}\tan^{-1}(x) + C.$$

Check:

$$\frac{d}{dx}\left(\frac{-x}{2(x^2+1)} + \frac{1}{2}\tan^{-1}(x) + C\right)$$

$$= \frac{d}{dx}\left(-\frac{x}{2}(x^2+1)^{-1} + \tan^{-1}(x) + C\right)$$

$$= -\frac{1}{2}(x^2+1)^{-1} + \frac{x}{2}(2x)(x^2+1)^{-2} + \frac{1}{2}\frac{1}{(x^2+1)}$$

$$= \frac{x^2}{(x^2+1)^2}.$$

Integrating using trigonometric identities

There are many possible ways of using trigonometric identities in order to perform integration. We shall just look at examples of how to deal with powers of trigonometric functions. For even powers of a trigonometric function, the double angle formula may be used. For odd powers of a trigonometric function a method involving the substitution

$$\cos^2(x) = 1 - \sin^2(x) \quad \text{or} \quad \sin^2(x) = 1 - \cos^2(x)$$

is used.

Example 7.13   Find $\int \sin^2(x)\,dx$.

Solution   As $\cos(2x) = 1 - 2\sin^2(x)$,

$$\sin^2(x) = \frac{1 - \cos(2x)}{2}.$$

The integral becomes

$$\int \sin^2(x)\,dx = \int \frac{1 - \cos(2x)}{2}\,dx = \frac{1}{2}x - \frac{1}{4}\sin(2x) + C.$$

Check:

$$\frac{d}{dx} = \left(\frac{1}{2}x - \frac{1}{4}\sin(2x) + C\right) = \frac{1}{2} - \frac{1}{4} \cdot 2\cos(2x)$$

$$= \frac{1}{2}(1 - \cos(2x))$$

$$= \sin^2(x) \quad \text{(from the double angle formula)}.$$

Example 7.14   Find $\int \cos^3(x)\,dx$.

Solution   As $\cos^2(x) + \sin^2(x) = 1$, we have $\cos^2(x) = 1 - \sin^2(x)$.

$$\int \cos^3(x)\,dx = \int \cos(x)\cos^2(x)\,dx$$

$$= \int \cos(x)(1 - \sin^2(x))\,dx$$

$$= \int \cos(x) - \cos(x)\sin^2(x)\,dx$$

$$= \int \cos(x)\,dx - \int \cos(x)\sin^2(x)\,dx.$$

The second part of this integral is of the form

$$\int f(u)\frac{du}{dx}\,dx.$$

Substitute $u = \sin(x)$; then

$$du = \cos(x)\,dx \quad \Rightarrow \quad dx = \frac{du}{\cos(x)}.$$

Hence,

$$\int \cos(x)\sin^2(x)\,dx = \int \cos(x)\,u^2\,\frac{du}{\cos(x)} = \int u^2\,du = \frac{u^3}{3} + C.$$

Re-substituting $u = \sin(x)$ gives

$$\int \cos(x)\sin^2(x)\,dx = \frac{1}{3}\sin^3(x) + C.$$

Therefore,

$$\int \cos^3(x)\,dx = \int \cos(x) - \cos(x)\sin^2(x)\,dx$$

$$= \sin(x) - \frac{1}{3}\sin^3(x) + C.$$

Check:

$$\frac{d}{dx}(\sin(x) - \frac{1}{3}\sin^3(x) + C)$$

$$= \cos(x) - \frac{3}{3}\sin^2(x)\cos(x) = \cos(x)(1 - \sin^2(x))$$

$$= \cos(x)\cos^2(x) = \cos^3(x).$$

7.4 Applications of integration

In Section 6.7, we listed the applications of differentiation that are important in engineering. Here, we list the equivalent relationships using integration.

Mechanics

$x = \int v \, dt$, where $v =$ velocity, $x =$ distance, $t =$ time

$v = \int a \, dt$, where $a =$ acceleration, $v =$ velocity, $t =$ time

$W = \int F \, dx$, where $F =$ force, $W =$ work done (or energy used), $x =$ distance moved in the direction of the force

$p = \int F \, dt$, where $F =$ force, $p =$ momentum, $t =$ time

$W = \int P \, dt$, where $P =$ power, $W =$ work done (or energy used), $t =$ time

$p = \int E \, dv$, where $E =$ kinetic energy, $v =$ velocity, $p =$ momentum.

Gases

$p = \int W \, dV$, where $p =$ pressure, $W =$ work done under isothermal expansion, $V =$ volume.

Electrical circuits

$Q = \int I \, dt$, where $I =$ current, $Q =$ charge, $t =$ time

$I = (1/L) \int V \, dt$, where $V =$ voltage drop across an inductor, $L =$ inductance, $I =$ current, $t =$ time.

Electrostatics

$V = - \int E \, dx$, where $V =$ potential, $E =$ electric field, $x =$ distance.

Example 7.15   A car moving with a velocity of $12 \, \mathrm{m \, s^{-1}}$ accelerated uniformly for $10 \, \mathrm{s}$ at $1 \, \mathrm{m \, s^{-2}}$ and then kept a constant velocity. Calculate:

(a)   the distance travelled during the acceleration,
(b)   the velocity reached after $20 \, \mathrm{m}$,
(c)   the time taken to travel $100 \, \mathrm{m}$ from the time that the acceleration first started.

Solution   Take as time 0 the time when the car begins to accelerate. From $t = 0$ to $t = 10$, the acceleration is $1 \, \mathrm{m \, s^{-2}} \, dv/dt = 1$. Therefore, $v = \int 1 \, dt = t + C$. For $0 \leq t \leq 10$, this gives $v = t + C$. To find the constant C, we need to use other information given in the problem. We know that at $t = 0$ the velocity is $12 \, \mathrm{m \, s^{-1}}$. Substituting this into $v = t + C$ gives

$$12 = 0 + C \quad \Leftrightarrow \quad C = 12$$

so, $v = t + 12$. For $t > 10$, the velocity is constant, therefore $v = 10 + 12 = 22 \, \mathrm{m \, s^{-1}}$ for $t > 10$. The velocity function is therefore

$$v = \begin{cases} t + 12 & 0 \leq t \leq 10 \\ 22 & t > 10. \end{cases}$$

To find the distance travelled we need to integrate one more time

$$v = \frac{dx}{dt} = t + 12 \quad \text{for } 0 \leq t \leq 10$$

$$x = t^2 + 12t + C \quad \text{for } 0 \leq t \leq 10.$$

To find the value of C, consider that the distance travelled is 0 at $t = 0$. Hence, $0 = C$ and therefore

$$x = t^2 + 12t \quad \text{for } 0 \leq t \leq 10.$$

For $t > 10$, we have a different expression for the velocity

$$v = \frac{dx}{dt} = 22 \quad \text{for } t > 10$$

$$x = 22t + C \quad \text{for } t > 10.$$

To find the value of C is this expression, we need some information about the distance travelled, for instance, at $t = 10$. Using $x = t^2 + 12t$, we get that at $t = 10, x = 100 + 120 = 220$ m. Substituting this into $x = 22t + C$ gives $220 = 22 \times 10 + C$, which gives $C = 0$, so $x = 22t$ for $t \geq 10$. The function for x is therefore

$$x = \begin{cases} t^2 + 12t & 0 \leq t \leq 10 \\ 22t & t > 10. \end{cases}$$

As we now have expressions for v and x, we are in a position to answer the questions.

(a) The distance travelled during the acceleration is the distance after 10 s.

$$x = (10)^2 + 12(10) = 220 \, \text{m}.$$

(b) To find the velocity reached after 20 m, we need first to find the time taken to travel 20 m

$$x = 20 \Rightarrow 20 = t^2 + 12t$$

$$t^2 + 12t - 20 = 0.$$

Using the formula for solving a quadratic equation:

$$at^2 + bt + c = 0 \quad \Leftrightarrow \quad t = \frac{-b \pm \sqrt{b^2 - 4ac}}{2a}$$

$$t = \frac{-12 \pm \sqrt{144 + 80}}{2} = \frac{-12 \pm \sqrt{224}}{2}$$

$t \approx -13.5$ or $t \approx 1.5$ s.

As t cannot be negative, $t \approx 1.5$ s. Substituting the value for t into the expression for v gives

$$v = t + 12 = 1.5 + 12 = 13.5 \, \text{m s}^{-1}.$$

So the velocity after 20 m is 13.5 ms^{-1}.

(c) The time taken to travel $100\,\mathrm{m}$ from when the acceleration first started can be found from $x = 100$

$$100 = t^2 + 12t \quad \Leftrightarrow \quad t^2 + 12t - 100 = 0.$$

Using the quadratic formula gives

$$t = \frac{-12 \pm \sqrt{144 + 400}}{2} \approx \frac{-12 \pm 23.2}{2}$$

$t \approx -17.66$ or $t \approx 5.66$.
The time taken to travel $100\,\mathrm{m}$ is $5.66\,\mathrm{s}$.

Example 7.16 The current across a $1\,\mu\mathrm{F}$ capacitor from time $t = 3\,\mathrm{ms}$ to $t = 4\,\mathrm{ms}$ is given by $I(t) = -2t$. Find the voltage across the capacitor during that period of time given that $V = -0.1\,\mathrm{V}$ when $t = 3\,\mathrm{ms}$.

Solution For a capacitor $V = Q/C$, where $Q = \int I\,\mathrm{d}t$ and C is the capacitance, V the voltage drop across the capacitor, Q the charge, and I the current. So

$$V = \frac{1}{1 \times 10^{-6}} \int -2t\,\mathrm{d}t$$
$$= 10^6(-t^2) + C$$

when $t = 3\,\mathrm{ms}$, $V = 0.1$, so $V = 0.1$ when $t = 3 \times 10^{-3}$

$$0.1 = 10^6(-10^{-6} \times 9) + C \quad \Leftrightarrow \quad C = 0.1 + 9 \quad \Leftrightarrow \quad C = 9.1.$$

So $V = -10^6 t^2 + 9.1$.

7.5 The definite integral

The definite integral from $x = a$ to $x = b$ is defined as the area under the curve between those two points. In the graph in Figure 7.1, the area under the graph has been approximated by dividing it into rectangles. The height of each is the value of y and if each rectangle is the same width then the area of the rectangle is $y\delta x$.

If the rectangle is very thin, then y will not vary very much over its width and the area can reasonably be approximated as the sum of all of these rectangles.

The symbol for a sum is Σ (read as capital Greek letter sigma). The area under the graph is approximately

$$A = y_1\delta x + y_2\delta x + y_3\delta x + y_4\delta x + \cdots = \sum_{x=a}^{x=b-\delta x} y\delta x.$$

We would assume that if δx is made smaller, the approximation to the exact area would improve. An example is given for the function $y = x$ in Figure 7.2. Between the values of 1 and 2, we divide the area into strips, first of width 0.1, then 0.01, then 0.001.

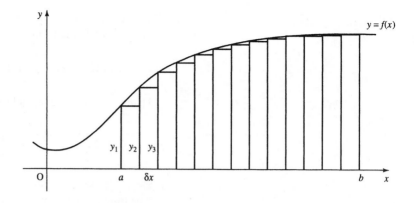

Figure 7.1 *A graph of*
y = f(x) and the area under
the graph from x = a to
x = b. This is approximated
by splitting the area into strips
of width δx.

When $\delta x = 0.1$, the approximate calculation gives

$$1 \times 0.1 + 1.1 \times 0.1 + 1.2 \times 0.1 + 1.3 \times 0.1 + 1.4 \times 0.1 + 1.5 \times 0.1$$
$$+ 1.6 \times 0.1 + 1.7 \times 0.1 + 1.8 \times 0.1 + 1.9 \times 0.1 = 1.45$$

When $\delta x = 0.01$, the calculation gives

$$1 \times 0.01 + 1.01 \times 0.01 + 1.02 \times 0.01 + \cdots$$
$$+ 1.98 \times 0.01 + 1.99 \times 0.01 = 1.495$$

When $\delta x = 0.001$, the calculation gives

$$1 \times 0.001 + 1.001 \times 0.001 + 1.002 \times 0.001 + \cdots$$
$$+ 1.998 \times 0.001 + 1.999 \times 0.001 = 1.4995$$

The exact answer is given by the area of a trapezoid which is equal to the average length of the parallel sides multiplied by the width. In this case for $y_a = 1$ and $y_b = 2$ we get

$$\text{Area} = \tfrac{1}{2}(1 + 2)1 = 1.5$$

We can see that the smaller the strips the nearer the area approximates to the exact area of 1.5. Therefore, as the width of the strips gets smaller and smaller, then there is a better approximation to the area, and we say that in the limit, as the width tends to zero, we have the exact area, which is called the definite integral.

The area under the curve, $y = f(x)$ between $x = a$ and $x = b$ is found as

$$\int_a^b y \, dx = \lim_{\delta x \to 0} \sum_{x=a}^{x=b-\delta x} y \, \delta x$$

which is read as 'The definite integral of y from $x = a$ to $x = b$ equals the limit as δx tends to 0 of the sum of y times δx for all x from $x = a$ to $x = b - \delta x$.

This is the definition of the definite integral which gives a number as its result, not a function.

We need to show that our two ways of defining integration (the indefinite integral as the inverse process to differentiation and the definite integral as the area under the curve) are consistent. To do this, consider

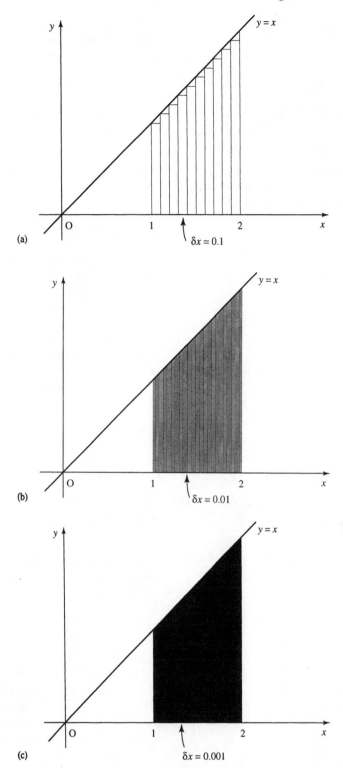

Figure 7.2 *The area under the graph y = x between x = 1 and x = 2: (a) divided into strips of width 0.1 gives 1.45; (b) divided into strips of width 0.01 gives 1.495; (c) divided into strips of width 0.001 gives 1.4995.*

an integral of y from some starting point, a, up to any point x. Then, the area A is

$$A = \int_a^x y\, dx$$

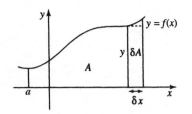

Figure 7.3 *The area A is given by the definite integral $\int_a^x y \, dx$ and the increase in the area, $\delta A = y \delta x$.*

Notice that as we move the final point x, A will change. Now consider moving the final value by a small amount, δx, this will increase the area by δA and δA is approximately the area of a rectangle of height y and width δx. This is shown in Figure 7.3.

So, we have $\delta A \approx y \delta x$, that is,

$$\frac{\delta A}{\delta x} \approx y.$$

Taking the limit as δx tends to 0 gives

$$y = \frac{dA}{dx}.$$

This shows that finding the area under the graph does in fact give a function which, when differentiated, gives back the function of the original graph, that is, the area function gives the 'inverse' of differentiation. The area function, A, is not unique because different functions will be found by moving the position of the starting point for the area, however in each case dA/dx will be the original function.

This is illustrated for the area under $y = \frac{1}{2}t$ in Figure 7.4, where there are two area functions, one starting from $t = -1$ and the other from $t = 0$.

The first area function is

$$A = \frac{t^2}{4} - \frac{1}{4}$$

and the second is

$$A = \frac{t^2}{4}$$

The definite integral, the area under a particular section of the graph, can be found, as in Figure 7.5, by subtracting the areas.

In practice, we do not need to worry about the starting value for finding the area. The effect of any constant of integration will cancel out.

Example 7.17 Find $\int_2^3 2t \, dt$.

This is the area under the graph from $t = 2$ to $t = 3$. As $\int 2t \, dt = t^2 + C$, the area up to 2 is $(2)^2 + C = 4 + C$ and the area up to 3 is $(3)^2 + C = 9 + C$. The difference in the areas is $9 + C - (4 + C) = 9 - 4 = 5$. Therefore, $\int_2^3 2t \, dt = 5$.

The working of a definite integral is usually laid out as follows

$$\int_2^3 2t \, dt = \left[t^2\right]_2^3 = (3)^2 - (2)^2 = 5.$$

The square brackets indicate that the function should be evaluated at the top value, in this case 3, and then have its value at the bottom value, in this case 2, subtracted.

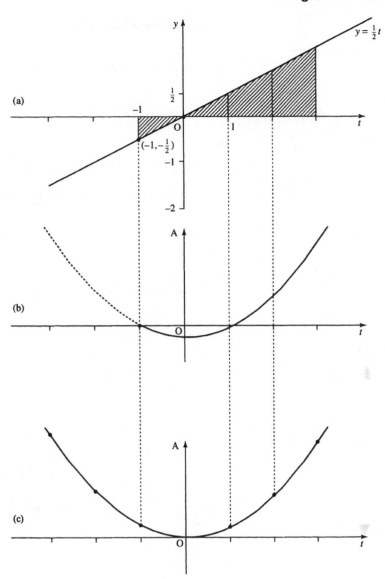

Figure 7.4   *(a) The graph of*
y = t/2. (b) The area under
the graph of y = t/2 starting
from t = −1. (c) The area
under the graph of y = t/2
starting from t = 0.

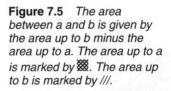

Figure 7.5   *The area*
between a and b is given by
the area up to b minus the
area up to a. The area up to a
is marked by   .   *The area up*
to b is marked by ///.

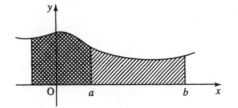

Example 7.18   Find

$$\int_{-1}^{1} 3x^2 + 2x - 1 \, dx.$$

Solution

$$\int_{-1}^{1} 3x^2 + 2x - 1 \, dx = \left[x^3 + x^2 - x \right]_{-1}^{1}$$
$$= (1^3 + 1^2 - 1) - ((-1)^3 + (-1)^2 - (-1))$$
$$= 1 - (-1 + 1 + 1) = 1 - 1 = 0.$$

Example 7.19   Find

$$\int_{0}^{\pi/6} \sin(3x + 2) \, dx.$$

Solution

$$\int_{0}^{\pi/6} \sin(3x + 2) \, dx = \left[-\frac{1}{3} \cos(3x + 2) \right]_{0}^{\pi/6}$$
$$= \frac{1}{3} \cos\left(3\frac{\pi}{6} + 2\right) - \left(-\frac{1}{3} \cos(2)\right)$$
$$= \frac{1}{3} \cos\left(\frac{\pi}{2} + 2\right) + \frac{1}{3} \cos(2) \approx 0.1644.$$

Example 7.20   Find the shaded area in Figure 7.6, where $y = -x^2 + 6x - 5$.

Solution   First, we find where the curve crosses the x-axis, that is, when $y = 0$

$$0 = -x^2 + 6x - 5 \quad \Leftrightarrow \quad x^2 - 6x + 5 = 0$$
$$\Leftrightarrow \quad (x - 5)(x - 1) = 0 \quad \Leftrightarrow \quad x = 5 \lor x = 1$$

Figure 7.6   *The shaded area is bound by the graph of $y = -x^2 + 6x - 5$ and the x-axis.*

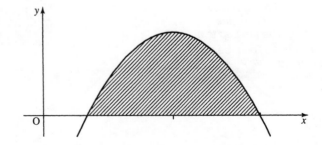

This has given the limits of the integration. Now we integrate:

$$\int_1^5 -x^2 + 6x - 5\,dx = \left[-\frac{x^3}{3} + \frac{6x^2}{2} - 5x\right]_1^5$$

$$= -\frac{(5)^3}{3} + \frac{6(5)^2}{2} - 5(5)$$

$$- \left(-\frac{(1)^3}{3} + \frac{6(1)^2}{2} - 5(1)\right)$$

$$= -\frac{125}{3} + 75 - 25 + \frac{1}{3} - 3 + 5 = \frac{32}{3} = 10\frac{2}{3}$$

Therefore, the shaded area is $10\frac{2}{3}$ units^2.

Finding the area when the integral is negative

The integral can be negative if the curve is below the x-axis as in Figure 7.7, where the area under the curve $y = \sin(x)$ from $x = \pi$ to $x = 3\pi/2$ is illustrated.

$$\int_\pi^{3\pi/2} \sin(x)\,dx = \left[-\cos(x)\right]_\pi^{3\pi/2} = -\cos\left(\frac{3\pi}{2}\right) + \cos(\pi) = -1$$

The integral is negative because the values of y are negative in that region. In the case where all of that portion of the curve is below the x-axis to find the area we just take the modulus. Therefore, the shaded area $A = 1$.

This is important because negative and positive areas can cancel out giving an integral of 0. In Figure 7.8, the area under the curve $y = \sin(x)$ from $x = 0$ to $x = 2\pi$ is pictured. The area under the curve has a positive part from 0 to π and an equal negative part from π to 2π.

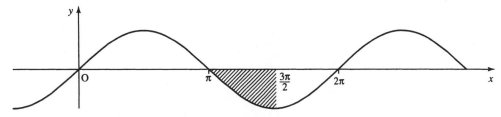

Figure 7.7   *The area under the curve given by $\int_\pi^{3\pi/2} \sin(x)\,dx$.*

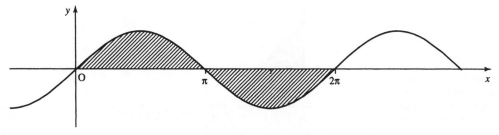

Figure 7.8   *The area under the graph $y = \sin(x)$ from $x = 0$ to 2π.*

The following gives an integral of 0

$$\int_0^{2\pi} \sin(x)\,dx = \left[-\cos(x)\right]_0^{2\pi} = -\cos(2\pi) - (-\cos(0))$$

$$= -1 - (-1) = 0$$

To prevent cancellation of the positive and negative parts of the integration, we find the total shaded area in two stages

$$\int_0^{\pi} \sin(x)\,dx = \left[-\cos(x)\right]_0^{\pi} = -\cos(\pi) - (-\cos(0)) = 2$$

and

$$\int_{\pi}^{2\pi} \sin(x)\,dx = \left[-\cos(x)\right]_{\pi}^{2\pi} = -\cos(2\pi) - (-\cos(\pi)) = -2$$

So, the total area is $2 + |-2| = 4$.

We have seen that if we wish to find the area bounded by a curve which crosses the x-axis, then we must find where it crosses the x-axis first and perform the integration in stages.

Example 7.21   Find the area bounded by the curve $y = x^2 - x$ and the x-axis and the lines $x = -1$ and $x = 1$.

Solution   First, we find if the curve crosses the x-axis. $x^2 - x = 0 \Leftrightarrow x(x - 1) = 0 \Leftrightarrow x = 0$ or $x = 1$. The sketch of the graph with the required area shaded is given in Figure 7.9.

Therefore, the area is the sum of A_1 and A_2. We find A_1 by integrating from -1 to 0

$$\int_{-1}^0 (x^2 - x)\,dx = \left[\frac{x^3}{3} - \frac{x^2}{2}\right]_{-1}^0 = 0 - \left(\frac{(-1)^3}{3} - \frac{(-1)^2}{2}\right)$$

$$= \frac{1}{3} + \frac{1}{2} = \frac{5}{6}$$

therefore, $A_1 = \frac{5}{6}$.

Figure 7.9   *Sketch of $y = x(x - 1)$, with the area bounded by the x-axis and $x = -1$ and $x = 1$ marked. The area above the x-axis is marked as A_1 and the area below the x-axis is marked as A_2.*

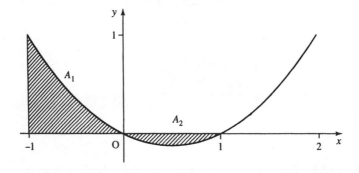

Find A_2 by integrating from 0 to 1 and taking the modulus

$$\int_0^1 (x^2 - x)\,dx = \left[\frac{x^3}{3} - \frac{x^2}{2}\right]_0^1 = \frac{1}{3} - \frac{1}{2} = -\frac{1}{6}$$

Therefore, $A_2 = \frac{1}{6}$.

Then, the total area is $A_1 + A_2 = \frac{5}{6} + \frac{1}{6} = 1$.

7.6 The mean value and r.m.s. value

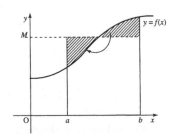

Figure 7.10 *The mean value of a function is the value it would take if it were constant over the range but with the same integral.*

The mean value of a function is the value it would have if it were constant over the range but with the same area under the graph, that is, with the same integral (see Figure 7.10).

The formula for the mean value is

$$M = \frac{1}{b-a}\int_a^b y\,dx.$$

Example 7.22   Find the mean value of $i(t) = 20 + 2\sin(\pi t)$ for $t = 0$ to 0.5.

Solution   Using the formula $a = 0, b = 0.5$ gives

$$M = \frac{1}{0.5 - 0}\int_0^{0.5} 20 + 2\sin(\pi t)\,dt$$

$$2\left[20t - \frac{2}{\pi}\cos(\pi t)\right]_0^{0.5} = 2(10 - 0 - \left(0 - \frac{2}{\pi}(1)\right) \approx 21.27$$

The root mean squared (r.m.s) value

The 'root mean squared value' (r.m.s. value) means the square root of the mean value of the square of y. The formula for the r.m.s. value of y between $x = a$ and $x = b$ is

$$\text{r.m.s.}(y) = \sqrt{\frac{1}{b-a}\int_a^b y^2\,dx}$$

The advantage of the r.m.s.value is that as all the values for y are squared, they are positive, so the r.m.s.value will not give 0 unless we are considering the zero function. If the function represents the voltage then the r.m.s. value can be used to calculate the average power in the signal. In contrast the mean value gives zero if calculated for the sine or cosine over a complete cycle, giving no additional useful information.

Example 7.23   Find the r.m.s. value of $y = x^2 - 3$ between $x = 1$ and $x = 3$.

$$(\text{r.m.s.}(y))^2 = \frac{1}{3-1}\int_1^3 (x^2-3)^2\,dx = \frac{1}{2}\int_1^3 (x^4 - 6x^2 + 9)\,dx$$

$$= \frac{1}{2}\left[\frac{x^5}{5} - \frac{6x^3}{3} + 9x\right]_1^3$$

$$= \frac{1}{2}\left(\left(\frac{243}{5} - 54 + 27\right) - \left(\frac{1}{5} - 2 + 9\right)\right) = 7.2$$

Therefore, the r.m.s value is $\sqrt{7.2} \approx 2.683$.

7.7 Numerical Methods of Integration

Many problems may be difficult to solve analytically. In such cases numerical methods may be used. This is often necessary in order to perform integrations. The following integrals could not be solved by the methods of integration we have met so far:

$$\int_2^3 \frac{\sin(x)}{x^2 + 1}\, dx$$

$$\int_{-3}^2 2^{x^2}\, dx$$

Numerical methods can usually only give an approximate answer.

General method

We wish to approximate the integral

$$\int_a^b f(x)\, dx$$

Formulae for numerical integration are obtained by considering the area under the graph and splitting the area into strips, as in Figure 7.11. The area of the strips can be approximated using the trapezoidal rule or Simpson's rule. In each case, we assume that the thickness of each strip is h and that there are N strips, so that

$$h = \frac{(b - a)}{N}$$

Numerical methods are obviously to be used with a computer or possibly a programmable calculator. However, it is a good idea to be able to check some simple numerical results, which needs some understanding of the algorithms used.

The trapezoidal rule

The strips are approximated to trapeziums with parallel sides of length y_{r-1} and y_r as in Figure 7.12. The area of each strip is $(h/2)(y_{r-1} + y_r)$.

Figure 7.11 *Numerical integration is performed by splitting the area into strips of width h. The area of the strips is approximated using the trapezoidal rule or Simpson's rule.*

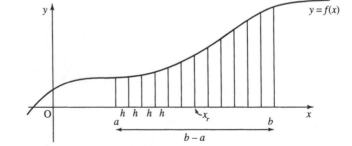

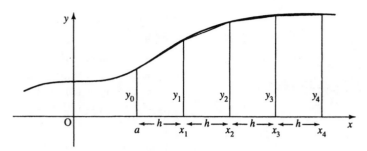

Figure 7.12 *The trapezoidal rule is found by approximating each of the strips as a trapezium.*

The formula becomes:

$$A = h \left(\tfrac{1}{2} y_0 + y_1 + y_2 + \cdots + y_{N-1} + \tfrac{1}{2} y_N \right)$$

where $x_r = a + rh$.

$$y_r = f(x_r)$$
$$N = \frac{(b-a)}{h}.$$

A computer program would more likely use the equivalent recurrence relation, where A_r is the area up to rth strip (at $x = x_r$)

$$A_r = A_{r-1} + \frac{h}{2}(y_{r-1} + y_r)$$

for $r = 1$ to N and $A_0 = 0$.

 This is simply stating that the area is found by adding on the area of one strip at a time to the previously found area.

Example 7.24   We wish to approximate

$$\int_1^3 x^2 \, dx.$$

The limits of the integration are 1 and 3, so $a = 1$ and $b = 3$. We choose a step size of 0.5, therefore,

$$N = \frac{(b-a)}{h} = \frac{(3-1)}{0.5} = 4.$$

Using $x_r = a + rh$ and $y_r = f(x_r)$, which in this case gives $y_r = x_r^2$ we get

$$
\begin{aligned}
x_0 &= 1 & y_0 &= (1)^2 = 1 \\
x_1 &= 1.5 & y_1 &= (1.5)^2 = 2.25 \\
x_2 &= 2 & y_2 &= (2)^2 = 4 \\
x_3 &= 2.5 & y_3 &= (2.5)^2 = 6.25 \\
x_4 &= 3 & y_4 &= (3)^2 = 9.
\end{aligned}
$$

Using the formula for the trapezoidal rule:

$$A = h(\tfrac{1}{2}y_0 + y_1 + y_2 + \cdots + y_{N-1} + \tfrac{1}{2}y_N)$$

we get

$$A = 0.5(0.5 + 2.25 + 4 + 6.25 + 4.5) = 8.75.$$

Hence, by the trapezoidal rule:

$$\int_1^3 x^2 \, \mathrm{d}x \approx 8.75.$$

Simpson's rule

For Simpson's rule, the area of each strip is approximated by drawing a parabola through three adjacent points (see Figure 7.13). Notice that the number of strips must be even.

The area of the strips in this case is not obvious as in the case of the trapezoidal rule. Three strips together have an area of:

$$\frac{h}{3}(y_{2n-2} + 4y_{2n-1} + y_{2n})$$

where $r = 2n$. The formula then becomes

$$A = \frac{h}{3}(y_0 + 4y_1 + 2y_2 + 4y_3 + 2y_4 + \cdots + 2y_{N-2} + 4y_{N-1} + y_N)$$

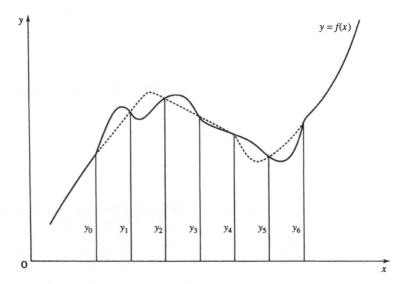

Figure 7.13 *Simpson's rule is found by approximating the areas of the strips by drawing a parabola through three adjacent points. To do this the total number of strips must be even.*

where

$$x_r = a + rh, \quad y_r = f(x_r), \quad N = \frac{(b-a)}{h}$$

as before.

Again, a computer program would more likely use the recurrence relation to define the area

$$A_{2n} = A_{2(n-1)} + \frac{h}{3}(y_{2n-2} + 4y_{2n-1} + y_{2n})$$

where $r = 1, \ldots, N/2$ and $A_0 = 0$.

Example 7.25 Find $\int_1^3 x^2 \, dx$ using Simpson's rule with $h = 0.5$.

Solution From the limits of the integral we find that $a = 1$ and $b = 3$. So,

$$N = \frac{(b-a)}{h} = \frac{(3-1)}{0.5} = 4.$$

Using $x_r = a + rh$ and $y_r = f(x_r)$, which in this case gives

$$y_r = x_r^2$$

we get

$$
\begin{array}{ll}
x_0 = 1 & y_0 = (1)^2 = 1 \\
x_1 = 1.5 & y_1 = (1.5)^2 = 2.25 \\
x_2 = 2 & y_2 = (2)^2 = 4 \\
x_3 = 2.5 & y_3 = (2.5)^2 = 6.25 \\
x_4 = 3 & y_4 = (3)^2 = 9.
\end{array}
$$

Hence,

$$A = \frac{0.5}{3}(1 + 4(2.25) + 2(4) + 4(6.25) + 9) \approx 8.66667.$$

In this case, as we are integrating a parabola the result is exact (except for rounding errors).

7.8 Summary

1. Integration can be defined as the inverse process of differentiation. If $y = f(x)$ then

$$\frac{dy}{dx} = f'(x) \quad \Leftrightarrow \quad y = \int \frac{dy}{dx} \, dx = f(x) + C$$

or equivalently

$$\int \frac{dy}{dx} \, dx = y + C.$$

This is called indefinite integration and C is the constant of integration.

2. A table of standard integrals can be found as in Table 7.1 by swapping the columns of Table 6.3, rearranging them in a more convenient form and adding the constant of integration.

3. Integrals of combinations of the functions given in Table 7.1 cannot always be found but some methods can be tried as follows.
   (a) Substitute $u = ax + b$ to find $\int f(ax + b) \, dx$.

(b)   Substitute when the integral is of the form $\int f(u)(du/dx)\,dx$.

(c)   Use integration by parts when the integral is of the form $\int u\,dv$, using the formula $\int u\,dv = uv - \int v\,du$.

(d)   Use trigonometric identities to integrate powers of $\cos(x)$ and $\sin(x)$.

4.   Integration has many applications, some of which are listed in section 7.4.

5.   The definite integral, from $x = a$ to $x = b$, is defined as the area under the curve between those two values. This is written as $\int_a^b f(x)\,dx$.

6.   The mean value of a function is the value the functions would have if it were constant over the range but with the same area under the graph. The mean value from $x = a$ to $x = b$ of y is

$$M = \frac{1}{b-a} \int_a^b y\,dx.$$

7.   The root mean squared value (r.m.s. value) is the square root of the mean value of the square of y. The r.m.s. value from $x = a$ to $x = b$ is given by

$$\text{r.m.s.}(y) = \sqrt{\frac{1}{b-a} \int_a^b y^2\,dx}.$$

8.   Two methods of numerical integration are: trapezoidal rule and Simpson's rule, where

$$A \approx \int f(x)\,dx.$$

Trapezoidal rule

$$A = h\left(\tfrac{1}{2}y_0 + y_1 + y_2 + \cdots + y_{N-1} + \tfrac{1}{2}y_N\right)$$

Simpson's rule

$$A = \frac{h}{3}(y_0 + 4y_1 + 2y_2 + 4y_3 + 2y_4 + \cdots + 2y_{N-2} + 4y_{N-1} + y_N).$$

In both cases, $x_r = a + rh$ and $N = (b - a)/h$. For Simpson's rule, N must be even. h is called the step size and N is the number of steps.

7.9 Exercises

7.1. Find the following integrals

(a) $\int (x^3 + x^2)\,dx$

(b) $\int 2\sin(x) + \sec^2(x)\,dx$

(c) $\int \frac{1}{x^2}\,dx$

(d) $\int (1 + x^2 + 3x^3)\,dx$

(e) $\int (1 - 5x)\,dx$

(f) $\int \cos(2 - 4x)\,dx$

(g) $\int \sqrt{2x - 1}\,dx$

(h) $\int \frac{1}{\sqrt{x+2}}\,dx$

(i) $\int x(x^2 - 4)^3\,dx$

(j) $\int x\sqrt{(1 + x^2)}\,dx$

(k) $\int \frac{\cos(x)}{(1 + \sin(x))^2}\,dx$

(l) $\int (x^2 + x - 6)(2x + 1)\,dx$

(m) $\int \frac{4x^2}{(x^2 - 7)^2}\,dx$

(n) $\int_1^2 x\cos(x)\,dx$

(o) $\int x^2 \cos(x)\,dx$

(p) $\int_2^4 x\sqrt{x - 1}\,dx$

(q) $\int x(2x - 3)^4\,dx$

(r) $\int_0^{\pi/2} \sin^5(x)\,dx$

(s) $\int \cos^4(x)\,dx$

(t) $\int \sin(3x)\cos(5x)\,dx$.

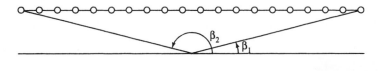

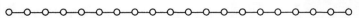

Figure 7.14   *Field on the axis of a solenoid for Exercise 7.7.*

7.2. Given that $v = ds/dt = 3 - t$, find s in terms of t if $s = 5$ when $t = 0$. What is the value of s when $t = 2$?

7.3. Find the equation of the curve with the gradient $dy/dt = -5$ which passes through the origin.

7.4. A curve, $y = f(x)$ passes through the point $(0,1)$ and its gradient at any point is $1 - 2x^2$. Find the function.

7.5. The voltage across an inductor of inductance $3\,\text{H}$ is measured as $V = 2\sin(2t - (\pi/6))$.
The current at $t = 0$ is 0. Given that $V = L(di/dt)$ find the current after $10\,\text{s}$.

7.6. The velocity of a spring is found to be $V = 6\sin(3\pi t)$. Assuming that the spring is perfect, so that $v = (1/k)(dF/dt)$, where k is the spring constant (known to be 0.5), v the velocity, and F the force operating on the spring, find the force, given that it is $0\,\text{N}$ initially.

7.7. The magnitude of the magnetic flux density at the midpoint of the axis of a solenoid, as in Figure 7.14, can be found by the integral

$$B = \int_{\beta_1}^{\beta_2} \frac{\mu_0 n I}{2} \sin(\beta)\,d\beta$$

where μ_0 is the permeability of free space ($\approx 4\pi \times 10^{-7}\,\text{H m}^{-1}$), n is the number of turns and I is the current. If the solenoid is so long that $\beta_1 \approx 0$ and $\beta_2 \approx \pi$, show that $B = \mu_0 n I$.

7.8. Find the area under the curve $y = x + x^2$ between the lines $x = 1$ and $x = 4$.

7.9. Find the area bounded by the x-axis and the portion of the curve $y = 2(x - 1)(x - 4)$ which lies below it.

7.10. Find the total area bounded by the curve $y = 2x - x^2$, the x-axis and the lines $x = -1$ and $x = 1$.

7.11. Find the mean value of $i(t) = 5 - \cos(t/2)$ for $t = 0$ to $t = 5$.

7.12. Calculate the r.m.s. value of $i = 3\cos(50\pi t)$ between $t = 0$ and $t = 0.01$.

7.13. Approximate:

$$\int_0^1 \frac{\sin(x)}{x}\,dx$$

(a) using the trapezoidal rule with $h = 0.2$ and $h = 0.1$; (b) using Simpson's rule with $h = 0.5$ and $h = 0.25$.

7.14. Find an approximate value of

$$A = \int_1^3 \frac{dx}{x}$$

(a) by the trapezoidal rule with $N = 6$; (b) by Simpson's rule with $N = 6$.

7.15. Approximate $\int_0^1 x^5\,dx$ using Simpson's rule with $N = 10$.

8 The exponential function

8.1 Introduction

In Example 2.11, we looked at an acoustical absorption problem. We found that after a single note on a trombone had been played, the sound intensity decayed according to the expression $I = 10^{-11t}$. This is a dying exponential function. Many other physical situations involve decay or growth in an exponential fashion; for instance, population growth or the decay of charge on a discharging capacitor. The functions $y = a^t$, where a can be any positive number, are called exponential functions. In this chapter, we shall look at how they describe this particular type of growth or decay, when the growth or decay is proportional to the current size of the 'population', y. This situation can be described by a differential equation of the form $dy/dt = ky$. The special case where $k = 1$, giving the equation $dy/dt = y$, leads us to define the number $e \approx 2.7182818$ and the function $y = e^t$, which is called *the* exponential function. The inverse function is $y = \log_e(t)$, which is also refered to as $y = \ln(t)$ and called the natural or Napierian logarithm.

The function $y = e^t$ is neither even nor odd; however, it is possible to split any real function into an even part and an odd part and in this case we find that this gives the hyperbolic functions. These hyperbolic functions have properties that are surprisingly similar to the properties of the trigonometric functions and hence have similar names, $\cosh(t)$, $\sinh(t)$ (the hyperbolic sine and hyperbolic cosine) from which we can define also $\tanh(t)$, the hyperbolic tangent. We also look at differentiation and integration problems involving the exponential and hyperbolic functions.

8.2 Exponential growth and decay

Supposing, following some deed of heroism, the police offered you the choice of the following rewards:

(a) Tomorrow you receive 1 c and the following day 2 c and after that 4 c, then 8 c and each day the amount doubles for the next month.
(b) Tomorrow you receive €2, the following day €4 and the day after €6, then €8 so that each day you receive €2 more than the day before. Again you receive payments on every day for the next month.

If, although not motivated by personal greed, you wish to receive the highest possible reward (in order, presumably, to donate the amount to charity), which reward should you accept? Option (b) superficially appears to be the best because at least it starts off with enough money to buy a small sandwich. However, a closer look reveals the that if you choose option (a), on the last day (assuming there are 31 days in the month) you receive in excess of €10 000 000 with the total reward exceeding €20 million. However, option (b) only reaps €62 on the final day with a total award of only €992.

Both options are examples of growth. Option (b) gives a constant growth rate of €2. If y_n is the amount received on day n, the way that this grows could be expressed by $y_{n+1} = y_n + 2$. This is expressing the fact that the amount received on day $n+1$ is €2 more than the amount received on day n. This also means that the change each day in the amount received is 2, which can be expressed as $\Delta y_n = 2$, where $\Delta y_n = y_{n+1} - y_n$ and represents the change in y from day n to day $n + 1$.

Option (a) is an example of exponential growth (or geometric growth). The amount received each day is proportional to the amount received the day before, in this case twice as much. y_n can be expressed by $y_{n+1} = 2y_n$. The change in y is equal to the value of y itself, $\Delta y_n = y_n$.

Because of the nature of exponential growth it is unlikely that you would be given such an attractive award as option (a) represents. However, exponential growth is not beyond the reach of the everyday person as savings accounts offer this opportunity. Unfortunately, the amount you receive does not increase as quickly as doubling each day but it is based on how much you have already in the bank; hence, it is exponential. Supposing you opened an account that paid an annual interest of 6% and the annual rate of inflation was 3%, then the real rate of growth is approximately 3% per annum. If y_n is the value of the amount you have in the bank after n years then $y_{n+1} = 1.03y_n$. We can also express this by saying that the interest received each year, that is, the change in y_n, Δy_n, is 3% of y_n, that is, $\Delta y_n = 0.03y_n$, where $\Delta y_n = y_{n+1} - y_n$. If the rate of interest remains constant then if you deposit €1 tomorrow then your descendents, in only 500 years time, will receive an amount worth over €2.5 million in real terms.

The models of growth that we have discussed so far give examples of recurrence relations, also called difference equations. Their solutions are not difficult to find. For instance, if $y_{n+1} = 2y_n$ and we know that on day 1 we received 1 c, that is, $y_1 = 0.01$ then we can substitute $n = 1, 2, 3, \ldots$ (as we did in Section 1.4) to find values of the function giving

$$0.01, 0.02, 0.04, 0.08, 0.16, 0.32, \ldots$$

Clearly, there is a power of 2 involved in the expression for y_n, so we can guess that $y = 0.01\ (2^{n-1})$. By checking a few values of n we can confirm that this is indeed the amount received each day. When we deposit €1 in the bank at a real rate of growth of 3% we get the recurrence relation $y_{n+1} = 1.03y_n$, where y_n is the current day value of the amount in the savings account after n years. Substituting a few values beginning with $y_0 = 1$ (the amount we initially deposit) we then get $1, 1.03, 1.06, 1.09, 1.13, 1.16, 1.19, \ldots$ (to the nearest cent). Each time we multiply the amount by 1.03, there must be a power of 1.03 in the solution for y. We can guess the solution as $y_n = (1.03)^n$. By checking a few values of n we can confirm that this is, in fact, the amount in the bank after n years.

The models we have looked at so far are discrete models. In the case of the money in the bank the increase occurs at the end of each year. However, if we consider population growth, for instance, then it is not possible to say that the population grows at the end of a certain period, the growth could happen at any moment of time. In this case, providing the population is large enough, it is easier to model the situation continuously, using a differential equation. Such models take the form of $dy/dt = ky$. dy/dt is the rate of growth, if k is positive, or the rate of decay, if k is negative. The equation states that the rate of growth or decay of a population of size y is proportional to the size of the population.

Example 8.1   A malfunctioning fridge maintains a temperature of 6°C which allows a population of bacteria to reproduce such that, on average, each bacteria divides every 20 min. Assuming no bacteria die in the time under consideration find a differential equation to describe the population growth.

Solution   If the population at time t is given by p then the rate of change of the population is $\mathrm{d}p/\mathrm{d}t$. The increase in the population is such that it approximately doubles every 20 min, that is, it increases by p in 20×60 s. That gives a rate of increase as $p/(20 \times 60)$ per second. Hence, the differential equation describing the population is

$$\frac{\mathrm{d}p}{\mathrm{d}t} = \frac{p}{1200}.$$

Figure 8.1   *A closed RC circuit.*

Example 8.2   A capacitor, in an RC circuit, has been charged to a charge of Q_0. The voltage source has been removed and the circuit closed as in Figure 8.1. Find a differential equation that describes the rate of discharge of the capacitor if $C = 0.001\,\mu\mathrm{F}$ and $R = 10\,\mathrm{M}\Omega$.

Solution   The voltage across a capacitor is given by Q/C where C is the capacitance and Q is the charge on the capacitor. The voltage across the resistor is given by Ohm's Law as IR. From Kirchoff's voltage law, the sum of the voltage drops in the circuit must be 0; therefore, as the circuit is closed, we get voltage across the resistor + voltage across the capacitor = 0:

$$\Rightarrow IR + \frac{Q}{C} = 0.$$

By definition, the current is the rate of change of charge with respect to time, that is $I = \mathrm{d}Q/\mathrm{d}t$ giving the differential equation

$$R\frac{\mathrm{d}Q}{\mathrm{d}t} + \frac{Q}{C} = 0.$$

We can rearrange this equation as

$$R\frac{\mathrm{d}Q}{\mathrm{d}t} = -\frac{Q}{C} \quad \Leftrightarrow \quad \frac{\mathrm{d}Q}{\mathrm{d}t} = -\frac{Q}{RC}.$$

We can see that this is an equation for exponential decay. The rate of change of the charge on the capacitor is proportional to the remaining charge at any point in time with a constant of proportionality given by $1/RC$. In this case as $R = 10\,\mathrm{M}\Omega$ and $C = 0.001\,\mu\mathrm{F}$, we get

$$\frac{\mathrm{d}Q}{\mathrm{d}t} = -100Q.$$

Example 8.3   Radioactivity is the emission of α- or β-particles and γ-rays due to the disintegration of the nuclei of atoms. The rate of disintegration is proportional to the number of atoms at any point in time and the constant of proportionality is called the radioactivity decay constant. The radioactive decay constant for Radium B is approximately $4.3 \times 10^{-4}\,\mathrm{s}^{-1}$. Give a differential equation that describes the decay of the number of particles N in a piece of Radium B.

Solution If the number of particles at time t is N, then the rate of change is dN/dt. The decay is proportional to the number of atoms, and we are given that the constant of proportionality is $4.3 \times 10^{-4}\,\text{s}^{-1}$ so that we have

$$\frac{dN}{dt} = -4.3 \times 10^{-4} N$$

as the equation which describes the decay.

Example 8.4 An object is heated so that its temperature is $400\,\text{K}$ and the temperature of its surroundings is $300\,\text{K}$, and then it is left to cool. Newton's law of cooling states that the rate of heat loss is proportional to the excess temperature over the surroundings. Furthermore, if m is the mass of the object and c is its specific heat capacity then the rate of change of heat is proportional to the rate of fall of temperature of the body, and is given by

$$\frac{dQ}{dt} = -mc\frac{d\phi}{dt}$$

where Q is the heat in the body and ϕ is its temperature. Find a differential equation for the temperature that describes the way the body cools.

Solution Newton's law of cooling gives

$$\frac{dQ}{dt} = A(\phi - \phi_s)$$

where A is some constant of proportionality, Q is the heat in the body, ϕ is its temperature, and ϕ_s is the temperature of its surroundings. As we also know that

$$\frac{dQ}{dt} = -mc\frac{d\phi}{dt}$$

this can be substituted in our first equation giving:

$$-mc\frac{d\phi}{dt} = A(\phi - \phi_s) \quad \Leftrightarrow \quad \frac{d\phi}{dt} = -\frac{A}{mc}(\phi - \phi_s)$$

$A/(mc)$ can be replaced by a constant k, giving

$$\frac{d\phi}{dt} = -k(\phi - \phi_s).$$

In this case, the temperature of the surroundings is known to be $300\,\text{K}$, so the equation describing the rate of change of temperature is

$$\frac{d\phi}{dt} = -k(\phi - 300).$$

We have established that there are a number of important physical situations that can be described by the equation $dy/dt = ky$. The rate of change of y is proportional to its value. We would like to solve this equation, that is, find y explicitly as a function of t. In Chapter 7, we solved simple differential equations such as $dy/dt = 3t$ by integrating both sides with respect to t, for example,

$$\frac{dy}{dt} = 3t \quad \Leftrightarrow \quad y = \int 3t \, dt \quad \Leftrightarrow \quad y = \frac{3t^2}{2} + C.$$

However, we cannot solve the equation $dy/dt = ky$ in this way because the right-hand side is a function of y not of t. If we integrate both sides with respect to t we get

$$\frac{dy}{dt} = ky \quad \Leftrightarrow \quad y = \int ky \, dt.$$

Although this is true we are no nearer solving for y as we need to know y as a function of t in order to find $\int ky \, dt$.

When we solved equations in Chapter 3 of the Background Mathematics notes available on the companion website for this book, we said that one method was to guess a solution and substitute that value for the unknown into the equation to see if it gave a true statement. This would be a very long method to use unless we are able to make an informed guess. We can use this method with this differential equation as we know from our experience with problems involving discrete growth that a solution should involve an exponential function of the form $y = a^t$. The main problem is to find the value of a, which will go with any particular equation. To do this we begin by looking for an exponential function that would solve the equation $dy/dt = y$, that is, we want to find the function whose derivative is equal to itself.

8.3 The exponential function $y = e^t$

Figures 8.2(a) and 8.3(a) give graphs of $y = 2^t$ and $y = 3^t$, which are two exponential functions. We can sketch their derivative functions by drawing tangents to the graph and measuring the gradient of the tangent at various different points. The derivative functions are pictured in Figure 8.2(b), dy/dt where $y = 2^t$, and in Figure 8.3(b), dy/dt where $y = 3^t$.

We can see that for these exponential functions the derivative has the same shape as the original function but has been scaled in the y-direction, that is, multiplied by a constant, k, so that $dy/dt = ky$ as we expected:

$$\frac{d}{dt}(2^t) = (C)(2^t) \quad \text{and} \quad \frac{d}{dt}(3^t) = (D)(3^t)$$

where C and D are constants. We can see from the graphs that $C < 1$ and $D > 1$. Thus, the derivative of 2^t gives a squashed version of the original graph and the derivative of 3^t gives a stretched version of the original graph.

It would seem reasonable that there would be a number somewhere between 2 and 3 that we can call e, which has the property that the derivative of e^t is exactly the same as the original graph. That is,

$$\frac{d}{dt}(e^t) = e^t.$$

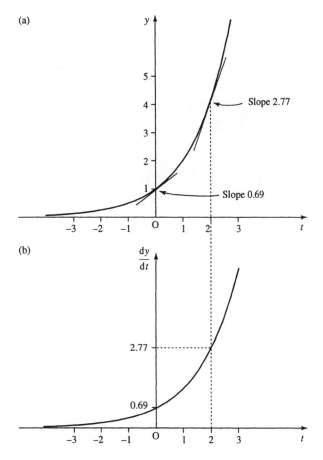

Figure 8.2 *(a) The graph of $y = 2^t$ with some tangents marked. (b) The graph of the derivative (the gradient of the tangent at any point on $y = 2^t$ plotted against t).*

Finding the value of e

There are various methods for finding the value of e and a graphical investigation into finding e to one decimal place is given in the projects and investigations available on the companion website for this book. An alternative, numerical, method is to look at the gradient of the chord for the function $y = e^t$ at $t = 0$:

$$\frac{\delta y}{\delta t} = \frac{f(t + \delta t) - \delta t}{\delta t} = \frac{e^{t + \delta t} - e^t}{\delta t}$$

At $t = 0$ all functions $y = a^t$ have value 1 so that the gradient of the chord at $t = 0$ is

$$\frac{\delta y}{\delta t} = \frac{e^{\delta t} - 1}{\delta t}$$

We defined e as the number for which $\dfrac{d}{dt}(e^t) = e^t$ so that at the point $t = 0$ the gradient of the tangent is given by $dy/dt = 1$. For small δt the gradient of the tangent is approximately equal to the gradient of the chord

$$\frac{dy}{dt} \approx \frac{\delta y}{\delta t}$$

and therefore

$$1 \approx \frac{e^{\delta t} - 1}{\delta t}$$

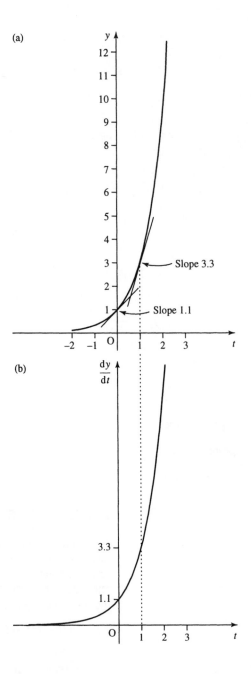

Figure 8.3 *(a) The graph of y = 3^t with some tangents marked. (b) The graph of the derivative (the gradient of the tangent at any point on y = 3^t plotted against t).*

Rearranging this equation gives $e \approx (1+\delta t)^{1/\delta t}$ for small δt. Replacing δt by $1/n$, with n large, gives

$$e \approx \left(1 + \frac{1}{n}\right)^n$$

for large n.

Let n tend to infinity and this gives the well-known limit

$$e = \lim_{n \to \infty} \left(1 + \frac{1}{n}\right)^n.$$

We can use the expression

$$e \approx \left(1 + \frac{1}{n}\right)^n$$

Table 8.1 *Estimating* e *using* $(1 + 1/n)^n$ *gives* e $= 2.71828$ *to five decimal places*

n	$(1 + (1/n))^n$ to five decimal places
1000	2.71692
10 000	2.71815
100 000	2.71827
1 000 000	2.71828
10 000 000	2.71828

for large n to calculate e on a calculator. This is done in Table 8.1 to five decimal places.

We have shown that e $= 2.71828$ to five decimal places. e is an irrational number which means that it cannot be written exactly as a fraction (or as a decimal). The function e^t is often referred to as $\exp(t)$. e^t and its derivative are shown in Figure 8.4.

By definition of the logarithm (as given in Chapter 4 of the Background Mathematics notes available on the companion website for this book), we know that the inverse function to e^t is $\log_e(t)$ (log, base e, of t). This is often represented by the short hand of $\ln(t)$ and called the natural or Napierian logarithm.

We are now able to solve the differential equation $dy/dt = y$ as we know that one solution is $y = e^t$ because the derivative of $y = e^t$ is e^t. When we discussed differential equations in Chapter 7, we noticed that there was an arbitrary constant that was involved in the solution of a differential equation. In this case the constant represents the initial size of the population, or the initial charge or the initial number of atoms or the initial temperature. The general solution to $dy/dt = y$ is $y = y_0e^t$ where y_0 is the value of y at time $t = 0$. We can show that this is, in fact, the general solution by substituting into the differential equation.

Example 8.5

(a)   Show that any function of the form $y = y_0e^t$, where y_0 is a constant, is a solution to the equation

$$\frac{dy}{dt} = y$$

(b)   Show that in the function $y = y_0e^t$, $y = y_0$ when $t = 0$.

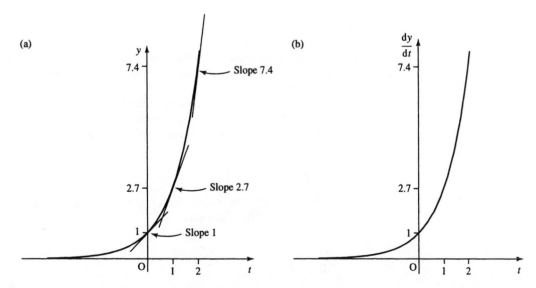

Figure 8.4   *(a) The graph of $y = e^t$ with some tangents marked. (b) The graph of the derivative (the gradient of the tangent at any point on e^t plotted against t).*

Solution

(a)   To show that $y = y_0 e^t$ are solutions, we first differentiate

$$\frac{d}{dt}(y_0 e^t) = y_0 e^t$$

(as y_0 is a constant and $d(e^t)/dt = e^t$).
   Substitute for dy/dt and for y into the differential equation and we get $y_0 e^t = y_0 e^t$, which is a true statement for all t. Hence the solutions to $dy/dt = y$ are $y = y_0 e^t$.

(b)   Substitute $t = 0$ in the function $y = y_0 e^t$ and we get $y = y_0 e^0$. As any number raised to the power of 0 is 1, we have $y = y_0$. Hence y_0 is the value of y at $t = 0$.
   Using the function of a function rule we can find the derivative of e^{kt}, where k is some constant, and show that this function can be used to solve differential equations of the form $dy/dt = ky$.

The derivative of e^{kt}

To find the derivative of $y = e^{kt}$ where k is a constant substitute $u = kt$ so that $y = e^u$

$$\frac{du}{dt} = k \quad \text{and} \quad \frac{dy}{du} = e^u$$

therefore, using the chain rule,

$$\frac{dy}{dt} = \frac{du}{dt}\frac{dy}{du} = k\,e^u = k\,e^{kt}$$

(resubstituting $u = kt$). Therefore,

$$\frac{d}{dt}(e^{kt}) = k\,e^{kt}$$

Notice that if we substitute y for e^{kt} into $d(e^{kt})/dt = k\,e^{kt}$ we get $dy/dt = ky$, which was the differential equation we set out to solve for our growth or decay problems.
   This tells us that one solution to the equation $dy/dt = ky$ is $y = e^{kt}$. The general solution must involve a constant, so we try $y = y_0 e^{kt}$ where y_0 is the initial size of the population, or initial temperature, etc.

Example 8.6

(a)   Show that any function of the form $y = y_0 e^{kt}$, where y_0 is a constant, is a solution to the equation

$$\frac{dy}{dt} = ky$$

(b)   Show that in the function $y = y_0 e^{kt}$, $y = y_0$ when $t = 0$.

Solution

(a) To show that $y = y_0 e^{kt}$ are solutions we first differentiate:

$$\frac{d}{dt}(y_0 e^{kt}) = y_0 k e^{kt}$$

as y_0 is a constant and $d(e^{kt})/dt = k e^{kt}$. Substitute for dy/dt and for y into the differential equation and we get

$$y_0 k e^{kt} = k y_0 e^{kt}$$

which is a true statement for all values of t. Hence the solutions to $dy/dt = ky$ are $y = y_0 e^{kt}$.

(b) Substitute $t = 0$ in the function $y = y_0 e^t$ and we get $y = y_0 e^0$. As any number raised to the power of 0 is 1, we have $y = y_0$. Hence, y_0 is the value of y at $t = 0$.

Example 8.7 Solve the differential equation given in Example 8.2, describing the discharge of a capacitor in a closed RC circuit with $R = 10\,\text{M}\Omega$ and $C = 0.001\,\mu\text{F}$:

$$\frac{dQ}{dt} = -100Q$$

and find a particular solution given that at $t = 0$ the voltage drop against the capacitor was 1000 V.

Solution We have discovered that the solution to a differential equation of the form $dy/dt = kt$ is given by $y = y_0 e^{kt}$ where y_0 is the initial value of y.

Comparing $dy/dt = ky$ with $dQ/dt = -100Q$, and replacing y by Q and k by -100, we get the solution

$$Q = Q_0 e^{-100t}$$

To find the value of Q_0 we need to find the value of Q when $t = 0$. We are told that the initial value of the voltage across the capacitor was 1000 V and we know that the voltage drop across a capacitor is Q/C. Therefore, we have

$$\frac{Q_0}{0.001 \times 10^{-6}} = 1000 \quad \Leftrightarrow \quad Q_0 = 1000 \times 0.001 \times 10^{-6}$$

$$\Leftrightarrow \quad Q_0 = 10^{-6}\,\text{C}.$$

Therefore, the equation that describes the charge as the capacitor discharges is $Q = 10^{-6} e^{-100t}\,\text{C}$ at time t s.

The derivative of a^t

The derivative of $y = 2^t$ can now be found by observing that $2 = e^{(\ln(2))}$. Therefore, $y = 2^t = (e^{\ln(2)})^t = e^{\ln(2)t}$. This is of the form e^{kt} with

$k = \ln(2)$. As

$$\frac{\mathrm{d}}{\mathrm{d}t}(\mathrm{e}^{kt}) = k\mathrm{e}^{kt}$$

then

$$\frac{\mathrm{d}}{\mathrm{d}t}(\mathrm{e}^{\ln(2)^t}) = \ln(2)\mathrm{e}^{t\,\ln(2)}$$

Using again the fact that $\mathrm{e}^{\ln(2)t} = (\mathrm{e}^{\ln(2)})^t = 2^t$ we get

$$\frac{\mathrm{d}}{\mathrm{d}t}(2^t) = \frac{\mathrm{d}}{\mathrm{d}t}(\mathrm{e}^{\ln(2)t}) = \ln(2)\mathrm{e}^{\ln(2)t} = \ln(2)2^t$$

that is

$$\frac{\mathrm{d}}{\mathrm{d}t}(2^t) = \ln(2)2^t$$

Compare this result to that which we found by sketching the derivative of $y = 2^t$ in Figure 8.2(b). We said that the derivative graph was a squashed version of the original graph. This result tells us that the scaling factor is $\ln(2) \approx 0.693$, which confirms our observation that the scaling factor, $C < 1$.

Using the same argument for any exponential function $y = a^t$ we find that $\mathrm{d}y/\mathrm{d}t = \ln(a)a^t$.

In finding these results we have used the fact that an exponential function, to whatever base, a, can be written as e^{kt} where $k = \ln(a)$.

The derivative of $y = \ln(x)$

$y = \ln(x)$ is the inverse function of $f(x) = \mathrm{e}^x$, and therefore we can find the derivative in a manner similar to that used to find the derivatives of the inverse trigonometric functions in Chapter 5.

$y = \ln(x)$ where $x > 0$

$\Leftrightarrow \quad \mathrm{e}^y = \mathrm{e}^{\ln(x)}$ (take the exponential of both sides)

$\Leftrightarrow \quad \mathrm{e}^y = x$ (as exp is the inverse function to ln, $\mathrm{e}^{\ln(x)} = x$)

We wish to differentiate both sides with respect to x but the left-hand side is a function of y, so we use the chain rule, setting $w = \mathrm{e}^y$, thus, equation $\mathrm{e}^y = x$ becomes $w = x$ and $\mathrm{d}w/\mathrm{d}y = \mathrm{e}^y$.

Differentiating both sides of $w = x$ with respect to x gives $\mathrm{d}w/\mathrm{d}x = 1$, where

$$\frac{\mathrm{d}w}{\mathrm{d}x} = \frac{\mathrm{d}w}{\mathrm{d}y}\frac{\mathrm{d}y}{\mathrm{d}x}$$

from the chain rule. So

$$e^y \frac{dy}{dx} = 1$$

and resubstituting $x = e^y$ we get

$$x \frac{dy}{dx} = 1 \quad \Leftrightarrow \quad \frac{dy}{dx} = \frac{1}{x}$$

(we can divide by x as $x > 0$). Hence,

$$\frac{d}{dx}(\ln x) = \frac{1}{x}.$$

The derivative of the log, of whatever the base, can be found using the change of base rule for logarithms as given in Chapter 4 of the Background Mathematics notes available on the companion website for this book. We can write

$$\log_a(x) = \frac{\ln(x)}{\ln(a)}.$$

Therefore

$$\frac{d}{dx}(\log_a(x)) = \frac{d}{dx}\left(\frac{\ln(x)}{\ln(a)}\right) = \frac{1}{\ln(a)x}.$$

8.4 The hyperbolic functions

Any function defined for both positive and negative values of x can be written as the sum of an even and odd function. That is, for any function $y = f(x)$ we can write

$$f(x) = f_e(x) + f_o(x)$$

where

$$f_e(x) = \frac{f(x) + f(-x)}{2}$$

and

$$f_o(x) = \frac{f(x) - f(-x)}{2}.$$

The even and odd parts of the function e^x are given the names of hyperbolic cosine and hyperbolic sine. The names of the functions are usually shortened to $\cosh(x)$ (read as 'cosh of x') and $\sinh(x)$ (read as 'shine of x').

$$e^x = \cosh(x) + \sinh(x)$$

and

$$\cosh(x) = \frac{e^x + e^{-x}}{2}, \quad \sinh(x) = \frac{e^x - e^{-x}}{2}.$$

They are called the hyperbolic sine and cosine because they bear the same sort of relationship to the hyperbola as the sine and cosine do to the circle. When we introduced the trigonometric functions in Chapter 5 we used a rotating rod of length r. The horizontal and vertical positions of the tip of the rod as it travels around the circle defines the cosine and sine function, respectively. A point (x, y) on the circle can be defined using $x = r\cos(\alpha)$, $y = r\sin(\alpha)$. These are called parametric equations for the circle and α is the parameter. If the parameter is eliminated then we get the equation of the circle

$$\frac{x^2}{r^2} + \frac{y^2}{r^2} = 1$$

(shown in Figure 8.4(a)). Any point on a hyperbola can similarly be defined in terms of a parameter, α, and thus we get $x = a\cosh(\alpha)$ and $y = b\sinh(\alpha)$.

If the parameter is eliminated from the equations we get the equation for the hyperbola as

$$\frac{x^2}{a^2} - \frac{y^2}{b^2} = 1$$

Figure 8.5(b) shows the graph of the hyperbola.

The function $y = \tanh(x)$ is defined, similarly to the $\tan(x)$, as

$$\tanh(x) = \frac{\sinh(x)}{\cosh(x)}$$

and the reciprocal of these three main functions may be defined as

$$\operatorname{cosech}(x) = \frac{1}{\sinh(x)} \quad \text{(the hyperbolic cosecant)}$$

$$\operatorname{sech}(x) = \frac{1}{\cosh(x)} \quad \text{(the hyperbolic secant)}$$

$$\coth(x) = \frac{1}{\tanh(x)} \quad \text{(the hyperbolic cotangent)}$$

The graphs of $\cosh(x)$, $\sinh(x)$, and $\tanh(x)$ are shown in Figure 8.6.

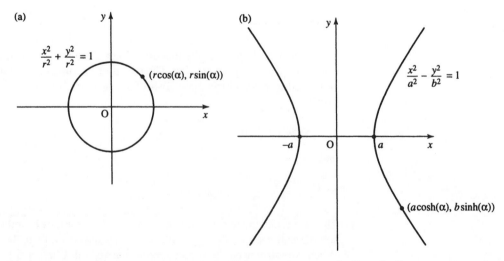

Figure 8.5 *(a)* $x = r\cos(\alpha)$, $y = r\sin(\alpha)$ *defines a point on the circle* $x^2/r^2 + y^2/r^2 = 1$. *(b)* $x = a\cosh(\alpha)$ *and* $y = b\sinh(\alpha)$ *defines a point on the hyperbola* $x^2/a^2 - y^2/b^2 = 1$.

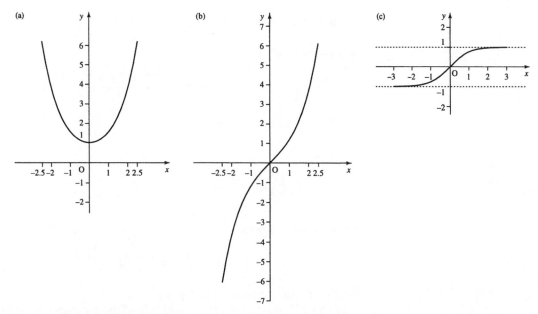

Figure 8.6   *(a) The graph of y = cosh(x). (b) The graph of y = sinh(x). (c) The graph of y = tanh(x).*

Table 8.2   *Summary of important hyperbolic identities*

$$\cosh(x) = (e^x + e^{-x})/2$$
$$\sinh(x) = (e^x - e^{-x})/2$$
$$\tanh(x) = \sinh(x)/\cosh(x) = (e^x - e^{-x})/(e^x + e^{-x})$$
$$\cosh(x) + \sinh(x) = e^x$$
$$\cosh(x) - \sinh(x) = e^{-x}$$
$$\cosh(A \pm B) = \cosh(A)\cosh(B) \pm \sinh(A)\sinh(B)$$
$$\sinh(A \pm B) = \sinh(A)\cosh(B) \pm \cosh(A)\sinh(B)$$
$$\tanh(A \pm B) = (\tanh(A) \pm \tanh(B))/(1 \pm \tanh(A)\tanh(B))$$

Hyperbolic identities

The hyperbolic identities are similar to those for trigonometric functions. A list of the more important ones is given in Table 8.2.

Example 8.8   Show that $\cosh(A + B) = \cosh(A)\cosh(B) + \sinh(A)\sinh(B)$.

Solution   Substitute

$$\cosh(A) = \frac{e^A + e^{-A}}{2}$$

$$\sinh(A) = \frac{e^A - e^{-A}}{2}$$

$$\cosh(B) = \frac{e^B + e^{-B}}{2}$$

$$\sinh(B) = \frac{e^B - e^{-B}}{2}$$

into the right-hand side of the expression

$$\cosh(A)\cosh(B) + \sinh(A)\sinh(B)$$

$$= \frac{(e^A + e^{-A})}{2} \frac{(e^B + e^{-B})}{2} \frac{(e^A - e^{-A})}{2} \frac{(e^B - e^{-B})}{2}.$$

Multiplying out the brackets gives

$$\tfrac{1}{4}\left(e^{A+B} + e^{A-B} + e^{-A+B} + e^{-(A+B)}\right.$$

$$\left. + (e^{A+B} - e^{A-B} - e^{-A+B} + e^{-(A+B)})\right).$$

Simplifying then gives

$$\tfrac{1}{4}(2e^{A+B} + 2e^{-(A+B)}) = \tfrac{1}{2}(e^{A+B} + e^{-(A+B)})$$

which is the definition of $\cosh(A + B)$.

We have shown that the right-hand side of the expression is equal to the left-hand side, and therefore

$$\cosh(A + B) = \cosh(A)\cosh(B) + \sinh(A)\sinh(B).$$

Inverse hyperbolic functions

The graphs of the inverse hyperbolic functions $\sinh^{-1}(x)$, $\cosh^{-1}(x)$, and $\tanh^{-1}(x)$ are given in Figure 8.7.

As $\cosh(x)$ is not a one-to-one function, it has no true inverse. However, if we limit x to zero or positive values only then $\cosh^{-1}(x)$ is indeed the inverse function and $\cosh^{-1}(\cosh(x)) = x$. The $\sinh^{-1}(x)$ function is defined for all values of x, but $\cosh^{-1}(x)$ is defined for $x \geqslant 1$ only and $\tanh^{-1}(x)$ is defined for $-1 < x < 1$.

As the hyperbolic functions are defined in terms of the exponential function we might suspect that the inverse would be defined in terms of the logarithm. The logarithmic equivalences are

$$\sinh^{-1}(x) = \ln\left(x + \sqrt{x^2 + 1}\right) \quad \text{for all } x$$

$$\cosh^{-1}(x) = \ln\left(x + \sqrt{x^2 - 1}\right) \quad x \geqslant 1$$

$$\tanh^{-1}(x) = \frac{1}{2}\ln\left(\frac{1+x}{1-x}\right) \quad -1 < x < 1$$

Example 8.9   Show that $\sinh^{-1}(x) = \ln(x + \sqrt{x^2 + 1})$ using the definitions

$$y = \sinh^{-1}(x) \quad \Leftrightarrow \quad \sinh(y) = x$$

and

$$\sinh(y) = \frac{e^y - e^{-y}}{2}$$

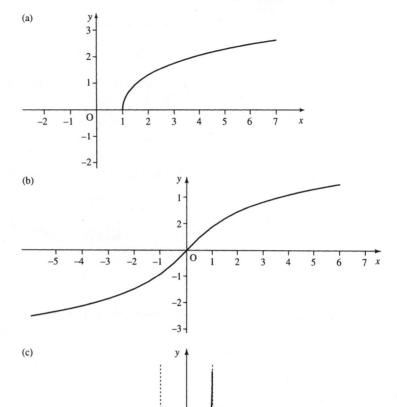

Figure 8.7 *(a) The graph of*
$y = \cosh^{-1}(x), \ x \geqslant 1.$ *(b)*
The graph of $y = \sinh^{-1}(x).$
(c) The graph of
$\tanh^{-1}(x), -1 \leqslant x \leqslant 1.$

Solution

$$y = \sinh^{-1}(x) \quad \Leftrightarrow \quad \sinh(y) = x$$

Using

$$\sinh(y) = \frac{e^y - e^{-y}}{2}$$

to substitute on the left-hand side, we get

$$\frac{e^y - e^{-y}}{2} = x$$

$$\Leftrightarrow \quad e^y - e^{-y} = 2x \quad \text{(multiplying by 2)}$$

$$\Leftrightarrow \quad e^{2y} - 1 = 2xe^y \quad \text{(multiplying by } e^y \text{ and using properties of}$$
$$\text{powers to write } e^y \cdot e^y = e^{2y})$$

$$\Leftrightarrow \quad e^{2y} - 2xe^y - 1 = 0 \quad \text{(subtracting } 2xe^y \text{ from both sides)}$$

This is now a quadratic equation in e^y

$$(e^y)^2 - 2xe^y - 1 = 0.$$

Using the formula for solving a quadratic equation, where $a = 1, b = -2x, c = -1$ gives

$$e^y = \frac{2x \pm \sqrt{4x^2 + 4}}{2}.$$

Dividing the top and bottom lines by 2 gives

$$e^y = x \pm \sqrt{x^2 + 1}.$$

Taking ln of both sides and using $\ln(e^y) = y$ (ln is the inverse function of exp) we get

$$y = \ln(x \pm \sqrt{x^2 + 1}).$$

We discount the negative sign inside the logarithm, as this would lead to a negative values, for which the logarithm is not defined. So finally

$$\sinh^{-1}(x) = \ln(x + \sqrt{x^2 + 1}).$$

Calculations

The hyperbolic and inverse hyperbolic functions are often not given in a calculator. To calculate a hyperbolic function then use the definitions

$$\cosh(x) = \frac{e^x + e^{-x}}{2}$$

$$\sinh(x) = \frac{e^x - e^{-x}}{2}$$

$$\tanh(x) = \frac{\sinh(x)}{\cosh(x)} = \frac{e^x - e^{-x}}{e^x + e^{-x}}$$

To calculate the inverse hyperbolic functions use their logarithmic equivalences.

Example 8.10   Calculate the following, and where possible use the appropriate inverse function to check your result:

(a) $\sinh(1.444)$ (b) $\tanh^{-1}(-0.5)$ (c) $\cosh(-1)$
(d) $\cosh^{-1}(3)$ (e) $\cosh^{-1}(0)$

Solution   From the definition

$$\sinh(1.444) = \frac{e^{1.444} - e^{-1.444}}{2}$$

$$\approx 2.0008152$$

$$= 2.001 \text{ to } 4 \text{ s.f.}$$

Check: Use the inverse function of the sinh, that is, find $\sinh^{-1}(2.0008152)$. From the logarithmic equivalence

(a) $\sinh^{-1}(2.0008152) = \ln(2.0008152 + \sqrt{(2.0008152)^2 + 1})$

$$= 1.444$$

As this is the original number input into the sinh function we have found $\sinh^{-1}(\sinh(1.444)) = 1.444$, which confirms the accuracy of our calculation.

(b) To calculate $\tanh^{-1}(-0.5)$ use the logarithmic equivalence giving

$$\tanh^{-1}(0.5) = \frac{1}{2}\ln\left(\frac{1 + (-0.5)}{1 - (-0.5)}\right) = \frac{1}{2}\ln\left(\tfrac{1}{3}\right) \approx -0.5493061$$

$$= -0.5493 \text{ to 4 s.f.}$$

Check: Use the inverse function of $\tanh^{-1}$, that is, find

$$\tanh(-0.5493061) = \frac{e^{-0.5493061} - e^{0.5493061}}{e^{-0.5493061} + e^{0.5493061}} \approx -0.5$$

This is the original number input to the $\tanh^{-1}$ function and this confirms the accuracy of our calculation.

(c) $\cosh(-1) = \dfrac{e^{-1} + e^{1}}{2} \approx 1.5430806$

$$= 1.543 \text{ to 4 s.f.}$$

Check: Use the inverse function of cosh, that is, $\cosh^{-1}$:

$$\cosh^{-1}(1.5430806) = \ln(1.5430806 + \sqrt{1.5430806^2 - 1}) \approx 1$$

This is not the number that we first started with, which was -1. However, we know that $\cosh^{-1}(x)$ is only a true inverse of $\cosh(x)$ if the domain of $\cosh(x)$ is limited to positive values and zero. We did not expect the inverse to 'work' in this case where we started with a negative value.

(d) To calculate $\cosh^{-1}(3)$, use

$$\cosh^{-1}(3) = \ln(3 + \sqrt{3^2 - 1}) = \ln(3 + \sqrt{8})$$

$$\approx 1.7627472$$

$$= 1.763 \text{ to 4 s.f.}$$

Check: The inverse function to $\cosh^{-1}$ is cosh, so we find

$$\cosh(1.7627472) = \frac{e^{1.7627472} + e^{-1.7627472}}{2} \approx 3$$

This confirms the accuracy of our calculation as we have shown $\cosh(\cosh^{-1}(3)) = 3$.

(e) Using the logarithmic definition of $\cosh^{-1}$ leads to an attempt to take the square root of a negative number. This confirms that $\cosh^{-1}(0)$ is not defined in $\mathbb{R}$.

Derivatives

Derivatives of the hyperbolic functions can be found by reverting to their definitions in terms of the exponential function.

Example 8.11　Show that

$$\frac{d}{dx}(\sinh(x)) = \cosh(x)$$

Solution　As

$$\sinh(x) = \frac{e^x - e^{-x}}{2}$$

then

$$\frac{d}{dx}(\sinh(x)) = \frac{d}{dx}\left(\frac{e^x - e^{-x}}{2}\right)$$

$$= \frac{e^x - (-1)e^{-x}}{2} = \frac{e^x + e^{-x}}{2} = \cosh(x).$$

Therefore

$$\frac{d}{dx}(\sinh(x)) = \cosh(x).$$

In a way similar to Example 8.6, we can find

$$\frac{d}{dx}(\cosh(x)) = \sinh(x)$$

and

$$\frac{d}{dx}(\tanh(x)) = \text{sech}^2(x).$$

The derivatives of the inverse hyperbolic functions can be found using the same method as given for the derivatives of the inverse trigonometric functions (in Chapter 5) and give

$$\frac{d}{dx}(\sinh^{-1}(x)) = \frac{1}{\sqrt{1 + x^2}}$$

$$\frac{d}{dx}(\cosh^{-1}(x)) = \frac{1}{\sqrt{x^2 - 1}}$$

$$\frac{d}{dx}(\tanh^{-1}(x)) = \frac{1}{1 - x^2}.$$

8.5 More differentiation and integration

We are now able to add the functions $y = e^x$ and $y = \ln(x)$, $y = a^x$, $y = \log_a(x)$, and the hyperbolic and inverse hyperbolic functions to the list of functions (Table 8.3). By swapping the columns and rearranging some of the terms in a more convenient fashion, and adding the constant of integration, we get a list of integrals (Table 8.4).

The methods of differentiation and integration of combined functions, discussed in Chapters 12 and 13, can equally be applied to exponential and logarithmic functions.

Table 8.3 *The derivatives of some simple functions*

$f(x)$	$f'(x)$
C	0
x^n	nx^{n-1}
$\cos(x)$	$-\sin(x)$
$\sin(x)$	$\cos(x)$
$\tan(x)$	$\sec^2(x)$
$\sin^{-1}(x)$	$1/\sqrt{1-x^2}$
$\cos^{-1}(x)$	$-1/\sqrt{1-x^2}$
$\tan^{-1}(x)$	$1/(1+x^2)$
e^x	e^x
a^x	$(\ln(a)a^x)$
$\ln(x)$	$1/x$
$\log_a(x)$	$1/(\ln(a)x)$
$\cosh(x)$	$\sinh(x)$
$\sinh(x)$	$\cosh(x)$
$\tanh(x)$	$\sech^2(x)$
$\sinh^{-1}(x)$	$1/\sqrt{1+x^2}$
$\cosh^{-1}(x)$	$1/\sqrt{x^2-1}$
$\tanh^{-1}(x)$	$1/(1-x^2)$

Table 8.4 *Some standard integrals*

$f(x)$	$\int f(x)dxf(x)$
1	$x+C$
$x^n (n \neq -1)$	$(x^{n+1})/(n+1)+C$
$\sin(x)$	$-\cos(x)+C$
$\cos(x)$	$\sin(x)+C$
$\sec^2(x)$	$\tan(x)+C$
$1/\sqrt{1-x^2}$	$\sin^{-1}(x)+C$
$-1/\sqrt{1-x^2}$	$\cos^{-1}(x)+C$
$1/(1+x^2)$	$\tan^{-1}(x)+C$
e^x	e^x+C
a^x	$(a^x/\ln(a))+C$
$1/x$	$\ln(x)+C$
$\cosh(x)$	$\sinh(x)+C$
$\sinh(x)$	$\cosh(x)+C$
$\sech^2(x)$	$\tanh(x)+C$
$1/\sqrt{1+x^2}$	$\sinh^{-1}(x)+C$
$1/\sqrt{x^2-1}$	$\cosh^{-1}(x)+C$
$1/(1-x^2)$	$\tanh^{-1}(x)+C$

Example 8.12   Find derivatives of the following:

(a) $y = e^{-2t^2+3}$ (b) $x = e^{-t}\cos(3t)$ (c) $y = \dfrac{\sinh(x)}{x}$ $x \neq 0$.

Solution   (a) To differentiate $y = e^{-2t^2+3}$ using the function of a function rule think of this as $y = e^{()}$ ('$y = e$ to the bracket').

Now differentiate y with respect to $()$ and multiply by the derivative of $()$ with respect to t. That is, use

$$\frac{dy}{dt} = \frac{dy}{d()}\frac{d()}{dt}$$

where $()$ represents the expression in the bracket

$$\frac{dy}{dt} = e^{-2t^2+3}\frac{d}{dt}(-2t^2+3) = e^{-2t^2+3}(-4t) = -4te^{-2t^2+3}.$$

(b) To find the derivative of $x = e^{-t}\cos(3t)$, write $x = uv$ so that $u = e^{-t}$ and $v = \cos(3t)$; then,

$$\frac{du}{dt} = -e^{-t} \qquad \frac{dv}{dt} = -3\sin(3t)$$

where we have used the chain rule to find both these derivatives.

Now use the product rule

$$\frac{dx}{dt} = u\frac{dv}{dt} + v\frac{du}{dt}$$

$$\frac{dx}{dt} = -e^{-t}\cos(3t) - e^{-t}3\sin(3t) = -e^{-t}\cos(3t) - 3e^{-t}\sin(3t).$$

(c) To find the derivative of

$$y = \frac{\sinh(x)}{x}$$

we use the formula for the quotient of two functions where $y = u/v, u = \sinh(x), v = x$, and

$$\frac{dy}{dx} = \frac{v(du/dx) - u(dv/dx)}{v^2}$$

Hence, we get

$$\frac{d}{dx}\left(\frac{\sinh(x)}{x}\right) = \frac{x\cosh(x) - \sinh(x)\cdot 1}{x^2}$$

$$= \frac{x\cosh(x) - \sinh(x)}{x^2}.$$

Example 8.13   Find the following integrals:

(a) $\int xe^{x^2+2}\,dx$ (b) $\int \sinh(t)\cosh^2(t)\,dt$ (c) $\int xe^x\,dx$

(d) $\int_1^2 \ln(x)\,dx$ (e) $\int\left(\dfrac{3x^2+2x}{x^3+x^2+2}\right)dx$

Solution (a) $\int x e^{x^2+2}\, dx$. Here, we have a function of a function $e^{(x^2+2)}$ multiplied by a term that is something like the derivative of the term in the bracket.

Try a substitution, $u = x^2 + 2$

$$\Rightarrow \frac{du}{dx} = 2x \Rightarrow dx = \frac{du}{2x}$$

then

$$\int x e^{x^2+2}\, dx = \int x e^u \frac{du}{2x} = \int \tfrac{1}{2} e^u\, du = \tfrac{1}{2} e^u + C$$

resubstituting $u = x^2 + 2$ gives

$$\int x e^{x^2+2} dx = \tfrac{1}{2} e^{x^2+2} + C$$

(b) To find $\int \sinh(t) \cosh^2(t)\, dt$ we remember that $\cosh^2(t) = (\cosh(t))^2$, so

$$\int \sinh(t) \cosh^2(t)\, dt = \int \sinh(t)(\cosh(t))^2 dt$$

$\sinh(t)$ is the derivative of the function in the bracket, $\cosh(t)$, so a substitution, $u = \cosh(t)$, should work:

$$u = \cosh(t) \Rightarrow \frac{du}{dt} = \sinh(t)$$

$$\Rightarrow dt = \frac{du}{\sinh(t)}$$

$$\int \sinh(t) \cosh^2(t)\, dt = \int \sinh(t) u^2 \frac{du}{\sinh(t)} = \int u^2\, du$$

$$= u^3 + C$$

resubstituting $u = \cosh(t)$ gives

$$\int \sinh(t) \cosh^2(t)\, dt = \frac{\cosh^3(t)}{3} + C$$

(c) $\int x e^x dx$. Use integration by parts

$$\int u\, dv = uv - \int v\, du$$

and choose $u = x$ and $dv = e^x dx$ giving

$$\frac{du}{dx} = 1 \quad \text{and} \quad v = \int e^x dx = e^x$$

Then

$$\int x e^x dx = x e^x - \int e^x dx$$

$$= x e^x - e^x + C$$

(d) $\int_1^2 \ln(x)dx$. Write $\ln(x) = 1\ln(x)$ and use integration by parts with

$$u = \ln(x) \quad \text{and} \quad dv = 1\,dx$$

$$\Rightarrow du = \frac{dx}{x} \quad v = \int 1\,dx = x$$

$$\int_1^2 \ln(x)dx = [x\ln(x)]_1^2 - \int_1^2 x\frac{1}{x}\,dx$$

$$= 2\ln(2) - 1\ln(1) - \int_1^2 1\,dx$$

$$= 2\ln(2) - [x]_1^2$$

$$= 2\ln(2) - (2-1) \approx 0.3863 \text{ to 4 s.f.}$$

(e) We rewrite

$$\int \frac{3x^2 + 2x}{x^3 + x^2 + 2}\,dx = \int (3x^2 + 2x)(x^3 + x^2 + 2)^{-1}\,dx.$$

Notice that there are two brackets. To decide what to substitute we notice that

$$\frac{d}{dx}(x^3 + x^2 + 2) = 3x^2 + 2x.$$

so it should work to substitute $u = x^3 + x^2 + 2$

$$\Rightarrow \frac{du}{dx} = 3x^2 + 2x \Rightarrow dx = \frac{dx}{3x^2 + 2x}.$$

The integral becomes

$$\int (3x^2 + 2x)u^{-1}\frac{du}{3x^2 + 2x}$$

$$\int \frac{du}{u} = \ln(u) + C.$$

Resubstituting for u gives

$$\int \frac{3x^2 + 2x}{x^3 + x^2 + 2}\,dx = \ln(x^3 + x^2 + 2) + C.$$

Integration using partial fractions

The fact that expressions like $1/(3x + 2)$ can be integrated using a substitution which results in an integral of the form:

$$\int \frac{1}{u}\,du = \ln(u) + C$$

is exploited when we perform the integration of fractional expressions like

$$\frac{2x - 1}{(x - 3)(x + 1)}.$$

We first rewrite the function to be integrated using partial fractions.

Example 8.14   Find

(a) $\displaystyle \int \frac{2x-1}{(x-3)(x+1)} \, dx$

(b) $\displaystyle \int \frac{x^2}{(x+2)(2x-1)^2} \, dx.$

Solution

(a) $\displaystyle \int \frac{2x-1}{(x-3)(x+1)} \, dx.$

Rewrite the expression using partial fractions. We need to find A and B so that:

$$\frac{2x-1}{(x-3)(x+1)} = \frac{A}{(x-3)} + \frac{B}{(x+1)}$$

where this should be true for all values of x.

Multiplying by $(x-3)(x+1)$ gives $2x - 1 = A(x+1) + B(x-3)$. This is an identity, so we can substitute values for x:

substitute $x = -1$ giving $-3 = B(-4)$ $\Leftrightarrow$ $B = 3/4$

substitute $x = 3$ giving $5 = A(4)$ $\Leftrightarrow$ $A = 5/4$

Hence,

$$\frac{2x-1}{(x-3)(x+1)} = \frac{5}{4(x-3)} + \frac{3}{4(x+1)}.$$

So

$$\int \frac{2x-1}{(x-3)(x+1)} \, dx = \int \frac{5}{4(x-3)} + \frac{3}{4(x+1)} \, dx.$$

As $(x-3)$ and $(x+1)$ are linear functions of x, we can find each part of this integral using substitutions of $u = x - 3$ and $u = x + 1$:

$$\int \frac{5}{4(x-3)} + \frac{3}{4(x+1)} \, dx = \frac{5}{4} \ln(x-3) + \frac{3}{4} \ln(x+1) + C$$

$$\int \frac{2x-1}{(x-3)(x+1)} \, dx = \frac{5}{4} \ln(x-3) + \frac{3}{4} \ln(x+1) + C.$$

Check:

$$\frac{d}{dx} \left(\frac{5}{4} \ln(x-3) + \frac{3}{4} \ln(x+1) + C \right) = \frac{5}{4(x-3)} + \frac{3}{4(x+1)}$$

writing this over a common denominator gives

$$\frac{d}{dx} \left(\frac{5}{4} \ln(x-3) + \frac{3}{4} \ln(x+1) + C \right) = \frac{5(x+1) + 3(x-3)}{4(x-3)(x+1)}$$

$$= \frac{5x + 5 + 3x - 9}{4(x-3)(x+1)}$$

$$= \frac{8x - 4}{4(x-3)(x+1)}$$

$$= \frac{2x - 1}{(x-3)(x+1)}$$

(b) $\displaystyle \int \frac{x^2}{(x+2)(2x-1)^2} \, dx$

Again, we can use partial fractions. Because of the repeated factor in the denominator we use both a linear and a squared term in that factor. We need to find A, B, and C so that

$$\frac{x^2}{(x+2)(2x-1)^2} = \frac{A}{(x+2)} + \frac{B}{(2x-1)} + \frac{C}{(2x-1)^2}$$

where this should be true for all values of x.

Multiply by $(x+2)(2x-1)^2$ to get

$$x^2 = A(2x-1)^2 + B(2x-1)(x+2) + C(x+2)$$

This is an identity, so we can substitute values for x

substitute $x = \frac{1}{2}$ giving $0.25 = C(2.5)$ $\Leftrightarrow$ $C = 0.1$

substitute $x = -2$ giving $4 = A(-5)^2$ $\Leftrightarrow$ $A = 4/25 = 0.16$

substitute $x = 0$ giving $0 = A + B(-1)(2) + C(2)$.

Using the fact that $A = 0.16$ and $C = 0.1$, we get

$$0 = 0.16 - 2B + 0.2 \quad \Leftrightarrow \quad 0 = 0.36 - 2B \quad \Leftrightarrow \quad 2B = 0.36 \quad \Leftrightarrow \quad B = 0.18.$$

Then we have

$$\int \frac{x^2}{(x+2)(2x-1)^2} dx = \int \frac{0.16}{(x+2)} + \frac{0.18}{(2x-1)} + \frac{0.1}{(2x-1)^2} dx$$

$$= 0.16 \ln(x+2) + \frac{0.18}{2} \ln(2x-1) - \frac{0.1}{2}(2x-1)^{-1} + C$$

$$= 0.16 \ln(x+2) + 0.09 \ln(2x-1) - \frac{0.05}{(2x-1)} + C.$$

Check:

$$\frac{d}{dx}\left(0.16 \ln(x+2) + 0.09 \ln(2x-1) - \frac{0.05}{2x-1} + C\right)$$

$$= \frac{d}{dx}(0.16 \ln(x+2) + 0.09 \ln(2x-1) - 0.05(2x-1)^{-1} + C)$$

$$= \frac{0.16}{(x+2)} + \frac{0.09}{(2x-1)}(2) + 0.05(2)(2x-1)^{-2}$$

$$= \frac{0.16}{(x+2)} + \frac{0.18}{(2x-1)} + \frac{0.1}{(2x-1)^2}.$$

Writing this over a common denominator gives

$$\frac{0.16(2x-1)^2 + 0.18(2x-1)(x+2) + 0.1(x+2)}{(x+2)(2x-1)^2}$$

$$= \frac{0.16(4x^2 - 4x + 1) + 0.18(2x^2 + 3x - 2) + 0.1x + 0.2}{(x+2)(2x-1)^2}$$

$$= \frac{0.64x^2 - 0.64x + 0.16 + 0.36x^2 + 0.54x - 0.36 + 0.1x + 0.2}{(x+2)(2x-1)^2}$$

$$= \frac{x^2}{(x+2)(2x-1)^2}.$$

8.6 Summary

1. Many physical situations involve exponential growth or decay where the rate of change of y is proportional to its current value.

2. All exponential functions, $y = a^t$, are such that $dy/dt = ky$, that is, the derivative of an exponential function is also an exponential function scaled by a factor k.

3. The exponential function $y = e^t$ has the property that $dy/dt = y$, that is, its derivative is equal to the original function:

$$\frac{d}{dt}(e^t) = e^t,$$

where $e \approx 2.71828$. The inverse function to e^t is $\log_e(t)$, which is abbreviated to $\ln(t)$. This is called the natural or Napierian logarithm.

4. The general solution to $dy/dt = ky$ is $y = y_0 e^{kt}$, where y_0 is the value of y at $t = 0$.

5. $\dfrac{d}{dt}(a^t) = \ln(a)a^t$ and

$$\frac{d}{dt}(\log_a(t)) = \frac{1}{\ln(a)t}.$$

6. The hyperbolic cosine (cosh) and hyperbolic sine (sinh) are the even and odd parts of the exponential function:

$$e^x = \cosh(x) + \sinh(x)$$

$$\cosh(x) = \frac{e^x + e^{-x}}{2}$$

$$\sinh(x) = \frac{e^x - e^{-x}}{2}$$

These functions get the name hyperbolic because of their relationship to a hyperbola. The hyperbolic tangent is defined by

$$\tanh(x) = \frac{\sinh(x)}{\cosh(x)} = \frac{e^x - e^{-x}}{e^x + e^{-x}}$$

There are various hyperbolic identities, which are similar to the trigonometric identities (Table 8.2).

7. The inverse hyperbolic functions $\cosh^{-1}(x)(x \geqslant 1), \sinh^{-1}(x), \tanh^{-1}(x)(-1 < x < 1)$ have the following logarithmic identities:

$$\sinh^{-1}(x) = \ln(x + \sqrt{x^2 + 1}) \quad \text{for all } x \in \mathbb{R}$$

$$\cosh^{-1}(x) = \ln(x + \sqrt{x^2 - 1}) \quad x \geqslant 1$$

$$\tanh^{-1}(x) = \frac{1}{2}\ln\left(\frac{1+x}{1-x}\right) \quad -1 < x < 1$$

$\cosh^{-1}(x)$ is the inverse of $\cosh(x)$ if the domain of $\cosh(x)$ is limited to the positive values of x and zero.

8. Adding the derivatives and integrals of the exponential, ln, hyperbolic and inverse hyperbolic functions to the tables of standard derivatives and integrals gives Tables 8.3 and 8.4.

9. Partial fractions can be used to integrate fractional functions such as

$$\frac{x + 1}{(x - 1)(x + 2)}.$$

8.7 Exercises

8.1. Using $\dfrac{d}{dt}(e^t) = e^t$ show that the function $2e^{3t}$ is a solution to the differential equation

$$\frac{dy}{dt} = 3y$$

8.2. Assuming $p = p_0e^{kt}$ find p_0 and k such that

$$\frac{dp}{dt} = \frac{p}{1200}$$

and $p = 1$ when $t = 0$.

8.3. Assuming $N = N_0e^{kt}$ find N_0 and k such that

$$\frac{dN}{dt} = -4.3 \times 10^{-4}N \quad \text{and} \quad N = 5 \times 10^6 \text{ at } t = 0.$$

8.4. Assuming $\phi = Ae^{kt} + 300$ find A and k such that

$$\frac{d\phi}{dt} = -0.1(\phi - 300) \quad \text{and} \quad \phi = 400 \text{ when } t = 0.$$

8.5. Using the definitions of

$$\cosh(x) = \frac{e^x + e^{-x}}{2}$$

and

$$\sinh(x) = \frac{e^x - e^{-x}}{2}$$

show that
(a) $\cosh^2(x) - \sinh^2(x) = 1$
(b) $\sinh(x - y) = \sinh(x)\cosh(y) - \cosh(x)\sinh(y)$

8.6. Using $y = \tanh^{-1}(x) \Leftrightarrow \tanh(y) = x$, where $-1 < x < 1$, and

$$\tanh(y) = \frac{e^y - e^{-y}}{e^y + e^{-y}}$$

show that

$$\tanh^{-1}(x) = \frac{1}{2}\ln\left(\frac{1+x}{1-x}\right)$$

where $-1 < x < 1$.

8.7. Calculate the following and where possible use the appropriate inverse functions to check your result:

(a) $\cosh(2.1)$ (b) $\tanh(3)$ (c) $\sinh^{-1}(0.6)$
(d) $\tanh^{-1}(1.5)$ (e) $\cosh^{-1}(-1.5)$

8.8. Differentiate the following:

(a) $z = e^{t^2-2}$ (b) $x = e^{-t}\cosh(2t)$

(c) $\dfrac{x^2 - 1}{\sinh(x)}$ (d) $\ln(x^3 - 3x)$

(e) $\log_2(2x)$ (f) a^{4t}

(g) $2^t t^2$ (h) $1/(e^{t-1})^2$

8.9. Find the following integrals:

(a) $\displaystyle\int e^{4t-3}dt$ (b) $\displaystyle\int_2^3 \frac{dt}{4t - 1}$

(c) $\displaystyle\int x\sinh(2x^2)dx$ (d) $\displaystyle\int x\ln(x)dx$

(e) $\displaystyle\int_0^1 e^x x^2\,dx$ (f) $\displaystyle\int \frac{\sinh(t)}{\cosh(t)}dt$

(g) $\displaystyle\int \frac{2(x-1)}{x^2 - 2x - 4}dx$ (h) $\displaystyle\int \frac{t+1}{(t-3)(t-1)}dt$

(i) $\displaystyle\int_2^4 \frac{-t}{t^2(t-1)}dt$

8.10. The charge on a discharging capacitor in an RC circuit decays according to the expression $Q = 0.001e^{-10t}$. Find an expression for the current using $I = dQ/dt$ and find after how long the current is half of its initial value.

8.11. A charging capacitor in an RC circuit with a d.c. voltage of 5 V charges according to the expression $q = 0.005(1 - e^{-0.5t})$. Given that the current $i = dq/dt$, calculate the current: (a) when $t = 0$; (b) after 10 s; and (c) after 20 s.

9 Vectors

9.1 Introduction

Many things can be represented by a simple number, for instance, time, distance, mass, which are then called scalar quantities. Others, however, are better represented by both their size, or magnitude, and a direction. Some of these are velocity, acceleration, and force. These quantities are called vector quantities because they are represented by vectors.

A simple example of a vector is one that describes displacement. Supposing someone is standing in a room with floor tiles (as in Figure 9.1) and moving from one position to another can be described by the number of tiles to the right and the number of tiles towards the top of the page.

In the example, to move from the door to the cupboard can be represented by (4, 2). This vector consists of two numbers, where the order of the numbers is important. Moving (4, 2) results in a different final position to that if we move (2, 4). The magnitude of the displacement can be found by drawing a straight line from the starting position to the final position and measuring the length. From Pythagoras theorem this can be found as $\sqrt{4^2 + 2^2} = \sqrt{20} \approx 4.47$. The direction can be described by an angle, for instance, the angle made to the wall with the window on it.

This example shows that a two-dimensional vector (2D) can be used to represent movement on a flat surface. A 2D vector is two numbers, where the order of the numbers is important.

If the room in Figure 9.1 also had wall tiles then we could represent a position above the floor by the number of tiles towards the ceiling. This three-dimensional (3D) vector can be represented by three numbers. It can also be represented by the distance travelled and the direction, angles made to the floor and the angle made to the wall.

Velocity is an example of a vector quantity. This can be described by two things, the speed, which is the rate of change of distance travelled with respect to time, and also the direction in which it is travelling. Similarly, force can be described by the size of magnitude of the force and also the direction in which it operates.

Vectors have their own rules for addition and subtraction. If two forces of equal magnitude operate on one object then the net effect will depend on the direction of the forces. If the forces operate in opposite directions

Figure 9.1 *A tiled room. To reach the cupboard from the door we need to move four tiles to the right and two tiles towards the top of the page. This can be represented by the vector (4, 2).*

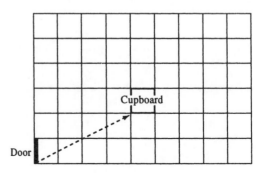

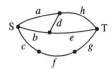

Figure 9.2 *A network consisting of sides a, b, c, d, e, f, g, and h.*

they could balance each other out, like two tug-of-war teams in a stalemate struggle. Alternatively, they could operate in the same direction or partially in the same direction and cause the object to have an acceleration.

For the examples of vectors given so far, the maximum dimension of the vector is three as there are only three spatial dimensions. However, there are many examples when vectors of higher dimension are useful. For instance, a path through the network given in Figure 9.2 can be represented by a list of 1s and 0s to indicate whether each of the edges is included in the path. A path from S to T can be represented by a vector, for instance:

a	b	c	d	e	f	g	h
0	1	0	0	1	0	0	0	represents the path *be*
0	0	1	0	0	1	1	0	represents the path *cfg*.

Although there are many other types of vectors we will concentrate on vectors of two or three dimensions, called spatial or geometrical vectors, used to represent physical quantities in space. Many of the ideas in this chapter are only true for geometrical vectors of two and three dimensional. As 3D vectors can only be correctly represented by making a 3D model, it is important to concentrate on understanding 2D vectors as they can be drawn on a piece of paper allowing results to be checked easily.

9.2 Vectors and vector quantities

A vector is a string of numbers, for example,

$(1, 2, -1)$

$(1, 0)$

$(3, -4, 2, -6, 8)$

$(2.6, 9, -1.2, 0.3)$.

The length of the string is called the dimension of the vector. For the examples given above, the dimensions are 3, 2, 5, and 4, respectively. The commas can be left out, so the examples given above can be written as

$(1\ 2\ -1)$

$(1\ 0)$

$(3\ -4\ 2\ -6\ 8)$

$(2.6\ 9\ -1.2\ 0.3)$.

Vectors may also be written as columns, giving:

$$\begin{pmatrix} 1 \\ 2 \\ -1 \end{pmatrix} \begin{pmatrix} 1 \\ 0 \end{pmatrix} \begin{pmatrix} 3 \\ -4 \\ 2 \\ -6 \\ 8 \end{pmatrix} \begin{pmatrix} 2.6 \\ 9 \\ -1.2 \\ 0.3 \end{pmatrix}.$$

Whether vectors are written as columns or rows only becomes important when we look at matrices (Chapter 13). However, the order of the numbers in the vector is important: $(0, 1)$ is a different vector from $(1, 0)$.

We will mainly deal with 2D or 3D vectors. Vectors are represented in a diagram by a line segment with an arrow as in Figure 9.3. In printed material vectors can be represented by bold letters: **a**. They are also represented by $\underline{a}$ or $\vec{a}$ or $\overrightarrow{AB}$, where A and B are points at either end of the vector.

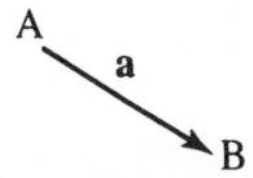

In the rest of this chapter, we will assume that we are dealing with 2D or 3D vectors represented in rectangular form, also called Cartesian form.

Figure 9.3 *A vector is drawn in a diagram as a line segment with an arrow to indicate its direction.*

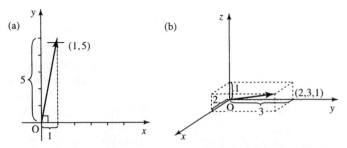

Figure 9.4 *(a) A two-dimensional rectangular set of axes and the vector (1, 5). The axes are at right angles and the numbers in the vector give the x, y translation it represents. (b) A three-dimensional rectangular set of axes and the vector (2, 3, 1). The axes are at right angles and the numbers in the vector correspond to the x, y, z translation it represents.*

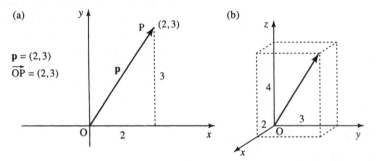

Figure 9.5 *(a) The position vector $\mathbf{p} = (2, 3)$ or $\overrightarrow{OP} = (2, 3)$ is used to represent a point in the plane. The point can be found by translating from the origin by 2 in the x-direction followed by 3 in the y-direction; hence, $\mathbf{p} = (2, 3)$. (b) The position vector $\mathbf{p} = (2, 3, 4)$ is used to represent a point in space. The point can be found by translating from the origin by 2 in the x-direction, followed by 3 in the y-direction, and by 4 in the z-direction; hence, $\mathbf{p} = (2, 3, 4)$.*

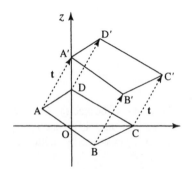

Figure 9.6 *(a) The vector $\mathbf{t} = (2, 3)$ is used to represent a translation of the figure ABCD. Each of the points defining the figure have been translated by (2,3).*

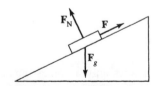

Figure 9.7 *The object is being pulled up the slope using a force $\mathbf{F}$ which has a direction parallel with the slope of the hill. There is also a force due to gravity $\mathbf{F}_g$ acting vertically downwards and a force at right angles to the plane, $\mathbf{F}_N$.*

This means that the numbers in the vectors correspond to the x, y, z values for a set of rectangular axes, as shown in Figure 9.4. This assumption is important for many of the geometrical interpretations presented here.

Position vectors and translation vectors

Vectors can represent points in a plane, as in Figure 9.5(a), or points in space, as in Figure 9.5(b). These are called position vectors. They can be thought of as representing a translation from the origin.

Vectors can represent a translation that can be applied to figures. In Figure 9.6, a four-sided figure ABCD has been translated through the vector (2, 3).

Vector quantities

Vectors can represent physical quantities that have both a magnitude and a direction. In Figure 9.7, there is an example of the forces acting on a body that is being pulled up a slope. By using vectors and vector addition the resultant force acting on the body can be found and therefore the direction in which the body will travel can be found together with the size of the acceleration. Other quantities with both magnitude and direction are velocity, acceleration, and moment. Quantities that only have magnitude and no direction are called scalar quantities and can be represented using a number, for example, mass and length.

9.3 Addition and subtraction of vectors

Addition

To add two vectors, add the corresponding elements of the vectors.

Example 9.1

$\mathbf{a} = (2, 3) \quad \mathbf{b} = (4, 2)$

$\mathbf{a} + \mathbf{b} = (2, 3) + (4, 2) = (2 + 4, 3 + 2) = (6, 5)$

$\mathbf{c} = (1, 3, 1.5) \quad \mathbf{d} = (5, -2, 1)$

$\mathbf{c} + \mathbf{d} = (1, 3, 1.5) + (5, -2, 1) = (1 + 5, 3 + (-2), 1.5 + 1) = (6, 1, 2.5).$

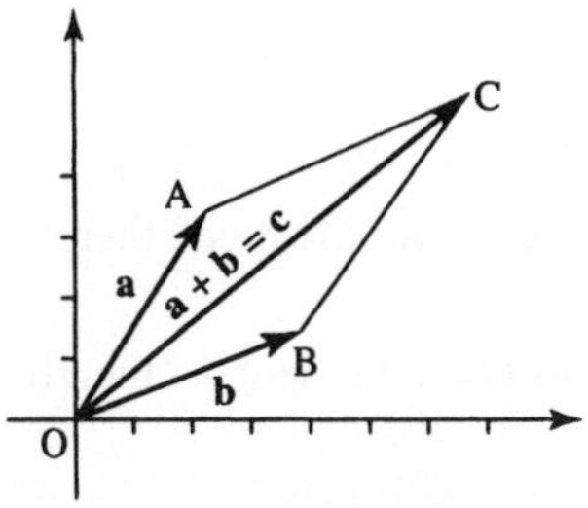

If the vectors are represented in the plane then the vector sum can be found using the parallelogram law, as in Figure 9.8. The resultant or vector sum of $\mathbf{a}$ and $\mathbf{b}$ is found by drawing vector $\mathbf{a}$ and then drawing vector $\mathbf{b}$ from the tip of vector $\mathbf{a}$ which gives the point C. Then $\mathbf{a} + \mathbf{b}$ can be found by drawing a line starting at O to the point C. If we imagine walking from O to A, along vector $\mathbf{a}$, and then from A to C, along vector $\mathbf{b}$, this has the same effect as walking direct from O to C, along vector $\mathbf{c}$. We can also use the parallelogram to show that $\mathbf{a} + \mathbf{b} = \mathbf{b} + \mathbf{a}$. To find $\mathbf{b} + \mathbf{a}$, start with vector $\mathbf{b}$ and draw vector $\mathbf{a}$ from the tip of vector $\mathbf{b}$; this also gives the point C. Then if we walk from O to B along vector $\mathbf{b}$ and then from B to C along vector $\mathbf{a}$, this has the same effect as walking along the other two sides of the parallelogram or walking direct from O to C. Hence

$$\mathbf{a} + \mathbf{b} = \mathbf{b} + \mathbf{a} = \mathbf{c}$$

Figure 9.8 *The resultant or vector sum of* **a** *and* **b**.

Subtraction

To subtract one vector from another, subtract the corresponding elements of the vectors.

Example 9.2

$\mathbf{a} = (2, 3) \quad \mathbf{b} = (4, 2)$

$\mathbf{a} - \mathbf{b} = (2, 3) - (4, 2) = (2 - 4, 3 - 2) = (-2, 1).$

Using a vector diagram we can perform vector subtraction in two ways. Draw $\mathbf{a}$ and $-\mathbf{b}$ and add as before or simply draw vectors $\mathbf{a}$ and $\mathbf{b}$ from the same point and the line joining the tip of $\mathbf{b}$ to the tip of $\mathbf{a}$ gives the vector $\mathbf{a} - \mathbf{b}$. These methods are explained in Figure 9.9.

Example 9.3 In Figure 9.10, $\overrightarrow{OA} = \mathbf{a}$ and $\overrightarrow{OB} = \mathbf{b}$. OB = BC, OA = EO, and AD is parallel to OC and EF. Write the following vectors in terms of $\mathbf{a}$ and $\mathbf{b}$:

(a) $\overrightarrow{OE}$ (b) $\overrightarrow{OC}$ (c) $\overrightarrow{BA}$ (d) $\overrightarrow{AB}$ (e) $\overrightarrow{AD}$ (f) $\overrightarrow{BE}$ (g) $\overrightarrow{BF}$.

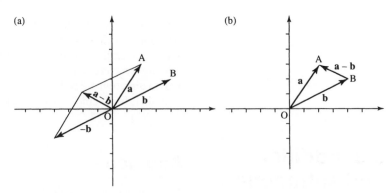

Figure 9.9　*(a) To find* **a** − **b** *by using addition, draw* **a** *and* **b**. *Then* −**b** *is the vector in the opposite direction. Then add* **a** *and* −**b** *by drawing a parallelogram as in Figure 9.8. (b) Use the triangle OAB.* $\overrightarrow{BA}$ *gives the vector* **a** − **b**. *To see this imagine walking directly from* B *to* A, *this is the same as walking from* B *to* O, *which is backwards along* **b** *and therefore is the vector* −**b**, *and then along* $\overrightarrow{OA}$ *which is the vector* **a**. *Hence,* $\overrightarrow{BA} = -\mathbf{b} + \mathbf{a} = \mathbf{a} - \mathbf{b}$.

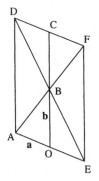

Figure 9.10　*Using vectors. See Example 9.3.*

Solution

(a) $\overrightarrow{OE}$ is the same length as **a** in the opposite direction; therefore, $\overrightarrow{OE} = -\mathbf{a}$.

(b) $\overrightarrow{OC}$ is in the same direction as **b**, but twice the length; therefore, $\overrightarrow{OC} = 2\mathbf{b}$.

(c) $\overrightarrow{BA}$ is in a triangle with **a** and **b**. To get from B to A we would walk in the reverse direction along **b** and then along **a**: $\overrightarrow{BA} = -\mathbf{b} + \mathbf{a} = \mathbf{a} - \mathbf{b}$.

(d) $\overrightarrow{AB} = -\overrightarrow{BA} = -(\mathbf{a} - \mathbf{b}) = \mathbf{b} - \mathbf{a}$.

(e) $\overrightarrow{AD}$ is parallel to $\overrightarrow{OC}$ in the same direction; therefore, as $\overrightarrow{OC} = 2\mathbf{b}$ then $\overrightarrow{AD} = 2\mathbf{b}$.

(f) To find $\overrightarrow{BE}$ we need to know $\overrightarrow{OE}.\overrightarrow{OE}$ is the same length as **a** in the opposite direction; therefore, $\overrightarrow{OE} = -\mathbf{a}$. To get from B to E we could go from B to O (−**b**) and then from O to E (−**a**); therefore, $\overrightarrow{BE} = -\mathbf{b} - \mathbf{a}$.

(g) $\overrightarrow{BF}$ is the same length as $\overrightarrow{AB}$ and in the same direction; therefore, $\overrightarrow{BF} = \overrightarrow{AB} = \mathbf{b} - \mathbf{a}$.

9.4 Magnitude and direction of a 2D vector – polar co-ordinates

We have already noted that a vector has magnitude and direction. A 2D vector can be represented by its length (also called magnitude or modulus), r (or |**r**|), and its angle to the x-axis, also called its argument, θ. If the vector is (x, y) then $r^2 = x^2 + y^2$, from Pythagoras's theorem The angle is given by $\tan^{-1}(y/x)$ if x is positive and by $\tan^{-1}(y/x) + \pi$ if x is negative. Hence, $\mathbf{r} = (r, \theta)$ in polar coordinates, and can also be written as $r\angle\theta$ so it is clear that the second number represents the angle. As it is usual to give the angle between $-\pi$ and $+\pi$, it may be necessary to subtract 2π from the angle given by this formula. (As a rotation of 2π is a complete rotation this will make no difference to the position of the vector.)

Example 9.4   Find the magnitude and direction of

(a) $(2, 3)$ (b) $(-1, -4)$ (c) $(1, -2.2)$ (d) $(-2, 5.6)$

Solution   To perform these conversions to polar form it is a good idea to draw a diagram of the vector in order to be able to check that the angle is of the correct size. Figure 9.11 shows the diagrams for each part of the example.

(a) $\mathbf{r} = (2, 3)$ has magnitude $\sqrt{2^2 + 3^2} \approx 3.606$ and angle given by $\tan^{-1}(3/2) \approx 0.983$ and therefore in polar coordinates $\mathbf{r}$ is $3.606 \angle 0.983$.

(b) $\mathbf{r} = (-1, -4)$ has magnitude $r = \sqrt{(-1)^2 + (-4)^2} \approx 4.123$, and angle given by $\tan^{-1}(-4/-1) + \pi \approx 1.326 + 3.142 = 4.467$. As this angle is bigger than π, subtract 2π (a complete revolution) to give

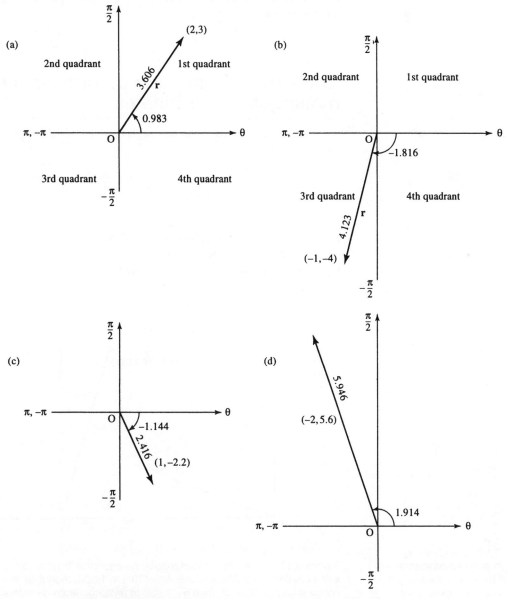

Figure 9.11   *Converting vectors to polar form: (a) $r = (2, 3) = 3.606 \angle 0.983$; (b) $r = (-1, -4) =$ 4.123 $\angle -1.816$; (c) $r = (1, -2.2) = 2.416 \angle -1.144$; (d) $r = (-2, 5.6) = 5.946 \angle 1.914$.*

−1.816. Therefore, in polar co-ordinates $\mathbf{r} = 4.123 \angle -1.816$. Note that the angle is between $-\pi$ and $-\pi/2$, meaning that the vector must lie in the third quadrant, which we can see is correct from the diagram.

(c) $\mathbf{r} = (1, -2.2)$ has magnitude $\mathbf{r} = \sqrt{(1)^2 + (-2.2)^2} \approx 2.416$ and the angle is given by $\tan^{-1}(-2.2/1) \approx -1.144$. Therefore, in polar co-ordinates $\mathbf{r} = 2.416 \angle -1.144$. Note that the angle is between $-\pi/2$ and 0, meaning that the vector must lie in the fourth quadrant.

(d) $\mathbf{r} = (-2, 5.6)$ has magnitude $r = \sqrt{(-2)^2 + (5.6)^2} \approx 5.946$ and angle given by $\tan^{-1}(5.6/-2) + \pi \approx 1.914$. Therefore, in polar co-ordinates $\mathbf{r} = 5.946 \angle 1.914$. Note that the angle is between $\pi/2$ and π, meaning that the vector must lie in the second quadrant.

Many calculators have a rectangular to polar conversion facility. Look this up on the instructions with your calculator and check the results. Remember, to get the result in radians you should first put your calculator into radian mode.

Conversion from polar co-ordinates to rectangular co-ordinates

If a vector is given by its length and angle to the x-axis, that is, $\mathbf{r} = r\angle\theta$, then

$$x = r\cos(\theta) \qquad y = r\sin(\theta)$$

Hence, in rectangular co-ordinates $\mathbf{r} = (r\cos(\theta), r\sin(\theta))$.

This result can easily be found from the triangle, as shown in Figure 9.12; examples are given in Figure 9.13.

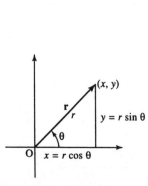

Figure 9.12 *If a vector, **r**, is known in polar co-ordinates, $\mathbf{r} = r\angle\theta$ then from the triangle $\cos(\theta) = x/r \Leftrightarrow x = r\cos(\theta)$ and $\sin(\theta) = y/r \Leftrightarrow y = r\sin(\theta)$.*

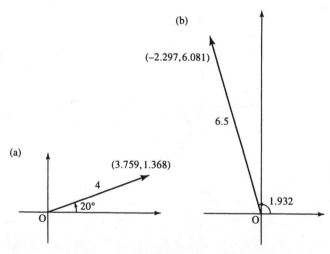

Figure 9.13 *(a) $4\angle 20°$ in rectangular co-ordinates is given by $x = 4\cos(20°) \approx 3.759$ and $y = 4\sin(20°) \approx 1.368$; therefore, the vector is (3.759, 1.368). (b) $6.5\angle 1.932$ in rectangular co-ordinates is given by $x = 6.5\cos(1.932) \approx -2.297$ and $y = 6.5\sin(1.932) \approx 6.081$; therefore, the vector is (−2.297, 6.081).*

Adding two vectors expressed in polar co-ordinates

To add two vectors expressed in polar co-ordinates, first express them in rectangular co-ordinates, then find the sum and then convert back into polar co-ordinates.

Example 9.5   Find $2 \angle 20° + 4 \angle 50°$.

Solution   Using $x = r \cos(\theta)$ and $y = r \sin(\theta)$, we can express the two vectors in rectangular co-ordinates, giving

$$2 \angle 20° \approx (1.879, 0.684)$$

$$4 \angle 50° \approx (2.5711, 3.064).$$

Therefore, $2 \angle 20° + 4 \angle 50° \approx (1.879, 0.684) + (2.571, 3.064) = (4.45, 3.748)$.

Finally, this can be represented in polar co-ordinates by using $r = \sqrt{x^2 + y^2}$ and $\theta = \tan^{-1}(y/x)$ ($+\pi$, if x is negative) giving $(4.45, 3.748) \approx 5.818 \angle 40.106°$.

Example 9.6   Find $4 \angle 1 + 2 \angle -1.6$.

Solution   Using $x = r \cos(\theta)$ and $y = r \sin(\theta)$, we can express the two vectors in rectangular co-ordinates, giving $4 \angle 1 \approx (2.161, 3.366)$:

$$2 \angle -1.6 \approx (-0.058, -1.999).$$

Therefore, $4 \angle 1 + 2 \angle - 1.6 \approx (2.161, 3.366) + (-0.058, -1.999) = (2.103, 1.367)$.

Finally, this can be represented in polar co-ordinates by using $r = \sqrt{x^2 + y^2}$ and $\theta = \tan^{-1}(y/x)$ ($+\pi$, if x is negative), giving $(2.103, 1.367) \approx 2.508 \angle 0.576$.

9.5 Application of vectors to represent waves (phasors)

In Section 5.3, we found the amplitude, phase, and cycle rate (frequency) of a wave. $f(t) = A \cos(\omega t + \phi)$ has amplitude A, angular frequency ω, and phase ϕ. Suppose we consider waves of a fixed frequency (say 50 Hz giving $\omega = 50 \times 2\pi \approx 314$); then, different waves can be represented by the amplitude and phase, giving $y = A \angle \phi$. The ideas of vectors can then be used to add and subtract waves and find their combined effect.

If a wave can be represented in polar form by $A \angle \phi$, then what does the rectangular form of the vector represent? We find that if the wave is split into cosine and sine terms by using the trigonometric identity $\cos(A + B) = \cos(A) \cos(B) - \sin(A) \sin(B)$, we get:

$$f(t) = A \cos(\omega t + \phi) = A \cos(\phi) \cos(\omega t) - A \sin(\phi) \sin(\omega t).$$

As $A \cos(\phi)$ is a constant, not involving an expression in t, this can be replaced by c and similarly $A \sin(\phi)$ can be replaced by d giving

$$f(t) = c \cos(\omega t) - d \sin(\omega t)$$

where $c = A \cos(\phi)$ and $d = A \sin(\phi)$.

So the vector (c, d) used to represent a wave represents the function $f(t) = c \cos(\omega t) - d \sin(\omega t)$ and if expressed in polar form $A \angle \phi$ it represents the equivalent expression

$$f(t) = A \cos(\omega t + \phi)$$

Example 9.7 Express the following as a single cosine term and give the amplitude and phase of the resultant function:

$$x = 3\cos(2t) + 2\sin(2t).$$

Solution Comparing $x = 3\cos(2t) + 2\sin(2t)$ with the expression $f(t) = c\cos(\omega t) - d\sin(\omega t)$ gives $c = 3, d = -2$ and $\omega = 2$. Expressing the vector $(3, -2)$ in polar form gives $3.605\angle-0.588$ and hence $x = 3.605\cos(2t - 0.588)$ giving the amplitude as 3.605 and phase as -0.588.

Check: Expand $x = 3.605\cos(2t - 0.588)$ using

$$\cos(A - B) = \cos(A)\cos(B) + \sin(A)\sin(B)$$
$$3.605\cos(2t - 0.588) = 3.605\cos(2t)\cos(0.588)$$
$$+ 3.605\sin(2t)\sin(0.588)$$
$$= 3\cos(2t) + 2\sin(2t)$$

which is the original expression.

Example 9.8 Express the following as a single cosine term and hence give the magnitude and phase of the resultant function:

$$y = -2\cos(t) - 4\sin(t)$$

Solution Comparing $y = -2\cos(t) - 4\sin(t)$ with the expression $f(t) = c\cos(\omega t) - d\sin(\omega t)$ gives $c = -2, d = 4$, and $\omega = 1$. Expressing the vector $(-2, 4)$ in polar form gives $4.472\angle 2.034$ and hence $y = 4.472\cos(t + 2.034)$.

Check: Expand $y = 4.472\cos(t + 2.034)$ using $\cos(A + B) = \cos(A)\cos(B) - \sin(A)\sin(B)$: $4.472\cos(t + 2.034) = 4.472\cos(t)\cos(2.034) - 4.472\sin(t)\sin(2.034) = -2\cos(t) - 4\sin(t)$.

Example 9.9 Express $x = 3\cos(20t + 5)$ as the sum of cosine and sine terms.

Solution On representing x as the phasor $3\angle 5$ with angular frequency 20, $3\angle 5$ converts to rectangular form as the vector $(0.851, -2.877)$ and this now gives the values of (c, d) in the expression $f(t) = c\cos(\omega t) - d\sin(\omega t)$, giving $x = 0.851\cos(20t) + 2.877\sin(20t)$.

Example 9.10 Find the resultant wave found from combining the following into one term:

$$f(t) = 3\cos(314t + 0.5) + 2\cos(314t + 0.9).$$

Solution As both terms are of the same angular frequency, $314\,\text{radians s}^{-1}$, we can express the two component parts by their amplitude and phase and then add the two vectors, giving $3\angle 0.5 + 2\angle 0.9$.

Expressing these in rectangular form gives $(2.633, 1.438) + (1.243, 1.567) = (3.876, 3.005)$.

Finally, expressing this again in polar form gives $4.904\angle 0.659$, so the resultant expression is $f(t) = 4.904\cos(314t + 0.659)$.

This method is a shorthand version of writing out all the trigonometric identities. It is even quicker if you use the polar – rectangular and rectangular – polar conversion facility on a calculator.

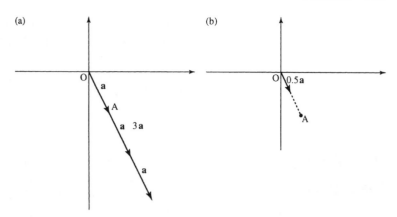

Figure 9.14 *The vector* $(1, -2)$*: (a) multiplied by 3; (b) multiplied by 0.5.*

9.6 Multiplication of a vector by a scalar and unit vectors

Multiplying a vector by a scalar has the effect of changing the length without affecting the direction. Each number in the vector is multiplied by the scalar

Example 9.11   If $\mathbf{a} = (1, -2)$ then

$$3\mathbf{a} = 3(1, -2) = (3 \times 1, 3 \times (-2)) = (3, -6)$$
$$0.5\mathbf{a} = 0.5(1, -2) = (0.5 \times 1, 0.5 \times (-2)) = (0.5, -1).$$

This is shown in Figure 9.14.

Unit vectors

Unit vectors have length 1. They are often represented by vectors with a cap on them $\hat{\mathbf{r}}$. Hence, $\hat{\mathbf{r}}$ means the unit vector in the same direction as $\mathbf{r}$.

To find the unit vector in the same direction as $\mathbf{r}$, divide $\mathbf{r}$ by its length: $\hat{\mathbf{r}} = \mathbf{r}/|\mathbf{r}|$, where $|\mathbf{r}|$ represents the magnitude of vector $\mathbf{r}$.

In Section 9.5, we found the length of a 2D vector (x, y) is $\sqrt{x^2 + y^2}$, similarly it can be shown that in three dimensions (x, y, z) the length is $\sqrt{x^2 + y^2 + z^2}$.

Example 9.12   Find unit vectors in the direction of the following vectors:

(a) $(1, -1)$ (b) $(3, 4)$ (c) $(0.5, 1, 0.2)$.

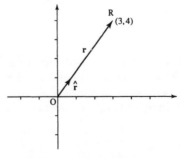

Figure 9.15 *The vector $\mathbf{r} = (3, 4)$ has modulus, or length $\mathbf{r} = \sqrt{3^2 + 4^2} = 5$. The unit vector in the same direction is found by dividing the vector $\mathbf{r}$ by its length giving $\hat{\mathbf{r}} = \frac{1}{5}(3, 4) = (0.6, 0.8)$.*

Solution   (a) Find the length of $(1, -1)$ given by $\sqrt{x^2 + y^2} = \sqrt{1^2 + 1^2} = \sqrt{2}$. Therefore, the unit vector is

$$\frac{1}{\sqrt{2}}(1, -1) \approx (0.707, -0.707)$$

(b) Find the length of $(3, 4)$ given by $\sqrt{x^2 + y^2} = \sqrt{3^2 + 4^2} = 5$. Therefore, the unit vector (see Figure 9.15) is

$$\frac{1}{5}(3, 4) = (0.6, 0.8).$$

(c) Find the length of $(0.5, 1, 0.2)$ given by $\sqrt{x^2 + y^2 + z^2} = \sqrt{(0.5)^2 + 1^2 + (0.2)^2} \approx 1.13578$. Therefore, the unit vector is

$$\frac{1}{1.13578}(0.5, 1, 0.2) \approx (0.44, 0.88, 1.176).$$

9.7 Basis vectors

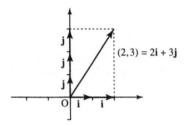

Figure 9.16 *Any vector in a plane can be expressed in terms of the vectors **i** and **j**; for instance, $(2, 3) = 2(1, 0) + 3(0, 1) = 2\mathbf{i} + 3\mathbf{j}$.*

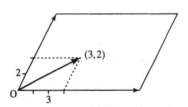

Figure 9.17 *A room shaped as a parallelogram. Any position in the room can be found by moving parallel to the sides. The basis vectors used are not at right angles.*

Vectors in a plane are made up of a part in the x-direction and a part in the y-direction, for example, $(2, 3) = (2, 0) + (0, 3)$. **i** and **j** are used to represent unit vectors in the x-direction and y-direction, that is, $\mathbf{i} = (1, 0)$ and $\mathbf{j} = (0, 1)$.

Any vector in the plane can be expressed in terms of **i** and **j**. **i** and **j** are called the Cartesian unit basis vectors, which is the name given to a co-ordinate system where the axes are at right angles to each other (orthogonal) (see Figure 9.16):

$$(2, 3) = 2(1, 0) + 3(0, 1) = 2\mathbf{i} + 3\mathbf{j}.$$

The unit vector in the z direction is often given the symbol **k**; and in three dimensions, using rectangular axes we have:

$$\mathbf{i} = (1, 0, 0) \quad \mathbf{j} = (0, 1, 0) \quad \mathbf{k} = (0, 0, 1)$$

that is

$$\begin{aligned}(5, -1, 2) &= (5, 0, 0) + (0, -1, 0) + (0, 0, 2) \\ &= 5(1, 0, 0) + (-1)(0, 1, 0) + 2(0, 0, 1) \\ &= 5\mathbf{i} - \mathbf{j} + 2\mathbf{k}.\end{aligned}$$

The vectors **i** and **j** form a basis set because all 2D geometrical vectors can be expressed in such terms. Similarly, all 3D vectors can be expressed in terms of **i**, **j**, and **k**. There are many other sets of vectors that can be used as a basis set: for instance, if we were in a room shaped like a parallelogram we could express any position in the room by moving parallel to one of the sides and then parallel to the other side. This is shown in Figure 9.17. Other basis sets are not as useful for interpreting spatial vectors as they do not give the same geometrical results. For instance, the interpretation of the scalar product, given in Section 9.8, relies on the fact that we use Cartesian basis vectors.

9.8 Products of vectors

There are two products of vectors that are commonly used: the scalar product, which results in a scalar, and the vector or cross product, which gives a vector as the result. However, both of these products are irreversible: they have no inverse operation. In other words, it is not possible to divide by a vector.

Scalar product

The scalar product of two vectors is defined by

$$\mathbf{a} \cdot \mathbf{b} = (\mathbf{a}_1, \mathbf{a}_2) \cdot (\mathbf{b}_1, \mathbf{b}_2) = a_1 b_1 + a_2 b_2.$$

Notice that the scalar product gives a simple number as the result.

Example 9.13   Find the scalar product of (2, 3) and (1, 2).

Solution   Using the definition

$$(2, 3) \cdot (1, 2) = (2)(1) + (3)(2) = 2 + 6 = 8.$$

Interpretation of the scalar product

The scalar product of **a** and **b** is related to the length of the vectors in the following way: $\mathbf{a} \cdot \mathbf{b} = ab\cos(\theta)$, where θ is the angle between the two vectors and a is the magnitude of **a** and b is the magnitude of **b** (see Figure 9.18).

The magnitude of a vector is the square root of the dot product with itself; hence, $\mathbf{a} \cdot \mathbf{a} = a^2$.

The scalar product can be used to find the angle between two vectors. It can also be used to find the length of a vector and can be used to test if two vectors are at right angles (orthogonal).

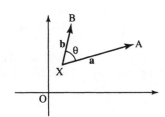

Figure 9.18   *The scalar product* $\mathbf{a} \cdot \mathbf{b} = ab\cos(\theta)$ *where* θ *is the angle between the two vectors.*

Example 9.14   Find the angle between (1, −1) and (3, 2).

Solution   If $\mathbf{a} = (1, -1)$ and $\mathbf{b} = (3, 2)$, then $\mathbf{a} \cdot \mathbf{b} = (1, -1) \cdot (3, 2) = (1)(3) + (-1)(2) = 3 - 2 = 1$.

We now use the relationship

$$\mathbf{a} \cdot \mathbf{b} = ab\cos(\theta) \quad \Leftrightarrow \quad \cos(\theta) = \frac{\mathbf{a} \cdot \mathbf{b}}{ab}$$

to find the angle between the vectors. We find the magnitude of **a** and the magnitude of **b**

$$a = \sqrt{1^2 + (-1)^2} \approx 1.414 \quad \text{and} \quad b = \sqrt{3^2 + 2^2} \approx 3.606.$$

Hence,

$$\cos(\theta) = \frac{\mathbf{a} \cdot \mathbf{b}}{ab}$$

becomes

$$\cos(\theta) = \frac{1}{1.414 \times 3.606} \approx 0.196$$

giving $\theta = \cos^{-1}(0.196) \approx 1.373$ radians.

Example 9.15   Show that (2, −1) and (−0.5, −1) are at right angles.

Solution   If two vectors are at an angle θ, with $\cos(\theta) = 0$, then $\theta = \pm 90°$, so the vectors are at right angles. Hence, if we find that $\mathbf{a} \cdot \mathbf{b} = 0$ this shows that **a** and **b** are at right angles (as long as one of the vectors is not the null vector (0, 0)). In this case the scalar product gives:

$$(2, -1) \cdot (-0.5, -1) = 2(-0.5) + (-1)(-1) = -1 + 1 = 0$$

As the scalar product of the two vectors is 0 the angle between them is 90°, so they are at right angles.

Example 9.16 Show that $(1, -1)$ and $(-2, -2)$ are at right angles.

Solution

$$(1, -1) \cdot (-2, -2) = (1)(-2) + (-1)(-2) = -2 + 2 = 0$$

Hence, they are at right angles.

Direction cosines

We have seen that the scalar product of two vectors a and b, $\mathbf{a} \cdot \mathbf{b} = ab \cos(\theta)$. We can use this result to show that the components of a unit vector are the direction cosines of the vector, that is,

$$\hat{\mathbf{r}} = (\cos(\alpha), \cos(\beta))$$

where α is the angle that the vector makes to the x-axis and β is the angle that the vector makes to the y-axis.

To show this, consider a unit vector $\hat{\mathbf{r}} = (x, y) = x\mathbf{i} + y\mathbf{j}$. If we take the scalar product with the unit vector along the x-axis, $\mathbf{i} = (1, 0)$, we will get

$$\hat{\mathbf{r}} \cdot \mathbf{i} = (x, y) \cdot (1, 0) = x$$

and we know that $\hat{\mathbf{r}} \cdot \mathbf{i} = |\hat{\mathbf{r}}||\mathbf{i}| \cos(\alpha)$, where $|\hat{\mathbf{r}}|$ and $|\mathbf{i}|$ are the magnitudes of $\hat{\mathbf{r}}$ and $\mathbf{i}$, respectively, and α is the angle between them. In this case, as we have two unit vectors their magnitudes are 1. This means that $\hat{\mathbf{r}} \cdot \mathbf{i} = \cos(\alpha)$, where α is the angle between the two vectors. In this case, α is the angle that the vector, $\hat{\mathbf{r}}$, makes to the x-axis. We have shown that $\hat{\mathbf{r}} \cdot \mathbf{i} = \cos(\alpha)$ and we know that $\hat{\mathbf{r}} \cdot \mathbf{i} = x$, so we have that $x = \cos(\alpha)$, where α is the angle to the x-axis. By considering $\hat{\mathbf{r}} \cdot \mathbf{j}$, we find that $y = \cos(\beta)$, where β is the angle that the vector $\hat{\mathbf{r}}$ makes to the y-axis. So we have that the components of any unit vector $\hat{\mathbf{r}}$ are the direction cosines of the vector.

If we consider any vector $\mathbf{r}$ we can find that a unit vector is the same direction as $\mathbf{r}$ by dividing by the magnitude of $\mathbf{r}$. Hence, we have that:

$$\hat{\mathbf{r}} = \frac{\mathbf{r}}{|\mathbf{r}|} = (\cos(\alpha), \cos(\beta))$$

where α is the angle the vector $\mathbf{r}$ makes to the x-axis and β is the angle the vector $\mathbf{r}$ makes to the y-axis (see Figure 9.19).

In three dimensions we get:

$$\hat{\mathbf{r}} = \frac{\mathbf{r}}{|\mathbf{r}|} = (\cos(\alpha), \cos(\beta), \cos(\gamma))$$

where α is the angle the vector $\mathbf{r}$ makes to the x-axis, β is the angle it makes to the y-axis, and γ the angle it makes to the z-axis.

Example 9.17 Find the angle that the following vectors make to the axes: (a) $(3, 6)$; (b) $(-1, -4, 5)$.

Solution (a) $\mathbf{r}/|\mathbf{r}| = (\cos(\alpha), \cos(\beta))$, where α and β are the angles made to the x and y axes. Therefore,

$$(\cos(\alpha), \cos(\beta)) = \frac{(3, 6)}{\sqrt{3^2 + 6^2}} = \frac{(3, 6)}{\sqrt{45}} \approx (0.44721, 0.89443).$$

The angle made to the x-axis is $\alpha = \cos^{-1}(0.44721) \approx 1.107$ and the angle made to the y-axis is $\beta = \cos^{-1}(0.89443) \approx 0.463$. The angles made to the x and y axes are 1.107 and 0.463 radians, respectively.

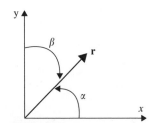

Figure 9.19
$\hat{\mathbf{r}} = (\cos(\alpha), \cos(\beta))$ *where* α
is the angle the vector $\mathbf{r}$
makes to the x-*axis and* β *is
the angle the vector* $\mathbf{r}$ *makes
to the* y-*axis.*

(b) $\mathbf{r}/|\mathbf{r}| = (\cos(\alpha), \cos(\beta), \cos(\gamma))$, where α, β and γ are the angles made to the x, y, and z axes. Therefore,

$$(\cos(\alpha), \cos(\beta), \cos(\gamma)) = \frac{(-1, -4, 5)}{\sqrt{(-1)^2 + (-4)^2 + 5^2}} = \frac{(-1, -4, 5)}{\sqrt{42}}$$
$$\approx (-0.1543, -0.61721, 0.77151).$$

The angles made to axes in the x, y, and z directions are found by taking the inverse cosines of the above: 1.726, 2.236, 0.6896 radians, respectively.

Vector components

The scalar product can be used to find the component of a vector in a given direction. This is a useful idea, for instance, if we are resolving forces and we want to add up all the forces acting in a certain direction. We can use the dot product with a unit vector in the direction of interest to find the component in that direction. The component of a vector $\mathbf{F}$ in the direction of a vector $\mathbf{r}$ is $\mathbf{F} \cdot (\mathbf{r}/|\mathbf{r}|)$

Example 9.18 A removal company wants to move a piano from the upstairs window of a small house. A smooth plank is placed against a wall near the window so that it touches the wall at a height of 4 m and the base of the plank is 1.5 m from the building. The piano, of mass 800 kg, is to be slid down the plank while attached by a rope. Taking the acceleration due to gravity to be $g \approx 9.81 \text{ m s}^{-2}$, what force is required on the other end of the rope to hold the piano steady while it is on the plank?

Solution We draw the situation as in Figure 9.20 using x and y axes. The vector that represents the plank goes from $(0, 0)$ to $(1.5, 4)$. This is the direction vector $\mathbf{p} = (1.5, 4) - (0, 0) = (1.5, 4)$.

The acceleration due to gravity is in a vertical direction and is $\mathbf{a} = (0, -g)$. From Newton's second law the force due to gravity is

$$\mathbf{F} = m\mathbf{a} = 800 \times (0, -9.81).$$

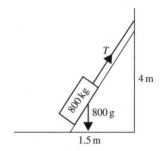

Figure 9.20 *Piano on a plank for Example 9.18.*

The component of the force due to gravity acting along the direction of the plank is the scalar product of the force with a unit vector in the direction of the plank, that is

$$\mathbf{F} \cdot \frac{\mathbf{p}}{|\mathbf{p}|} = 800 \frac{(0, -9.81).(1.5, 4)}{\sqrt{(1.5)^2 + 4^2}} = 800 \frac{-39.24}{\sqrt{18.25}} = -7348 \text{ } N \text{ to 4 s.f.}$$

The $-$ sign indicates that the component of the force due to gravity is in the opposite direction to the vector $\mathbf{p}$ along the plank. In order to hold the piano steady on the plank, we would need to have a force of equal magnitude in the opposing direction to be provided by the rope. That is, we would require a force of 7348 N on the rope.

Vector (or cross) product

The vector product of $\mathbf{a}$ and $\mathbf{b}$ is defined by $\mathbf{a} \times \mathbf{b} = ab \sin(\theta)\hat{\mathbf{n}}$, where $\hat{\mathbf{n}}$ is the unit vector normal to the plane of $\mathbf{a}$ and $\mathbf{b}$ and θ is the angle between $\mathbf{a}$ and $\mathbf{b}$.

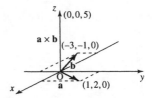

Figure 9.21 *The vector product of two vectors lying in the x, y plane,* $(1, 2, 0) \times (-3, -1, 0) = (0, 0, 5).$

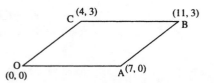

Figure 9.22 *The magnitude of the vector product of two vectors gives the area of the parallelogram formed by the vectors. The area of OABC is given by* $|(4, 3) \times (7, 0)| = |(4.0) - (3.7)| = |0 - 21| = 21$ *square units.*

If **a** and **b** are vectors that lie in the x, y plane, the vector product will be a vector normal to that plane, that is, wholly in the z-direction. It can be found from the following expression:

$$(a_1, a_2, 0) \times (b_1, b_2, 0) = (0, 0, a_1 b_2 - a_2 b_1) = (a_1 b_2 - a_2 b_1)\mathbf{k}$$

where **k** is the unit vector in the z-direction.

Example 9.19 $(1, 2, 0) \times (-3, -1, 0) = (0, 0, (1)(-1) - (2)(-3)) = (0, 0, 5)$. This is shown in Figure 9.21.

Applications of the vector product

The vector product can be used to find the area of a parallelogram with sides OA and OB (see Figure 9.22).

The area of a parallelogram is given by $ab \sin(\theta)$, where a and b are the lengths of the sides and θ is the angle between them. Consider vectors **a** and **b** representing the sides of the parallelogram. We know that $\mathbf{a} \times \mathbf{b} = ab \sin(\theta)\hat{\mathbf{n}}$, where $\hat{\mathbf{n}}$ is a unit vector normal to the plane of **a** and **b**. As $\hat{\mathbf{n}}$ is a unit vector it has a magnitude of 1, so $|\mathbf{a} \times \mathbf{b}| = ab \sin(\theta)$, which is exactly the same as the formula for the area of the parallelogram. Therefore, the area of the parallelogram can be found by taking the magnitude of the vector product of the vectors that define the sides of the parallelogram.

As $\sin(\theta) = 0$ when $\theta = 0$ or $\theta = 180°$, the vector product can also be used to test for parallel vectors (vectors pointing in the same direction or in exactly opposing directions). Again we only need to consider the magnitude of the vector product.

Example 9.20 Show that the vectors $(0.2, -5)$ and $(-1, 25)$ are parallel.

Solution Find $|(0.2, -5) \times (-1, 25)| = |(0.2 \times 25) - (-5 \times -1)| = |5 - 5| = 0$.

As we know that $|a \times b| = ab \sin(\theta)$, then as a and b are of non-zero length, $|\mathbf{a} \times \mathbf{b}| = 0 \Leftrightarrow \sin(\theta) = 0$. This shows that the vectors are parallel.

9.9 Vector equation of a line

In Chapter 2, we looked at the equation of a line that we found to be $y = mx + c$, where the gradient of the line is m and the line goes through the point $(0, c)$. We also found that the equation of a line that goes through two points, (x_1, y_1) and (x_2, y_2), is

$$\frac{y - y_1}{y_2 - y_1} = \frac{x - x_1}{x_2 - x_1}.$$

We would like to be able to express the equation of a line as a vector equation. If we know that the two points A and B represented by the position vectors **a** and **b** lie on the line, then a vector in the direction of the line will be a vector joining those two points, that is, **b** − **a**. As the line must go through A, we can see that any multiple of **b** − **a** added to the position vector **a** must lie on the line. This is shown in Figure 9.23

If we call the position vector of any point on the line **r** where **r** = (x, y), we now have the vector equation of the line as **r** = **a** + λ(**b** − **a**) where $\lambda \in \mathbb{R}$.

This can be rewritten as **r** = **a**$(1 − \lambda)$ + λ**b**.

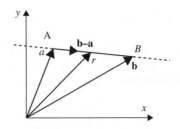

Figure 9.23   *The vector equation of the line. The vector* **r** *represents points on the line joining* A *and* B.

Example 9.21   Find the vector equation of a line through the points $(2, 4)$ and $(0, 6)$ and show that your result agrees with the equation of the line $y = −x + 6$, as found in Example 2.3.

Solution   Using **r** = **a**$(1 − \lambda)$ + λ**b** and **a** = $(2, 4)$, **b** = $(0, 6)$, we find

$$r = (2, 4)(1 − \lambda) + \lambda(0, 6) = (2(1 − \lambda) + \lambda(0), 4(1 − \lambda) + \lambda 6)$$

$$\mathbf{r} = (2 − 2\lambda, 4 + 2\lambda)$$

To show that this is the same as equation $y = −x + 6$ use **r** = (x, y)

$$(x, y) = (2 − 2\lambda, 4 + 2\lambda) \quad \Leftrightarrow \quad x = 2 − 2\lambda \text{ and } y = 4 + 2\lambda$$

This is a parametric equation for the line with parameter λ.

Eliminate λ by rewriting the equation for x so that λ is the subject and substitute into the equation for y:

$$x = 2 − 2\lambda \quad \Leftrightarrow \quad \lambda = \tfrac{1}{2}(2 − x)$$

Substituting for λ in $y = 4 + 2\lambda$ gives

$$y = 4 + 2\left(\tfrac{1}{2}\right)(2 − x)$$
$$\Leftrightarrow \quad y = 6 − x$$
$$\Leftrightarrow \quad y = −x + 6$$

This shows that the vector equation of the line is equivalent to $y = −x + 6$.

9.10 Summary

1. A vector is a string of numbers where the length of the string is called the dimension of the vector.
2. Vectors are used to represent points on a plane or in space, translations and physical quantities that have both magnitude and direction (called vector quantities).
3. The vector sum is found by adding corresponding elements of the vectors, or from a diagram by using a parallelogram.
4. To subtract vectors, subtract corresponding elements of the vectors. A triangle may be used to perform vector subtraction in a diagram.
5. Two-dimensional vectors **r** = (x, y) can be expressed in polar co-ordinates using

$$r = \sqrt{x^2 + y^2}$$

and $\theta = \tan^{-1}(y)$ $(+\pi$, if x is negative), so that $(x, y) = r\angle\theta$, where r or $|\mathbf{r}|$ is the magnitude, or length, of the vector and θ is the angle that the vector makes to the x-axis, also called its argument. To convert from polar to rectangular co-ordinates use:

$$x = r \cos(\theta) \quad \text{and} \quad y = r \sin(\theta).$$

To add vectors given in polar form they must first be converted to rectangular form.

6. Waves of a fixed frequency can be represented by phasors giving the amplitude and phase. $f(t) = A \cos(\omega t + \phi)$ can be represented by its amplitude and phase $A \angle \phi$. Converting this vector to rectangular form gives (c,d) where

$$f(t) = c \cos(\omega t) - d \sin(\omega t).$$

Using ideas of conversion from polar to rectangular form and vector addition, waves of the same frequency can be easily combined.

7. Unit vectors have length 1. To find the unit vector in the same direction as a vector $\mathbf{r}$ divide the vector by its length:

$$\hat{\mathbf{r}} = \frac{\mathbf{r}}{|\mathbf{r}|}.$$

8. Any vectors in the plane can be represented in terms of $\mathbf{i} = (1,0)$ and $\mathbf{j} = (0,1)$ and in three dimensions by $\mathbf{i} = (1,0,0), \mathbf{j} = (0,1,0)$, and $\mathbf{k} = (0,0,1)$. These are the Cartesian unit basis vectors and they are at right angles to each other.

9. Where $\mathbf{a} = (a_1, a_2)$ and $\mathbf{b} = (b_1, b_2)$ are two vectors, the scalar product is given by $\mathbf{a} \cdot \mathbf{b} = (a_1, a_2) \cdot (b_1, b_2) = a_1 b_1 + a_2 b_2$ and $\mathbf{a} \cdot \mathbf{b} = ab \cos(\theta)$, where a, b are the magnitudes of the vectors $\mathbf{a}$ and $\mathbf{b}$, and θ is the angle between them. The scalar product can be used to find the angle between two vectors.

10. The components of any unit vector give the cosines of the angles that the vector makes to each of the Cartesian axes. Then we have for any vector $\mathbf{r}$:

$$\hat{\mathbf{r}} = \frac{\mathbf{r}}{|\mathbf{r}|} = (\cos(\alpha), \cos(\beta))$$

where α is the angle the vector $\mathbf{r}$ makes to the x-axis and β is the angle it makes to the y-axis. In three dimensions:

$$\hat{\mathbf{r}} = \frac{\mathbf{r}}{|\mathbf{r}|} = (\cos(\alpha), \cos(\beta), \cos(\gamma))$$

where α is the angle the vector $\mathbf{r}$ makes to the x-axis, β is the angle it makes to the y-axis, and γ is the angle it makes to the z-axis. $\cos(\alpha), \cos(\beta)$, and $\cos(\gamma)$ are called the direction cosines of the vector.

11. The scalar product can be used to find the component of a vector in any given direction. Component of a vector $\mathbf{F}$ in the direction of a vector $\mathbf{r} = \mathbf{F} \cdot (\mathbf{r}/|\mathbf{r}|)$.

12. The vector product is given by $\mathbf{a} \times \mathbf{b} = ab \sin(\theta) \hat{\mathbf{n}}$, where a and b are the magnitudes of the vectors $\mathbf{a}$ and $\mathbf{b}$, θ is the angle between them, and $\hat{\mathbf{n}}$ is a unit vector normal to the plane of $\mathbf{a}$ and $\mathbf{b}$. If $\mathbf{a}$ and $\mathbf{b}$ are vectors in the x, y plane, that is, $\mathbf{a} = (a_1, a_2, 0)$ and $\mathbf{b} = (b_1, b_2, 0)$, we have

$$(a_1, a_2, 0) \times (b_1, b_2, 0) = (0, 0, a_1 b_2 - a_2 b_1) = (a_1 b_2 - a_2 b_1)\mathbf{k}$$

where $\mathbf{k}$ is the unit vector in the z-direction. The magnitude of the vector product can be used to find the area of a parallelogram.

13. The vector equation of a line passing through two points $\mathbf{a}$ and $\mathbf{b}$ is

$$\mathbf{r} = \mathbf{a}(1 - \lambda) + \lambda \mathbf{b} \quad \lambda \in \mathbb{R}$$

9.12 Exercises

9.1. In Figure 9.24, $\overrightarrow{OA} = \mathbf{a}$ and $\overrightarrow{OB} = \mathbf{b}$, OA = BC = OD, OB = AC = EO, EOB and DOA are straight lines. Write the following in terms of $\mathbf{a}$ and $\mathbf{b}$:

(a) $\overrightarrow{AB}$ (b) $\overrightarrow{BA}$ (c) $\overrightarrow{OC}$ (d) $\overrightarrow{OE}$ (e) $\overrightarrow{OD}$
(f) $\overrightarrow{ED}$ (g) $\overrightarrow{DE}$ (h) $\overrightarrow{DA}$ (i) $\overrightarrow{BE}$ (j) $\overrightarrow{EA}$

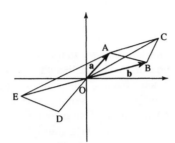

Figure 9.24   *Vectors for Exercise 9.1.*

9.2. Given $\mathbf{a} = (1, 3), \mathbf{b} = (-1, 2), \mathbf{c} = (3, 6, 2)$, and $\mathbf{d} = (6, 4, -1)$, find the following:

(a) $\mathbf{a} + \mathbf{b}$ (b) $\mathbf{a} - \mathbf{b}$ (c) $\mathbf{b} - \mathbf{a}$ (d) $-\mathbf{b} + \mathbf{a}$
(e) $2\mathbf{b}$ (f) $\mathbf{a} + 2\mathbf{b}$ (g) $3\mathbf{a} - \mathbf{b}$ (h) $\mathbf{c} - \mathbf{d}$
(i) $10\mathbf{c}$ (j) $\mathbf{c} + 6\mathbf{d}$ (k) $6\mathbf{c} - \mathbf{d}$

9.3. Express the following in its polar form, $r \angle \theta$, where r is the length of the vector and θ its angle to the x-axis:

(a) $(1, 3)$ (b) $(3, -1)$ (c) $(-1, -3)$ (d) $(5, -6)$

9.4. Express the following vectors $r \angle \theta$, where r is the modulus of the vector and θ the angle to the x-axis, in rectangular form. The angle is expressed in radians.

(a) $5 \angle \pi$ (b) $1 \angle -\pi$ (c) $\frac{1}{2} \angle \pi/4$ (d) $3 \angle \pi/3$

9.5. Express the following as a sum of a cosine and sine term in ωt:
(a) $f(t) = 3\cos(3t - 2)$
(b) $f(t) = 10\cos(20t + 5)$

9.6. Express the following as single cosine terms:
(a) $f(t) = 4\cos(10t) - 3\sin(10t)$
(b) $g(t) = -2\cos(157t) + 10\sin(157t)$

9.7. Express the following as a single wave:
(a) $6\cos(2t - 3) + 10\cos(2t + 2)$
(b) $\cos(t - \pi/2) + \cos(t + \pi/2)$
(c) $2\cos(628t - 1.57) - 6\cos(628t)$

9.8. Find the unit vectors in the same direction as the following:

(a) $(6, 8)$ (b) $(5, 12)$ (c) $(5, -12)$ (d) $(1, 1)$

(e) $(3, 2)$ (f) $(2, 0)$ (g) $(0, -3)$ (h) $(2, 4, 4)$

(i) $(1, -1, 2)$ (j) $(0.5, 0, -0.5)$

9.9. Express the following vectors in terms of $\mathbf{i} = (1, 0)$ and $\mathbf{j} = (0, 1)$ or in terms of $\mathbf{i} = (0, 0, 1), \mathbf{j} = (0, 1, 0)$, and $\mathbf{k} = (0, 0, 1)$ for 3D vectors:

(a) $(5, 2)$ (b) $(-1, -2)$ (c) $(-6, 2)$ (d) $(-1, 2, -3)$

(e) $(0.2, -1.6, 3.3)$

9.10. Find the following scalar products:

(a) $(1, -2) \cdot (3, 3)$ (b) $(9, 2) \cdot (-1, 6)$
(c) $(6, -1) \cdot (-1, -3)$

9.11. Find the angle between the following pairs of vectors:

(a) $(1, -2)$ and $(5, 1)$ (b) $(6, -1)$ and $(1, 6)$
(c) $(2, -1)$ and $(4, 9)$

9.12. Show that the following pairs of vectors are at right angles to each other:

(a) $(2, 1)$ and $(-1, 2)$ (b) $(-6, 3)$ and $(1, 2)$
(c) $(0.5, -2)$ and $(4, 1)$

9.13. Find the angles that the following vectors make to the axes:

(a) $(3, 6)$ (b) $(-1, -4, 5)$

9.14. (a) Find the component of the vector $(-1, 5)$ in the direction of the following vectors:

(i) $(0.5, 0.5)$ (ii) $(0.5, -0.5)$ (iii) $(-5, 1)$

(iv) $(1, -5)$ (v) $(8, 2)$

(b) Find the component of the vector $(-1, 2, 7)$ in the direction of the following vectors:

(i) $(1, 1, 1)$ (ii) $(6, 0, 2)$

9.15. Show that the following pairs of vectors are parallel:

(a) $(-3, 1)$ and $(1.5, -0.5)$ (b) $(6, 3)$ and $(18, 9)$

9.16. Find the area of the parallelogram OABC where two adjacent sides are:
(a) $\overrightarrow{OA} = (1, -1)$ and $\overrightarrow{OC} = (5, 2)$
(b) $\overrightarrow{OA} = (4, -1)$ and $\overrightarrow{OC} = (2, 2)$
(c) $\overrightarrow{OA} = (-3, 1)$ and $\overrightarrow{OC} = (2, 3)$

9.17 A straight line passes through the pair of points given. Find the vector equation of the line in each case:

(a) $(0, 1), (-1, 4)$ (b) $(1, 1), (-2, -4)$
(c) $(1, 1), (6, 3)$ (d) $(-1, -4), (-3, -4)$

10 Complex numbers

10.1 Introduction

In the previous chapter, we have shown that a single frequency wave can be represented by a phasor. We begin this chapter with a brief look at linear system theory. Such systems, when the input is a single frequency wave, produce an output at the same frequency which may be phase shifted with a scaled amplitude. Using complex numbers the system can be represented by a number which multiplies the input phasor having the effect of rotating the phasor and scaling the amplitude. We can define j as the number which rotates the phasor by $\pi/2$ without changing the amplitude. If this multiplication is repeated, hence rotating the phasor by $(\pi/2) + (\pi/2) = \pi$, then the system output will be inverted. In this way we can get the fundamental definition $j^2 = -1$. j is clearly not a real number as any real number squared is positive. j is called an imaginary number.

The introduction of imaginary numbers allows any quadratic equation to be solved. In previous chapters we said that the equation $ax^2 + bx + c = 0$ had no solutions when the formula leads to an attempt to take the square root of a negative number. The introduction of the number j makes square roots of negative numbers possible and in these cases the equation has complex roots. A complex number, z, has a real and imaginary part, $z = x + jy$ where x is the real part and y is the imaginary part. Real numbers are represented by points on a number line. Complex numbers need a whole plane to represent them.

We shall look at operations involving complex numbers, the conversion between polar and Cartesian (rectangular) form and the application of complex numbers to alternating current theory.

By looking at the problem of motion in a circle, we show the equivalence between polar and exponential form of complex numbers and represent a wave in complex exponential form. We can also obtain formulae for the sine and cosine in terms of complex exponentials, and we solve complex equations $z^n = c$, where c is a complex number.

10.2 Phasor rotation by $\pi/2$

In system theory, a system is represented (Figure 10.1) as a box with an input and an output. We think about the system after it has been in operation for a length of time, so any initial switching effects have disappeared.

Of particular importance are systems which, when the input is a single frequency wave, produce an output, at the same frequency that can be characterized by a phase shift and a change of amplitude of the wave. Examples of such systems are electrical circuits which are made up of lumped elements, that is, resistors, capacitors, and inductors. Here the input and output are voltages. Such a system is shown in Figure 10.2(a). Equivalent mechanical systems are made up of masses, springs and dampers, and the input and output is the external force applied and the

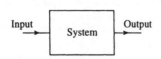

Figure 10.1 *A system is characterized by a box with an input and output.*

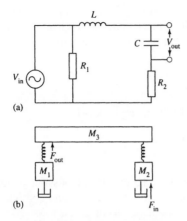

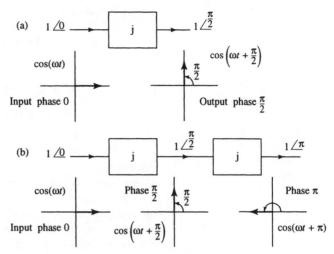

Figure 10.2 (a) An electrical system made up of resistors, capacitors, and inductors with voltage as input and output. (b) A mechanical system made up of masses, springs, and dampers. The input is the external force and the output is tension in the spring.

Figure 10.3 (a) A system which produces a phase shift of $\pi/2$, that is, rotates a phasor by $\pi/2$. This may be represented as a multiplication by j. (b) A system consisting of two sub-systems, both of which produce a phase shift of $\pi/2$ giving a combined shift of π. As a phase shift of π inverts a wave, that is, $\cos(\omega t + \pi) = -\cos(\omega t)$ this is equivalent to multiplication by -1. Hence, $j \times j = -1$.

tension in the spring. An example of such a mechanical system is shown in Figure 10.2(b).

We saw in Chapter 9 that a single frequency wave can be characterized by its amplitude and phase and these can be represented by vectors, called phasors. The advantage of complex numbers is that a phasor can be treated as a number and the system can be represented also by a number multiplying the input phasor.

Consider a single frequency input of 0 phase and amplitude 1. If there is a system which has the effect of simply shifting the phase by $\pi/2$, then we represent this by the imaginary number j. So $j \times 1\angle 0 = 1\angle \pi/2$. This system is shown in Figure 10.3(a). Supposing now we consider a system which can be broken into two components both of which shift the phase by $\frac{\pi}{2}$ as shown in Figure 10.3(b). The combined effect of the two systems is to multiply the input by $j \times j$. The final output wave, shifted now by π, is the $\cos(\omega t + \pi) = -\cos(\omega t)$ so it is -1 times the initial input. For this to be so then $j \times j = -1$.

This is the central definition for complex numbers:

$$j \times j = -1$$

meaning that $j = \sqrt{-1}$, where j is an operator which rotates a phasor by $\pi/2$.

We will return to these linear time-invariant systems in Chapter 16.

10.3 Complex numbers and operations

Complex numbers allow us to find solutions to all quadratic equations. Equations like $x^2 + 4 = 0$ do not have real roots because

$$x^2 + 4 = 0 \Leftrightarrow x^2 = -4$$

and there is no real number, which if squared will give -4.

If we introduce new numbers by using $j = \sqrt{-1}$, then a solution to $x^2 + 4 = 0$ is $x = j2$. $j2$ is an imaginary number. To check that $j2$ is in fact a solution to $x^2 + 4 = 0$, substitute it into the equation $x^2 + 4 = 0$, to give

$$(j2)^2 + 4 = 0 \iff j^2(2)^2 + 4 = 0$$
$$\iff (-1)(4) + 4 = 0 \text{ using } j^2 = -1$$
$$\iff 0 = 0$$

which is true.

Therefore, $x = j2$ is a solution. In order to solve all possible quadratic equations we need to use complex numbers, that is numbers that have both real and imaginary parts. Mathematicians often use i instead of j to represent $\sqrt{-1}$. However, j is used in engineering work to avoid confusion with the symbol for the current.

Real and imaginary parts and the complex plane

A complex number, z, can be written as the sum of its real and imaginary parts:

$$z = a + jb$$

where a and b are real numbers.

The real part of z is a ($\mathrm{Re}(z) = a$). The imaginary part of z is b ($\mathrm{Im}(z) = b$).

Complex numbers can be represented in the complex plane (often called an Argand diagram) as the points (x, y) where

$$z = x + jy$$

for example, $z = 1 - j2$ is shown in Figure 10.4. The methods used for visualizing and adding and subtracting complex numbers is the same as that used for two-dimensional vectors in Chapter 9.

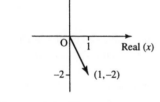

Figure 10.4   *The number $z = 1 - j2$. The real part is plotted along the x-axis and the imaginary part along the y-axis.*

Equality of two complex numbers

Two complex numbers can only be equal if their real parts are equal and their imaginary parts are equal.

Example 10.1   If $a - 2 + jb = 6 + j2$, where a and b are known to be real numbers, then find a and b

Solution

$$a - 2 + jb = 6 + j2$$

We know that a and b are real, so

$$a - 2 = 6 \quad \text{(real parts must be equal)}$$
$$\iff a = 8$$
$$b = 2 \quad \text{(imaginary parts must be equal)}$$

Check by substituting $a = 8$ and $b = 2$ into

$$a - 2 + jb = 6 + j2$$

which gives

$$8 - 2 + j2 = 6 + j2$$
$$\Leftrightarrow 6 + j2 = 6 + j2$$

which is correct

Addition of complex numbers

To add complex numbers, add the real parts and the imaginary parts.

Example 10.2 Given $z_1 = 3 + j4$ and $z_2 = 1 - j2$, find $z_1 + z_2$.

Solution

$$z_1 + z_2 = 3 + j4 + 1 - j2 = (3 + 1) + j(4 - 2) = 4 + j2$$

On the Argand diagram, the numbers add like vectors by the parallelogram law as in Figure 10.5.

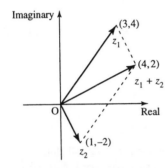

Figure 10.5 *Adding two complex numbers using the parallelogram law.*

Subtraction of complex numbers

To subtract complex numbers, we subtract the real and imaginary parts.

Example 10.3 Given $z_1 = 3 + j4$ and $z_2 = 1 - j2$, find $z_1 - z_2$.

Solution

$$z_1 - z_2 = 3 + j4 - (1 - j2) = 3 - 1 + j(4 - (-2)) = 2 + j6.$$

On the Argand diagram, reverse the vector z_1 to give $-z_2$ and then add z_1 and $-z_2$ as in Figure 10.6.

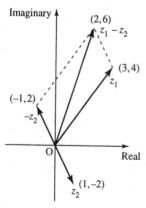

Figure 10.6 *To find $z_1 - z_2$ on an Argand diagram, reverse vector z_2 to give $-z_2$ and then add giving $z_1 + -z_2$.*

Multiplication of complex numbers

To multiply complex numbers multiply out the brackets, as for any other expression, and remember that $j^2 = -1$.

Example 10.4 Given $z_1 = 3 + j4$ and $z_2 = 1 - j2$, find $z_1 \cdot z_2$.

Solution

$$z_1 z_2 = (3 + j4)(1 - j2) = 3 + j4 + 3(-j2) + (j4)(-j2)$$
$$= 3 + j4 - j6 - j^2 8$$
$$= 3 - j2 + 8 \quad (\text{using } j^2 = -1)$$
$$= 11 - j2.$$

Example 10.5   Find $(4 - j2)(8 - j)$.

Solution   Multiplying as before gives

$$(4 - j2)(8 - j) = 32 - j16 - j4 + (-j2)(-j)$$
$$= 32 - j20 + j^2 2$$

Using $j^2 = -1$ gives $32 - j20 - 2 = 30 - j20$

The complex conjugate

The complex conjugate of a number, $z = x + jy$, is the number with equal real part and the imaginary part negated. This is represented by z^*:

$$z^* = x - jy$$

A number multiplied by its conjugate is always real and positive (or zero). For example, $z = 3 + j4 \Leftrightarrow z^* = 3 - j4$.

$$zz^* = (3 + j4)(3 - j4) = (3)(3) + (j4)3 + 3(-j4) + (j4)(-j4)$$
$$= 9 + j12 - j12 - j^2 16$$
$$= 9 - j^2 16 = 9 + 16 = 25 \quad (\text{using } j^2 = -1).$$

Note that the conjugate of the conjugate takes you back to the original number.

$$z = 3 + j4$$
$$z^* = 3 - j4$$
$$z^{**} = 3 + j4 = z$$

The conjugate of a number can be found on an Argand diagram by reflecting the position of the number in the real axis (see Figure 10.7).

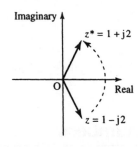

Figure 10.7   *The complex conjugate of a number can be found by reflecting the number in the real axis in the diagram are shown. The diagram shows $1 - j2$ and its conjugate $1 + j2$.*

Example 10.6   Find complex conjugates of the following and show that zz^* is real and positive, or zero, in each case

(a) $2 - j5$ (b) $-4 + j2$ (c) -5 (d) $j6$
(e) $a + jb$,   where a and b are real.

Solution   (a) The conjugate of $2 - j5$ is

$$(2 - j5)^* = 2 + j5$$

Hence

$$zz^* = (2 - j5)(2 + j5) = (2)(2) + (-j5)2 + 2(j5) + (-j5)(j5)$$
$$= 4 - j10 + j10 - j^2 25$$
$$= 4 - j^2 25 \quad (\text{using } j^2 = -1)$$
$$= 4 + 25 = 29$$

which is real and positive. We have shown that $2 - j5$ multiplied by its conjugate $2 + j5$ gives a real, positive number.

(b) $(-4 + j2)^* = -4 - j2$

$(-4 + j2)(-4 - j2) = (-4)(-4) + (j2)(-4)$
$$+ (-4)(-j2) + (j2)(-j2)$$
$$= 16 - j8 + j8 - j^2 4$$
$$= 16 - j^2 4 \quad (\text{using } j^2 = -1)$$
$$= 16 + 4 = 20$$

which is real and positive.

(c) -5 is a real number and therefore its complex conjugate is the same: -5. $(-5)^* = -5$ and $(-5)(-5) = 25$, which is real and positive.

(d) $(j6)^* = -j6$

$(j6)(-j6) = -j^2 36 = 36$

which is real and positive.

(e)
$$(a + jb)^* = a - jb$$
$(a + jb)(a - jb) = (a)(a) + (jb)(a) + (a)(-jb) + (jb)(-jb)$
$$= a^2 + jab - jab - j^2 b^2 = a^2 - j^2 b^2$$
$$\text{using } j^2 = -1, \text{ this gives } a^2 + b^2$$

As a and b are real, this must be a real number. Also, we know that the square of a real number is greater than or equal to 0. So $a^2 + b^2$ is real, and it is positive if $a \neq 0$, $b \neq 0$ or zero if both a and b are zero.

It is a good idea to remember this last result that $a + jb$ multiplied by its conjugate, $a - jb$, gives $a^2 + b^2$. That is, any number multiplied by its complex conjugate gives the sum of the square of the real part and the square of the imaginary part. This is the same as the value of the modulus of z squared, that is,

$zz^* = |z|^2.$

Division of complex numbers

To divide complex numbers, we use the fact that a number times its conjugate is real to transform the bottom line of the fraction to a real number. If we multiply the bottom line by its complex conjugate, we must also multiply the top line in order not to change the value of the number.

Example 10.7   Given $z_1 = 3 + j4$ and $z_2 = 1 - j2$, find z_1/z_2.

Solution

$$\frac{z_1}{z_2} = \frac{3 + j4}{1 - j2}$$
$$= \frac{(3 + j4)(1 + j2)}{(1 - j2)(1 + j2)}$$

Here, we have multiplied the top and bottom line by z_2^* to make the bottom line entirely real. Hence, we get

$$\frac{(3 + j4 + j6 + j^2 8)}{(1 + 2^2)} = \frac{(-5 + j10)}{5} = \frac{-5}{5} + \frac{j10}{5} = -1 + j2.$$

Example 10.8   Find

$$\frac{-3+j2}{10+j5}$$

in the form $x + jy$.

Solution   Multiply the top and bottom line by the complex conjugate of $10 + j5$ to make the bottom line real

$$\frac{-3+j2}{10+j5} = \frac{(-3+j2)(10-j5)}{(10+j5)(10-j5)}$$

$$= \frac{(-3)(10)+j2(10)+(-3)(-j5)+(j2)(-j5)}{(10^2+5^2)}$$

$$= \frac{-30+j20+j15-j^2 10}{125}$$

$$= \frac{-30+j20+j15+10}{125}$$

$$= \frac{-20+j35}{125} = \frac{-20}{125} + \frac{j35}{125} = -0.16 + j0.28.$$

10.4 Solution of quadratic equations

In Chapter 2 of the Background Mathematics Notes available on the companion website for this book, we looked at solutions of $ax^2 + bx + c = 0$ where a, b and c are real numbers and said that the solutions are given by the formula

$$x = \frac{-b \pm \sqrt{b^2 - 4ac}}{2a}$$

We discovered that there are no real solutions if $b^2 - 4ac < 0$ because we would need to take the square root of a negative number. We can now find complex solutions in this case by using $j = \sqrt{-1}$.

Example 10.9   Solve the following where $x \in \mathbb{C}$, the set of complex numbers:

(a) $x^2 + 3x + 5 = 0$ (b) $x^2 - x + 1 = 0$
(c) $4x^2 - 2x + 1 = 0$ (d) $4x^2 + 1 = 0$

Solution   (a) To solve $x^2 + 3x + 5 = 0$, compare with $ax^2 + bx + c = 0$. This gives $a = 1, b = 3$, and $c = 5$. Substitute in

$$x = \frac{-b \pm \sqrt{b^2 - 4ac}}{2a}$$

to give

$$x = \frac{-3 \pm \sqrt{3^2 - 4(1)(5)}}{2(1)} = \frac{-3 \pm \sqrt{-11}}{2}$$

We write $-11 = (-1) \cdot (11)$, so

$$\sqrt{-11} = \sqrt{-1}\sqrt{11} \approx j3.317 \quad \text{(using } j = \sqrt{-1})$$

Therefore,

$$x \approx \frac{-3 \pm j3.317}{2} \Rightarrow x \approx -1.5 + j1.658 \lor x \approx 1.5 - j1.658.$$

(b) To solve $x^2 - x + 1 = 0$, compare it with $ax^2 + bx + c = 0$. We all that $a = 1$, $b = -1$, and $c = 1$. Substitute in

$$x = \frac{-b \pm \sqrt{b^2 - 4ac}}{2a}$$

to give

$$x = \frac{-(-1) \pm \sqrt{(-1)^2 - 4(1)(1)}}{2(1)} = \frac{+1 \pm \sqrt{-3}}{2}$$

We write $-3 = (-1) \cdot 3$, so

$$\sqrt{-3} = \sqrt{-1}\sqrt{3} \approx j1.732 \quad (\text{using } j = \sqrt{-1})$$

Therefore,

$$x \approx \frac{1 \pm j1.732}{2} \Rightarrow x \approx 0.5 + j0.866 \lor x \approx 0.5 - j0.866.$$

(c) To solve $4x^2 - 2x + 1 = 0$, compare it with $ax^2 + bx + c = 0$. We get $a = 4$, $b = -2$, and $c = 1$. Substitute in

$$x = \frac{-b \pm \sqrt{b^2 - 4ac}}{2a}$$

to give

$$x = \frac{-(-2) \pm \sqrt{(-2)^2 - 4(4)(1)}}{2(4)} = \frac{2 \pm \sqrt{-12}}{8}$$

We write $-12 = (-1) \cdot 12$, so

$$\sqrt{-12} = \sqrt{-1}\sqrt{12} \approx j3.4641 \quad (\text{using } j = \sqrt{-1})$$

Therefore,

$$x \approx \frac{2 \pm j3.4641}{8} \Leftrightarrow x \approx 0.25 + j0.433 \lor x \approx 0.25 - j0.433.$$

(d) To solve $4x^2 + 1 = 0$, we could use the formula as in the other cases but it is quicker to do the following:

$$4x^2 + 1 \Leftrightarrow 4x^2 = -1 \quad (\text{subtracting 1 from both sides})$$

$$\Leftrightarrow x^2 = -\tfrac{1}{4} \quad (\text{dividing both sides by 4})$$

$$\Leftrightarrow x = \pm\sqrt{-0.25} \quad (\text{taking the square root of both sides})$$

as $\sqrt{-0.25} = \sqrt{-1}\sqrt{0.25} = j0.5$, we get $x = \pm j0.5 \Leftrightarrow x = j0.5 \lor x = -j0.5$.

If $ax^2 + bx + c = 0$ and the coefficients a, b, c are all real numbers, then we find that the two roots of the equation, if they are not entirely real, must be the complex conjugates of each other. This is true for all the cases we looked at in Example 10.9.

$$x^2 + 3x + 5 = 0$$

has solutions $x = -1.5 + j1.658$ and $x = -1.5 - j1.658$.

$$x^2 - x + 1 = 0$$

has solutions $x = 0.5 + j0.866$ and $x = 0.5 - j0.866$.

$$4x^2 + 1 = 0$$

has solutions $x = j0.5$ and $x = -j0.5$.

We can show that if the coefficients a, b, and c are real in the equation $ax^2 + bx + c = 0$, then the roots of the equation must either be real or complex conjugates of each other. We know from the formula that

$$ax^2 + bx + c = 0 \quad \Leftrightarrow \quad x = \frac{-b \pm \sqrt{b^2 - 4ac}}{2a}$$

If x has an imaginary part, then $b^2 - 4ac < 0$, so that $\sqrt{b^2 - 4ac}$ is an imaginary number. We can write this is terms of j as follows:

$$\sqrt{b^2 - 4ac} = \sqrt{-(4ac - b^2)} = \sqrt{-1}\sqrt{(4ac - b^2)}$$
$$= j\sqrt{(4ac - b^2)}$$

So the solutions, in the case, $b^2 - 4ac < 0$ are

$$x = \frac{-b \pm j\sqrt{4ac - b^2}}{2a}$$

which can be written, using

$$p = \frac{-b}{2a}$$

and

$$q = \frac{\sqrt{4ac - b^2}}{2a},$$

as

$$x = p \pm jq \Leftrightarrow x = p - jq \lor x = p + jq$$

where p and q are real.

This shows that the two solutions are complex conjugates of each other. This fact can be used to find the other root when one root is known.

Example 10.10   Given that the equation $x^2 - kx + 8 = 0$, where $k \in \mathbb{R}$ has one solution $x = 2 - j2$ then find the other solution and also the value of k.

Solution   We know that non-real solutions must be complex conjugates of each other so if one solution is $x = 2 - j2$ the other one must be $x = 2 + j2$. To find k, we use the result that if an equation has exactly two solutions x_1 and x_2, then the equation must be equivalent to $(x - x_1)(x - x_2) = 0$. We know that $x = 2 + j2$ or $x = 2 - j2$, therefore, the equation must be equivalent to

$$(x - (2 + j2))(x - (2 - j2)) = 0.$$

Multiplying out the brackets gives:

$$x^2 - x(2 + j2) - x(2 - j2) + (2 + j2)(2 - j2) = 0$$
$$\Leftrightarrow x^2 + x(-2 - j2 - 2 + j2) + (4 + 4) = 0$$
$$\Leftrightarrow x^2 - 4x + 8 = 0$$

Compare $x^2 - 4x + 8 = 0$ with $x^2 - kx + 8 = 0$. The coefficient of x^2 are equal in both cases, as are the constant terms, so the equations would be the same if $-k = -4 \Leftrightarrow k = 4$, giving the solution as $k = 4$.

10.5 Polar form of a complex number

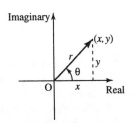

Figure 10.8   *The number $x + jy$ can be expressed in polar form by the length of the line representing the number, r, and the angle it makes to the x axis, θ; that is, $x + jy = r\angle\theta$.*

From the Argand diagram in Figure 10.8, we can see that a complex number can be expressed in terms of the length of the vector (the modulus) and the angle it makes with the x-axis (the argument). This is exactly the same process as that as in expressing vectors in polar coordinates as in Section 9.4.

If $z = x + jy$ then z can be represented in polar form as $r\angle\theta$ where

$$r^2 = x^2 + y^2$$
$$\tan(\theta) = \frac{y}{x}$$

Hence

$$r = \sqrt{x^2 + y^2}, \quad \theta = \tan^{-1}\left(\frac{y}{x}\right) \quad (+\pi \text{ if } x \text{ is negative})$$

We can write the complex number as

$$z = r\angle\theta$$

r, the modulus of z, is also written as $|z|$.

As it is usual to give the angle between $-\pi$ and π, it may be necessary to subtract 2π from the angle given by this formula. As 2π is a complete rotation, this will make no difference to the position of the complex number on the diagram.

Example 10.11   Express the following complex numbers in polar form

(a) $3 + j2$ (b) $-2 - j5$
(c) $-4 + j2$ (d) $4 - j2$

Solution   To perform these conversions to polar form, it is a good idea to draw a diagram of the number in order to check that the angle is of the correct size (see Figure 10.9).

(a) $3 + j2$ has modulus $r = \sqrt{3^2 + 2^2} \approx 3.61$ and the angle is given by $\tan^{-1}(2/3)$; therefore, in polar form $3 + j2 \approx 3.61\angle0.59$

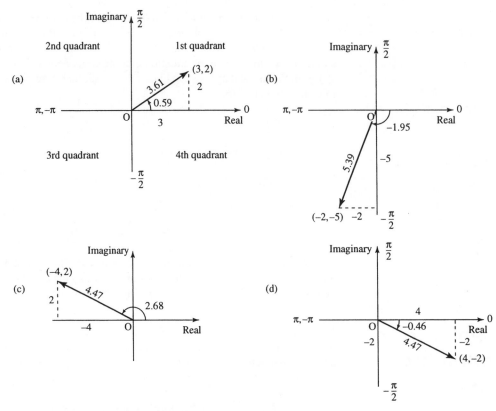

Figure 10.9 *Conversion to polar form:* (*a*) $3 + j2 \approx 3.61\angle0.59$; (*b*) $-2 - j5 \approx 5.39\angle-1.95$;
(*c*) $-4 + j2 \approx 4.47\angle2.68$; (*d*) $4 - j2 \approx 4.47\angle-0.46$.

(b) $-2 - j5$ has modulus $r = \sqrt{(-2)^2 + (-5)^2} \approx 5.39$ and the angle is given by $\tan^{-1}(-5/(-2)) + \pi \approx 4.332$. As this angle is bigger than 2π, subtract 2π (a complete revolution) to give -1.95. Therefore, in polar form $-2 - j5 \approx 5.39\angle - 1.95$. Note that the angle is between $-\pi$ and $-\pi/2$, meaning that the number must lie in the third quadrant, which we can see is correct from the diagram.

(c) $-4 + j2$ has modulus $r = \sqrt{(-4)^2 + 2^2} \approx 4.47$ and the angle is given by $\tan^{-1}(2/(-4)) + \pi \approx -0.46 + \pi \approx 2.68$. Therefore, in polar form $-4+j2 \approx 4.47\angle2.68$. Note that the angle is between $\pi/2$ and π, meaning that the number must lie in the second quadrant, which we can see is correct from the diagram.

(d) $4 - j2$ has modulus $r = \sqrt{(4)^2 + (-2)^2} \approx 4.47$ and the angle is given by $\tan^{-1}(-2/4) \approx -0.46$. Therefore, in polar form $4 - j2 \approx 4.47\angle - 0.46$. Note that the angle is between $-\pi/2$ and 0 meaning that the number must lie in the fourth quadrant, which we can see is correct from the diagram.

Check the calculations by using the rectangular to polar conversion facility on your calculator.

Conversion from polar form to Cartesian (rectangular) form

If a number is given by its modulus and argument, in polar form, $r\angle\theta$, we can convert back to Cartesian (rectangular) form using:

$$x = r\cos(\theta) \quad \text{and} \quad y = r\sin(\theta)$$

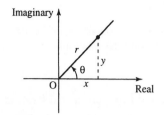

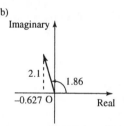

Figure 10.10 *The number*
$r\angle\theta$ *can be written as* $x + jy$.
Using the triangle,
$\cos(\theta) = x/r$ *giving*
$x = r\cos(\theta)$. *Also*
$\sin(\theta) = y/r$ *giving*
$y = r\sin(\theta)$.

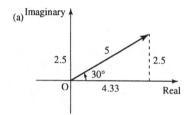

Figure 10.11   *(a)* $5\angle30°$ *in*
Cartesian (rectangular) form
is given by $x = 5\cos(30°) \approx$
4.33, $y = 5\sin(30°) = 2.5$,
therefore,
$5\angle30° \approx 4.33 + j2.5$.
(b) $2.2\angle1.86$ *in rectangular*
form is given by
$x = 2.2\cos(1.86) \approx -0.627$
and $y = 2.2\sin(1.86) \approx 2.1$,
therefore,
$2.2\angle1.86 \approx -0.627 + j2.1$.

This can be seen from Figure 10.10 and examples are given in Figure 10.11. As $z = x + jy$, $z = r\cos(\theta) + jr\sin(\theta) = r(\cos(\theta) + j\sin(\theta))$.

Addition, subtraction, multiplication, and division of complex numbers in polar form

To add and subtract two complex numbers, always express them first in rectangular form; that is, write as $z = a + jb$. The result of the addition or subtraction then can be converted back to polar form.

To multiply two numbers in polar form, multiply the moduli and add the arguments.

To divide two numbers in polar form divide the moduli and subtract the arguments.

Example 10.12   Given

$$z_1 = 3\angle\pi/6 \quad z_2 = 2\angle\pi/4$$

Find $z_1 + z_2, z_1 - z_2, z_1z_2$, and z_1/z_2.

Solution   To find $z_1 + z_2$ use $r\angle\theta = r(\cos(\theta) + j\sin(\theta))$

$$z_1 = 3(\cos(\pi/6) + j\sin(\pi/6)) \approx 2.5981 + j1.5$$
$$z_2 = 2(\cos(\pi/4) + j\sin(\pi/4)) \approx 1.4142 + j1.4142$$
$$z_1 + z_2 \approx 2.5981 + j1.5 + 1.4142 + j1.4142$$
$$= 4.0123 + j2.9142.$$

To express $z_1 + z_2$ back in polar form, use $r = \sqrt{x^2 + y^2}$ and $\theta = \tan^{-1}(y/x)$ ($+\pi$ if x is negative).

$$r = \sqrt{4.0123^2 + 2.9142^2} \approx 4.959,$$
$$\theta = \tan^{-1}(2.9142/4.0123) \approx 0.6282$$

Hence, $z_1 + z_2 \approx 4.959\angle0.6282$.

To find $z_1 - z_2$, we already have found (above) that $z_1 = 3\angle\pi/6 \approx 2.5981 + j1.5$ and $z_2 = 2\angle\pi/4 \approx 1.4142 + j1.4142$. So $z_1 - z_2 \approx 2.5981 + j1.5 - (1.4142 + j1.4142) = 1.1839 + j0.0858$.

To express $z_1 - z_2$ back in polar form use, $r = \sqrt{x^2 + y^2}$ and $\theta = \tan^{-1}(y/x)$ ($+\pi$ if x is negative).

Then $z_1 - z_2 \approx 1.1839 + j0.0858 \approx 1.187\angle0.0723$.

To find z_1z_2, multiply the moduli and add the arguments:

$$z_1z_2 = 3\angle\pi/6 \cdot 2\angle\pi/4 = (3) \cdot (2)\angle((\pi/6) + (\pi/4)) = 6\angle5\pi/12.$$

To find z_1/z_2, divide the moduli and subtract the arguments:

$$\frac{z_1}{z_2} = \frac{3\angle\pi/6}{2\angle\pi/4} = \frac{3}{2}\angle((\pi/6) - (\pi/4)) = 1.5\angle - \pi/12.$$

10.6 Applications of complex numbers to AC linear circuits

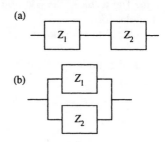

(a)

(b)

Figure 10.12
(a) Components in series.
(b) components in parallel.

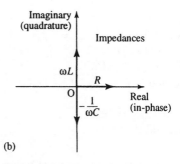

(a)

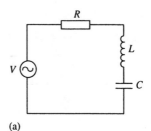

(b)

Figure 10.13 (a) The circuit for Example 10.13. (b) The impedances of the circuit elements shown on an Argand diagram. As the elements are in series, the resultant is found by taking the sum of the impedances of the components.

An alternating current (AC) electrical circuit consisting of resistors, capacitors, and inductors can be analysed using the relationship

$$V = ZI$$

where V is the voltage, Z the impedance, and I the current and V, Z, and I are all complex quantities.

Each component has a complex impedance associated with it and for two components in series, as in Figure 10.12(a), the resultant impedance Z_R is found by the formula

$$Z_R = Z_1 + Z_2$$

For two components in parallel, as in Figure 10.12(b), the resultant impedance is given by

$$\frac{1}{Z_R} = \frac{1}{Z_1} + \frac{1}{Z_2}$$

In the case of parallel circuit elements, it may be easier to calculate the admittance, the reciprocal of the impedance $Y = 1/Z$. Then use the fact that for circuit elements in parallel

$$Y_R = Y_1 + Y_2 \quad \text{and} \quad V = I/Y.$$

The real part of Z is called the resistance and the imaginary part is called the reactance.

$$Z = R + jS$$

where R is the resistance in ohms and S is the reactance in ohms.
The impedances of circuit elements are as follows:

Resistor	$Z = R$	No reactive element
Capacitor	$Z = 1/(j\omega C) = -j/(\omega C)$	Purely reactive
Inductor	$Z = j\omega L$	Purely reactive

where ω is the angular frequency of the source, $\omega = 2\pi f$, f is the frequency in Hz, R is the resistance (in ohms) , C is the capacitance (in farads), and L is the impedance (in henries).

Example 10.13 Find the impedance of the circuit shown in Figure 10.13(a) at 20 kHz where $L = 2\,\text{mH}$, $C = 100\,\mu\text{F}$, and $R = 2000\,\Omega$. Assuming a voltage amplitude of 300 V, calculate the current I and relative phase.

Solution The impedances of the elements are shown in Figure 10.13(b). As we are given that f, the frequency of the input, is 20 kHz, $\omega = 2\pi f = 2\pi \times 20 \times 10^3$. As the elements are in series, we can sum the impedances

$$Z = R + j\omega L - \frac{j}{\omega C}$$

$$= 2000 + j2\pi \times 20 \times 10^3 \times 2 \times 10^{-3}$$

$$- \frac{j}{2\pi \times 20 \times 10^3 \times 100 \times 10^{-6}}$$

giving

$$2000 + j\left(80 - \frac{1}{4\pi}\right) \approx 2000 + j251.2$$

Expressing this is polar form gives $Z \approx 2016\angle 7°$. Therefore, from $V = ZI$ and given $V = 300$,

$$I = \frac{300}{Z} = \frac{300}{2016\angle 7°} \approx 0.149\angle -7°.$$

giving a current of magnitude 0.149 A with a relative phase of $-7°$.

Example 10.14 One form of a 'tuned circuit' that can be used as a band-pass filter is given in Figure 10.14(a). Given that $R = 300\,\Omega$, $L = 2\,mH$, and $C = 10\,\mu F$, find the admittance of the circuit at 2 kHz. Given that the current source is of amplitude 12 A, find the voltage amplitude and its relative phase.

Solution The admittances of the elements are shown in Figure 10.14(b). As the elements are in parallel, we can sum the admittances

$$Y = \frac{1}{R} + j\omega C - \frac{j}{\omega L} \quad \text{where } \omega = 2 \times 10^3 \times 2\pi$$

giving

$$Y = \frac{1}{300} + j(2\pi \times 10^3 \times 2 \times 10 \times 10^{-6})$$

$$- \frac{j}{2 \times 10^3 \times 2\pi \times 2 \times 10^{-3}}$$

$$\approx 0.003333 + j0.0859.$$

Expressing this is polar form gives $Y = 0.086\angle 88°$. Therefore, from $V = ZI$ or $V = I/Y$, we have

$$V = \frac{12}{0.086\angle 88°} \approx 139.6\angle -88°$$

giving a voltage amplitude of 139.6 V with a relative phase of $-88°$.

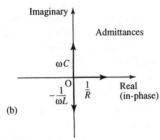
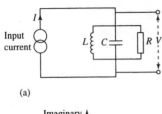

Figure 10.14 (a) The circuit for Example 10.14(b) The admittances of the circuit elements shown on an Argand diagram. As the elements are in parallel, the resultant is found by taking the sum of the admittances of the components.

10.7 Circular motion

When we introduced the sine and cosine function in Chapter 5, we used the example of a rotating rod of length r. We plotted the position of the rod, its height, and horizontal distance from the centre against the angle through which the rod had rotated. This defined the sine and cosine functions. We consider this problem again. This time we specify that the circular motion is at constant angular velocity ω. That is, the rate of change of the angle θ, $d\theta/dt$, is constant and equals ω. That is:

$$\frac{d\theta}{dt} = \omega \quad \text{(where } \omega \text{ is a constant)}$$

$$\Leftrightarrow \quad \theta = \omega t + \phi$$

where ϕ is the angle when $t = 0$. If we start with the rod horizontal, then $\phi = 0$ and we have $\theta = \omega t$. The (x, y) position of the tip of the rod of length r is given as a function of time by $x = r\cos(\omega t)$ and $y = r\sin(\omega t)$, where ω is the constant angular velocity, so the rotating vector is given by $\mathbf{r} = (r\cos(\omega t), r\sin(\omega t))$ with $\theta = \omega t$.

Consider now a ball on the end of a string with constant angular velocity ω. Can we obtain an expression for its acceleration? The acceleration is of particular importance because we know from Newton's second law

(a)   (b)

that the force required to maintain the circular motion can be found by using $\mathbf{F} = m\mathbf{a}$, where $\mathbf{F}$ is the force, m is the mass, and $\mathbf{a}$ is the acceleration. We assume, in this discussion, that the effects of gravity and air resistance are negligible. The ball is being rotated in a plane which is vertical to the ground.

First, imagine lying flat on the ground in a line with the x-axis and watching the ball. It appears to oscillate back and forth and its position is given by $x = r \cos(\omega t)$. This is pictured in Figure 10.15.

Differentiating with respect to t gives the component of the velocity in the x-direction

$$\frac{dx}{dt} = -r\omega \sin(\omega t).$$

The acceleration is the derivative of this velocity

$$\frac{d}{dt}\left(\frac{dx}{dt}\right).$$

This is also written as d^2x/dt^2 (read as 'd two x by dt squared') and it means the derivative of the derivative

$$\frac{dx}{dt} = -r\omega \sin(\omega t)$$

differentiating again gives

$$\frac{d^2x}{dt^2} = -r\omega^2 \cos(\omega t) = -\omega^2(r \cos(\omega t))$$

and as $x = r \cos(\omega t)$

$$\frac{d^2x}{dt^2} = -\omega^2 x.$$

This equation tells us that the horizontal acceleration is proportional to the horizontal distance from the origin in a direction towards the origin. This type of behaviour is called simple harmonic motion.

We can consider the movement in the y-direction by changing our point of view as in Figure 10.16. Again we can find the component of the acceleration, this time in the y-direction.

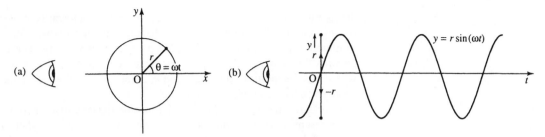

Figure 10.16 *If we observe the ball from the position of the eye in Figure 10.16(a), we can only see the motion in the vertical direction and the ball appears to oscillate up and down as shown in Figure 10.16(b). Also shown is the graph of y against time, t : y = r sin(ωt).*

We differentiate $y = r \sin(\omega t)$ to get the component of velocity in the y-direction:

$$\frac{\mathrm{d}y}{\mathrm{d}t} = r\omega \cos(\omega t)$$

Differentiate again to find the acceleration:

$$\frac{\mathrm{d}^2 y}{\mathrm{d}t^2} = -r\omega^2 \sin(\omega t)$$

and as $y = r \sin(\omega t)$

$$\frac{\mathrm{d}^2 y}{\mathrm{d}t^2} = -\omega^2 y$$

Notice that this is the same equation as we had for x. We can represent the motion, both in the x- and y-directions by using a complex number to represent the rotating vector. The real part of z represents the position in the x-direction and the imaginary part of z represents the position in the y-direction:

$$z = x + \mathrm{j}y = r \cos(\omega t) + \mathrm{j}r \sin(\omega t)$$

Then

$$\frac{\mathrm{d}z}{\mathrm{d}t} = -r\omega \sin(\omega t) + \mathrm{j}r\omega \cos(\omega t)$$

The real part of $\mathrm{d}z/\mathrm{d}t$ represents the component of velocity in the x-direction and the imaginary part represents the velocity in the y-direction. Again, we can differentiate to find the acceleration

$$\frac{\mathrm{d}^2 z}{\mathrm{d}t^2} = -r\omega^2 \cos(\omega t) - \mathrm{j}r\omega^2 \sin(\omega t)$$

$$= -\omega^2 (r \cos(\omega t) + \mathrm{j}r \sin(\omega t)) = -\omega^2 z$$

as

$$z = r \cos(\omega t) + \mathrm{j}r \sin(\omega t)$$

So, we get

$$\frac{\mathrm{d}^2 z}{\mathrm{d}t^2} = -\omega^2 z$$

This shows that the acceleration operates along the length of the vector z towards the origin and it must be of magnitude $|-\omega^2 z| = \omega^2 r$ where r is the radius of the circle. The ball is always accelerating towards the centre of the circle. This also tells us the force that the string must provide in order to maintain the circular motion at constant angular velocity. The force towards the centre, called the centripetal force, must be $|F| = m\omega^2 r$, where r is the radius of the circle and m is the mass of the ball. This has been given by Newton's second law $\mathbf{F} = m\mathbf{a}$.

We can use the equation for circular motion to show that it is possible to represent a complex number, z, in the form $z = r\,e^{j\theta}$, where r is the modulus and the argument. To do this we must first establish the conditions which determine a particular solution to the equation

$$\frac{d^2z}{dt^2} = -\omega^2 z$$

We know that one solution of the differential equation

$$\frac{d^2z}{dt^2} = -\omega^2 z$$

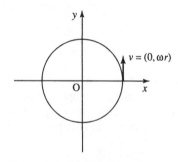

y

$v = (0, \omega r)$

O

x

Figure 10.17 *The initial velocity vector (as shown) must be in the positive y direction if the motion is anti-clockwise.*

with the condition that $z = r$ when $t = 0$, is given by $z = r\cos(\omega t) + jr\sin(\omega t)$. Unfortunately, there is at least one other solution, given by the case where the string travels clockwise rather than anti-clockwise, that is, $z = r\cos(-\omega t) + jr\sin(-\omega t)$. However, we can pin down the solution to the anti-clockwise direction of rotation by using the fact that we defined the angular velocity by $d\theta/dt = \omega$. This gives a condition on the initial velocity (at $t = 0$). From Figure 10.17 we can see that the velocity must be positive and only have a component in the y-direction at $t = 0$.

This discounts the possibility of the motion being clockwise as this would give a negative initial velocity. From $z = r\cos(\omega t) + jr\sin(\omega t)$

$$\frac{dz}{dt} = -r\omega\sin(\omega t) + jr\omega\cos(\omega t)dt$$

and at $t = 0$, $dz/dt = jr\omega$. We now have enough information to say that

$$\frac{d^2z}{dt^2} = -\omega^2 z \quad \text{and} \quad z = r \quad \text{when } t = 0$$

and

$$\frac{dz}{dt} = j\omega r \text{ when } t = 0 \quad \Leftrightarrow \quad z = r\cos(\omega t) + jr\sin(\omega t)$$

In Chapter 8 we looked at the exponential function and we found that $y = y_0\,e^{kt}$ is a solution to the equation $dy/dt = ky$. This equation models the situation where the rate of change of the population is proportional to its current size: the first derivative of y is proportional to y. The equation

$$\frac{d^2z}{dt^2} = -\omega^2 z$$

is similar only now the acceleration is related to z, that is the second derivative is proportional to z. As the exponential functions have the property that the derivative gives a scaled version of the original function, we must also get a scaled version of the original function if we differentiate twice. So we can try a solution of the form $z = r\,e^{kt}$ for the equation

$$\frac{d^2z}{dt^2} = -\omega^2 z$$

$$z = r\,e^{kt}$$

$$\Rightarrow \frac{dz}{dt} = rk\,e^{kt}$$

$$\Rightarrow \frac{d^2z}{dt^2} = rk^2\,e^{kt}$$

Substituting into

$$\frac{d^2z}{dt^2} = -\omega^2 z$$

we get

$$rk^2\,e^{kt} = -\omega^2 r\,e^{kt}$$

Dividing both sides by $r\,e^{kt}$ gives $k^2 = -\omega^2 \Leftrightarrow k = \pm j\omega$.

This gives two possible solutions: $z = r\,e^{j\omega t}$ when $k = j\omega$ and $z = r\,e^{-j\omega t}$ when $k = -j\omega$. Again, we can use the initial velocity to determine the solution. Using $z = r\,e^{j\omega t}$ we get

$$\frac{dz}{dt} = jr\omega\,e^{j\omega t}$$

at $t = 0$ then we get the velocity as $j\omega r$, which was one of the conditions we wanted to fulfil.

This shows that the two expression $z = r\,e^{j\omega t}$ and $z = r\cos(\omega t) + jr\sin(\omega t)$ both satisfy

$$\frac{d^2z}{dt^2} = -\omega^2 z \quad \text{and} \quad z = r \quad \text{when } t = 0$$

and

$$\frac{dz}{dt} = j\omega r \quad \text{when } t = 0.$$

We have stated that these initial conditions are enough to determine the solution of the differential equation. So, the only possibility is that

$$r\,e^{j\omega t} = r\cos(\omega t) + jr\sin(\omega t)$$

This shows the equivalence of the polar form of a complex number and the exponential form. Replacing ωt by θ, we get $r\,e^{j\theta} = r\cos(\theta) + jr\sin(\theta)$, which we recognize as the polar form for a complex number $z = r\angle\theta$ where r is the modulus and θ is the argument. We can represent any complex number $z = x + jy$ in the form $r\,e^{j\theta}$. r and θ are found, as given before for the polar form, by

$$r = \sqrt{x^2 + y^2}$$

$$\theta = \tan^{-1}\left(\frac{y}{x}\right) \quad (+\pi \text{ if } x \text{ is negative})$$

Conversely, to express a number given in exponential form in rectangular (Cartesian) form, we can use $r\,e^{j\theta} = r\cos(\theta) + jr\sin(\theta)$.

Example 10.15   Show that $z = 2\,e^{j3t}$ is a solution to $d^2z/dt^2 = -9z$ where $z = 2$ when $t = 0$ and $dz/dt = j6$ when $t = 0$.

Solution   If $z = 2\,e^{j3t}$, then when $t = 0$, $z = 2\,e^0 = 2$

$$\frac{dz}{dt} = 2(j3)e^{j3t} = j6e^{j3t}$$

when $t = 0$

$$\frac{dz}{dt} = j6e^0 = j6.$$

Hence

$$\frac{d^2z}{dt^2} = j6(j3)e^{j3t}$$

$$\Rightarrow \quad \frac{d^2z}{dt^2} = -18e^{j3t}$$

Substituting into $d^2z/dt^2 = -9z$ gives $-18\,e^{j3t} = -9(2\,e^{j3t}) \Leftrightarrow -18\,e^{j3t} = -18\,e^{j3t}$, which is true.

Example 10.16   Express $z = 3 + j4$ in exponential form.

Solution   The modulus r if given by

$$r = \sqrt{3^2 + 4^2} = \sqrt{25} = 5$$

The argument is $\tan^{-1}(4/3) \approx 0.9273$. Hence, $z \approx 5\,e^{j0.9273}$.

Example 10.17   Find the real and imaginary parts of the following

(a) $3\,e^{j(\pi/2)}$ (b) e^{-j} (c) e^{3+j2}
(d) $e^{-j(j-1)}$ (e) j^j

Solution   (a) Use $r\,e^{j\theta} = r\cos(\theta) + jr\sin(\theta) \cdot 3\,e^{j(\pi/2)}$ has $r = 3$ and $\theta = \pi/2$

$$3\,e^{j(\pi/2)} = 3\cos(\pi/2) + j3\sin(\pi/2)$$
$$= 3.0 + j(3)(1) = j3$$

The real part is 0 and the imaginary part is 3.
   (b) Comparing e^{-j} with $re^{j\theta}$ gives $r = 1$ and $\theta = -1$. Using $r\,e^{j\theta} = r\cos(\theta) + jr\sin(\theta)$

$$e^{-j} = 1\cos(-1) + j\sin(-1)$$
$$\approx 0.5403 - j0.8415$$

So the real part of e^{-j} is approximately 0.5403 and the imaginary part is approximately -0.8415.

(c) Notice that e^{3+j2} is not in the form $r\,e^{j\theta}$ because the exponent has a real part. We therefore split the exponent into its real and imaginary bits by using the rules of powers:

$$e^{3+j2} = e^3 e^{j2}$$

$e^3 \approx 20.09$ is a real number, and the remaining exponent j2 is purely imaginary:

$$e^3 e^{j2} \approx 20.09\,e^{j2}$$

Comparing $e^3 e^{j2}$ with $re^{j\theta}$ gives $r = e^3$ and $\theta = 2$. Using $re^{j\theta} = r\cos(\theta) + jr\sin(\theta)$ gives

$$e^3 e^{j2} = e^3 \cos(2) + je^3 \sin(2)$$
$$\approx -8.359 + j18.26$$

The real part of e^{3+j2} is approximately -8.359 and the imaginary part is approximately 18.26.

(d) For $e^{-j(j-1)}$ we need first to write the exponent in a form that allows us to split it into its real and imaginary bits. So, we remove the brackets to give

$$e^{-j(j-1)} = e^{-j^2+j}$$

Using $j^2 = -1$, this gives

$$e^{-j(j-1)} = e^{1+j}$$

Now, using the rules of powers we can write

$e^{1+j} = e^1 e^j$. e^1 is a real number and the exponent of e^j is purely imaginary. Comparing $e^1 e^j$ with $re^{j\theta}$ gives $r = e^1$ and $\theta = 1$. Using $re^{j\theta} = r\cos(\theta) + jr\sin(\theta)$ gives

$$e^1 e^{j1} = e^1 \cos(1) + je^1 \sin(1)$$
$$\approx 1.469 + j2.287$$

The real part of $e^{-j(j-1)}$ is approximately 1.469 and the imaginary part is approximately 2.287

(e) j^j looks a very confusing number as we have only dealt with complex powers when the base is e. We therefore begin by looking for a way of rewriting the expression so that its base is e. To do this, we write the base, j, in exponential form. From the Argand diagram in Figure 10.18, as j is represented by the point (0,1) we can see that that it has modulus 1 and argument $\pi/2$. Hence, $j = e^{j(\pi/2)}$.

Use this to replace the base in the expression j^j so that

$j^j = \left(e^{j(\pi/2)}\right)^j = e^{j^2(\pi/2)} = e^{-\pi/2}$. We can now see that j^j is in fact a real number!

$$j^j = e^{-\pi/2} \approx 0.2079.$$

The real part of j^j is approximately 0.2079 and the imaginary part is 0.

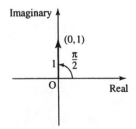

Figure 10.18 j *is represented by the point* (0, 1) *in the complex plane. Therefore, it has modulus 1 and argument* $\pi/2$, *that is,* $j = e^{j(\pi/2)}$.

10.8 The importance of being exponential

We have shown that

$$r\,e^{j\theta} = r\cos(\theta) + jr\sin(\theta) = r\angle\theta$$

Here r is the modulus of the complex number and θ is the argument. Therefore, the exponential form is simply another way of writing the polar form. The advantage of the exponential form is its simplicity. For circular motion, at constant angular velocity, it can represent the motion both in the real and imaginary (x and y) directions in one simple expression $r\,e^{j\omega t}$. The rules for multiplication and powers of complex numbers in exponential form are given by the rules of powers, as for any other number, given in Chapter 4 of the Background Notes in Mathematics available on the companion website for this book, thus confirming the rules that we gave for the polar form.

Multiplication:

$$r_1\,e^{j\theta_1} r_2\,e^{j\theta_2} = r_1 r_2\,e^{j(\theta_1+\theta_2)}$$

That is, we multiply the moduli and add the arguments.

Division:

$$\frac{r_1\,e^{j\theta_1}}{r_2\,e^{j\theta_2}} = \frac{r_1}{r_2}e^{(\theta_1-\theta_2)}$$

That is, we divide the moduli and subtract the arguments.

The complex conjugate is of a number $r\,e^{j\theta}$ is the number of same modulus and negative argument. That is, the complex conjugate of $r\,e^{j\theta}$ is $r\,e^{-j\theta}$

Powers:

$$(r\,e^{j\theta})^n = r\,e^{jn\theta}.$$

That is, to raise a complex number to the power n, take the nth power of its modulus and multiply its argument by n.

Example 10.18   Given $z_1 = 3\,e^{j(\pi/6)} z_2 = 2\,e^{j(\pi/4)}$, Find z_1+z_2, z_1-z_2, $z_1 z_2, z_1/z_2, z_1^* z_2^*$ and z_1^3.

Solution   To find $z_1 + z_2$, use $r\,e^{j\theta} = r(\cos(\theta) + j\sin(\theta))$

$$z_1 = 3\,e^{j(\pi/6)} = 3\left(\cos\left(\frac{\pi}{6}\right) + j\sin\left(\frac{\pi}{6}\right)\right) \approx 2.5981 + j1.5$$

$$z_2 = 2\,e^{j(\pi/4)} = 2\left(\cos\left(\frac{\pi}{4}\right) + j\sin\left(\frac{\pi}{4}\right)\right) \approx 1.4142 + j1.4142$$

Therefore,

$$z_1 + z_2 \approx 2.5981 + j1.5 + 1.4142 + j1.4142 = 4.0123 + j2.9142$$

To express $z_1 + z_2$ back in exponential form, use $r = \sqrt{x^2 + y^2}$ and $\theta = \tan^{-1}(y/x)$ ($+\pi$ if x is negative)

$$r = \sqrt{(4.0123)^2 + (2.9142)^2} \approx 4.959 \text{ and}$$

$$\theta = \tan^{-1}(2.9142/4.0123) \approx 0.6282$$

Hence, $z_1 + z_2 \approx 4.959\,e^{j0.6282}$

To find $z_1 - z_2$, we already have found (above) that

$$z_1 = 3\,e^{j(\pi/6)} \approx 2.5981 + j1.5 \text{ and}$$

$$z_2 = 2\,e^{j(\pi/4)} \approx 1.4142 + j1.4142$$

Therefore

$$z_1 - z_2 \approx 2.5981 + j1.5 - (1.4142 + j1.4142) = 1.1839 + j0.0858$$

To express $z_1 - z_2$ back in polar form, use $r = \sqrt{x^2 + y^2}$ and $\theta = \tan^{-1}(y/x)$ ($+\pi$ if x is negative):

$$z_1 - z_2 \approx 1.1839 + j0.0858 \approx 1.187\,e^{j0.0723}.$$

To find $z_1 z_2$, multiply the moduli and add the arguments

$$z_1 z_1 = 3\,e^{j(\pi/6)} 2e^{j(\pi/4)} = 3.2\,e^{j((\pi/6)+(\pi/4))} = 6\,e^{-j(5\pi/12)}.$$

To find z_1/z_2, divide the moduli and subtract the arguments

$$\frac{z_1}{z_2} = \frac{3\,e^{j(\pi/6)}}{2\,e^{j(\pi/4)}} = \frac{3}{2}e^{j((\pi/6)-(\pi/4))} = 1.5\,e^{-j(\pi/12)}.$$

To find $z_1^* z_2^*$, we find the complex conjugate of z_1 and z_2

$$z_1^* = (3\,e^{j(\pi/6)})^* = 3\,e^{-j(\pi/6)}$$

$$z_2^* = (2\,e^{j(\pi/4)})^* = 2\,e^{-j(\pi/4)}$$

$$z_1^* z_2^* = 3\,e^{-j(\pi/6)} 2\,e^{-j(\pi/4)} = 3.2\,e^{-j((\pi/6)+(\pi/4))} = 6\,e^{-j(5\pi/12)}.$$

To find z_1^3

$$z_1^3 = (3\,e^{j(\pi/6)})^3 = 3^3 (e^{j(\pi/6)})^3 = 27\,e^{j(\pi/6)\times 3} = 27\,e^{j(\pi/2)}.$$

Expressions for the trigonometric functions

From the exponential form, we can find expressions for the cosine or sine in terms of complex numbers. We begin with a complex number z of modulus 1 and its complex conjugate

$$e^{j\theta} = \cos(\theta) + j\sin(\theta) \tag{10.1}$$

$$e^{-j\theta} = \cos(\theta) - j\sin(\theta) \tag{10.2}$$

From these, we can find the expression for the cosine and sine in terms of the complex exponential. Adding Equation (10.1) and (10.2), we have

$$e^{j\theta} + e^{-j\theta} = \cos(\theta) + j\sin(\theta) + \cos(\theta) - j\sin(\theta)$$

$$\Leftrightarrow \quad e^{j\theta} + e^{-j\theta} = 2\cos(\theta)$$

Dividing both sides by 2 gives

$$\tfrac{1}{2}(e^{j\theta} + e^{-j\theta}) = \cos(\theta)$$

$$\Leftrightarrow \quad \cos(\theta) = \tfrac{1}{2}(e^{j\theta} + e^{-j\theta})$$

Now, subtracting Equation (10.2) from Equation (10.1), we get

$$e^{j\theta} - e^{-j\theta} = \cos(\theta) + j\sin(\theta) - (\cos(\theta) - j\sin(\theta))$$

$$\Leftrightarrow \quad e^{j\theta} - e^{-j\theta} = 2j\sin(\theta)$$

Dividing both sides by 2j gives

$$\frac{1}{2j}(e^{j\theta} - e^{-j\theta}) = \sin(\theta)$$

$$\Leftrightarrow \quad \sin(\theta) = \frac{1}{2j}(e^{j\theta} - e^{-j\theta})$$

So, we have

$$\cos(\theta) = \tfrac{1}{2}(e^{j\theta} + e^{-j\theta})$$

$$\sin(\theta) = \frac{1}{2j}(e^{j\theta} - e^{-j\theta})$$

Using $\tan(\theta) = \sin(\theta)/\cos(\theta)$, we get

$$\tan(\theta) = \frac{(1/2j)(e^{j\theta} - e^{-j\theta})}{(1/2)(e^{j\theta} + e^{-j\theta})} = \frac{1}{j}\left(\frac{e^{j\theta} - e^{-j\theta}}{e^{j\theta} + e^{-j\theta}}\right)$$

Compare these with the definition of the sinh, cosh, and tanh functions given in Chapter 8:

$$\cosh(\theta) = \tfrac{1}{2}(e^{\theta} + e^{-\theta})$$

$$\sinh(\theta) = \tfrac{1}{2}(e^{\theta} - e^{-\theta})$$

$$\tanh(x) = \frac{e^{\theta} - e^{-\theta}}{e^{\theta} + e^{-\theta}}$$

We see that:

$$\cos(j\theta) = \cosh(\theta)$$

$$\sin(j\theta) = j\sinh(\theta)$$

$$\tan(j\theta) = j\tanh(\theta)$$

De Moivre's theorem

Using the expression for the cmoplex number in terms of a sine and cosine, $re^{j\theta} = r(\cos(\theta) + j\sin(\theta))$, and using this in $r\,e^{j\theta n} = r\,e^{jn\theta}$, we get

$$(r(\cos(\theta) + j\sin(\theta))^{n} = r^{n}(\cos(n\theta) + j\sin(n\theta))$$

This is called De Moivre's theorem and can be used to obtain multiple angle formulae.

Example 10.19 Find $\sin(3\theta)$ in terms of powers of $\sin(\theta)$ and $\cos(\theta)$.

Solution   We use the fact that $\sin(3\theta) = \text{Im}(\cos(3\theta) + j\sin(3\theta))$, where Im( ) represents 'the imaginary part of'. Hence

$$\sin(3\theta) = \text{Im}(e^{j3\theta})$$

$$= \text{Im}((\cos(\theta) + j\sin(\theta))^3).$$

Expanding

$$(\cos(\theta) + j\sin(\theta))^3 = (\cos(\theta) + j\sin(\theta))(\cos(\theta) + j\sin(\theta))^2$$

$$= (\cos(\theta) + j\sin(\theta))(\cos^2(\theta) + 2j\cos(\theta)\sin(\theta) + j^2\sin^2(\theta))$$

$$= \cos^3(\theta) + j\sin(\theta)\cos^2(\theta) + 2j\cos^2(\theta)\sin(\theta)$$

$$+ j^2\cos(\theta)\sin^2(\theta) + 2j^2\cos(\theta)\sin^2(\theta) + j^3\sin^3(\theta)$$

$$= \cos^3(\theta) + 3j\sin(\theta)\cos^2(\theta) - 3\cos(\theta)\sin^2(\theta) - j\sin^3(\theta)$$

$$= \cos^3(\theta) - 3\cos(\theta)\sin^2(\theta) + j(3\sin(\theta)\cos^2(\theta) - \sin^3(\theta)).$$

As $\sin(3\theta) = \text{Im}((\cos(\theta) + j\sin(\theta))^3)$, we take the imaginary part of the expression we have found to get

$$\sin(3\theta) = 3\sin(\theta)\cos^2(\theta) - \sin^3(\theta).$$

Example 10.20   Express $\cos^3(\theta)$ in terms of cosines of multiples of θ.

Solution   Using $\cos(\theta) = (1/2)(e^{j\theta} + e^{-j\theta})$ and the expansion $(a+b)^3 = a^3 + 3a^2b + 3ab^2 + b^3$:

$$\cos^3(\theta) = \left(\frac{1}{2}(e^{j\theta} + e^{-j\theta})\right)^3$$

$$= \frac{1}{2^3}(e^{j3\theta} + 3e^{j\theta} + 3e^{-j\theta} + e^{-j3\theta})$$

$$\frac{1}{2^2}\left(\frac{1}{2}(e^{j3\theta} + e^{-j3\theta}) + \frac{3}{2}(e^{j\theta} + e^{-j\theta})\right).$$

As

$$\cos(\theta) = \tfrac{1}{2}(e^{j\theta} + e^{-j\theta}) \quad \text{and} \quad \cos(3\theta) = \tfrac{1}{2}(e^{j3\theta} + e^{-j3\theta})$$

we get

$$\cos^3(\theta) = \tfrac{1}{4}\cos(3\theta) + \tfrac{3}{4}\cos(\theta).$$

The exponential form can be used to solve complex equations of the form $z^n = c$, where c is a complex number. A particularly important example is the problem of finding all the solutions of $z^n = 1$, called the n roots of unity.

The *n* roots of unity

To solve the equation $z^n = 1$, we use the fact that 1 is a complex number with modulus 1 and argument 0, as can be seen in Figure 10.19(a). However, we can also use an argument of 2π, 4π, 6π, or any other multiple of 2π. As 2π is a complete revolution, adding 2π on to the argument of any complex number does not change the position of the vector representing it and therefore does not change the value of the number.

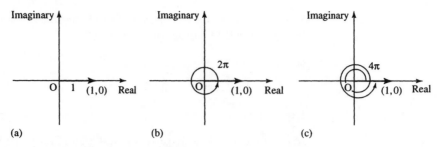

Figure 10.19   *(a)* 1 *is the complex number* e^{j0}, *that is, with a modulus of* 1 *and an argument of* 0. *(b)* 1 *can also be represented using an argument of* 2π, *that is,* $1 = e^{j2\pi}$. *(c)* 1 *represented with an argument of* 4π, *that is,* $1 = e^{j4\pi}$.

The equation $z^n = 1$ can be expressed as

$$z^n = e^{j2\pi N} \quad \text{where } N \in \mathbb{Z}$$

We can solve this equation by taking the nth root of both sides, which is the same as taking both sides to the power $1/n$.

$$(z^n)^{1/n} = e^{j2\pi N/n} \quad \text{where } N \in \mathbb{Z}$$

We can substitute some values for N to find the various solutions also using the fact that there should be n roots to the equation $z^n = 1$ so that we can stop after finding all n roots.

Example 10.21   Find all the solutions to $z^3 = 1$.

Solution   Write 1 as a complex number with argument $2\pi N$ giving the equation as

$$z^3 = e^{j2\pi N} \quad \text{where } N \in \mathbb{Z}.$$

Taking the cube root of both sides:

$$(z^3)^{1/3} = e^{j(2\pi N/3)} \quad \text{where } N \in \mathbb{Z}.$$

Substituting

$$N = 0: \quad z = e^{j2\pi 0} = 1$$

$$N = 1: \quad z = e^{j2\pi/3}$$

$$N = 2: \quad z = e^{j4\pi/3}.$$

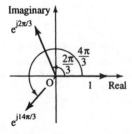

Figure 10.20   *The solutions to* $z^3 = 1$ *are* $z = 1$, $z = e^{j2\pi/3}$, *and* $z = e^{j4\pi/3}$. *Notice that one solution can be obtained from another by rotation through* $2\pi/3$.

There is no need to use any more values of N. We use the fact that there should be three roots of a cubic equation. If we continued to substitute values for N, then the values will begin to repeat. For example, substituting $N = 3$ gives $z = e^{j2\pi 3/3} = e^{j2\pi}$, which we know is the same as e^{j0} (subtracting 2π from the argument) which equals 1, which is a root that we have already found.

The solutions to $z^3 = 1$ are shown on an Argand diagram in Figure 10.20. The principal root of a complex equation is the one found nearest to the position of the positive x-axis. Notice that in the case of $z^3 = 1$, the principal root is 1 and the other solutions can be obtained from another by rotation through $2\pi/3$. Hence, another way of finding the n roots of $z^n = 1$ is to start with the principal root of $z = 1 = e^{j0}$ and add on multiples of $2\pi/n$ to the argument, in order to find the other roots.

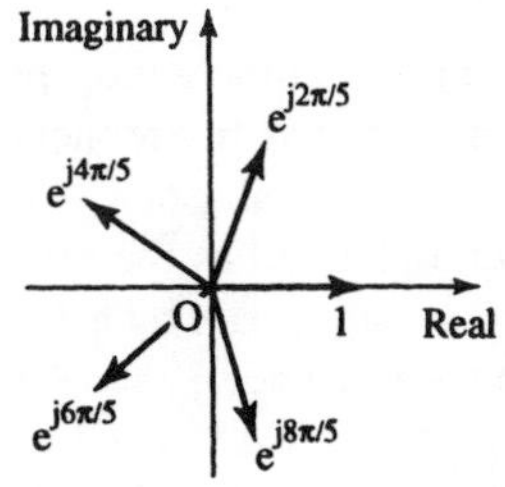

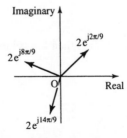

Figure 10.21 *The solutions to $z^5 = 1$ are $z=1$, $e^{j2\pi/5}$, $e^{j4\pi/5}$, $e^{j6\pi/5}$, and $e^{j8\pi/5}$. Notice that one solution can be obtained from another by rotation through $2\pi/5$.*

Example 10.22 Find all the roots of $z^5 = 1$

Solution One root, the principal root, is $z = 1 = e^{j0}$. The other roots can be found by rotating this around the complex plane by multiples of $2\pi/5$. Therefore, we have the solutions:

$$z = 1, \ e^{j2\pi/5}, \ e^{j4\pi/5}, \ e^{j6\pi/5}, \ e^{j8\pi/5}.$$

These are shown in Figure 10.21.

Solving some other complex equations

If we have the equation $z^n = c$, where c is any complex number, then we write the right-hand side of the equation in exponential form and use the fact that we can add a multiply of 2π to the argument without changing the value of the number. Write

$$c = r \, e^{j\theta} = r \, e^{j(\theta + 2\pi N)} \quad \text{where } N \in \mathbb{Z}$$
$$z^n = r \, e^{j(\theta + 2\pi N)}$$
$$\Leftrightarrow z = r^{(1/n)} e^{(j(\theta + 2\pi N)/n)}$$

taking the nth root of both sides.

Example 10.23 Solve $z^3 = -4 + j4\sqrt{3}$.

Solution Write $-4 + j4\sqrt{3}$ in exponential form, $r \, e^{j\theta}$

$$r = \sqrt{(-4)^2 + (4\sqrt{3})^2} = \sqrt{16 + 48} = \sqrt{64} = 8$$

$$= \tan^{-1}\left(-\frac{4\sqrt{3}}{4}\right) + \pi = 2\pi/3 \quad (\text{using } \tan^{-1}(\sqrt{3}) = \pi/3).$$

So, the equation becomes

$$z^3 = 8e^{j(2\pi/3 + 2\pi N)} \quad \text{where } N \in \mathbb{Z}$$
$$\Leftrightarrow \quad z = 8^{1/3} \, e^{j(2\pi/3 + 2\pi N)/3}$$
$$\Leftrightarrow \quad z = 2 \, e^{j(2\pi/3 + 2\pi N)/3}, \quad \text{where } N \in \mathbb{Z}$$

Substituting some values for N gives

$$N = 0: \quad z = 2 \, e^{j2\pi/9}$$
$$N = 1: \quad z = 2 \, e^{j(2\pi/9 + 2\pi/3)} = 2 \, e^{j8\pi/9}$$
$$N = 2: \quad z = 2 \, e^{j(2\pi/9 + 4\pi/3)} = 2 \, e^{j14\pi/9}.$$

The solutions are

$$z = 2 \, e^{j2\pi/9}, \ 2 \, e^{j8\pi/9}, \ 2 \, e^{j14\pi/9}.$$

These are shown in Figure 10.22.

Figure 10.22 *The solutions to $z^3 = -4 + j4/3$ are $z = 2e^{j2\pi/9}$, $2e^{j8\pi/9}$, and $2e^{j14\pi/9}$. Notice that one solution can be obtained from another by rotation through $2\pi/3$.*

10.9 Summary

1. Simple systems can be represented by a complex number multiplying a single frequency input. The output then has the same frequency as the input, with a modified amplitude and a shifted phase.

2. j is the number which when multiplying a phasor has the effect of rotating the phase by $\pi/2$ (or 90°). $j \times j$ rotates the phase by π (180°), which is equivalent to multiplication by -1. Hence $j^2 = -1$ and $j = \sqrt{-1}$. Sometimes i is used instead of j to represent $\sqrt{-1}$.

3. A complex number is any number that can be represented on the complex plane. It can be written as $z = x + jy$ (x and y real) where x is the real part of z ($\text{Re}(z) = x$) and y is the imaginary part of z ($\text{Im}(z) = y$). A complex number expressed in the form $z = x + jy$ is said to be in Cartesian or rectangular form.

4. The complex conjugate of $a + jb$ is $a - jb$; $(a + jb)^* = a - jb$. The product of a number and its complex conjugate is always a real number greater than or equal to 0: $(a + jb)(a + jb)^* = (a + jb)(a - jb) = a^2 + b^2$, which is real and $\geqslant 0$. $zz^* = |z|^2$, where z is any complex number: a number multiplied by its conjugate gives its modulus squared.

5. The operations of addition and subtraction of complex numbers are like those for vectors: simply add or subtract the real parts and then the imaginary parts. Multiply as follows, remembering $j^2 = -1$.

$$(1 + j2)(-3 - j3) = (1)(-3) + j2(-3) + 1(-j3) + (j2)(-j3)$$
$$= -3 - j6 - j3 - j^2 6$$
$$= -3 - j9 + 6 = 3 - j9$$

To divide multiply the top and bottom lines by the complex conjugate of the bottom line as follows:

$$\frac{1 + j2}{-3 - j3} = \frac{(1 + j2)(-3 + j3)}{(-3 - j3)(-3 + j3)}$$
$$= \frac{-3 - j6 + j3 - 6}{(-3)^2 + (3)^2} = \frac{-9 - j3}{18} = -\frac{1}{2} - \frac{j}{6}$$

6. All quadratic equations can now be solved if $x \in \mathbb{C}$, that is, x is a complex number. If $ax^2 + bx + c = 0$ where a, b, c are real numbers, then

$$x = \frac{-b \pm \sqrt{b^2 - 4ac}}{2a}$$

The solutions are real if $b^2 \geqslant 4ac$. If $b^2 < 4ac$, then the solutions can be written as $x = p \pm jq$, where

$$p = \frac{-b}{2a}$$

and

$$q = \frac{\sqrt{4ac - b^2}}{2a}, \quad p \text{ and } q \text{ are real}$$

and therefore, non-real roots are complex conjugates of each other.

7. Complex numbers can be written in polar form $r\angle\theta$, where r is the modulus of the number and θ is the argument. The modulus is the

length of the vector representing the complex number and θ is the angle made with the positive real axis.

$$z = x + \mathrm{j}y = r\angle\theta,$$

$$r = \sqrt{x^2 + y^2} \text{ and } \theta = \tan^{-1}\left(\frac{y}{x}\right)(+\pi \text{ if } x < 0)$$

$$x = r\cos(\theta) \text{ and } y = r\sin(\theta)$$

To add or subtract complex numbers expressed in polar form, first convert to rectangular form. To multiply, multiply the moduli and add the arguments and to divide, divide the moduli and subtract the arguments:

$$r_1\angle\theta_1 r_2\angle\theta_2 = r_1 r_2\angle(\theta_1 + \theta_2)$$

$$\frac{r_1\angle\theta_1}{r_2\angle\theta_2} = \frac{r_1}{r_2}\angle(\theta_1 - \theta_2)$$

8. Complex numbers are used in the analysis of alternating current (AC) circuits. ω is the angular frequency of the source. Resistors, capacitors, and inductors have associated impedances, Z, where for a resistor $Z = R$, for a capacitor $Z = 1/\mathrm{j}\omega C$, and for an inductor $Z = \mathrm{j}\omega L$, where R is the resistance, C is the capacitance, and L is the inductance. The voltage and the current are related by

$$V = ZI$$

and the impedances of circuit elements obey

$$Z_R = Z_1 + Z_2$$

for elements in series, and

$$\frac{1}{Z_R} = \frac{1}{Z_1} + \frac{1}{Z_2}$$

for elements in parallel, where Z_R is the resultant impedance. The admittance Y is the reciprocal of the impedance: $Y = 1/Z$.

9. $\mathrm{d}^2 y/\mathrm{d}t^2 = -\omega^2 y$ is the differential equation that defines waves as a function of time. This is an equation of motion where the acceleration is proportional to the distance from the origin. This is called simple harmonic motion. By examining the case of circular motion, at constant angular velocity, where the rotating vector $z = x + \mathrm{j}y = r\cos(\omega t) + \mathrm{j}r\sin(\omega t)$ obeys this equation, we can show the equivalence of the polar representation of a complex wave and the exponential form:

$$r\,\mathrm{e}^{\mathrm{j}\omega t} = r\cos(\omega t) + \mathrm{j}r\sin(\omega t) = r\angle\omega t$$

as $\theta = \omega t$, we have

$$r\,\mathrm{e}^{\mathrm{j}\theta} = r\cos(\theta) + \mathrm{j}r\sin(\theta) = r\angle\theta$$

Here, r is the modulus of the complex number and θ is the argument.

Replacing θ by $-\theta$, we get

$$r\,e^{-j\theta} = r\cos(\theta) - j\sin(\theta) = r\angle-\theta$$

From these, using the case where $r = 1$, we can find the expression for the cosine and sine in terms of the complex exponential:

$$\cos(\theta) = \tfrac{1}{2}(e^{j\theta} + e^{-j\theta})$$

$$\sin(\theta) = \frac{1}{2j}(e^{j\theta} - e^{-j\theta})$$

$$\tan(\theta) = \frac{1}{j}\left(\frac{e^{j\theta} - e^{-j\theta}}{e^{j\theta} + e^{-j\theta}}\right)$$

and by comparing these with the definition of the sinh, cosh, and tanh function we see that:

$$\cos(j\theta) = \cosh(\theta)$$

$$\sin(j\theta) = j\sinh(\theta)$$

$$\tan(j\theta) = j\tanh(\theta)$$

10. The advantage of the exponential form is its simplicity. For circular motion, at constant angular velocity, it can represent the motion both in the real and imaginary (x and y) directions in one simple expression $r\,e^{j\omega t}$. The rules for multiplication and powers of complex numbers in exponential form are given by the rules of powers, as for any other number, given in Chapter 4 of the Background Mathematics Notes on the companion website for this book.

   Multiplication:

   $$r_1\,e^{j\theta_1} r_2\,e^{j\theta_2} = r_1 r_2\,e^{j(\theta_1 + \theta_2)}$$

   that is, we multiply the moduli and add the arguments.

   Division:

   $$\frac{r_1\,e^{j\theta_1}}{r_2\,e^{j\theta_2}} = \frac{r_1}{r_2}e^{(\theta_1 - \theta_2)}$$

   that is, we divide the moduli and subtract the arguments.

   Powers:

   $$(r\,e^{j\theta})^n = r\,e^{jn\theta}$$

   This last relationship can be used to show De Moivre's theorem. Using the expression for the complex number in terms of a sine and cosine, $r\,e^{j\theta} = r(\cos(\theta) + j\sin(\theta))$, and using this in the expression above, we get

   $$(r(\cos(\theta) + j\sin(\theta))^n = r^n(\cos(n\theta) + j\sin(n\theta))$$

   The complex conjugate of $r\,e^{j\theta}$ is $r\,e^{-j\theta}$.

   The derivative of a complex exponential is easy to find. As

   $$\frac{d}{dt}(e^t) = e^t$$

   therefore

   $$\frac{d}{dt}(e^{j\omega t}) = j\omega e^{j\omega t}.$$

10.10 Exercises

10.1. Given $z_1 = 1 - 2j, z_2 = 3 + 3j$, and $z_3 = -1 + 4j$.
  (a) Represent z_1, z_2, and z_3, on an Argand diagram.
  (b) Find the following and show the results on the Argand diagram

    (i) $z_1 + z_2$ (ii) $z_3 - z_1$ (iii) z_1^*

  (c) Calculate

    (i) $z_1 + z_2 + z_3$ (ii) $z_1 - z_3 + z_2$ (iii) $z_1 z_1^*$
    (iv) z_1/z_2 (v) $z_1 z_3$.

10.2. Simplify

  (a) j^8 (b) j^{11} (c) j^{28}

10.3. Find each of the following complex numbers in the form $a + jb$, where a and b are real:

  (a) $(3 - 7j)(2 + j4)$ (b) $(-1 + 2j)^2$ (c) $\dfrac{4 - j3}{5 - j}$

  (d) $\dfrac{5 + j3}{j(4 - j9)} - \dfrac{6}{j}$

10.4. Find the real and imaginary parts of $z^2 + 1/z^2$, where $z = (3 + j)/(2 - j)$

10.5. Given that x and y are real and that $2x - 3 + j(y - x) = x + j2$, find x and y.

10.6. Find the roots x_1 and x_2 of the following quadratic equations. In each case, find the product $(x - x_1)(x - x_2)$ and show that the original equation is equivalent to $(x - x_1)(x - x_2) = 0$.

  (a) $x^2 - 3x + 2 = 0$ (b) $-6 + 2x - x^2 = 0$
  (c) $3x^2 - x + 1 = 0$ (d) $4x^2 - 7x - 2 = 0$
  (e) $2x^2 + 3 = 0$.

10.7. The equation $x^2 + bx + c = 0$ where b and c are real numbers, has one complex root, $x = -1 + j3$.

  (a) What is the other root?
  (b) Find b and c.

10.8. Convert the following to polar form:

  (a) $3 + j5$ (b) $-6 + j3$
  (c) $-4 - j5$ (d) $-5 - j3$.

10.9. Express in rectangular (Cartesian) form

  (a) $5\angle 225°$ (b) $4\angle 330°$
  (c) $2\angle 2.723$ (d) $5\angle -0.646$.

10.10. If x and y are real and $2x + y + j(2x - y) = 15 + j6$, find x and y

10.11. If $z_1 = 12\angle 3\pi/4$ and $z_2 = 3\angle 2\pi/5$, find:

  (a) $z_1 z_2$ (b) z_1/z_2 (c) $z_1 + z_2$
  (d) $z_2 - z_1$ (e) z_1^* (f) z_2^2.

  giving the results in polar form.

10.12. If $z = 2\angle 0.8$, find z^4.

10.13. Find the impedance of the circuit shown in Figure 10.23(a) at 90 kHz, where $L = 4\,\text{mH}$, $C = 2\,\text{pF}$, and $R = 400\,\text{k}\Omega$. Assuming a current source of amplitude 5 A, calculate the voltage V and its relative phase.

10.14. Find the admittance of the circuit given in Figure 10.23(b) at 20 kHz given that $R = 250\,\text{k}\Omega$, $L = 20\,\text{mH}$, and $C = 50\,\text{pF}$. Given that the voltage source has amplitude 10 V find the current, I, and its relative phase.

10.15. Feedback is applied to an amplifier such that

$$A' = \frac{A}{1 - \beta A}$$

  where A', A, and β are complex quantities. A is the amplifier gain, A' the gain with feedback, and β the

Figure 10.23 *(a) Circuit for Exercise 10.13. (b) Circuit for Exercise 10.14.*

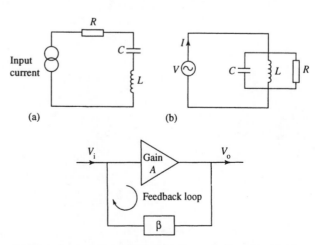

Figure 10.24 *An amplifier with feedback, as in Exercise 10.15.*

proportion of the output which has been fed back (see Figure 10.24)

(a) If at 30 Hz $A = 500\angle 180°$ and $\beta = 0.005\angle 160°$, calculate A'.

(b) At a particular frequency it is desired to have $A' = 300\angle 100°$ where it is known that $A = 400\angle 110°$. Find the value of β necessary to achieve this gain modification.

10.16. Write the following in exponential form:

(a) $3 + j5$ (b) $-6 + j3$
(c) $-4 - j5$ (d) $8\angle 22°$
(e) $3\angle -4.15$ (f) $6(\cos(1.9) + j\sin(1.9))$

10.17. Express the following in polar form and in the form $a + jb$:

(a) $4\,e^{j2}$ (b) $e^{-j(\pi/2)}$
(c) $2\,e^{-j\pi}$ (d) $-6\,e^{j4}$
(e) $\dfrac{1}{2}\,e^{-j5}$ (f) $e^{j(\pi/6)}\,3\,e^{j3(\pi/4)}$
(g) $e^{j(\pi/6)} + 3\,e^{j(3\pi/4)}$

10.18. Find the real and imaginary parts of the following:

(a) $2\,e^{-j\pi}$ (b) $-3\,e^{j0.5}$ (c) $2.5\,e^{-2+j}$
(d) $5\,e^{j(3+j)}$ (e) $(3 - j4)^{2+j}$

10.19. Given $z_1 = 12\,e^{j(3\pi/4)}$ and $z_2 = 3\,e^{j(2\pi/5)}$, find:

(a) $z_1 z_2$ (b) z_1/z_2 (c) $z_1 + z_2$ (d) $z_2 - z_1$
(e) z_1^* (f) z_1^*/z_2 (g) $z_1 z_1^*$

giving the results in exponential form.

10.20. If $z = 3\,e^{j0.46}$, find z^3 in polar and exponential forms.

10.21. Find all the solutions of the following and show them on an Argand diagram

(a) $z^4 = 1$ (b) $z^6 = -1$
(c) $z^5 + 32 = 0$ (d) $3z^3 + 2 = 0$.

10.22. Find $\cos(3\theta)$ in powers of $\sin(\theta)$ and $\cos(\theta)$.

10.23. Express $\sin^3(\theta)$ in terms of sines of multiples of θ.

10.24. Find the fifth roots of $-2 + j3$ and represent the results on an Argand diagram.

10.25. Solve the following equations:

(a) $z^2 + 2jz - 2 = 0$ (b) $z^2 - 3jz = j$.

10.26. Show that the following are solutions to the differential equation $d^2z/dt^2 = -\omega^2 z$ and find the value of ω in each case:

(a) $z = 2\,e^{-j4t}$ (b) $z = 4\,e^{j0.5t}$.

11 Maxima and minima and sketching functions

11.1 Introduction

Differentiation can be used to examine the behaviour of a function and find regions where it is increasing or decreasing, and where it has maximum and minimum values. For instance, we may be interested in finding the maximum height, maximum power, or generating the maximum profit, or in finding ways to use the minimum amount of energy or minimum use of materials. Maximum and minimum points can also help in the process of sketching a function.

11.2 Stationary points, local maxima and minima

Example 11.1 Throw a stone in the air and initially it will have a positive velocity as the height, s, increases; that is, $ds/dt > 0$. At some point it will start to fall back to the ground, the distance from the ground is then decreasing, and the velocity is negative, $ds/dt < 0$. In order to go from a positive velocity to a negative velocity there must be a turning point, where the stone is at its maximum height and the velocity is zero. If the stone has initial velocity $20 \, \text{ms}^{-1}$, how can we find the maximum height that it reaches?

In order to express the velocity of the stone we can make the assumption that air resistance is negligible and use the relationship between distance and time for motion under constant acceleration, giving

$$s = ut + \tfrac{1}{2}at^2$$

where s is the distance travelled, u the initial velocity, t is time, and a the acceleration. In this case, $u = 20 \, \text{ms}^{-1}$ and $a = -g$ (acceleration due to gravity $\approx 10 \, \text{ms}^{-2}$), so $s = 20t - 5t^2$.

At the maximum height, the rate of change of distance with time must be 0, that is, the velocity is 0. Therefore, we differentiate to find the velocity:

$$v = \frac{ds}{dt} = 20 - 10t$$

Putting $v = 0$ gives

$$0 = 20 - 10t \quad \Leftrightarrow \quad 10t = 20 \quad \Leftrightarrow \quad t = 2$$

We have shown that the maximum height is reached after 2 s. But what is that height? Substituting $t = 2$ into the equation for s gives

$$s = 20(2) - 5(2)^2 = 20\,\text{m}$$

giving the maximum value of s as 20 m.

This example illustrates the important step in finding maximum and minimum values of a function, $y = f(x)$. That is, we differentiate and solve

$$\frac{\mathrm{d}y}{\mathrm{d}x} = 0$$

This may give various values of x. The points where $\mathrm{d}y/\mathrm{d}x = 0$ are called the stationary points but having found these we still need a way of deciding whether they could be maximum or minimum values. In the example, we knew that a stone thrown into the air must reach a maximum height and then return to the ground, and so by solving $\mathrm{d}s/\mathrm{d}t = 0$ we would find the time at the maximum. Other problems may not be so clear cut and thus we need a method of distinguishing between different types of stationary points.

A stationary point is classified as either a local maximum, a local minimum, or a point of inflexion. The plural of maximum is maxima and the plural of minimum is minima. The word 'local' is used in the description, because local maxima or local minima do not necessarily give the overall maximum or minimum values of the function. For instance, in Figure 11.1 there is a local maximum at B, but the value of y at $x = x_1$ is actually bigger; hence, the overall maximum value of the function in the range is given by y at x_1.

To see how to classify stationary points, examine Figure 11.1, where points A, B, and C are all stationary points.

In order to analyse the slope of the function, imagine the function as representing the cross-section of a mountain range and we are crossing it from left to right.

At points A, B, and C in Figure 11.1, the gradient of the tangent to the curve is zero, that is, $\mathrm{d}y/\mathrm{d}x = 0$.

At A there is a local minimum, where the graph changes from going downhill to going uphill.

At B there is a local maximum, where the graph changes from going uphill to going downhill.

Figure 11.1 *A graph of some function $y = f(x)$ plotted from $x = x_1$ to $x = x_2$. Points A, B, and C in the graph are stationary points. They are points where the gradient of the tangent to the curve is zero, that is, $\mathrm{d}y/\mathrm{d}x = 0$.*

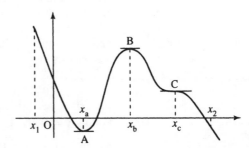

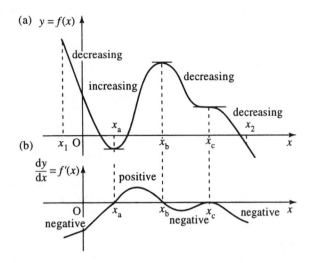

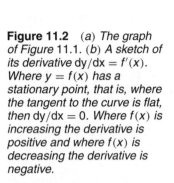

Figure 11.2 (a) The graph of Figure 11.1. (b) A sketch of its derivative dy/dx = $f'(x)$. Where $y = f(x)$ has a stationary point, that is, where the tangent to the curve is flat, then dy/dx = 0. Where $f(x)$ is increasing the derivative is positive and where $f(x)$ is decreasing the derivative is negative.

At C there is a point of inflexion, where the graph goes flat briefly before resuming its descent.

Local maxima and minima are also called turning points because the function is changing from increasing to decreasing or vice versa.

By looking at where the graph is going uphill, that is, $dy/dx > 0$, at where it is going downhill, that is, $dy/dx < 0$, and especially remembering to mark the points where $dy/dx = 0$, the stationary points, we can draw a very rough sketch of the derivative of any function just by looking at its graph. This we do for our example graph in Figure 11.2. By examining the graph of dy/dx we can see that at a local minimum point $dy/dx = 0$ and dy/dx is negative just before the minimum and positive afterwards. For a local maximum point $dy/dx = 0$ and dy/dx is positive just before the maximum and negative afterwards.

If at the point where $dy/dx = 0$ and the derivative has the same sign on either side of the stationary point then it must be a point of inflexion. (At point C, dy/dx is negative just before and just after $x = c$.)

Analysing the sign of dy/dx on either side of a stationary point is one way of classifying whether it is a maximum, minimum, or point of inflexion. Another, sometimes quicker way, is to use the derivative of the derivative, the second derivative, d^2y/dx^2, also referred to as $f''(x)$.

To understand how to use the second derivative, examine Figure 11.2(b). We can see that at x_a, $f'(x)$ is heading uphill; hence, its slope, $f''(x)$, is positive. At x_b, $f'(x)$ is heading downhill; hence, its slope, $f''(x)$, is negative. At x_c, $f'(x)$ has zero slope and therefore $f''(x)$ is 0.

We can now summarize the steps involved in finding and classifying the stationary points of a function $y = f(x)$ and in finding the overall maximum or minimum value of a function.

Step 1   Find the values of x at the stationary points. First, find dy/dx and then solve for x such that $dy/dx = 0$.

Step 2   To classify the stationary points there is a choice of method, although Method 2 does not always give a conclusive result.

   Method 1   For each of the values of x found in Step 1, find out whether the derivative is positive or negative just before the stationary point and just after the stationary point. The classifications are summarized below.

dy/dx before	→	0	→	dy/dx after	Type of stationary point
+	→	0	→	−	Maximum point
−	→	0	→	+	Minimum point
+	→	0	→	+	Point of inflexion
−	→	0	→	−	Point of inflexion

These results are summarized in Figure 11.3.

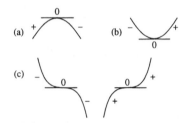

Figure 11.3 *Distinguishing stationary points:* (a) *a maximum point;* (b) *a minimum point;* (c) *two points of inflexion. The sign of the slope of the curve, given by* dy/dx, *are marked on either side of the stationary point.*

Method 2 Find the second derivative, d^2y/dx^2 and substitute in turn each of the values of x found in Step 1 (the values of x at the stationary points).

If d^2y/dx^2 is negative, this indicates that there is a maximum point.

If d^2y/dx^2 is positive, this indicates that there is a minimum point.

However, if d^2y/d$x^2 = 0$, then the test is inconclusive and we must revert to Method 1.

Step 3 Find the values of y at the maximum and minimum points to give the co-ordinates of the turning points.

Step 4 The overall maximum or minimum values of the function can be found in the following way, as long the function is continuous over the values of x of interest. Substitute the boundary values of x into the function to find the corresponding values for y. Compare the values of y found in Step 3 to these boundary values to find the overall maximum and minimum.

Example 11.2   Find and classify the stationary points of $y = x^3 - 9x^2 + 24x + 3$ and find the overall maximum and minimum value of the function in the range $x = 0$ to $x = 5$.

Solution
Step 1. First, we must solve dy/d$x = 0$

$$y = x^3 - 9x^2 + 24x + 3 \implies \frac{dy}{dx} = 3x^2 - 18x + 24$$

So we put

$$3x^2 - 18x + 24 = 0$$

$$\Leftrightarrow \quad x^2 - 6x + 8 = 0 \quad \text{(dividing by 3)}$$

$$\Leftrightarrow \quad (x - 2)(x - 4) = 0$$

(factorizing to find the roots, although we could also use the formula for solving a quadratic equation if no factorization can easily be found)

$$\Leftrightarrow \quad x - 2 = 0 \quad \text{or} \quad x - 4 = 0$$
$$\Leftrightarrow \quad x = 2 \quad \text{or} \quad x = 4$$

Therefore, the stationary points occur when $x = 2$ or $x = 4$.

Step 2. To classify these we use Method 2 as outlined above and look at the second derivative, that is, we differentiate dy/dx

$$\frac{dy}{dx} = 3x^2 - 18x + 24 \quad \Rightarrow \quad \frac{d^2y}{dx^2} = 6x - 18$$

At $x = 2$, $d^2y/dx^2 = 12 - 18 = -6$, which is negative, showing that at $x = 2$, $f''(x)$ is negative and we therefore have a local maximum.

At $x = 4$, $d^2y/dx^2 = 24 - 18 = 6$, which is positive, showing that at $x = 4$, $f''(x)$ is positive and we therefore have a local minimum.

Step 3. We still need to know the function value, the value of y at these stationary points. To find the value of y we substitute into the original expression. At $x = 2$, we get

$$y = (2)^3 - 9(2)^2 + 24(2) + 3 = 8 - 36 + 48 + 3 = 23$$

Therefore, the local maximum occurs at the point (2,23). At $x = 4$, we get

$$y = (4)^3 - 9(4)^2 + 24(4) + 3 = 64 - 144 + 96 + 3 = 19$$

Hence, the local minimum occurs at the point (4,19).

Step 4. To find the overall maximum value and minimum value, substitute the boundary values for x. These are given as $x = 0$ and $x = 5$.

At $x = 0$, $y = (0)^3 - 9(0)^2 + 24(0) + 3 = 0 - 0 + 0 + 3 = 3$
At $x = 5$, $y = (5)^3 - 9(5)^2 + 24(5) + 3 = 125 - 225 + 120 + 3 = 23$
So the boundary points are $(0, 3)$ and $(5, 23)$.

Comparing the numbers 3 and 23 with the values of the function at the maximum and minimum in Step 3, that is, 23 and 19, we can see that the overall maximum value occurs at $x = 5$ and $x = 2$, where $y = 23$. The overall minimum value occurs at $x = 0$, where $y = 3$. These findings are confirmed by the sketch of the function, which we now have sufficient information to make, as in Figure 11.4.

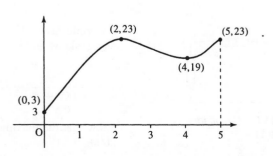

Figure 11.4 *Sketch of* $y = x^3 - 9x^2 + 24x + 3$.

Example 11.3   Find and classify the stationary points of $y = -(2-x)^4$.

Solution
  Step 1

$$y = -(2 - x)^4 \quad \Rightarrow \quad \frac{dy}{dx} = 4(2 - x)^3$$

Stationary points occur where $dy/dx = 0$:

$$4(2 - x)^3 = 0 \quad \Leftrightarrow \quad x = 2$$

Step 2. To classify this stationary point, we differentiate again:

$$\frac{d^2 y}{dx^2} = -12(2 - x)^2$$

At $x = 2$, we get $d^2y/dx^2 = -12(2 - 2)^2 = 0$. So the second derivative is zero.

We cannot use the second derivative test to classify the stationary point because a zero value is inconclusive, so we go back to the first derivative and examine its sign at a value of x just less than $x = 2$ and just greater than $x = 2$. This can be done with the help of a table. Choose any values of x less than $x = 2$ and greater than $x = 2$, and here we choose $x = 1$ and $x = 3$. Be careful if the function is discontinuous at any point not to cross the discontinuity

x	1	2	3
$dy/dx = 4(2 - x)^3$	4	0	-4

At $x = 1, dy/dx = 4(2 - 1)^3 = 4$ (positive); at $x = 3, dy/dx = 4(2-3)^3 = -4$ (negative). Therefore, near the point $x = 2$ the derivative goes from positive to zero to negative. Therefore, the graph of the function goes from travelling uphill to travelling downhill, showing that we have a maximum value.

Step 3. Finally, we find the value of the function at the maximum point. At $x = 2, y = 0$, that is, there is a maximum at $(2, 0)$.

Applications of maximum and minimum values of a function

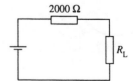

2000 Ω

R_L

Figure 11.5 *Circuit for Example* 11.4.

Example 11.4   The power delivered to the load resistance R_L for the circuit shown in Figure 11.5 is defined by

$$P = \frac{25 R_L}{(2000 + R_L)^2}$$

Show that the maximum power delivered to the load occurs for $R_L = 2000$.

Solution For a maximum value $dP/dR_L = 0$

$$P = \frac{25R_L}{(2000 + R_L)^2}$$

Using the quotient rule to find the derivative we get

$$\frac{dP}{dR_L} = \frac{25(2000 + R_L)^2 - 25R_L(2)(2000 + R_L)}{(2000 + R_L)^4}$$

$$\Leftrightarrow \quad \frac{dP}{dR_L} = \frac{(2000 + R_L)(50\,000 + 25R_L - 50R_L)}{(2000 + R_L)^4}$$

As R_L is positive $2000 + R_L$ is non-zero so we can cancel the common factor of $2000 + R_L$. Also simplifying the top line gives

$$\frac{dP}{dR_L} = \frac{50\,000 - 25R_L}{(2000 + R_L)^3}$$

Setting $dP/dR_L = 0$ gives

$$\frac{50\,000 - 25R_L}{(2000 + R_L)^3} = 0$$

multiplying by $(2000 + R_L)^3$, we get

$$50\,000 - 25R_L = 0 \quad \Leftrightarrow \quad R_L = 2000$$

We have shown that there is a stationary value of the function P when $R_L = 2000$ but now we need to check that it is in fact a maximum value. To do this we substitute values above and below $R_L = 2000$ (say $R_L = 1000$ and $R_L = 3000$) into dP/dR_L:

$$\frac{dP}{dR_L} = \frac{50\,000 - 25R_L}{(2000 + R_L)^3}$$

when $R_L = 1000$ the top line is positive and the bottom line is positive, so the sign of dP/dR_L is $+/+$, which is positive. When $R_L = 3000$ the top line is negative and the bottom line is positive, so the sign of dP/dR_L is $-/+$, which is negative.

So the derivative of the power with respect to the load resistance goes from positive to 0 to negative when $R_L = 2000$, indicating a maximum point.

We have shown that the maximum power to the load occurs when $R_L = 2000$.

Example 11.5 A rectangular field is to be surrounded by a fence of length 400 m. What is the dimensions of the field such that it has maximum area?

Solution The field is shown in Figure 11.6. Call the length of the sides a m and b m. Then the area is given by $A = ab$.

The perimeter is given by $P = 2a + 2b$ and as the length of the fence is 400 m we get:

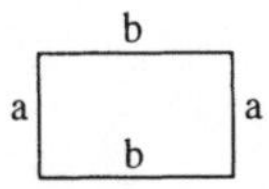

$$400 = 2a + 2b \quad \Leftrightarrow \quad 200 = a + b$$

We wish to find the maximum area, and to do this we need to be able to differentiate A in terms of one of the variables, a or b. We use the

Figure 11.6 *A rectangular field.*

information given about the length of the perimeter to express a in terms of b;

$$a = b - 200$$

and substitute this into the expression for the area, giving

$$A = b(200 - b) = 200b - b^2$$

To find the maximum value for A differentiate with respect to b:

$$\frac{dA}{db} = 200 - 2b$$

and solve

$$\frac{dA}{db} = 0 \quad \Rightarrow \quad 200 - 2b = 0 \quad \Leftrightarrow \quad 2b = 200 \quad \Leftrightarrow \quad b = 100$$

Check that this is indeed a local maximum value by differentiating a second time: $d^2 A/db^2 = -2$, which is negative.

Now we find the length of the other side of the field

$$a = 200 - b = 100$$

Therefore, the field with maximum area is a square of side 100. The area of the field is $100 \times 100 = 10\,000\,\text{m}^2$.

To check that this agrees with the original condition that $2a + 2b = 400$, substitute $a = 100$ and $b = 100$ to get $200 + 200 = 400$, which is correct.

11.3 Graph sketching by analysing the function behaviour

To sketch the graph of any function $y = f(x)$, we first analyse the main features of the function's behaviour. Look at the graph in Figure 11.7 and list the 'important features' of the graph. List these in a way that would enable someone else to sketch the graph from your description.

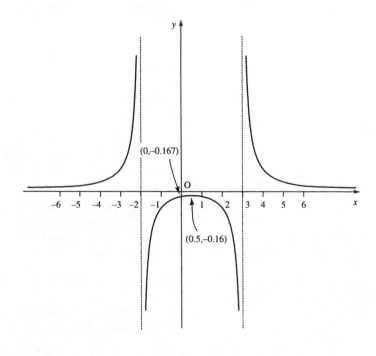

Figure 11.7 *Exercise in graph sketching.*

Here are some of the important features. The list is not exhaustive but should be enough to enable someone to reproduce the graph:

1. The graph is positive for $x < -2$ and $x > 3$ and negative for x between -2 and $+3$. There are no values of x for which y is zero and when $x = 0, y = -0.167$.
2. The graph is discontinuous at $x = -2$ and $x = 3$. y keeps getting larger as x approaches -2 from the left. A more precise way of expressing this is to say that as x tends to -2, with x less than -2, y tends to infinity. The symbol for infinity is ∞ and the symbol for 'tends to' is $\rightarrow$. x tends to -2, with x less than -2 can be expressed more briefly as $x \rightarrow 2^-$. This gives

   as $x \rightarrow -2^-, \ y \rightarrow \infty$.

   Similarly

   as $x \rightarrow -2^+, \ y \rightarrow -\infty$.

   For the discontinuity at $x = 3$ we have:

   as $x \rightarrow 3^-, \quad y \ \rightarrow -\infty$

   as $x \rightarrow 3^+, \quad y \rightarrow \infty$

3. There is a local maximum at $x = 0.5$, where $y = -0.16$.
4. For large values of x, y gets nearer to 0. This can be expressed as

   as $x \rightarrow \infty, \ y \rightarrow 0^+$

   Similarly:

   as $x \rightarrow -\infty, \ y \rightarrow 0^+$

This list of important features indicates the steps that should be taken in order to sketch a graph:

Step 1. Find the value of y when the graph crosses the y-axis; that is, when $x = 0$. If possible find where the graph crosses the x-axis; that is, the values of x where $y = 0$.

Step 2. Find any discontinuities in the function, that is, are there values of x where there is no value of y? Infinite discontinuities (a 'divide by zero') will lead to vertical asymptotes. These are vertical lines which the function approaches but does not meet. We must decide whether the function is positive or negative on either side of the asymptote.

Step 3. Find the co-ordinates of the maxima and minima.

Step 4. Find the behaviour of the function as x tends to plus and minus infinity.

Step 5. Mark these features, found from steps 1 to 4 on the graph and join them, where appropriate, to give the sketch of the graph.

Example 11.6 Sketch the curve whose equation is

$$y = \frac{2x + 1}{(x + 1)(x - 5)}$$

Table 11.1 *Finding the sign on either side of the asymptotes*

x	-2	-1	-0.75	-0.5	4	5	6
$y = 2x + 1/[(x+1)(x-5)]$	-0.4286	Not defined	0.348	0	-1.8	Not defined	0.93
Sign of y	$-$	Not defined	$+$	0	$-$	Not defined	$+$

Solution

Step 1. When $x = 0$:

$$y = \frac{2.0 + 1}{(0+1)(0-5)} = \frac{1}{-5} = -0.2$$

When $y = 0$,

$$\frac{2x + 1}{(x+1)(x-5)} = 0$$

then $2x + 1 = 0 \Leftrightarrow x = -0.5$.

Step 2. If the bottom line of the function were 0 this would lead to a 'divide by zero' which is undefined. This happens when $x + 1 = 0$, that is, $x = -1$ and when $x - 5 = 0$, that is, $x = 5$.

These are infinite discontinuities, that is, y will tend to plus or minus infinity as x approaches these values. The lines at $x = -1$ and $x = 5$ are the asymptotes.

To find the sign of y on either side of the asymptotes substitute values of x on either side of them (avoiding including any values where $y = 0$). This can be done using a table, as in Table 11.1.

Using Table 11.1, we can conclude that as y is negative to the left of the asymptote at $x = -1$ and positive to the right, then

as $x \to -1^-$, $y \to -\infty$

as $x \to -1^+$, $y \to +\infty$

Similarly, as y is negative to the left of the asymptote at $x = 5$ and positive to the right of it, then

as $x \to 5^-$, $y \to -\infty$

as $x \to 5^+$, $y \to +\infty$

Step 3. To find the turning points look for points where $dy/dx = 0$

$$y = \frac{2x + 1}{(x+1)(x-5)}$$

To differentiate, first multiply out the brackets on the bottom line of the expression and then use the formula for finding the derivative of a quotient.

$$y = \frac{2x + 1}{(x+1)(x-5)} \Leftrightarrow y = \frac{2x + 1}{x^2 - 4x - 5}$$

This gives

$$\frac{dy}{dx} = \frac{2(x^2 - 4x - 5) - (2x + 1)(2x - 4)}{(x^2 - 4x - 5)^2}$$

$$\frac{dy}{dx} = \frac{2x^2 - 8x - 10 - 4x^2 + 6x + 4}{(x^2 - 4x - 5)^2}$$

$$\frac{dy}{dx} = \frac{-2x^2 - 2x - 6}{(x^2 - 4x - 5)^2}$$

We now solve $dy/dx = 0$

$$\frac{-2x^2 - 2x - 6}{(x^2 - 4x - 5)^2} = 0$$

$$\Rightarrow -2x^2 - 2x - 6 = 0 \quad \Rightarrow \quad x^2 + x + 3 = 0$$

Using the formula to solve the quadratic equation gives

$$x = \frac{-1 \pm \sqrt{1^2 - 12}}{2} = \frac{-1 \pm \sqrt{-11}}{2}$$

Because of the square root of a negative number in this expression we can see that there are no real solutions; here, there are no turning points.

Step 4. When x is large in magnitude then the highest powers of x on the top and bottom lines of the function expression will dominate. In this case ignore all the other terms. Considering

$$y = \frac{2x + 1}{(x + 1)(x - 5)} = \frac{2x + 1}{x^2 - 4x - 5}$$

For x large in magnitude

$$y \sim \frac{2x}{x^2} = \frac{2}{x}$$

which tends to 0 and is positive as $x \to \infty$ and tends to 0 and is negative as $x \to -\infty$. We can say that as $x \to \infty$, $y \to 0^+$ and as $x \to -\infty$, $y \to 0^-$.

Step 5. Sketch the graph. This is done in two stages, as shown in Figure 11.8.

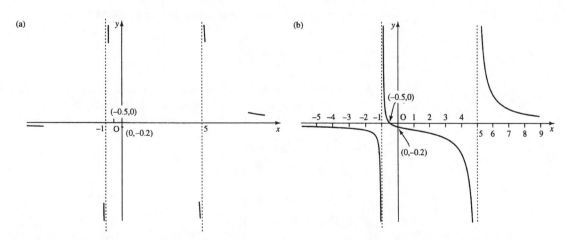

Figure 11.8 *Sketching the graph of $y = (2x + 1)/((x + 1)(x - 5))$: (a) first mark the important points as found in Example* 11.6; *(b) join up these points, where relevant, to give the sketch of the function.*

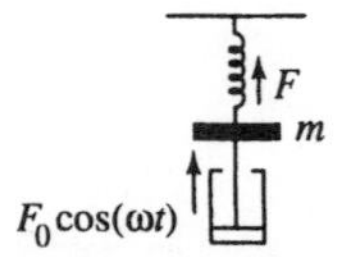

Figure 11.9 *A spring subjected to damped, forced motion.*

Example 11.7 A spring of modulus of elasticity k has a mass, m, suspended from it and is subjected to a oscillating force $F = F_0 \cos(\omega t)$, where $\omega > 0$. The motion of the mass is damped by the use of a dashpot of damping constant c. This is displayed in Figure 11.9. After some time the spring force is found to have oscillations of angular frequency ω and of magnitude

$$F = \frac{kF_0}{\sqrt{\omega^2 c^2 + (\omega^2 m - k)^2}}$$

Taking $m = 1$ and $k = 1$, sketch the graph of F/F_0 against ω for

(a) $c = 2$ (b) $c = \frac{1}{2}$ (c) $c = \frac{1}{4}$ (d) $c = 0$

Solution   We are given that

$$F = \frac{kF_0}{\sqrt{\omega^2 c^2 + (\omega^2 m - k)^2}}$$

Therefore,

$$\frac{F}{F_0} = \frac{k}{\sqrt{\omega^2 c^2 + (\omega^2 m - k)^2}}$$

Substituting $m = 1$ and $k = 1$ gives

$$\frac{F}{F_0} = \frac{1}{\sqrt{\omega^2 c^2 + (\omega^2 - 1)^2}}$$

Call this function H.

One method to solve this problem would be to consider cases (a), (b), (c), and (d) separately. However, it is quicker to leave c as an unknown constant and substitute in for the particular cases later.

Step 1.

$$H = \frac{1}{\sqrt{\omega^2 c^2 + (\omega^2 - 1)^2}}$$

when $\omega = 0, H = 1$. The denominator is a positive square root, which means that H is always $\geqslant 0$ where it is defined.

Step 2. Consider any points where H is not defined

$$H = \frac{1}{\sqrt{\omega^2 c^2 + (\omega^2 - 1)^2}}$$

As the term inside the square root in the expression for H is a sum of squares it is always, $\geqslant 0$. This means that H is always defined, except where the denominator is 0. H is not defined when

$$\sqrt{\omega^2 c^2 + (\omega^2 - 1)^2} = 0$$

$$\Rightarrow \omega^2 c^2 + (\omega^2 - 1)^2 = 0$$

$$\omega^2 c^2 + \omega^4 - 2\omega^2 + 1 = 0$$

$$\omega^4 + \omega^2 (c^2 - 2) + 1 = 0$$

By substituting the values of c of interest we see that in case (a) when $c = 2$, in case (b) when $c = 1/2$, and in case (c) when $c = 1/4$ there are no real solutions of this equation for ω. This means that there are no points where H is undefined. However, for case (d), where $c = 0$ we get the equation

$$\omega^4 - 2\omega^2 + 1 = 0 \iff (\omega^2 - 1)^2 = 0$$

$$\omega^2 - 1 = 0 \iff \omega^2 = 1 \iff \omega = 1 \text{ or } \omega = -1$$

As the frequency of the forcing function is positive then we just have the one value where H is discontinuous, at $\omega = 1$. For case (d), where $c = 0$, H is undefined when $\omega = 1$. This is an infinite 'divide by zero' discontinuity. We have already noted that H is always positive where it is defined and so we know that as $\omega \to 1^-, H \to +\infty$ and as $\omega \to 1^+, H \to +\infty$.

Step 3. To find the stationary points solve $\mathrm{d}H/\mathrm{d}\omega = 0$

$$H = \frac{1}{\sqrt{\omega^2 c^2 + (\omega^2 - 1)^2}}$$

is easier to differentiate if we use the rules of powers to give

$$H = (\omega^2 c^2 + (\omega^2 - 1)^2)^{-1/2}$$

Using the function of a function rule we get

$$\frac{\mathrm{d}H}{\mathrm{d}\omega} = -\frac{1}{2}(\omega^2 c^2 + (\omega^2 - 1)^2)^{-3/2}(2\omega c^2 + 2(\omega^2 - 1)(2\omega))$$

$$\frac{\mathrm{d}H}{\mathrm{d}\omega} = -\frac{2\omega c^2 + 2(\omega^2 - 1)(2\omega)}{2(\omega^2 c^2 + (\omega^2 - 1)^2)^{3/2}} = -\frac{2\omega c^2 + 4\omega^3 - 4\omega}{2(\omega^2 c^2 + (\omega^2 - 1)^2)^{3/2}}$$

Dividing the top and bottom lines by 2 and rearranging the terms on the top line gives

$$\frac{\mathrm{d}H}{\mathrm{d}\omega} = -\frac{2\omega^3 + \omega(c^2 - 2)}{(\omega^2 c^2 + (\omega^2 - 1)^2)^{3/2}}$$

Setting $\mathrm{d}H/\mathrm{d}\omega = 0$ gives

$$-\frac{2\omega^3 + \omega(c^2 - 2)}{(\omega^2 c^2 + (\omega^2 - 1)^2)^{3/2}} = 0$$

multiplying both sides by -1 times the denominator of the left-hand side gives

$$2\omega^3 + \omega(c^2 - 2) = 0$$

$$\iff \quad \omega(2\omega^2 + c^2 - 2) = 0$$

$$\iff \quad \omega = 0 \vee 2\omega^2 + c^2 - 2 = 0$$

$$\iff \quad \omega = 0 \vee 2\omega^2 = 2 - c^2$$

$$\iff \quad \omega = 0 \vee \omega^2 = (2 - c^2)/2$$

$$\iff \quad \omega = 0 \vee \omega = \pm\sqrt{(2 - c^2)/2}$$

As we assume that the frequency is positive (or zero) there are two possibilities and $\omega = 0$ and $\omega = \sqrt{(2 - c^2)/2}$. The second case does not give a real solution if $c = 2$.

We wait until specific values of c are substituted in order to analyse the type of stationary points and the value of H at these points. Also note that this analysis is not valid for the case $c = 0$ as this completely changes the nature of the function H.

Step 4. When ω is large in magnitude

$$H \sim \frac{1}{\sqrt{\omega^4}}$$

and this tends to 0 for large ω. Therefore, as $\omega \to \infty$, $H \to 0^+$.

Step 5. Now we can sketch the graph using the information found and after substituting the various values of c.

Case (a): $c = 2$. On substituting $c = 2$, we have

$$H = \frac{1}{\sqrt{4\omega^2 + (\omega^2 - 1)^2}}.$$

The graph passes through $(0, 1)$ (Step 1), and the function is defined for all $\omega \geqslant 0$ (Step 2). The stationary point is at $\omega = 0$. In this case

$$\frac{dH}{d\omega} = -\frac{2\omega^3 + 2\omega}{(4\omega^2 + (\omega^2 - 1)^2)^{3/2}} = -\frac{2\omega(\omega^2 + 1)}{(4\omega^2 + (\omega^2 - 1)^2)^{3/2}}$$

This is positive for $\omega < 0$ and negative for $\omega > 0$ and therefore there is a maximum value at $\omega = 0$.

Case (b): $c = \frac{1}{2}$. Here

$$H = \frac{1}{\sqrt{\omega^2/4 + (\omega^2 - 1)^2}}$$

The graph passes through $(0, 1)$ (Step 1), and there are no discontinuities (Step 2). The stationary points are at $\omega = 0$ and $\omega = \sqrt{(2 - c^2)/2} = \sqrt{7/8} \approx 0.935$ (Step 3):

$$\frac{dH}{d\omega} = \frac{-\omega(2\omega^2 - 7/4)}{(\omega^2/4 + (\omega^2 - 1)^2)^{3/2}}$$

$dH/d\omega < 0$ for $\omega < 0$ and $dH/d\omega > 0$ for $\omega > 0$; therefore, there is a minimum value at $\omega = 0$,

$dH/d\omega > 0$ for ω just less than $\sqrt{7/8}$,

$dH/d\omega < 0$ for ω just greater than $\sqrt{7/8}$.

Therefore there is a maximum value of H at $\omega = \sqrt{7/8} \approx 0.935$.
At the maximum

$$H = \frac{1}{\sqrt{7/32 + 1/64}} = \frac{1}{\sqrt{15/64}} \approx 2.06$$

Case (c): $c = \frac{1}{4}$.

$$H = \frac{1}{\sqrt{\omega^2/16 + (\omega^2 - 1)^2}}.$$

The graph passes through $(0, 1)$ (Step 1) and there are no discontinuities (Step 2). The stationary points are at $\omega = 0$ and $\omega = \sqrt{(2 - c^2)/2} = \sqrt{31/32} \approx 0.98$. There is a minimum value at $\omega = 0$ and a maximum at $\omega = \sqrt{31/32}$, where

$$H = \frac{1}{\sqrt{31/512 + 1/1024}} \approx 4.03$$

Case (d): $c = 0$

$$H = \frac{1}{\sqrt{(\omega^2 - 1)^2}} = \begin{cases} \dfrac{1}{1 - \omega^2} & \text{for } \omega < 1 \\ \dfrac{1}{\omega^2 - 1} & \text{for } \omega > 1 \end{cases}$$

The graph passes through $(0, 1)$ (Step 1). There is an infinite discontinuity at $\omega = 1$, and for

$$\omega \to 1^-, \quad H \to +\infty$$

$$\omega \to 1^+, \quad H \to +\infty$$

The stationary points have to be analysed separately as the general case always assumed $c > 0$. Differentiating H we get

$$\frac{\mathrm{d}H}{\mathrm{d}\omega} = \frac{2\omega}{(\omega^2 - 1)^2} \quad \omega < 1$$

$$\frac{\mathrm{d}H}{\mathrm{d}\omega} = \frac{-2\omega}{(\omega^2 - 1)^2} \quad \omega > 1$$

which has a zero value at $\omega = 0$.

$\mathrm{d}H/\mathrm{d}\omega$ is negative for ω just less than 0 and positive for ω just greater than 0. Therefore, there is a minimum point at $\omega = 0$. We can now sketch the graphs as in Figure 11.10.

11.4 Summary

1. To find and classify stationary values of a function $y = f(x)$, then

   Step 1. find $\mathrm{d}y/\mathrm{d}x$ and solve for x such that $\mathrm{d}y/\mathrm{d}x = 0$
   Step 2. classify the stationary points.
   Method 1. By examining the sign of $\mathrm{d}y/\mathrm{d}x$ near the point, in which case $+ \to 0 \to -$ indicates a local maximum point, $- \to 0 \to +$ indicates a local minimum point, and $+ \to 0 \to +$ or $- \to 0 \to -$ indicates a point of inflexion.
   Method 2. By finding $\mathrm{d}^2y/\mathrm{d}x^2$ at the point.

   If $\mathrm{d}^2y/\mathrm{d}x^2 < 0$ then there is a local maximum point

   If $\mathrm{d}^2y/\mathrm{d}x^2 > 0$ then there is a local minimum point

   However, if $\mathrm{d}^2y/\mathrm{d}x^2 = 0$ then this test is inconclusive and Method 1 must be used instead.

   Substitute the values of x at the stationary points to find the relevant values of y. Local maxima and minima are also called turning points.
2. If the function, defined for a range of values of x, is continuous, then the overall maximum and minimum values can be found by finding the values of y at the local maxima and minima and the values of y at the boundary points. The maximum of all of these is the global maximum and the minimum of all of these is the global minimum value.

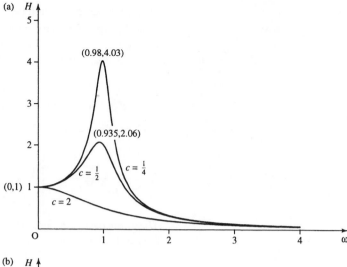

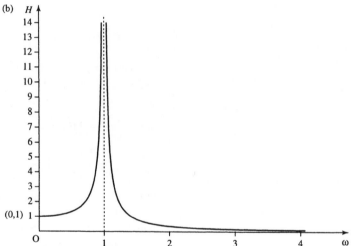

Figure 11.10   (a) The graph
of $H = 1/\sqrt{\omega^2 c^2 + (\omega^2 - 1)^2}$,
as in Example 11.7 for
$c = 2, c = \frac{1}{2}$, and $c = \frac{1}{4}$. (b)
The graph of H for $c = 0$.
$H = 1/(1 - \omega^2)$ for $\omega < 1$ and
$H = 1/(\omega^2 - 1)$ for $\omega > 1$.

3.   There are many practical problems that involve the need to find the
maximum or minimum value of a function.

4.   The stationary values are used when sketching a graph. We also
look for: (a) values where the graph crosses the axes; (b) points of
discontinuity and the behaviour near discontinuities; (c) behaviour
as x tends to $\pm\infty$.

11.5 Exercises

11.1.   Find and classify the stationary points of the following
functions:

(a) $y = x^2 - 5x + 2$ (b) $y = -3x^2 + 4x$

(c) $y = 3x^3 - x$ (d) $x = 2t + \dfrac{200}{t}$

(e) $w = z^4 + 4z^3 - 8z^2 + 2$

11.2.   Find the overall maximum and minimum value of
$x/(2x^2 + 1)$ in the range $x \in [-1, 1]$

11.3.   Sketch the graphs of the following functions:

(a) $y = \dfrac{(x - 3)(x + 5)}{x + 2}$ (b) $y = x + \dfrac{1}{x}$

(c) $y = x^3 - 3x - 1$ (d) $y = \dfrac{(x - 1)(x + 4)}{(x - 2)(x - 3)}$

11.4.   Sketch the graph of $y = 2\sin(x) - \sin(2x)$ for x
between -2π and 2π.

11.5. Find the overall maximum and minimum value of $y = x^3(x - 1)$ in the range $x \in [0, 2]$.

11.6. An open box of variable height and width is to have a length of 3 m. It should not use more than a total of $20\,\mathrm{m}^2$ of surface area. Find the height and width that gives maximum volume.

11.7. An LRC series circuit has an impedance of magnitude

$$Z = \sqrt{R^2 + \left(\omega L - \frac{1}{\omega C}\right)^2}$$

where R is the resistance, L the inductance, C the capacitance and ω is the angular frequency of the voltage source. Sketch Z against ω for the case where $R = 200\,\Omega$, $C = 0.03\,\mu\mathrm{F}$, $L = 2\,\mathrm{mH}$.

11.8. A crank is used to drive a piston as in Figure 11.11. The angular velocity of the crank shaft is the rate of change of the angle θ, $\omega = \mathrm{d}\theta/\mathrm{d}t$. The piston moves horizontally with velocity v_p and acceleration a_p. The crankpin performs circular motion with a velocity of v_c and centripetal acceleration of $\omega^2 r$. The acceleration a_p of the piston varies with θ and is related by

$$a_\mathrm{p} = \omega^2 r \left(\cos(\theta) + \frac{r\cos(2\theta)}{l}\right)$$

where r is the length of the crank and l is the length of the connecting rod. Substituting $r = 150\,\mathrm{mm}$ and $l = 375\,\mathrm{mm}$ find the maximum and minimum values of the acceleration a_p.

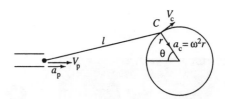

Figure 11.11   *Crank used to drive a piston (Exercise 11.8).*

11.9. A water wheel is constructed with symmetrical curved vanes of angle of curvature θ. Assuming that friction can be taken as negligible, the efficiency, η, that is, the ratio of output power to input power, is calculated as

$$\eta = \frac{2(V - v)(1 + \cos(\theta))v}{V^2}$$

where V is the velocity of the jet of water as it strikes the vane, v is the velocity of the vane in the direction of the jet, and θ is constant. Find the ratio v/V that gives maximum efficiency and find the maximum efficiency.

11.10. Power is transmitted by a fluid of density ρ moving with positive velocity V along a pipeline of constant cross-section area A. Assuming that the loss of power is mainly attributable to friction and that the friction coefficient f can be taken to be a constant, then the power transmitted is given by P, $P = \rho A(hV - cV^3)$, where, g is acceleration due to gravity and h is the head (the energy per unit weight). $c = 4fl/2gd$ where l is the length of the pipe and d is the diameter of the pipe. Assuming h is a constant find the value of V which gives a maximum value for P, and given the input power is $P_i = \rho g A V h$, find the maximum efficiency.

12 Sequences and series

12.1 Introduction

Sequences have two main applications: serving as a digital representation of a signal after analog to digital (A/D) conversion or as a method of solving a numerical problem by getting a sequence of answers, each one being closer to the exact, correct solution.

The advance of digital communications has resulted from the increased accuracy in reproduction of the stored or transmitted signal in digital form. For instance, the reproduction of the stored signal by a compact disc player is far superior to the analog reproduction of the old vinyl records. It is also very convenient to be able to apply filters and other processing techniques in digital form using computers or dedicated microprocessors.

Sequences are often defined in the form of a recurrence relation, special sorts of which are also called difference equations. Recurrence relations can be found which will solve certain problems numerically or they may be derived by modelling the physical processes in a digital system.

The sum of a sequence of terms is called a series. An important example of a series is the Taylor series which can be used to approximate a function. Later in the book, we will look at other examples of series such as z transforms and Fourier series.

Many problems involving sequences and series are solved using a computer. However, it is useful to be able to solve a few simple cases without the aid of a computer, as this can often help check a result for some special cases. Some examples of special sequences are the arithmetic progression and the geometric progression.

12.2 Sequences and series definitions

A sequence is a collection of objects (not necessarily all different) arranged in a definite order. Some examples of sequences are:

1. The numbers 1 to 10, that is, 1, 2, 3, 4, 5, 6, 7, 8, 9, 10
2. Red, Red and Amber, Green, Amber, Red
3. 3, 6, 9, 12, 15, 18.

When a sequence follows some 'obvious' rule then three dots (...) are used to indicate 'and so on', for example, list 1 above may be rewritten as 1, 2, 3, ..., 10.

The examples so far have all been finite sequences. Infinite sequences may use dots at the end, meaning carry on indefinitely in the same fashion, for example,

2, 4, 6, 8, 10, ...
1, 2, 4, 8, 16, ...
4, 9, 16, 25, 36, ...
1, 1, 2, 3, 5, 8, ...

and the (...) indicates that there is no end to the sequence of values and that they carry on in the same fashion.

The elements of a sequence can be represented using letters, for example,

$$a_1, a_2, a_3, a_4, \ldots, a_n, \ldots$$

The first term is called a_1, the second a_2 etc. (Sometimes it is more convenient to say that a sequence begins with a zeroth term, a_0).

If a rule exists by which any term in the sequence can be found then this may be expressed by the 'general term' of the sequence, usually called a_n or a_r. This rule may be expressed in the form of a recurrence relation, giving a_{n+1} in terms of $a_n, a_{n-1}, \ldots$. In this case, it may be quite difficult to find the explicit function definition, that is to solve the recurrence relation. We look at solving recurrence relations in Chapter 14.

A sequence is a function of natural numbers, or integers. The function expression is given by the general term.

Example 12.1 Find the general term of the sequence of numbers from 1 to 10

Solution 1, 2, 3, 4, 5, 6, 7, 8, 9, 10 has the general term $a_n = n$, where $n = 1$ to 10.

We can also write this in 'standard' function notation as

$$a(n) = n, \quad \text{where } n = 1 \text{ to } 10.$$

Check: To check that the correct general term or function expression has been found, reproduce a few of the members of the sequence by substituting values for n in the general term and check that the sequence found is the same as the given values. Wherever n occurs in the function expression or general term replace it by a value.

$$a_n = n$$

for $n = 1$ gives $a_1 = 1$
for $n = 2$ gives $a_2 = 2$, etc.

Example 12.2 Find the general term of the sequence $1, 4, 9, 16, 25, 36, \ldots$ and also define the sequence in terms of a recurrence relation.

Solution Notice that each term in the sequence is a complete square. The first term is 1^2, the second term 2^2, etc. We therefore speculate that the general term is

$$a_n = n^2, \quad \text{where } n = 1 \text{ to } \infty.$$

In function notation this is

$$a(n) = n^2$$

To define the sequence in terms of a recurrence relation means that we must find a way of getting to the $n + 1$th term if we know the nth term. There is no prescribed way of doing this: we merely have to try out a few ideas as to how to see a pattern in the sequence. In this case, we can best see the pattern with the aid of a diagram where we represent the 'square numbers' using a square as in Figure 12.1. Here we can see that to get from 2^2 to 3^2 we need to add a row of two dots and a column of three dots. To get from 3^2 to 4^2 we need to add a row of three dots and a column of four dots. In general, to get from n^2 to $(n + 1)^2$, we need to add a row of n dots and a column of $n + 1$ dots.

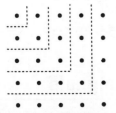

Figure 12.1 *The 'square numbers'.*

As n^2 is a_n and $(n+1)^2$ is a_{n+1}, the rule can be expressed as

$$a_{n+1} = a_n + n + n + 1 \quad \Leftrightarrow \quad a_{n+1} = a_n + 2n + 1.$$

However, we also need to give the starting value in order to define the sequence using a recurrence relation, so we can say that

$$a_{n+1} = a_n + 2n + 1, \quad \text{where } a_1 = 1.$$

Check: To check, we substitute a few values into both the explicit definition and the recurrence relation to see if we correctly reproduce the terms in the sequence.

Substitute $n = 1$, $n = 2$, $n = 3$, $n = 4$, and $n = 5$ into $a_n = n^2$. We get 1, 4, 9, 16, 25, correctly reproducing the first five terms of the sequence.

Substituting $n = 1$, $n = 2$, $n = 3$ and $n = 4$, into $a_{n+1} = a_n + 2n + 1$, where $a_1 = 1$ gives the following.

$n = 1$: $a_2 = a_1 + 2 + 1$; as $a_1 = 1$, this gives $a_2 = 1 + 2 + 1 = 4$, which is correct,

$n = 2$: $a_3 = a_2 + 4 + 1$, as $a_2 = 4$, this gives $a_3 = 4 + 4 + 1 = 9$, which is correct,

$n = 3$: $a_4 = a_3 + 6 + 1$, as $a_3 = 9$, this gives $a_4 = 9 + 6 + 1 = 16$, which is correct,

$n = 4$: $a_5 = a_4 + 8 + 1$, as $a_4 = 16$, this gives $a_5 = 16 + 8 + 1 = 25$, which is correct.

Example 12.3   Find a recurrence relation to define the Fibonacci sequence: $1, 1, 2, 3, 5, 8, 13, 21, \ldots$

Solution   After some trial and error attempts to spot the rule, we should be able to see that the way to get the next number is to add up the last two numbers, so the next term after 13, 21 is $13 + 21 = 34$ and the next is $21 + 34 = 55$, etc.

The recurrence relation is therefore

$$a_{n+1} = a_n + a_{n-1}$$

and because we have two previous values of the sequence used in the recurrence relation then we also need to give two initial values. So we define that $a_1 = 1$ and $a_2 = 1$.

The recurrence relation definition of the Fibonacci sequence is

$$a_{n+1} = a_n + a_{n-1}, \quad \text{where } a_1 = 1 \text{ and } a_2 = 1.$$

Check: Substitute a few values for n into the recurrence relation to see if it correctly reproduces the given values of the sequence

$n = 2$: $a_3 = a_2 + a_1$, where $a_1 = 1$ and $a_2 = 1$, so $a_3 = 1 + 1 = 2$, which is correct,

$n = 3$: $a_4 = a_3 + a_2$, where $a_2 = 1$ and $a_3 = 2$, so $a_4 = 2 + 1 = 3$, which is correct,

$n = 4$: $a_5 = a_4 + a_3$, where $a_4 = 3$ and $a_2 = 1$, so $a_5 = 3 + 2 = 5$, which is correct.

Digital representation of signals

Supposing we would like to give a digital representation of the sine wave of angular frequency 3: $f(t) = \sin(3t)$, then we might choose $a(n) = \sin(3n)$ to give the sequence of values.

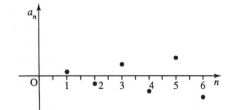

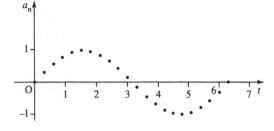

Figure 12.2   *The sequence given by* $a(n) = \sin(3n)$.

Figure 12.3   *The function* $f(t) = \sin(3t)$ *sampled at an interval of* $T = 0.1$, *giving the sequence* $a(n) = \sin(0.3n)$.

Substituting some values for n gives the sequence (to 2 significant figures or s.f.)

$n = 0$: $a(0) = \sin(0) = 0$
$n = 1$: $a(1) = \sin(3) = 0.14$
$n = 2$: $a(2) = \sin(6) = -0.27$
$n = 3$: $a(3) = \sin(9) = 0.41$

This sequence is shown on a graph in Figure 12.2.

This graph does not look like the sine wave it is supposed to represent. This is due to 'undersampling'. $a(n) = \sin(3n)$ is the function $f(t) = \sin(3t)$ sampled at a sampling rate of 1, which is very inadequate to see a good representation. Digital signals are usually expressed in terms of the sampling interval T so that a suitable sampling interval can be chosen. For the function $f(t) = \sin(3t)$, this gives the sequence

$$a(n) = \sin(3Tn) \quad \text{where } n = 0, 1, 2, 3, \ldots$$

The original variable, usually time, t, can be given by $t = Tn$. Choosing $T = 0.1$, for instance, gives

$$a(n) = \sin(3 \times 0.1n) = \sin(0.3n) \quad \text{where } t = nT = 0.1n.$$

Substituting a few values for n gives

$n = 0$: $a(0) = 0$, $t = 0$
$n = 1$: $a(1) = 0.3$, $t = 0.1$
$n = 2$: $a(2) = 0.56$, $t = 0.2$
$n = 3$: $a(3) = 0.78$, $t = 0.3$
$n = 4$: $a(4) = 0.93$, $t = 0.4$
$n = 5$: $a(5) = 1$, $t = 0.5$
$n = 6$: $a(6) = 0.97$, $t = 0.6$.

The values are plotted against t in Figure 12.3.

We can see that the picture in Figure 12.3 is a reasonable representation of the function. The digital representation of $f(t) = \sin(3t)$ is therefore $f(nT) = \sin(3nT)$, where n is an integer.

The problem of undersampling, which we saw in Figure 12.2 leads to a phenomenon called aliasing. Instead of looking like line $\sin(3t)$, Figure 12.2 looks like a sine wave of much lower frequency. This same phenomenon is the one that makes car wheels, pictured on the television, apparently rotate backwards and at the wrong frequency. The television picture is scanned 30 times per second whereas the wheel on the car is probably revolving in excess of 30 times per second. The sample rate is insufficient to give a good representation of the movement of the wheel. The sampling theorem states that the sampling interval must be less than $T = 1/(2f)$ seconds in order to be able to represent a frequency of f hertz.

Example 12.4   Represent the function $y = 4\cos(10\pi t)$ as a sequence using a sampling interval of $T = 0.01$. What is the maximum sampling interval that could be used to represent this signal?

Solution   The digital representation of $y = 4\cos(10\pi t)$ is given by $y(0.01n) = 4\cos(0.1\pi n)$.

Substituting some values for n gives (to 3 s.f.)

n	0	1	2	3	4	5	6	7	8	9	10
y	1	0.951	0.809	0.588	0.309	0	−0.309	−0.588	−0.809	−0.951	−1

The maximum sampling interval that could be used is $1/(2f)$ where f, the frequency in this case, is 5, giving $T = 1/(2 \times 5) = 0.1$.

Example 12.5   A triangular wave of period 2 is given by the function

$$f(t) = t \qquad 0 \leqslant t < 1$$

$$f(t) = 2 - t \quad 1 \leqslant t < 2.$$

Draw a graph of the function and give a sequence of values for $t \geqslant 0$ at a sampling interval of 0.1.

Solution   To draw the continuous function, use the definition $y = t$ between $t = 0$ and 1 and draw the function $y = 2 - t$ in the region where t lies between 1 and 2. The function is of period 2 so that section of the graph is repeated between $t = 2$ and 4, $t = 4$ and 6, etc.

The sequence of values found by using a sampling interval of 0.1 is given by substituting $t = Tn = 0.1n$ into the function definition, giving

$$a(n) = 0.1n \qquad 0 \leqslant 0.1n < 1 \text{ (for n between 0 and 10)}$$

$$a(n) = 2 - 0.1n \quad 1 \leqslant 0.1n < 2 \text{ (for n between 10 and 20)}.$$

The sequence then repeats periodically.

This gives the sequence:

0, 0.1, 0.2, 0.3, 0.4, 0.5, 0.6, 0.7, 0.8, 0.9, 1, 0.9, 0.8, 0.7, 0.6, 0.5, 0.4, 0.3, 0.2, 0.1, 0, 0.1, 0.2, . . .

The continuous function is plotted in Figure 12.4(a) and the digital function in Figure 12.4(b).

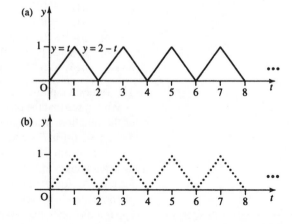

Figure 12.4   *(a) A triangular wave of period 2 given by $f(t) = t, 0 < t \leqslant 1, f(t) = 2 - t, 1 < t < 2$. (b) The function sampled at a sampling interval of 0.1.*

Series

A series is the sum of a sequence of numbers or of functions. If the series contains a finite number of terms then it is a finite series otherwise it is infinite. For example,

$$1 + 2 + 3 + 4 + 5 + 6 + 7 + 8 + 10$$

is a finite series, while

$$1 + 1/2 + 1/4 + 1/8 + 1/16 + \cdots + 1/2^n + \cdots$$

is an infinite series.

To represent series, we may use the sigma notation Σ.

$$\sum_{n=0}^{n=10} \frac{1}{2^n}$$

means 'sum all the terms $1/2^n$ for n from 0 to 10'.

Example 12.6   Express the following series in sigma notation

$$-1 + 4 - 9 + 16 + \cdots + 256.$$

Solution   To write in sigma notation, we need to first express the general term in the sequence. We notice here that the pattern is that each term is a complete square with every other term multiplied by -1. The general term is, therefore

$$(-1)^n n^2.$$

The $(-1)^n$ part of this will just cause the sign of the term to be negative or positive depending on whether n is odd or even.

We can now write

$$-1 + 4 - 9 + 16 + \cdots + 256 = \sum_{n=1}^{n=16} (-1)^n n^2.$$

The limits of the summation are found by considering the value of n to use for the first and last terms. Check that the expression is correct by substituting a few values for n which should recover terms in the original series.

We now look at two commonly encountered types of sequences and series, the arithmetic and geometric progression.

12.3 Arithmetic progression

An arithmetic progression (AP) is a sequence where each term is found by adding a fixed amount on to the previous term. This fixed amount is called the common difference. Some examples of arithmetic progressions are:

1. $-1, 3, 7, 11, 15, 19, 23, 27, \ldots$

Notice that successive terms can be found by adding 4 to the previous term

$$-1 + 4 = 3$$
$$3 + 4 = 7$$
$$7 + 4 = 11$$

$$\vdots$$

showing that the common difference is 4.

2. $25, 15, 5, -5, -15, \ldots$

Notice that successive terms can be found by adding -10 to the previous term

$$25 - 10 = 15$$
$$15 - 10 = 5$$
$$5 - 10 = -5$$
$$-5 - 10 = -15$$

showing that the common difference is -10.

It is not difficult to obtain the recurrence relation for the arithmetic progression. If we call the common difference d, then the $(n+1)$th term can be found by the previous term, the nth, by adding on d, that is

$$a_{n+1} = a_n + d.$$

This can also be expressed using the difference operator, Δ (the Greek capital letter delta) so that $\Delta a_n = d$, where $\Delta a_n = a_{n+1} - a_n$.

If the first term is a and the common difference is d then the sequence is

$$a, a + d, a + 2d, a + 3d, a + 4d, a + 5d, a + 6d, \ldots$$

The second term is $a + d$, the fourth is $a + 3d$, the seventh term is $a + 6d$ and the general term

$$a_n = a + (n - 1)d.$$

Example 12.7 The seventh term of an AP is 11 and the sixteenth term is 29. Find the common difference, the first term of the sequence, and the nth term.

Solution If the first term of the sequence is a and the common difference is d, then the seventh term is given by

$$a_n = a + (n - 1)d \quad \text{with } n = 7$$

so

$$a + 6d = 11. \tag{12.1}$$

Similarly, the sixteenth term is $a + 15d$ and as we are given that this is 29, we have

$$a + 15d = 29. \tag{12.2}$$

Subtracting Equation (12.1) from Equation (12.2) gives $9d = 18 \Leftrightarrow d = 2$, and substituting this into Equation (12.1) gives

$$a + 6 \times 2 = 11 \quad \Leftrightarrow \quad a = 11 - 12 \Leftrightarrow a = -1.$$

That is, the first term is -1 and the common difference is 2. Hence, the nth term is $a + (n - 1)d = -1 + (n - 1)2 = -1 + 2n - 2$ giving, $a_n = 2n - 3$.

Check: To check that the general term is correct for this sequence substitute $n = 7$ giving $a_7 = 2(7) - 3 = 14 - 3 = 11$; substitute $n = 16$ giving $a_{16} = 2(16) - 3 = 29$, which are the values given in the problem.

The sum of *n* terms of an arithmetic progression

There are some simple formulae which can be used to find the sum of the first n terms of an AP. These can be found by writing out all the terms of a general AP from the first term to the last term, ℓ, and then adding on the same series again but this time reversing it. We will begin by finding the sum of 20 terms of an AP with first term 1 and common difference 3, giving the general term as $1 + (n - 1) \times 3$ and the last term as $1 + 19 \times 3$

$$S_{20} = 1 + (1 + 3) + (1 + 2 \times 3) + \cdots + (1 + 18 \times 3) + (1 + 19 \times 3)$$

on reversing, we have

$$S_{20} = (1 + 19 \times 3) + (1 + 18 \times 3) + (1 + 17 \times 3) + \cdots + (1 + 1 \times 3) + 1$$

on adding we have $2S_{20} = (2 + 19 \times 3) + (2 + 19 \times 3) + (2 + 19 \times 3) + \cdots + (2 + 19 \times 3) + (2 + 19 \times 3)$.

Notice that each term in the last line is the same, and is equal to the sum of the first term (1) and the last term $(1 + 19 \times 3)$ giving $2 + 19 \times 3$. As there are 20 terms, we have

$$2S_{20} = 20 \times (2 + 19 \times 3)$$

$$S_{20} = 10 \times (2 + 19 \times 3) = 590.$$

It would obviously be simpler to be able to use a formula to calculate this rather than having to repeat this process for every AP. Therefore, we go through the same process for an AP of first term a and common difference d with last term l.

The sum of the first n terms of an AP is given by:-

$$S_n = a + (a + d) + (a + 2d) + \cdots + (a + (n - 2)d)$$
$$+ (a + (n - 1)d)$$
$$S_n = (a + (n - 1)d) + (a + (n - 2)d) + (a + (n - 3)d)$$
$$+ \cdots + (a + d) + a$$
$$2S_n = (2a + (n - 1)d) + (2a + (n - 1)d) + (2a + (n - 1)d)$$
$$+ \cdots + (2a + (n - 1)d) + (2a + (n - 1)d).$$

Using the fact that there are n terms, we have

$$2S_n = n(2a + (n - 1)d)$$
$$S_n = \frac{n}{2}(2a + (n - 1)d)$$

which gives the first of two formulae that can be used to find the sum of n terms of an AP.

An alternative formula is found by noticing that the sum is the given by the number of terms multiplied by the average term. The average term is half the sum of the first term and the last term: average term $= (a + l)/2$. This gives the sum of n terms as

$$S_n = \frac{n}{2}(a + l).$$

This is the second of the two formulae that may be used to find the sum of the first n terms of an AP.

Example 12.8 Find the sum of an AP whose first term is 3 and has 12 terms ending with -15.

Using the formula $S_n = \dfrac{n}{2}(a+l)$, and then substituting $n = 12$, $a = 3$, and $l = -15$:

$$S_n = \frac{12}{2}(3 - 15) = 6(-12) = -72.$$

Example 12.9 Find

$$\sum_{r=1}^{r=9} \left(1 - \frac{r}{4}\right).$$

Write out the series by substituting values for r

$$\sum_{r=1}^{r=9} \left(1 - \frac{r}{4}\right) = \left(1 - \frac{1}{4}\right) + \left(1 - \frac{2}{4}\right) + \left(1 - \frac{3}{4}\right) + \left(1 - \frac{4}{4}\right)$$

$$+ \left(1 - \frac{5}{4}\right) + \left(1 - \frac{6}{4}\right) + \cdots + \left(1 - \frac{9}{4}\right)$$

$$= \frac{3}{4} + \frac{1}{2} + \frac{1}{4} + 0 - \frac{1}{4} - \frac{1}{2} \cdots - \frac{5}{4}$$

This is the sum of an arithmetic progression with nine terms where $a = \frac{3}{4}$ and $d = -\frac{1}{4}$. Using

$$S_n = \frac{n}{2}(2a + (n-1)d)$$

gives $S_n = \frac{9}{2}\left(2 \times \frac{3}{4} + (9-1)\left(-\frac{1}{4}\right)\right) = \frac{9}{2}\left(\frac{6}{4} - \frac{8}{4}\right) = -\frac{9}{4}$.

12.4 Geometric progression

A geometric progression (GP) is a sequence where each term is found by multiplying the previous term by a fixed number. This fixed number is called the common ratio, r. We have already come across examples of geometric progressions in Chapter 8, where we looked at exponential growth. There we had the example of €1 deposited in a bank with a real rate of growth of 3% so we get the sequence 1, 1.03, 1.09, 1.13, 1.16, 1.19, ... (expressed to the nearest cent) where each year the amount in the bank is multiplied by 1.03.

Some more examples of GPs are

1. $16, 8, 4, 2, 1, 0.5, 0.25, 0.125, \ldots$

Notice that successive terms can be found by multiplying the previous term by 0.5:

$$16 \times 0.5 = 8$$
$$8 \times 0.5 = 4$$
$$4 \times 0.5 = 2$$

$$\vdots$$

showing that the common ratio is 0.5.

2. $1, 3, 9, 27, 81, \ldots$

Notice that successive terms can be found by multiplying the previous term by 3:

$1 \times 3 = 3$

$3 \times 3 = 9$

$9 \times 3 = 27$

$$\vdots$$

showing that the common ratio is 3.

3. $-1, 2, -4, 8, -16, \ldots$

Notice that successive terms can be found by multiplying the previous term by -2:

$(-1) \times (-2) = 2$

$2 \times (-2) = -4$

$(-4) \times (-2) = 8$

$$\vdots$$

showing that the common ratio is -2.

It is not difficult to obtain the recurrence relation for the geometric progression. If we call the common ratio r then the $(n+1)$th term can be found by the previous term, the nth, by multiplying by r, that is

$$a_{n+1} = r a_n.$$

This can also be expressed using the difference operator, Δ, so that $\Delta a_n = (r-1)a_n$, where $\Delta a_n = a_{n+1} - a_n$. If a GP has first term, a, and common ratio, r, then the sequence is

$$a, ar, ar^2, ar^2, ar^4, \ldots, ar^{n-1}, \ldots$$

The second term is ar, the fourth term is ar^3 and the seventh term is ar^6; the general term is given by

$$a_n = ar^{n-1}$$

Example 12.10   Find the general term of the GP
$16, 8, 4, 2, 1, 0.5, 0.25, 0.125, \ldots$

Solution   This GP has first term 16. The common ratio is found by taking the ratio of any two successive terms. Take the ratio of the first two terms (second term divided by the first term) to give

$$r = 8/16 = 0.5$$

The general term is given by $ar^{n-1} = 16(0.5)^{n-1}$.

Example 12.11   A GP has third term 12 and fifth term 48. Find the first term and the common ratio.

Solution   Call the first term a and the common ratio r. We know that the nth term is given by $a_n = ar^{n-1}$. The fact that the third term is 12 gives the equation

$$ar^2 = 12 \tag{12.3}$$

and the fact that the fifth term is 48 gives the equation

$$ar^4 = 48. \tag{12.4}$$

Dividing Equation (12.4) by Equation (12.3) gives

$$r^2 = 4.$$

This means that there are two possible values for the common ratio: either 2 or -2. To find the first term, substitute for r into Equation (18.3), to get

$$ar^2 = 12 \quad \text{and} \quad r = \pm 2 \quad \Rightarrow \quad 4a = 12 \quad \Leftrightarrow \quad a = 3$$

So the first term is 3.

The sum of a geometric progression

Consider the sum of the first six terms of the GP with first term 2 and common ratio 4. To try to find the sum we first write out the original series and then multiply the whole series by the common ratio, as this will reproduce the same terms in the series, only shifted up one place. We can then subtract the two expressions:

$$S_6 = 2 + 8 + 32 + 128 + 512 + 2048$$
$$4S_6 = \qquad 8 + 32 + 128 + 512 + 2048 + 9192$$
$$S_6 - 4S_6 = 2 \qquad\qquad\qquad\qquad\qquad - 9192$$

So

$$S_6 = \frac{2 - 9192}{1 - 4}$$

as $9192 = 2 \times 4^6$, this gives the sum of the first six terms as

$$S_6 = \frac{2(1 - 4^6)}{1 - 4}.$$

Applying this process to a general GP gives a formula for the sum of the first n terms. Consider the sum, S_n, of the first n terms of a GP whose first term is a and whose common ratio is r. Multiply this by r and subtract.

$$S_n = a + ar + ar^2 + \cdots + ar^{n-2} + ar^{n-1}$$
$$rS_n = \qquad ar + ar^2 + \cdots + ar^{n-2} + ar^{n-1} + ar^n$$
$$S_n - rS_n = a \qquad\qquad\qquad\qquad\qquad\qquad - ar^n.$$

This gives

$$S_n(1 - r) = a - ar^n = a(1 - r^n) \quad \Leftrightarrow \quad S_n = \frac{a(1 - r^n)}{1 - r}.$$

If $r > 1$, it may be more convenient to write

$$S_n = \frac{a(r^n - 1)}{r - 1}.$$

Example 12.12 The second term of a GP is 2, and the fifth term is 0.03125. Find the first term, the common ratio, and the sum of the first 10 terms.

Solution Let the first term be a and the common ratio r, then the nth term is ar^{n+1}. The second term is 2 giving the equation

$$ar = 2. \tag{12.5}$$

The fifth term is 0.03125 giving the equation

$$ar^4 = 0.03125. \tag{12.6}$$

Dividing Equation (12.6) by Equation (12.5) gives

$$\frac{ar^4}{ar} = \frac{0.03125}{2} \Leftrightarrow r^3 = 0.015625 \Leftrightarrow r = (0.015625)^{1/3} \Leftrightarrow r = 0.25.$$

Substituting this value for r into Equaton (21.5) gives

$$a(0.25) = 2 \Leftrightarrow a = 2/0.25 \Leftrightarrow a = 8.$$

Therefore, the first term is 8 and the common ratio is 0.25.
   The sum of the first n terms is given by

$$S_n = \frac{a(1 - r^n)}{1 - r}.$$

Substituting $a = 8$, $r = 0.25$, and $n = 10$ gives

$$S_{10} = \frac{a(1 - (0.25)^{10})}{1 - 0.25} = 13.33 \text{ to 4 s.f.}$$

Example 12.13 The general term of a series is given by

$$a_n = \frac{2^{n+1}}{3^n}.$$

Show that the terms of the series form a GP and find the sum of the first n terms.

Solution To show that this is a GP, we must show that consecutive terms have a common ratio. Take two terms, the mth term and the $(m + 1)$th term. The mth term is

$$\frac{2^{m+1}}{3^m} = a_m$$

and the $(m + 1)$th term is found by substituting $n = m + 1$ into the expression for the general term, which gives

$$a_{m+1} = \frac{2^{(m+1)+1}}{3^{(m+1)}} = \frac{2^{m+2}}{3^{m+1}}.$$

We can now spot that $a_{m+1} = \frac{2}{3}a_m$, meaning that the $(m + 1)$th term is found by multiplying the mth term by $\frac{2}{3}$.
   Alternatively, we could divide the $(m + 1)$th term by the mth term

$$\frac{a_{m+1}}{a_m} = \frac{2^{m+2}/3^{m+1}}{2^{m+1}/3^m} = \frac{2^{m+2}}{3^{m+1}} \times \frac{3^m}{2^{m+1}} = \frac{2}{3}$$

giving the common ratio as $\frac{2}{3}$.
   To find the sum of n terms, we need to know the first term. To find this, substitute $n = 1$ into $2^{n+1}/3^n$, giving $2^2/3 = \frac{4}{3}$. Thus, the sum of n terms is given by:

$$\frac{\frac{4}{3}\left(1 - \left(\frac{2}{3}\right)^n\right)}{1 - \frac{2}{3}} = \frac{\frac{4}{3}\left(1 - \left(\frac{2}{3}\right)^n\right)}{\frac{1}{3}} = 4\left(1 - \left(\frac{2}{3}\right)^n\right).$$

The sum to infinity of a geometric progression

Consider Example 12.13 concerning the series whose general term is $2^{n+1}/3^n$. This can be written as

$$\frac{4}{3} + \frac{4}{3}\left(\frac{2}{3}\right) + \frac{4}{3}\left(\frac{2}{3}\right)^2 + \frac{4}{3}\left(\frac{2}{3}\right)^3 + \frac{4}{3}\left(\frac{2}{3}\right)^4 + \cdots + \left(\frac{2^{n+1}}{3^n}\right) + \cdots$$

and the sum of n terms is $4(1 - (2/3)^n)$.

We can write out this sum for various values of n to 7 s.f. as in Table 12.1.

After 40 terms, the sum has become 4 to 7 s.f. and however many more terms are considered the sum is found to be 4 to 7 s.f. This shows that the limit of the sum is 4 to 7 s.f. The limit of the sum of n terms as n tends to infinity is called the sum to infinity.

We can see that the limit is exactly 4 in this case by looking what happens to $4(1 - (2/3)^n)$ as n tends to infinity.

$$4(1 - (2/3)^n) = 4 - 4(2/3)^n.$$

The second term becomes smaller and smaller as n gets bigger and bigger, and we can see that $(2/3)^n \to 0$ as $n \to \infty$, therefore

$$S_\infty = \lim_{n\to\infty} S_n = \lim_{n\to\infty} (4 - 4(2/3)^n) = 4.$$

This approach can also be applied to the general GP, where

$$S_n = \frac{a(1 - r^n)}{1 - r} = \frac{a}{1 - r} - \frac{ar^n}{1 - r}.$$

If $|r| < 1$, we have

$$\lim_{n\to\infty} (r^n) = 0$$

which gives

$$\lim_{n\to\infty} S_n = \lim_{n\to\infty} \frac{a}{1 - r} - \lim_{n\to\infty} \frac{ar^n}{1 - r} = \frac{a}{1 - r}.$$

We can write

$$S_\infty = \frac{a}{1 - r} \quad |r| < 1.$$

Example 12.14 Find the sum to infinity of a GP with first term -10 and common ratio 0.1.

Table 12.1 *The sum of the first n terms of the GP expressed to 7 s.f., for various values of n*

n	5	10	15	20	25	30	35	40	45	50
$4(1 - (2/3)^n)$	3.47325	3.930634	3.990865	3.998797	3.999842	3.999979	3.999997	4	4	4

Solution   The formula for the sum to infinity gives

$$S_\infty = \frac{a}{1 - r}.$$

Substituting $a = -10$ and $r = 0.1$ gives

$$S_\infty = \frac{-10}{1 - 0.1} = \frac{-10}{0.9} = -11\frac{1}{9}.$$

Example 12.15   Express the recurring decimal $0.0\dot{2} = (0.0222222\ldots)$ as a fraction.

Solution

$$0.0\dot{2} = 0.02 + 0.002 + 0.0002 + \cdots$$

This is the sum to infinity of a GP with first term 0.02 and common ratio 0.1. The sum to infinity is therefore

$$S_\infty = \frac{0.02}{1 - 0.1} = \frac{0.02}{0.9} = \frac{2}{90}$$

giving $0.0\dot{2} = 2/90$.

Example 12.16   Find the sum to infinity of

$$1 + z - z^2 + z^3 - z^4 \cdots$$

where $0 < z < 1$.

Solution   The first term of this series is 1 and the common ratio is $-z$ giving the sum to infinity as

$$S_\infty = \frac{1}{1 - (-z)} = \frac{1}{1 + z}.$$

Note that in the case of a GP with common ratio $|r| \geqslant 1$, the series sum will not tend to a finite limit. For instance, the sum of the GP

$$2 + 4 + 8 + 16 + 32 + \cdots + 2^n$$

gets much larger each time a new term is added. We say that the sum of this series tends to infinity.

12.5 Pascal's triangle and the binomial series

Expressions like $(3 + 2y)^5$ are called binomial expressions. Expanding these expressions can be very tedious as we need to multiply out $(3 + 2y)$ $(3 + 2y)(3 + 2y)(3 + 2y)(3 + 2y)$ term by term. To speed up this process, we analyse the coefficients of the terms in the expansion and find that they make a triangular pattern, called Pascal's triangle.

Pascal's triangle

Consider a simpler case of $(1+x)^5$. To work this out, we would first start with $(1+x)$ and multiply by $(1+x)$ to get $(1+x)^2$. We then multiply $(1+x)^2$ by $(1+x)$ to get $(1+x)^3$, etc. continuing this process gives the following

$$1 + x \qquad\qquad\qquad\qquad\qquad\qquad (1)$$
$$1 + 2x + x^2 \qquad\qquad\qquad\qquad\quad (2)$$
$$1 + 3x + 3x^2 + x^3 \qquad\qquad\qquad (3)$$
$$1 + 4x + 6x^2 + 4x^3 + x^4 \qquad\quad (4)$$
$$1 + 5x + 10x^2 + 10x^3 + 5x^4 + x^5 \quad (5)$$

If we write out the coefficients of each line of this in a triangular form, we get

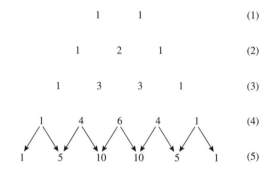

Notice that each line can be found from the line above by adding pairs of numbers, where the outer numbers are always 1. That is, looking at line 5, the first number is 1, the others are found by adding the two numbers above, $5 = 1 + 4$, $10 = 4 + 6$, $10 = 6 + 4$, $5 = 4 + 1$, and the last number is 1.

Example 12.17   Expand $(1 + x)^7$ in powers of x.

Solution   Write out the first seven lines of the Pascal's triangle in order to find the coefficients in the expansion, giving

$$\begin{array}{ll} 1\,1 & (1) \\ 1\,2\,1 & (2) \\ 1\,3\,3\,1 & (3) \\ 1\,4\,6\,4\,1 & (4) \\ 1\,5\,10\,10\,5\,1 & (5) \\ 1\,6\,15\,20\,15\,6\,1 & (6) \\ 1\,7\,21\,35\,35\,21\,7\,1 & (7) \end{array}$$

Now write out the expansion with the powers of x, giving

$$1 + 7x + 21x^2 + 35x^3 + 35x^4 + 21x^5 + 7x^6 + x^7.$$

As we can now easily find expansions of the form $(1+x)^n$, we now move on to the more difficult problem of expressions such as $(1 + 2y)^5$. We can find this by substituting $x = 2y$ into the expression for $(1 + x)^5$

$$(1 + 2y)^5 = 1 + 5(2y) + 10(2y)^2 + 10(2y)^3 + 5(2y)^4 + (2y)^5.$$

Remembering to take the powers of 2 as well as y this gives

$$(1 + 2y)^5 = 1 + 10y + 40y^2 + 80y^3 + 80y^4 + 32y^5.$$

Finally, we look at $(3 + 2y)^5$, and to do this we need to be able to expand expressions like $(a + b)^5$. To find $(a + b)^5$, start by dividing inside the bracket by a to give

$$(a + b)^5 = a^5 \left(1 + \frac{b}{a}\right)^5$$

and substitute $x = b/a$ and use the expansion for $(1 + x)^5$:

$$a^5 \left(1 + \frac{b}{a}\right)^5 = a^5 \left(1 + 5\left(\frac{b}{a}\right) + 10\left(\frac{b}{a}\right)^2\right.$$
$$\left. +10\left(\frac{b}{a}\right)^3 + 5\left(\frac{b}{a}\right)^4 + \left(\frac{b}{a}\right)^5\right).$$

Multiplying out gives

$$(a + b)^5 = a^5 + 5a^4b + 10a^3b^2 + 10a^2b^3 + 5ab^4 + b^5.$$

Notice the pattern on the powers of a and b. The powers of a are decreasing term by term as the powers of b are increasing. Always, the sum of the power of a and power of b is 5.

We can now expand $(3 + 2x)^5$ by using

$$(a + b)^5 = a^5 + 5a^4b + 10a^3b^2 + 10a^2b^3 + 5ab^4 + b^5$$

and substitute $a = 3$ and $b = 2x$, giving

$$(3 + 2x)^5 = (3)^5 + 5(3)^4(2x) + 10(3)^3(2x)^2$$
$$+ 10(3)^2(2x)^3 + 5(3)(2x)^4 + (2x)^5$$
$$= 243 + 810x + 1080x^2 + 720x^3 + 240x^4 + 32x^5.$$

Example 12.18   Expand $(2x - y)^6$.

Solution   Find the sixth row of Pascal's triangle

$$
\begin{array}{ll}
1\ 1 & (1) \\
1\ 2\ 1 & (2) \\
1\ 3\ 3\ 1 & (3) \\
1\ 4\ 6\ 4\ 1 & (4) \\
1\ 5\ 10\ 10\ 5\ 1 & (5) \\
1\ 6\ 15\ 20\ 15\ 6\ 1 & (6)
\end{array}
$$

This gives the expansion of $(a + b)^6$ as

$$a^6 + 6a^5b + 15a^4b^2 + 20a^3b^3 + 15a^2b^4 + 6ab^5 + b^6.$$

Now, substitute $a = 2x$ and $b = -y$

$$(2x - y)^6 = (2x)^6 + 6(2x)^5(-y) + 15(2x)^4(-y)^2 + 20(2x)^3(-y)^3$$
$$+ 15(2x)^2(-y)^4 + 6(2x)(-y)^5 + (-y)^6$$
$$(2x - y)^6 = 64x^6 - 192x^5y + 240x^4y^2 - 160x^3y^3$$
$$+ 60x^2y^4 - 12xy^5 + y^6.$$

Example 12.19 Expand

$$\left(x + \frac{1}{x}\right)^3.$$

Solution Find the third row of Pascal's triangle

$$
\begin{array}{ll}
1\ 1 & (1) \\
1\ 2\ 1 & (2) \\
1\ 3\ 3\ 1 & (3)
\end{array}
$$

This gives the expansion of $(a + b)^3$ as

$$(a + b)^3 = a^3 + 3a^2b + 3ab^2 + b^3.$$

Now, substitute $a = x$ and $b = 1/x$ to give

$$\left(x + \frac{1}{x}\right)^3 = x^3 + 3x^2\left(\frac{1}{x}\right) + 3x\left(\frac{1}{x}\right)^2 + \left(\frac{1}{x}\right)^3$$

$$\left(x + \frac{1}{x}\right)^3 = x^3 + 3x + \frac{3}{x} + \frac{1}{x^3}.$$

Example 12.20 Expand $(e^x - e^{-x})^4$.

Solution Find the fourth row of Pascal's triangle

$$
\begin{array}{ll}
1\ 1 & (1) \\
1\ 2\ 1 & (2) \\
1\ 3\ 3\ 1 & (3) \\
1\ 4\ 6\ 4\ 1 & (4)
\end{array}
$$

This gives the expansion of $(a + b)^4$ as

$$(a + b)^4 = a^4 + 4a^3b + 6a^2b^2 + 4ab^3 + b^4.$$

Now, substitute $a = e^x$ and $b = -e^{-x}$ to give

$$(e^x - e^{-x})^4 = (e^x)^4 + 4(e^x)^3(-e^{-x}) + 6(e^x)^2(-e^{-x})^2$$
$$+ 4(e^x)(-e^{-x})^3 + (-e^{-x})^4$$
$$= e^{4x} - 4e^{2x} + 6 - 4e^{-2x} + e^{-4x}.$$

The binomial theorem

The binomial theorem gives a way of writing the terms which we have found for the binomial expansion without having to write out all the lines of Pascal's triangle to find the coefficients. The rth coefficient in the binomial expansion of $(1 + x)^n$ is expressed by ^nC_r or

$$\binom{n}{r} = \frac{n!}{(n-r)!r!}$$

where '!' is the factorial sign. The factorial function is defined by

$$n! = n(n-1)(n-2)\cdots 1$$

For example,

$$3! = 3 \times 2 \times 1 = 6, \quad 6! = 6 \times 5 \times 4 \times 3 \times 2 \times 1 = 720.$$

The binomial expansion then gives

$$(1 + x)^n = 1 + \binom{n}{1}x + \binom{n}{2}x^2 + \binom{n}{3}x^3 + \binom{n}{4}x^4$$
$$+ \cdots + \binom{n}{r}x^r + \cdots + x^n$$

and

$$(a + b)^n = a^n + \binom{n}{1}a^{n-1}b + \binom{n}{2}a^{n-2}b^2 + \binom{n}{3}a^{n-3}b^3$$
$$+ \binom{n}{4}b^4 + \cdots + \binom{n}{r}a^{n-r}b^r + \cdots + b^n.$$

This can also be written as

$$(1 + x)^n = 1 + nx + \frac{n(n-1)}{2!}x^2 + \frac{n(n-1)(n-2)}{3!}x^3 + \cdots + x^n$$

and the expansion for $(a + b)^n$ becomes

$$(a + b)^n = a^n + na^{n-1}b + \frac{n(n-1)}{2!}a^{n-2}b^2$$
$$+ \frac{n(n-1)(n-2)}{3!}a^{n-3}b^3 + \cdots + b^n.$$

Example 12.21   Expand $(1 + x)^4$.

Solution   Using the binomial expansion

$$(1 + x)^n = 1 + nx + \frac{n(n-1)}{2!}x^2 + \frac{n(n-1)(n-2)}{3!}x^3 + \cdots + x^n.$$

Substituting $n = 4$ gives

$$(1 + x)^4 = 1 + 4x + \frac{4(3)}{2!}x^2 + \frac{4(3)(2)}{3!}x^3 + x^4$$
$$= 1 + 4x + 6x^2 + 6x^3 + x^4.$$

Example 12.22   Expand

$$\left(2 - \frac{1}{x}\right)^5.$$

Solution   Find the binomial expansion of $(a + b)^n$

$$(a + b)^n = a^n + na^{n-1}b + \frac{n(n-1)}{2!}a^{n-2}b^2$$
$$+ \frac{n(n-1)(n-2)}{3!}a^{n-3}b^3 + \cdots + b^n.$$

This gives the expansion of $(a + b)^5$ as

$$(a + b)^5 = a^5 + 5a^4b + \frac{5(4)}{2!}a^3b^2 + \frac{5(4)(3)}{3!}a^2b^3$$
$$+ \frac{5(4)(3)(2)}{4!}ab^4 + b^5$$
$$= a^5 + 5a^4b + 10a^3b^2 + 10a^2b^3 + 5ab^4 + b^5.$$

Now, substitute $a = 2$ and $b = -1/x$ to give

$$= (2)^5 + 5(2)^4(-1/x) + 10(2)^3(-1/x)^2 + 10(2)^2(-1/x)^3$$
$$+ 5(2)(-1/x)^4 + (-1/x)^5$$
$$= 32 - \frac{80}{x} + \frac{80}{x^2} - \frac{40}{x^3} + \frac{10}{x^4} - \frac{1}{x^5}.$$

Example 12.23   Find to 4 s.f without using a calculator: $(2.95)^4$

Solution   Write $2.95 = 3 - 0.05$ so we need to find $(3 - 0.05)^4$. Using the expansion

$$(a + b)^4 = a^4 + 4a^3b + \frac{4(3)}{2!}a^2b^2 + \frac{4(3)(2)}{3!}ab^3 + b^4.$$

Substitute $a = 3$ and $b = -0.05$:

$$(3 - 0.5)^4 = (3)^4 + 4(3)^3(-0.5) + \frac{4(3)}{2!}(3)^2(-0.5)^2$$
$$+ \frac{4(3)(2)}{3!}(3)(-0.5)^3 + (-0.5)^4$$
$$= 81 - 5.4 + 0.135 - 0.0015 + 0.00000625$$
$$= 75.73 \text{ to 4 s.f.}$$

12.6 Power series

A power series is of the form

$$a_0 + a_1x + a_2x^2 + a_3x^3 + a_4x^4 + \cdots + a_nx^n + \cdots$$

Many functions can be approximated by a power series. To find a series, we use repeated differentiation. Supposing we wanted to find a power series for $\sin(x)$ we could write:

$$\sin(x) = a_0 + a_1 x + a_2 x^2 + a_3 x^3 + a_4 x^4 + a_5 x^5 + \cdots$$
$$+ a_n x^n + \cdots \tag{12.7}$$

For $x = 0$, we know that $\sin(0) = 0$, hence substituting $x = 0$ in Equation (12.7) we find

$$0 = a_0.$$

To find a_1, we differentiate both sides of Equation (12.7) to give

$$\cos(x) = a_1 + 2a_2 x + 3a_3 x^2 + 4a_4 x^3 + 5a_5 x^4 + \cdots$$
$$+ na_n x^{n-1} + \cdots \tag{12.8}$$

Substititute $x = 0$ and as $\cos(0) = 1$, this gives:

$$1 = a_1.$$

Differentiating Equation (12.8) we get:

$$-\sin(x) = 2a_2 + 3.2a_3 x^1 + 4.3a_4 x^2 + 5.4a_5 x^3 + \cdots$$
$$+ n(n-1)a_n x^{n-2} + \cdots \tag{12.9}$$

Therefore, at $x = 0$

$$0 = 2a_2 \quad \Leftrightarrow \quad a_2 = 0.$$

Differentiating Equation (12.9) we get:

$$-\cos(x) = 3.2.1a_3 + 4.3.2a_4 x + \cdots$$
$$+ n(n-1)(n-2)a_n x^{n-3} + \cdots \tag{12.10}$$

Substituting $x = 0$ gives:

$$-1 = 3!a_3 \Leftrightarrow a_3 = -1/3!$$

A pattern is emerging, so that we can write:

$$\sin(x) = x - \frac{x^3}{3!} + \frac{x^5}{5!} - \frac{x^7}{7!} \cdots$$

This is a power series for $\sin(x)$ which we have found by expanding around $x = 0$.

When we expand around $x = 0$, we find a special case of the Taylor series expansion called a Maclaurin series.

Maclaurin series: definition

If a function $f(x)$ is defined for values of x around $x = 0$, within some radius R, that is, for $-R < x < R$ (or $|x| < R$) and if all its derivatives are defined then:

$$f(x) = f(0) + f'(0)x + \frac{f''(0)}{2!}x^2 + \frac{f'''(0)}{3!}x^3 + \cdots$$

$$+ \frac{f^{(n)}(0)}{n!}x^n + \cdots \tag{12.11}$$

Notice that this is a power series with coefficient sequence:

$$a_n = \frac{f^{(n)}(0)}{n!}$$

where $f^{(n)}(0)$ is found by finding the nth derivative of $f(x)$ with respect to x and then substituting $x = 0$.

Example 12.24 Find the Maclaurin series for $f(x) = e^x$ and give the range of values of x for which the series is valid.

Solution Find all order derivatives of $f(x)$ and substitute $x = 0$

$$f(x) = e^x, \ f'(x) = e^x, \ f''(x) = e^x, \ f'''(x) = e^x, \ f^{(iv)}(x) = e^x$$

$$f(0) = e^0 = 1, \ f'(0) = 1, \ f''(0) = 1, \ f'''(0) = 1, \ f^{(iv)}(0) = 1.$$

Substituting into Equation (12.11), the Maclaurin series is

$$e^x = 1 + x + \frac{x^2}{2!} + \frac{x^3}{3!} + \frac{x^4}{4!} + \cdots + \frac{x^n}{n!} + \cdots$$

As e^x exists for all values of x, we can use this series for all values of x.

Example 12.25 Find a power series for $\sinh(x)$ and give the values of x for which it is valid.

Solution Find all order derivatives of $\sinh(x)$ and substitute $x = 0$ in each one.

$$f(x) = \sinh(x), \ f'(x) = \cosh(x), \ f''(x) = \sinh(x), \ f'''(x) = \cosh(x),$$

etc.

$$f(0) = \sinh(0) = 0, \ f'(0) = 1, \ f''(0) = 0, \ f'''(0) = 1, \ f^{(iv)}(0) = 0$$

Then

$$\sinh(x) = x + \frac{x^3}{3!} + \frac{x^5}{5!} + \cdots$$

As $\sinh(x)$ is defined for all values of x then the series is valid for all values of x.

Example 12.26   Expand $f(x) = 1/(1 + x)$ in powers of x.

Solution   Find all order derivatives of $f(x)$ and substitute $x = 0$

$$f(x) = \frac{1}{1 + x} = (1 + x)^{-1}, \quad f'(x) = (-1)(1 + x)^{-2}$$

$$f''(x) = (-1)(-2)(1 + x)^{-3}, \quad f'''(x) = (-1)(-2)(-3)(1 + x)^{-4}$$

$$f(0) = 1 \quad f'(0) = -1 \quad f''(0) = 2! \quad f'''(0) = -3!.$$

Then

$$\frac{1}{1 + x} = 1 - x + \frac{2!}{2!}x^2 - \frac{3!}{3!}x^3 \cdots$$

$$= 1 - x + x^2 - x^3 \cdots (-1)x^n \cdots$$

As $1/(1 + x)$ is not defined at $x = -1$ we can only use this series for $|x| < 1$.

The binomial theorem revisited

The binomial theorem, as stated in the previous section, was only given for n as a whole positive number. We can now find the binomial expansion for $(1 + x)^n$ for all values of n using the Maclaurin series.

$$f(x) = (1 + x)^n.$$

Then

$$f'(x) = n(1 + x)^{n-1}$$

$$f''(x) = n(n - 1)(1 + x)^{n-2}$$

$$f'''(x) = n(n - 1)(n - 2)(1 + x)^{n-3}.$$

Substituting $x = 0$, we get

$$f(0) = 1, \quad f'(0) = n, \quad f''(0) = n(n-1), \quad f'''(0) = n(n-1)(n-2).$$

Therefore, using Equation (12.11) for the Maclaurin series, we find:

$$(1 + x)^n = 1 + nx + \frac{n(n - 1)}{2!}x^2 + \frac{n(n - 1)(n - 2)}{3!}x^3 \cdots$$

Notice that n can take fractional or negative values, but if n is negative, $|x| < 1$ (as found in Example 12.26). For many fractional values of n we also need to keep to the restriction $|x| < 1$.

Example 12.27 Expand $(1 + x)^{1/2}$ in powers of x.

Solution Using the binomial expansion

$$(1 + x)^n = 1 + nx + \frac{n(n - 1)}{2!}x^2 + \frac{n(n - 1)(n - 2)}{3!}x^3 \cdots \qquad (12.12)$$

and substituting $n = 1/2$ gives

$$(1 + x)^{1/2} = 1 + \frac{1}{2}x + \frac{\frac{1}{2}\left(\frac{1}{2} - 1\right)}{2!}x^2$$

$$+ \frac{\frac{1}{2}\left(\frac{1}{2} - 1\right)\left(\frac{1}{2} - 2\right)}{3!}x^3$$

$$+ \frac{\frac{1}{2}\left(\frac{1}{2} - 1\right)\left(\frac{1}{2} - 2\right)\left(\frac{1}{2} - 3\right)}{4!}x^4 + \cdots$$

$$= 1 + \frac{1}{2}x - \frac{1}{8}x^2 + \frac{1}{16}x^3 - \frac{5}{128}x^4 + \cdots$$

Notice that $(1 + x)^{1/2} = \sqrt{1 + x}$ is not defined for $x < -1$, so the series is only valid for $|x| < 1$.

Series to represent products and quotients

Example 12.28 Find the Maclaurin series up to the term in x^3 for the function

$$f(x) = \frac{(1 + x)^{1/2}}{1 - x}.$$

Solution As this function would be difficult to differentiate three times (to use the Maclaurin series directly), we use

$$f(x) = (1 + x)^{1/2}(1 - x)^{-1}$$

and find series for the two terms in the product then multiply them together.

$$(1 + x)^{1/2} = 1 + \frac{1}{2}x - \frac{1}{8}x^2 + \frac{1}{16}x^3 - \frac{5}{128}x^4 + \cdots$$

$$(1 - x)^{-1} = 1 + x + \frac{(-1)(-2)}{2!}(-x)^2$$

$$- \frac{(-1)(-2)(-3)}{3!}(-x)^3 \cdots$$

$$(1 - x)^{-1} = 1 + x + x^2 + x^3 + \cdots + x^n + \cdots$$

Then

$$(1 + x)^{1/2}(1 - x)^{-1} = \left(1 + \frac{1}{2}x - \frac{1}{8}x^2 + \frac{1}{16}x^3 - \frac{5}{128}x^4 + \cdots\right)$$

$$\times (1 + x + x^2 + x^3 + \cdots)$$

Up to the term in x^3:

$$(1+x)^{1/2}(1-x)^{-1} = 1 + x + x^2 + x^3 + \frac{1}{2}x + \frac{1}{2}x^2 + \frac{1}{2}x^3$$

$$- \frac{1}{8}x^2 - \frac{1}{8}x^3 + \frac{1}{16}x^3$$

$$= 1 + \left(1 + \frac{1}{2}\right)x + \left(1 + \frac{1}{2} - \frac{1}{8}\right)x^2$$

$$+ \left(1 + \frac{1}{2} - \frac{1}{8} + \frac{1}{16}\right)x^3$$

$$= 1 + \frac{3}{2}x + \left(\frac{11}{8}\right)x^2 + \left(\frac{23}{16}\right)x^3.$$

Approximation

Power series can be used for approximations.

Example 12.29 Use a series expansion to find $\sqrt{1.06}$ correct to 5 s.f.

Solution We need to write $\sqrt{1.06}$ in a way that we could use the binomial expansion, so we use $\sqrt{1.06} = \sqrt{1 + 0.06} = (1 + 0.06)^{1/2}$

When doing this it is important that the second term, in this case 0.06, should be a small number so that its higher powers will tend towards zero.

We can now use the binomial expansion, taking terms up to x^3, as we estimate that terms beyond that will be very small. Using Equation (12.12), we find

$$(1+x)^{1/2} = 1 + \frac{1}{2}x - \frac{1}{8}x^2 + \frac{1}{16}x^3 - \cdots$$

Now substitute $x = 0.06$ giving

$$\sqrt{1.06} \approx 1 + \frac{0.06}{2} - \frac{(0.06)^2}{8} + \frac{(0.06)^3}{16}$$

$$= 1 + 0.03 - 0.00045 + 0.0000135 = 1.0295635$$

$$\Rightarrow \sqrt{1.06} = 1.0296 \text{ to 5 s.f.}$$

Example 12.30 Find $\sin(0.1)$ correct to five decimal places by using a power series expansion.

Solution At the beginning of this section, we found that the power series expansion for $\sin(x)$ was as follows:

$$\sin(x) = x - \frac{x^3}{3!} + \frac{x^5}{5!} - \cdots$$

We substitute $x = 0.1$ to find $\sin(0.1)$ and continue until the next term is small compared to 0.000005 which means that it would not effect the result when calculated to five decimal places:

$$\sin(0.1) = 0.1 - \frac{(0.1)^3}{3!} + \frac{(0.1)^5}{5!} - \cdots$$

$$= 0.1 - 0.0001\dot{6} + 0.0000008\dot{3} - \ldots$$

$$= 0.09983 \text{ (to five decimal places)}.$$

Taylor series

Maclaurin's series is just a special case of Taylor series. A Taylor series is a series expansion of a function not necessarily taken around $x = 0$. This is given by:

If a function $f(x)$ is defined for values of x around $x = a$, within some radius R, that is, for $a - R < x < a + R$ (or $|x - a| < R$) and if all its derivatives are defined, then:

$$f(x) = f(a) + f'(a)(x - a) + \frac{f''(a)}{2!}(x - a)^2 + \frac{f'''(a)}{3!}(x - a)^3$$

$$+ \cdots + \frac{f^{(n)}(a)}{n!}(x - a)^n + \cdots \tag{12.13}$$

or substituting $x = a + h$, where h is usually considered to be a small value, this gives

$$f(a + h) = f(a) + f'(a)h + \frac{f''(a)}{2!}h^2 + \frac{f'''(a)}{3!}h^3$$

$$+ \cdots + \frac{f^{(n)}(a)}{n!}h^n + \cdots \tag{12.14}$$

Example 12.31   Given $\sin(45°) = 1/\sqrt{2}$ and $\cos(45°) = 1/\sqrt{2}$, approximate $\sin(44°)$ by using a power series expansion.

Solution

$$\sin(44°) = \sin(45° - 1°).$$

Remember that the sine function is defined as a function of radians so we must convert the angles to radians in order to use the Taylor series: $45° = \pi/4$ and $1° = \pi/180$.

Expand using the Taylor series for $\sin(a+h)$ where $a = \pi/4$ and using Equation (12.14)

$$f(a + h) = f(a) + f'(a)h + \frac{f''(a)}{2!}h^2 + \frac{f'''(a)}{3!}h^3$$

$$+ \cdots + \frac{f^{(n)}(a)}{n!}h^n + \cdots$$

$$f(x) = \sin(x), \quad f'(x) = \cos(x),$$

$$f''(x) = -\sin(x), \quad f'''(x) = \cos(x)$$

$$f(\pi/4) = \frac{1}{\sqrt{2}}, \quad f'(\pi/4) = \frac{1}{\sqrt{2}},$$

$$f''(\pi/4) = \frac{1}{\sqrt{2}}, \quad f'''(\pi/4) = \frac{1}{\sqrt{2}},$$

So

$$\sin((\pi/4)+h) = \frac{1}{\sqrt{2}} + \frac{1}{\sqrt{2}}h + \frac{1}{\sqrt{2}}\frac{h^2}{2!} + \frac{1}{\sqrt{2}}\frac{h^3}{3!} + \cdots$$

Substituting $h = -\pi/180$ radians, we get

$$\sin(44°) = \frac{1}{\sqrt{2}}\left(1 - \frac{\pi}{180} + \frac{1}{2}\left(\frac{\pi}{180}\right)^2 + \frac{1}{6}\left(\frac{\pi}{180}\right)^3 + \cdots\right)$$

$$= \frac{1}{\sqrt{2}}(1 - 0.01745 - 0.0001523 + 0.0000009 + \cdots)$$

$$= 0.69466 \text{ to five decimal places.}$$

L'Hopital's rule

When sketching graphs of functions in Chapter 11, we looked at graphs where the function is undefined for some values of x. The function $f(x) = 1/x$, for instance, is not defined when $x = 0$ and tends to $-\infty$ as $x \to 0^-$ and tends to $+\infty$ as $x \to 0^+$. Not all functions that have undefined points tend to $\pm\infty$ near the point where they are undefined. For example, consider the function $f(x) = \sin(x)/x$. The graph of this function is shown in Figure 12.5. The function is not defined for $x = 0$, which we can see by substituting $x = 0$ into the function expression. This gives a zero in the denominator and hence an attempt to divide by 0 which is undefined. However, we can see from the graph that the function, rather than tending to plus or minus infinity as $x \to 0$, just tends to 1. This is very useful because we are able to 'patch' the function by giving it a value at $x = 0$ and the new function is defined for all values of x.

We can define a new function. This particular function is quite famous, and is called the sinc function

$$\text{sinc}(x) = \begin{cases} \dfrac{\sin(x)}{x} & \text{where } x \neq 0 \\ 1 & \text{where } x = 0 \end{cases}$$

The points where functions may tend to a finite limit can be identified by looking out for points which lead to 0/0. These are called indeterminate points, indicating that they are a special type of undefined point.

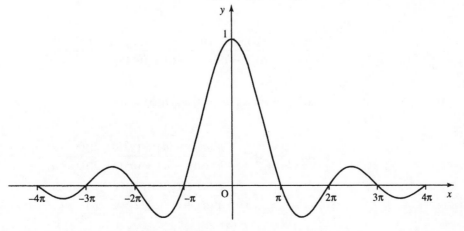

Figure 12.5   *The graph of the function $f(x) = \sin(x)/x$.*

We can examine what happens to the function near the indeterminate point by using the power series expansions of the denominator and numerator. Substituting the series for $\sin(x)/x$ into the expression $\sin(x)$ gives

$$\frac{\sin(x)}{x} = \frac{1}{x}\left(x - \frac{x^3}{3!} + \frac{x^5}{5!} - \frac{x^7}{7!}\cdots\right)$$

and therefore

$$\lim_{x\to 0}\frac{\sin(x)}{x} = \lim_{x\to 0}\frac{1}{x}\left(x - \frac{x^3}{3!} + \frac{x^5}{5!} - \cdots\right)$$

$$= \lim_{x\to 0}\left(1 - \frac{x^2}{3!} + \frac{x^4}{5!} - \cdots\right)$$

The last expression is defined at $x = 0$, so we can substitute $x = 0$ in order to find the limit. This gives the value 1. L'Hopital's rule is a quick way of finding this limit without needing to write out the series specifically.

L'Hopital's rule states that if a function $f(x) = g(x)/h(x)$ is indeterminate at $x = a$ then:

$$\lim_{x\to a}\frac{g(x)}{h(x)} = \lim_{x\to a}\frac{g'(x)}{h'(x)}.$$

If $g'(x)/h'(x)$ is defined at $x = a$, we can then use

$$\lim_{x\to a}\frac{g'(x)}{h'(x)} = \frac{g'(a)}{h'(a)}$$

and if $g'(x)/h'(x)$ is indeterminate at $x = a$, we can use the rule again.

We can show this to be true by using Equation (12.13) for the Taylor series expansion about a:

$$\frac{g(x)}{h(x)} = \frac{g(a) + g'(a)(x - a) + (g''(a)/2!)(x - a)^2 + \cdots}{h(a) + h'(a)(x - a) + (h''(a)/2!)(x - a)^2 + \cdots}$$

and given that $g(a) = 0$ and $h(a) = 0$

$$\lim_{x\to a}\frac{g(x)}{h(x)} = \lim_{x\to a}\frac{g'(a)(x - a) + (g''(a)/2!)(x - a)^2 + \cdots}{h'(a)(x - a) + (h''(a)/2!)(x - a)^2 + \cdots}$$

$$= \lim_{x\to a}\frac{g'(a) + (g''(a)/2!)(x - a) + \cdots}{h'(a) + (h''(a)/2!)(x - a) + \cdots}$$

$$= \frac{g'(a)}{h'(a)} \quad (\text{if } h'(a) \neq 0).$$

Example 12.32 Find

$$\lim_{x \to 1} \frac{x^3 - 2x^2 + 4x - 3}{4x^2 - 5x + 1}.$$

Solution Substituting $x = 1$ into

$$\frac{x^3 - 2x^2 + 4x - 3}{4x^2 - 5x + 1}$$

gives 0/0, which is indeterminate. Using L'Hopital's rule, we differentiate the top and bottom lines:

$$\lim_{x \to 1} \frac{x^3 - 2x^2 + 4x - 3}{4x^2 - 5x + 1} = \lim_{x \to 1} \frac{3x^2 - 4x + 4}{8x - 5}.$$

We find that the new expression is defined at $x = 1$, so

$$\lim_{x \to 1} \frac{3x^2 - 4x + 4}{8x - 5} = \frac{3 - 4 + 4}{8 - 5} = \frac{3}{3} = 1$$

Example 12.33 Find

$$\lim_{x \to 0} \frac{\cos(x) - 1}{x^2}.$$

Solution Substituting $x = 0$ into

$$\frac{\cos(x) - 1}{x^2}$$

gives 0/0, which is indeterminate. Therefore, using L'Hopital's rule, we differentiate the top and bottom lines:

$$\lim_{x \to 0} \frac{\cos(x) - 1}{x^2} = \lim_{x \to 0} \frac{-\sin(x)}{2x}.$$

We find that the new expression is also indeterminate at $x = 0$, so we use L'Hopital's rule again:

$$\lim_{x \to 0} \frac{-\sin(x)}{2x} = \lim_{x \to 0} \frac{-\cos(x)}{2}.$$

The last expression is defined at $x = 0$ so we can substitute $x = 0$ to give

$$\lim_{x \to 0} \frac{-\cos(x)}{2} = -\frac{1}{2}.$$

Hence,

$$\lim_{x \to 0} \frac{\cos(x) - 1}{x^2} = -\frac{1}{2}.$$

12.7 Limits and convergence

We have already briefly mentioned ideas of limits in various contexts. We will now look at the idea in more detail. When we looked at the sum to infinity of a geometric progression in Table 12.1 we looked at $S_n = 4(1 - (2/3)^n)$ as n was made larger and discovered that $S_{40} = 4$ to 7 s.f. and all values of $n > 40$ also gave $S_n = 4$ to 7 s.f. This can give us an idea of how to find out if a sequence tends to a limit:

1. Choose a number of significant figures
2. Write all terms in the sequence to that number of significant figures.
3. The sequence tends to a limit if the terms in the sequence, when expressed to the agreed number of significant figures, become constant, that is, do not change after some value of n.

Theoretically, this procedure must work for all possible numbers of significant figures. As my calculator only displays 8 s.f. I cannot go through this process for more than 7 s.f. A computer using double precision arithmetic could perform the calculations to far more (usually up to 18 s.f.).

Consider the series

$$1 + z + z^2 + z^3 + \cdots$$

which is a GP with first term 1 and common ratio z, the sum to n terms gives

$$\frac{1 - z^n}{1 - z}.$$

For $z = 1/2$ this gives the series

$$1 + \tfrac{1}{2} + \left(\tfrac{1}{2}\right)^2 + \left(\tfrac{1}{2}\right)^3 + \cdots$$

and the sum of n terms gives

$$S_n = \frac{1 - \left(\tfrac{1}{2}\right)^n}{1 - \tfrac{1}{2}} = 2\left(1 - \left(\frac{1}{2}\right)^n\right).$$

We can write, S_n, as a sequence of values to 3 s.f., 5 s.f., and 7 s.f. as is done in Table 12.2.

From these results we can see that the limit appears to be 2. The limit is reached to 3 s.f for $n = 9$, to 5 s.f for $n = 15$, and to 7 s.f for $n = 22$.

The more terms taken in a sequence which converges, then the nearer we will get to the limit. However we can only get as near as the number of significant figures, usually limited by the calculator, permits. When using a numerical method to solve a problem we use these ideas about convergence.

Table 12.2 *The values in the sequence $S_n =$ expressed to 3 s.f., 5 s.f., and 7 s.f. Notice the sequence becomes constant after $n = 9$ for 3 s.f, after $n = 15$ for 5 s.f and after $n = 22$ for 7 s.f.*

n	S_n (3 s.f.)	S_n (5 s.f.)	S_n (7 s.f.)
1	1	1	1
2	1.5	1.5	1.5
3	1.75	1.75	1.75
4	1.88	1.875	1.875
5	1.94	1.9375	1.9375
6	1.97	1.9688	1.96875
7	1.98	1.9844	1.984375
8	1.99	1.9922	1.992188
9	2	1.9961	1.996094
10	2	1.998	1.998047
11	2	1.999	1.999023
12	2	1.9995	1.999512
13	2	1.9998	1.999756
14	2	1.9999	1.999878
15	2	2	1.999939
16	2	2	1.999969
17	2	2	1.999985
18	2	2	1.999992
19	2	2	1.999996
20	2	2	1.999998
21	2	2	1.999999
22	2	2	2
23	2	2	2

12.8 Newton–Raphson method for solving equations

Although we already know how to solve linear equations and quadratic equations other equations may need to be solved by using a numerical method. One such method is the Newton–Raphson method. The method consists of an algorithm which can be expressed as follows:

Step 1: take an equation and write it in the form $f(x) = 0$, then,

Step 2: take a guess at a solution

Step 3: calculate a new value for x using

$$x \leftarrow x - \frac{f(x)}{f'(x)}$$

Step 4: Repeat Step 3 until come convergence criterion has been satisfied or until it is decided that the method has failed to find a solution. Here, the '$\leftarrow$' symbol has been used to represent the 'assignment operator'.

$$x \leftarrow x - \frac{f(x)}{f'(x)}$$

means replace x by a value found by taking the old value of x and calculating $x - f(x)/f'(x)$. We will return to the problems in Steps 1 and 4 later; first we will look at a simple example of using the Newton–Raphson method.

Example 12.34 Use Newton–Raphson method to find $\sqrt{5}$ correct to 7 s.f.

Solution $\sqrt{5}$ is one solution to the equation $x^2 = 5$

Step 1: Write the equation in the form $f(x) = 0$

$$x^2 = 5 \quad \Leftrightarrow \quad x^2 - 5 = 0$$

Step 2: Take a guess at the solution. We know that $\sqrt{5}$ is slightly bigger than $\sqrt{4}$ so take a first guess as $x = 2$.

Steps 3 and 4: Calculate

$$x \leftarrow x - \frac{f(x)}{f'(x)}$$

until some convergence criterion is satisfied.
    As $f(x) = x^2 - 5$, $f'(x) = 2x$

$$x \leftarrow x - \frac{f(x)}{f'(x)}$$

gives

$$x \leftarrow x - \frac{x^2 - 5}{2x}.$$

This can be written over a common denominator, giving

$$x \leftarrow \frac{x^2 + 5}{2x}$$

which is the Newton–Raphson formula for solving $x^2 - 5 = 0$.

Start with $x = 2$	$x \leftarrow \dfrac{4 + 5}{4}$	$x = 2.25$
Substitute $x = 2.25$	$x \leftarrow \dfrac{(2.25)^2 + 5}{2(2.25)}$	$x = 2.2361111$
Substitute $x = 2.2361111$	$x \leftarrow \dfrac{(2.2361111)^2 + 5}{2(2.2361111)}$	$x = 2.236068$
Substitute $x = 2.236068$	$x \leftarrow \dfrac{(2.236068)^2 + 5}{2(2.236068)}$	$x = 2.236068$

We notice that in the last iteration there has been no change in the value of x, so we assume that the algorithm has converged, giving

$$\sqrt{5} = 2.236068 \text{ to 7 s.f.}$$

The sequence of values we have found is:
    2, 2.25, 2.2361111, 2.236068, 2.236068, and we need to stop at this point because the value of x has not changed in the last iteration.

Table 12.3   *The function* $f(x) = x^3 - 3x^2 + 2x + 1$ *evaluated for a few values of* x

x	−1	−0.5	0	0.5	1	1.5	2
$f(x) = x^3 - 3x^2 + 2x + 1$	−7	−0.875	1	1.375	1	0.625	1

Example 12.35   Find a solution to the equation $x^3 - 3x^2 + 2x + 1 = 0$

Solution

Step 1: The equation is already expressed in the correct form.

Step 2: We need to find a first guess for the solution and to do this we could sketch the graph to see roughly where it crosses the x-axis or we could try substituting a few values into the function $f(x) = x^3 - 3x^2 + 2x + 1$ and look for a change of sign, which we have done in Table 12.3. As the function is continuous, the function must pass through zero in order to change from positive to negative, or vice versa. There is a change of sign between $x = -0.5$ and $x = 0$, so we take as a first guess a point half way between these two values, giving $x = -0.25$.

Step 3: Using the Newton–Raphson formula

$$x \leftarrow x - \frac{f(x)}{f'(x)}$$

and substituting $f(x) = x^3 - 3x^2 + 2x + 1$ gives

$$x \leftarrow x - \frac{x^3 - 3x^2 + 2x + 1}{3x^2 - 6x + 2}$$

and simplifying gives

$$x \leftarrow \frac{2x^3 - 3x^2 - 1}{3x^2 - 6x + 2}.$$

Starting by substituting $x = -0.25$ gives

$$x \leftarrow \frac{2(-0.25)^3 - 3(-0.25)^2 - 1}{3(-0.25)^2 - 6(-0.25) + 2} = \frac{-1.21875}{3.6875} = -0.3305084$$

Now substitute $x = -0.3305084$ giving

$$x \leftarrow \frac{2(-0.3305084)^3 - 3(-0.3305084)^2 - 1}{3(-0.3305084)^2 - 6(-0.3305084) + 2} = -0.3247489.$$

Substitute $x = -0.3247489$ giving

$$x \leftarrow \frac{2(-0.3247489)^3 - 3(-0.3247489)^2 - 1}{3(-0.3247489)^2 - 6(-0.3247489) + 2} = -0.3247179.$$

Substitute $x = -0.3247179$ giving

$$x \leftarrow \frac{2(-0.3247179)^3 - 3(-0.3247179)^2 - 1}{3(-0.3247179)^2 - 6(-0.3247179) + 2} = -0.3247179.$$

As the last two numbers are the same to the degree of accuracy we have used, there is no point in continuing. We have thus obtained the sequence of values:

$-0.25, -0.3305084, -0.3247489, -0.3247179, -0.3247179.$

Finally, we can check that we have found a good approximation to a solution of the equation by substituting $x = -0.3247179$ into the function $f(x) = x^3 - 3x^2 + 2x + 1$, which gives 2.441×10^{-7}. As this value is very close to 0 this confirms that we have found a reasonable approximation to a solution of the equation $f(x) = 0$.

The convergence criterion

In Examples 12.34 and 12.35 we decided to stop the calculation when the last two values found were equal. We had found the limit of the recurrence relation to 7 s.f. In a computer algorithm, we could test if the last two calculated values of x differ by a very small amount.

Example 12.36 A convergent sequence is defined by a recurrence relation. The calculation should stop when the limit has been found to an accuracy of at least three decimal places. Give a condition that could be used in this case.

Solution Assuming two consecutive terms are x_{n-1} and x_n, then the absolute difference between them is given by $|x_n - x_{n-1}|$. To test whether this is small enough to accept x_n as the limit to three decimal places we use the fact that a number expressed to three decimal places could have an absolute error of just less than 0.0005. So the condition we can use to stop the algorithm could be $|x_n - x_{n-1}| < 0.0005$. If this condition is satisfied we could then assume that x_n is the limit to three decimal places. To be on the safe side, however, it is better to perform the calculation one final time and check that it is also true that $|x_{n+1} - x_n| < 0.0005$ and then use x_{n+1} as the value of the limit which should be accurate to at least three decimal places.

Example 12.37 A convergent sequence is defined by a recurrence relation. The calculation should stop when the limit has been found to an accuracy of at least 4 s.f. Give a condition that could be used in this case.

Solution Assuming two consecutive terms are x_{n-1} and x_n then the absolute difference between them is given by $|x_n - x_{n-1}|$. To test whether this is small enough to accept x_n as the limit to 4 s.f. use the fact that a number expressed to 4 s.f. can have an absolute relative error of just less than 0.00005. As we do not know that value of the limit we must approximate it by the last value calculated in the sequence, so the absolute relative error is approximately

$$\frac{|x_n - x_{n-1}|}{|x_n|}$$

so an appropriate condition would be

$$\frac{|x_n - x_{n-1}|}{|x_n|} < 0.00005$$

or

$$|x_n - x_{n-1}| < 0.00005|x_n|.$$

As in the previous example it would be preferable to test that this condition holds on at least two successive iterations. Hence, we could also check that

$$|x_{n+1} - x_n| < 0.00005|x_{n+1}|$$

and take x_{n+1} as the limit of the sequence correct to four significant figures.

Divergence

A divergent sequence is one that does not tend to a finite limit. Some examples of divergent sequences are:

(i) $1, 0, 1, 0, 1, 0, 1, 0 \ldots$ which is an oscillating sequence,
(ii) $1, 2, 4, 8, 16 \ldots$ which tends to plus infinity,
(iii) $-1, -3, -5, \ldots$ which tends to minus infinity.

Recurrence relations that are used for some numerical method may not always converge, particularly if the initial value is chosen inappropriately. To check for this eventuality, it is usual to stop the algorithm after some finite number of steps, maybe 100 or 1000 iterations, depending on the problem. If no convergence has been found after that number of iterations then it is considered that sequence is failing to converge.

12.9 Summary

1. A sequence is a collection of objects arranged in a definite order. The elements of a sequence can be represented by $a_1, a_2, a_3, \ldots, a_n, \ldots$
2. If a rule exists by which any term in the sequence can be found then this may be used to express the general term of the sequence, usually represented by a_n or $a(n)$. This rule can also be expressed in the form of a recurrence relation where a_{n+1} is expressed in terms of $a_n, a_{n-1}, a_{n-2}, \ldots$
3. During analog to digital (A/D) conversion, a signal is sampled and can then be represented by a sequence of numbers. $f(t)$ can be represented by $a(n) = f(nT)$, where T is the sampling interval and $t = nT$. The sampling theorem states that the sampling interval must be less than $T = 1/(2f)$ seconds in order to be able to represent a frequency of f Hz.
4. A series is the sum of a sequence. To represent series we may use sigma notation, using the capital Greek letter sigma, Σ, to indicate the summation process, for example,

$$\sum_{n=0}^{n=10} \frac{1}{2^n}$$

means 'sum all the terms $1/2^n$ for n from 0 to 10'.
5. An arithmetic progression (AP) is a sequence where each term is found by adding a fixed amount, called the common difference, to the previous term. If the first term is a and the common difference is d, then the general term is $a_n = a + (n - 1)d$ and the sum of the first n terms is given by

$$S_n = \frac{n}{2}(2a + (n - 1)d) \quad \text{or} \quad S_n = \frac{n}{2}(a + l)$$

where l is the last term in the sequence and n is the number of terms.
6. An geometric progression (GP) is a sequence where each term is found by multiplying the previous term by a fixed amount, called

the common ratio. If the first term is a and the common ratio is r, then the general term is $a_n = ar^{n-1}$, and the sum of the first n terms is given by

$$S_n = \frac{a(1 - r^n)}{1 - r}.$$

The sum to infinity of a GP can be found if $|r| < 1$ and is given by

$$S_\infty = \frac{a}{1 - r}.$$

7. The binomial expansion gives

$$(a + b)^n = a^n + na^{n-1}b + \frac{n(n - 1)}{2!}a^{n-2}b^2$$

$$+ \frac{n(n - 1)(n - 2)}{3!}a^{n-3}b^3 + \cdots$$

where n can be a whole number or a fraction.

8. The Maclaurin series is a series expansion of a function about $x = 0$. If a function $f(x)$ is defined for values of x around $x = 0$, within some radius R, that is, for $-R < x < R$ (or $|x| < R$) and if all its derivatives are defined then:

$$f(x) = f(0) + f'(0)x + \frac{f''(0)}{2!}x^2 + \frac{f'''(0)}{3!}x^3$$

$$+ \cdots + \frac{f^{(n)}(0)}{n!}x^n + \cdots$$

This gives a power series with coefficient sequence:

$$a_n = \frac{f^{(n)}(0)}{n!}$$

where $f^{(n)}(0)$ is found by calculating the nth derivative of $f(x)$ with respect to x and then substituting $x = 0$.

9. Maclaurin's series is just a special case of Taylor series. A Taylor series is a series expansion of a function not necessarily taken around $x = 0$. If a function $f(x)$ is defined for values of x around $x = a$, within some radius R, that is, for $a - R < x < a + R$ (or $|x - a| < R$) and if all its derivatives are defined, then:

$$f(x) = f(a) + f'(a)(x - a) + \frac{f''(a)}{2!}(x - a)^2$$

$$+ \frac{f'''(a)}{3!}(x - a)^3 + \cdots + \frac{f^{(n)}(a)}{n!}(x - a)^n + \cdots$$

or, substituting $x = a + h$, where h is usually considered to be a small value, we get

$$f(a + h) = f(a) + f'(a)h + \frac{f''(a)}{2!}h^2 + \frac{f'''(a)}{3!}h^3$$

$$+ \cdots + \frac{f^{(n)}(a)}{n!}h^n + \cdots.$$

10. L'Hopital's rule is a way of finding the limit of a function at a point where it is undetermined (i.e. it gives 0/0 at the point). The

rule states that if a function $f(x) = g(x)/h(x)$ is indeterminate at $x = a$ then:

$$\lim_{x \to a} \frac{g(x)}{h(x)} = \lim_{x \to a} \frac{g'(x)}{h'(x)}.$$

If $g'(a)/h'(a)$ is defined, we can then use

$$\lim_{x \to a} \frac{g'(x)}{h'(x)} = \frac{g'(a)}{h'(a)}$$

and if $g'(a)/h'(a)$ is indeterminate, we can use the rule again.

11. To test if a sequence tends to a limit follow the following procedure:

(a) Choose a number of significant figures.

(b) Write all the terms in the sequence to that number of significant figures.

(c) The sequence tends to a limit if the terms in the sequence, when expressed to the agreed number of significant figures becomes constant, that is, do not change after some value of n.

This procedure must theoretically work for any chosen number of significant figures.

12. The algorithm for solving an equation using Newton–Raphson method can be described as

Step 1: Take an equation and write it in the form $f(x) = 0$.

Step 2: Take a guess at a solution

Step 3: Calculate a new value for x using

$$x \leftarrow x - \frac{f(x)}{f'(x)}$$

Step 4: Repeat Step 3 until some convergence criterion has been satisfied or until it is decided that the method has failed to find a solution.

13. Convergence criteria can either be based on the testing the size of the absolute error or the relative absolute error. To find the limit of a convergent sequence defined by a recurrence relation, correct to three decimal places, we can test for $|x_n - x_{n-1}| < 0.0005$ and to be correct to three significant figures we could test for $|x_n - x_{n-1}| < 0.0005|x_n|$. It is also necessary to put a limit on the number of iterations of some algorithm to check for the eventuality that the sequence fails to converge (is divergent).

12.10 Exercises

12.1. Find the next three terms in the following sequences. In each case, express the rule for the sequence as a recurrence relation.

(a) $-3, 1, 5, 9, 13, 17, \ldots$

(b) $8, 4, 2, 1, 0.5, \ldots$

(c) $18, 15, 12, 9, 6, 3, \ldots$

(d) $6, -6, 6, -6, \ldots$

(e) $10, 8, 6, 4, \ldots$

(f) $1, 2, 4, 7, 11, 16, 22, \ldots$

(g) $1, 3, 6, 10, 15, \ldots$

12.2. Given the following definitions of sequences write out the first five terms

(a) $a_n = 3n - 1$

(b) $x_n = 720/n$

(c) $b_n = 1 - n^2$

(d) $a_{n+1} = a_n + 2; a_1 = 6$

(e) $a_{n+1} = 3a_n; a_1 = 2$

(f) $a_{n+1} = -2a_n; a_1 = -1$

(g) $b_{n+1} = 2b_n - b_{n-1}; b_1 = 1/2, b_2 = 1$

(h) $\Delta y_n = 3; y_0 = 2$

(i) $\Delta y_n = 2y_n; y_0 = 1$.

12.3. Express the following using sigma notation

(a) $1 + x + x^2 + x^3 + \cdots + x^{10}$

(b) $-2 + 4 - 8 + 16 - \cdots + 256$

(c) $1 + 8 + 27 + 64 + 125 + 216$

(d) $-\frac{1}{3} + \frac{1}{9} - \frac{1}{27} + \cdots - \frac{1}{6561}$

(e) $\frac{1}{4} + \frac{1}{9} + \frac{1}{16} + \frac{1}{25} + \frac{1}{36} + \cdots + \frac{1}{100}$

(f) $-4 - 1 - \frac{1}{4} - \frac{1}{16} - \cdots - \frac{1}{4096}$.

12.4. Sketch the following functions and give the first 10 terms of their sequence representation ($t \geqslant 0$) at the sampling interval T given:

(a) $f(t) = \sin(2t), \quad T = 0.1$

(b) $f(t) = \cos(30t), \quad T = 0.01$

(c) $f(t)$ is the periodic function of period 16, defined for $0 \leqslant t < 16$ by

$$f(t) = \begin{cases} 2t & 0 \leq t \leq 4 \\ 16 - 2t & 4 < t \leq 12 \\ 2t - 32 & 12 < t < 16 \end{cases}$$

with sample interval $T = 1$.

(d) The square wave of period 2 given for $0 < t < 2$ by

$$f(t) = \begin{cases} 1 & 0 \leq t < 1 \\ -1 & 1 \leq t < 2 \end{cases}$$

with a sampling interval of $T = 0.25$.

12.5. The following are arithmetic progressions. Find the fifth, tenth, and general term of the sequence in each case.

(a) $6, 10, 14, \ldots$

(b) $3, 2.5, 2, \ldots$

(c) $-7, -1, 5, \ldots$

12.6. a_n is an arithmetic progression. Given the terms indicated, find the general term and find the sum of the first 20 terms in each case:

(a) $a_5 = 6, a_{10} = 26$

(b) $a_7 = -2, a_{16} = 2.5$

(c) $a_6 = 10, a_{12} = -8$.

12.7. The sum of the first 10 terms of an arithmetic progression is 50 and the first term is 2. Find the common difference and the general term and list the first six terms of the sequence.

12.8. How many terms are required in the arithmetic series $2 + 4 + 6 + 8 + \cdots$ to make a sum of 1056?

12.9. The following are geometric progressions. Find the fourth, eighth, and general term in each case:

(a) $1, 2, 4, \ldots$

(b) $1/3, 1/12, 1/48, \ldots$

(c) $-9, 3, -1, \ldots$

(d) $15, 18.75, 23.4375, \ldots$

12.10. a_n is a geometric progression. Given the terms indicated, find the general term and find the sum of the first 8 terms in each case.

(a) $a_3 = 8, a_6 = 1000$

(b) $a_6 = 54, a_9 = -486$

(c) $a_2 = -32, a_7 = 1$.

12.11. How many terms are required in the geometric series $8 + 4 + 2 + \cdots$ to make a sum of 15.9375?

12.12. A loan of €40 000 is repaid by annual instalments of €5000, except in the final year when the outstanding debt (if less than €5000) is repaid. Interest is charged at 10% per year, calculated at the end of each year on the outstanding amount of the debt. The first repayment is 1 year after the loan was taken out. Calculate the number of years required to repay the loan.

12.13. Evaluate the following

(a) $\displaystyle\sum_{n=1}^{n=4} 2^n$

(b) $\displaystyle\sum_{r=0}^{r=8} \frac{1}{2^r}$

(c) $\displaystyle\sum_{j=1}^{j=10} (-1)^j \left(\frac{1}{3}\right)^{j-2}$.

12.14. Find the sum of the first n terms of the following:

(a) $1 + z + z^2 + z^3 + \cdots$

(b) $1 - y^2 + y^4 - \cdots$

(c) $2x + \dfrac{4}{x} + \dfrac{8}{x^2} + \cdots$

12.15. State whether the following series are convergent and if they are find the sum to infinity.

(a) $2 + 1 + \frac{1}{2} + \frac{1}{4} + \cdots$

(b) $3 + 0 - 3 - 6 \cdots$

(c) $27 - 9 + 3 - 1 \cdots$

(d) $0.3 + 0.03 + 0.003 \cdots$

12.16. Find the following recurring decimals as fractions:

(a) $0.\dot{4}$ (b) $0.1\dot{6}$ (c) $0.0\dot{2}$.

12.17. Expand the following expressions

(a) $\left(1 + \frac{1}{2}x\right)^3$ (b) $(1 - x)^4$

(c) $(x - 1)^3$ (d) $(1 - 2y)^4$

(e) $(1 + x)^8$ (f) $(2x + 1)^3$

(g) $(2a + b)^3$ (h) $(x + (1/x))^7$ (i) $(a - 2b)^4$

12.18. Find the following using the expansion indicated:

(a) $(1.1)^3$ using $(1 + 0.1)^3$

(b) $(0.9)^4$ using $(1 - 0.1)^4$

(c) $(2.01)^3$ using $(2 + 0.01)^3$.

12.19. Give the first 4 terms in the binomial expansion of the following:

(a) $(1 + 2x)^5$ (b) $(1 - 3x)^8$

(c) $(2 + z)^6$ (d) $\left(1 + \frac{1}{2}x\right)^{16}$

(e) $(1 - x)^6$ (f) $(1 - 2x)^5$

12.20. Use $\sin(5\theta) = \text{Im}(e^{j5\theta}) = \text{Im}((\cos(\theta)+j\sin(\theta))^5)$ to find $\sin(5\theta)$ in terms of powers of $\cos(\theta)$ and $\sin(\theta)$.

12.21. Find the real and imaginary parts of the following:
(a) $(1-j)^6$ (b) $(1+j2)^4$ (c) $(3+j)^5$

12.22. Use a binomial expansion to find the following correct to four decimal places:

(a) $(0.99)^8$ (b) $(1.01)^7$ (c) $(2.05)^6$.

12.23. Find the first 4 non-zero terms in a power series expansion of the following functions and state for what values of x they are valid in each case.

(a) $\cos(x)$ (b) $\cosh(x)$ (c) $\ln(1+x)$
(d) $(1+x)^{1.5}$ (e) $(1+x)^{-2}$.

12.24. Find the first 4 non-zero terms in a power series expansion for the following functions:

(a) $\cos^2(x)$ (b) $\tan^{-1}(x)$ (c) $e^x \sin(x)$
(d) $(1-x)^{1.5}/(1+x)$.

12.25. Using a series expansion find the following correct to 4 significant figures:

(a) $\sqrt{1.05}$ (b) $\tan^{-1}(0.1)$ (c) $\sin(0.03)$
(d) $1/\sqrt{1.06}$

12.26. Using a series expansion and the given value of the function at $x = a$, evaluate the following correct to four significant figures:

(a) $\cos(7\pi/16)$ using $\cos(\pi/2) = 0$,
(b) $\sqrt{4.02}$ using $\sqrt{4} = 2$.

12.27. Find the following limits:
(a) $\lim_{x \to 2}((x^2 - x - 2)/(4x^3 - 4x - 7x - 2))$
(b) $\lim_{x \to 0}(x^3/(x - \sin(x)))$
(c) $\lim_{x \to -3}((x^2 + 6x + 9)/(4x^2 + 11x - 3))$
(d) $\lim_{x \to \pi/2}((\pi/2 - x)/\cos(x))$
(e) $\lim_{x \to 0}(\tan(x))/x)$
(f) $\lim_{x \to 0}(\sin(x-2)/(x^2 - 4x + 4))$.

12.28. Use the Newton–Raphson method to find a solution to the following equations correct to six significant figures:

(a) $x^3 - 2x = 1$ (b) $x^4 = 5$ (c) $\cos(x) = 2x$.

12.29. Suggest a test for convergence that could be used in a computer program so that the limit of a sequence, defined by some recurrence relation, could be assumed to be correct to

(a) six decimal places,
(b) six significant figures.

Part 2   Systems

13 Systems of linear equations, matrices, and determinants

13.1 Introduction

The widespread use of computers to solve engineering problems means that it is important to be able to represent problems in a form suitable for solution by a computer. Matrices are used to represent: systems of linear equations; transformations used in computer graphics or for robotic control; road, electrical and communication networks, and stresses and strains in materials. A matrix is a rectangular array of numbers of dimension $m \times n$ where m is the number of rows and n is the number of columns in the matrix. Matrices are also useful because they enable us to consider an array of numbers as a single object, represent it by a single symbol, and manipulate these symbols conveniently. In this chapter, we look at applications of matrices and arithmetic operations on matrices and some common numerical methods. We shall also look at the problem of solving systems of linear equations. The methods of solving linear systems of equations are well understood and we only need to be able to solve simple cases of such problems 'by hand'. However, it is important to be able to express a problem in matrix form and also appreciate situations where no solution exists or where more than one solution exists. This allows to analyse the problems of ill-conditioning of systems of equations, which can lead to instability in the solution and the problem of over- or under-determinacy, where either we have too much information, leading to possibly contradictory conditions, or we have not got enough to produce a single set of solutions for the unknowns.

We shall also look at the eigenvalue problem. The technique of finding eigenvalues will become particularly important when applied to systems of differential equations which we meet in Chapter 14.

13.2 Matrices

A matrix is a rectangular array of numbers. They may also be used as a simple store of information as in the following example.

Every weekday a household orders pints of milk, loaves of bread, and yoghurt from a milk lorry. The orders for the week can be displayed as follows:

	Milk	Bread	Yoghurt
Monday	3	2	4
Tuesday	4	1	0
Wednesday	2	2	4
Thursday	5	1	0
Friday	1	1	4

This information forms a matrix.

Transformations in a plane can be represented by using matrices, for example, a reflection about the x-axis can be represented by the matrix

$$\begin{pmatrix} 1 & 0 \\ 0 & -1 \end{pmatrix}$$

and rotation through the angle by

$$\begin{pmatrix} \cos(\theta) & -\sin(\theta) \\ \sin(\theta) & \cos(\theta) \end{pmatrix}.$$

We shall return to these examples later. Also in the chapter we will see that linear equations can be written in matrix form.

Notation

A matrix is represented by a capital letter **A** (bold) or by $[a_{ij}]$ where a_{ij} represents a typical element in the ith row and jth column of the matrix. We represent a general matrix in the following form:

$$\begin{array}{c} \\ \text{row number} \end{array} \begin{array}{c} \\ \begin{array}{c} 1 \\ 2 \\ 3 \\ \vdots \\ m \end{array} \end{array} \overset{\begin{array}{c} \text{column number} \\ \begin{array}{ccccc} 1 & 2 & 3 & \dots & n \end{array} \end{array}}{\begin{pmatrix} a_{11} & a_{12} & a_{13} & \dots & a_{1n} \\ a_{21} & a_{22} & a_{23} & \dots & a_{2n} \\ a_{31} & a_{32} & a_{33} & \dots & a_{2n} \\ \vdots & \vdots & \vdots & \ddots & \vdots \\ a_{m1} & a_{m2} & a_{m3} & \dots & a_{mn} \end{pmatrix}}$$

In order to refer to the element which is in the third row and the second column we can say a_{32}. The matrix

$$\begin{pmatrix} 3 & 2 \\ 6 & 1 \\ 8 & 2 \end{pmatrix}$$

is a 3×2 matrix (read as 3 by 2) as it has 3 rows and 2 columns.

The sum and difference of matrices

The sum and difference of matrices is found by adding or subtracting corresponding elements of the matrix. Only matrices of exactly the same dimension can be added or subtracted.

Example 13.1

$$\mathbf{A} = (2 \quad 1) \qquad \mathbf{B} = (6 \quad 2)$$

$$\mathbf{C} = \begin{pmatrix} 3 & 7 \\ 8 & 3 \end{pmatrix}$$

$$\mathbf{D} = \begin{pmatrix} 1 & 2 \\ 2 & 1 \end{pmatrix}$$

$$\mathbf{E} = \begin{pmatrix} 8 & 2 & 1 \\ 6 & 1 & 3 \end{pmatrix}$$

$$\mathbf{F} = \begin{pmatrix} 2 & 6 & 3 \\ 12 & -2 & -6 \end{pmatrix}$$

Find where possible: (a) $\mathbf{A} + \mathbf{B}$, (b) $\mathbf{C} + \mathbf{D}$, (c) $\mathbf{E} - \mathbf{F}$, (d) $\mathbf{A} + \mathbf{D}$.

Solution

(a) $\mathbf{A} + \mathbf{B} = (2 \quad 1) + (6 \quad 2) = (8 \quad 3)$

(b) $\mathbf{C} + \mathbf{D} = \begin{pmatrix} 3 & 7 \\ 8 & 3 \end{pmatrix} + \begin{pmatrix} 1 & 2 \\ 2 & 1 \end{pmatrix} = \begin{pmatrix} 4 & 9 \\ 10 & 4 \end{pmatrix}$

(c) $\mathbf{E} - \mathbf{F} = \begin{pmatrix} 8 & 2 & 1 \\ 6 & 1 & 3 \end{pmatrix} - \begin{pmatrix} 2 & 6 & 3 \\ 12 & -2 & -6 \end{pmatrix}$

$$= \begin{pmatrix} 6 & -4 & -2 \\ -6 & 3 & 9 \end{pmatrix}$$

(d) $\mathbf{A} + \mathbf{D}$ cannot be found because the two matrices are of different dimensions.

Multiplication of a matrix by a scalar

To multiply a matrix by a scalar, every element is multiplied by the scalar.

Example 13.2

If $\mathbf{A} = \begin{pmatrix} 2 & 5 \\ 6 & 1 \end{pmatrix}$

find $2\mathbf{A}$ and $\frac{1}{3}\mathbf{A}$

Solution

$$2\mathbf{A} = 2 \begin{pmatrix} 2 & 5 \\ 6 & 1 \end{pmatrix} = \begin{pmatrix} 4 & 10 \\ 12 & 2 \end{pmatrix}$$

$$\frac{1}{3}\mathbf{A} = \frac{1}{3} \begin{pmatrix} 2 & 5 \\ 6 & 1 \end{pmatrix} = \begin{pmatrix} \frac{2}{3} & \frac{5}{3} \\ 2 & \frac{1}{3} \end{pmatrix}$$

Multiplication of two matrices

To multiply two matrices, every row is multiplied by every column. For instance, if $\mathbf{C} = \mathbf{AB}$, to find the element in the second row and the third column of the product, $\mathbf{C}$, we take the second row of $\mathbf{A}$ and the third column of **B** and multiply them together, like taking the scalar product of two vectors. Multiplication is only possible if the number of columns in **A** is the same as the number of rows in **B**. For instance, if **A** is 2×3 it can only multiply matrices that are $3 \times n$ where n could be any dimension. The result of a 2×3 multiplying a 3×4 is a 2×4 matrix. Notice the pattern:

$$\underbrace{(2 \times 3)}_{\text{Must be equal}} \text{ multiplying } (3 \times 4) \text{ gives } 2 \times 4$$

Example 13.3

$$\mathbf{A} = \begin{pmatrix} 1 & -1 \\ 3 & 1 \end{pmatrix}$$

$$\mathbf{B} = \begin{pmatrix} 6 & 0 & -1 \\ 2 & 2 & 3 \end{pmatrix}$$

Find, if possible, **AB** and **BA**

$$\mathbf{AB} = \begin{pmatrix} 1 & -1 \\ 3 & 1 \end{pmatrix} \begin{pmatrix} 6 & 0 & -1 \\ 2 & 2 & 3 \end{pmatrix}$$

$$\begin{pmatrix} 1 \cdot 6 + (-1) \cdot 2 & 1 \cdot 0 + (-1) \cdot 2 & 1 \cdot (-1) + (-1) \cdot (3) \\ 3 \cdot 6 + 1 \cdot 2 & 3 \cdot 0 + 1 \cdot 2 & 3 \cdot (-1) + 1 \cdot 3 \end{pmatrix}$$

$$= \begin{pmatrix} 4 & -2 & -4 \\ 20 & 2 & 0 \end{pmatrix}$$

BA cannot be found because the number of columns in **B** is not equal to the number of rows in **A**.

We can justify the practical reasons for this method of matrix multiplication as in the following two examples. In the first, we return to our household shopping example.

Example 13.4   Every weekday a household orders pints of milk, loaves of bread and yoghurt from a milk lorry. The orders for the week are as follows:

	Milk	Bread	Yoghurt
Monday	3	2	4
Tuesday	4	1	0
Wednesday	2	2	4
Thursday	5	1	0
Friday	1	1	4

Next week, the dairy introduces a special offer and reduces its prices. The prices for this week and the next are as follows:

	This week	Next week
Milk	0.34	0.32
Bread	0.60	0.50
Yoghurt	0.33	0.30

Calculate the cost each day for this week and the next.

Solution   The cost each day is made up of the number of pints of milk times the cost of a pint plus the number of loaves of bread times the cost of a loaf plus the number of cartons of yoghurt times the cost of the yoghurt. In other words, we can find the cost each day by performing matrix multiplication

$$
\begin{pmatrix} 3 & 2 & 4 \\ 4 & 1 & 0 \\ 2 & 2 & 4 \\ 5 & 1 & 0 \\ 1 & 1 & 4 \end{pmatrix}
\begin{pmatrix} 0.34 & 0.32 \\ 0.60 & 0.50 \\ 0.33 & 0.30 \end{pmatrix}
$$

$$
= \begin{pmatrix}
3 \times (0.34) + 2 \times (0.60) + 4 \times (0.33) & 3 \times (0.32) + 2 \times (0.50) + 4 \times (0.30) \\
4 \times (0.34) + 1 \times (0.60) + 0 \times (0.33) & 4 \times (0.32) + 1 \times (0.50) + 0 \times (0.30) \\
2 \times (0.34) + 2 \times (0.60) + 4 \times (0.33) & 2 \times (0.32) + 2 \times (0.50) + 4 \times (0.30) \\
5 \times (0.34) + 1 \times (0.60) + 0 \times (0.33) & 5 \times (0.32) + 1 \times (0.50) + 0 \times (0.30) \\
1 \times (0.34) + 1 \times (0.60) + 4 \times (0.33) & 1 \times (0.32) + 1 \times (0.50) + 4 \times (0.30)
\end{pmatrix}
$$

$$
= \begin{pmatrix}
3.54 & 3.16 \\
1.96 & 1.78 \\
3.20 & 2.84 \\
2.30 & 2.10 \\
2.26 & 2.02
\end{pmatrix}
$$

The rows now represent the days of the week and the columns represent this week and the next week. Hence, for instance, the cost for Thursday of next week is given by the element $a_{42} = 2.10$.

Example 13.5   Figure 13.1 represents a communication network where the vertices a,b,f,g represent offices and vertices c,d,e represent switching centres. The numbers marked along the edges represent the number of connections between any two vertices. Calculate the number of routes from a,b to f,g.

Solution   The number of routes from a to f can be calculated by taking the number via c plus the number via d plus the number via e. In each case, this is given by multiplying the number of connections along the edges connecting a to c, c to f, etc giving the number of routes from a to f as: $3 \times 2 + 4 \times 6 + 1 \times 1$.

We can see that we can get the number of routes by matrix multiplication. The network from ab to cde is represented by:

$$
\begin{array}{c}
\quad \begin{array}{ccc} c & d & e \end{array} \\
\begin{array}{c} a \\ b \end{array} \begin{pmatrix} 3 & 4 & 1 \\ 2 & 1 & 3 \end{pmatrix}
\end{array}
$$

and from cde to fg by

$$
\begin{array}{c}
\quad \begin{array}{cc} f & g \end{array} \\
\begin{array}{c} c \\ d \\ e \end{array} \begin{pmatrix} 2 & 1 \\ 6 & 3 \\ 1 & 2 \end{pmatrix}
\end{array}
$$

So, the total number of routes is given by

$$
\begin{pmatrix} 3 & 4 & 1 \\ 2 & 1 & 3 \end{pmatrix}
\begin{pmatrix} 2 & 1 \\ 6 & 3 \\ 1 & 2 \end{pmatrix}
$$

$$
= \begin{pmatrix}
3 \times 2 + 4 \times 6 + 1 \times 1 & 3 \times 1 + 4 \times 3 + 1 \times 2 \\
2 \times 2 + 1 \times 6 + 3 \times 1 & 2 \times 1 + 3 \times 1 + 3 \times 2
\end{pmatrix}
$$

$$
\begin{array}{c}
\quad \begin{array}{cc} f & g \end{array} \\
\begin{array}{c} a \\ b \end{array} \begin{pmatrix} 31 & 17 \\ 13 & 11 \end{pmatrix}
\end{array}
$$

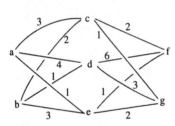

Figure 13.1   *A representation of a communication network.*

Hence, by interpreting the rows and columns of the resulting matrix we can see that there are 31 routes from a to f, 17 from a to g, 13 from b to f and 11 from b to g.

The unit matrix

The unit matrix is a square matrix which leaves any matrix, $\mathbf{A}$, unchanged under multiplication. If $\mathbf{A}$ is a square matrix, then

$$\mathbf{AI} = \mathbf{IA} = \mathbf{A}$$

The unit matrix has 1s on its leading diagonal and 0s elsewhere. In two dimensions

$$\mathbf{I} = \begin{pmatrix} 1 & 0 \\ 0 & 1 \end{pmatrix}$$

In three dimensions

$$\mathbf{I} = \begin{pmatrix} 1 & 0 & 0 \\ 0 & 1 & 0 \\ 0 & 0 & 1 \end{pmatrix}.$$

Example 13.6

$$A = \begin{pmatrix} 2 & -1 \\ 0 & 1 \end{pmatrix}, \quad \mathbf{B} = \begin{pmatrix} 3 \\ 2 \end{pmatrix}$$

Show that $\mathbf{AI} = \mathbf{IA} = \mathbf{A}$ and $\mathbf{IB} = \mathbf{B}$.

Solution

$$\mathbf{AI} = \begin{pmatrix} 2 & -1 \\ 0 & 1 \end{pmatrix} \begin{pmatrix} 1 & 0 \\ 0 & 1 \end{pmatrix}$$

$$= \begin{pmatrix} 2 \times 1 + (-1) \times 0 & 2 \times 0 + (-1) \times 1 \\ 0 \times 1 + 1 \times 0 & 0 \times 0 + 1 \times 1 \end{pmatrix}$$

$$= \begin{pmatrix} 2 & -1 \\ 0 & 1 \end{pmatrix} = \mathbf{A}$$

$$\mathbf{IA} = \begin{pmatrix} 1 & 0 \\ 0 & 1 \end{pmatrix} \begin{pmatrix} 2 & -1 \\ 0 & 1 \end{pmatrix}$$

$$= \begin{pmatrix} 1 \times 2 + 0 \times 0 & 1 \times (-1) + 0 \times 1 \\ 0 \times 2 + 1 \times 0 & 0 \times (-1) + 1 \times 1 \end{pmatrix}$$

$$= \begin{pmatrix} 2 & -1 \\ 0 & 1 \end{pmatrix} = \mathbf{A}$$

$$\mathbf{IB} = \begin{pmatrix} 1 & 0 \\ 0 & 1 \end{pmatrix} \begin{pmatrix} 3 \\ 2 \end{pmatrix}$$

$$= \begin{pmatrix} 1 \times 3 + 0 \times 2 \\ 0 \times 3 + 1 \times 2 \end{pmatrix}$$

$$= \begin{pmatrix} 3 \\ 2 \end{pmatrix} = \mathbf{B}$$

The transpose of a matrix

The transpose of a matrix is obtained by interchanging the rows and the columns. The transpose of a matrix $\mathbf{A}$ is represented by $\mathbf{A}^{\mathrm{T}}$.

Example 13.7 Given

$$\mathbf{A} = \begin{pmatrix} 2 & -1 \\ 6 & 3 \end{pmatrix}, \quad \mathbf{B} = \begin{pmatrix} 2 & 1 & 8 \\ -1 & 0 & 1 \end{pmatrix}$$

find $\mathbf{A}^T$ and $\mathbf{B}^T$

Solution The first row of $\mathbf{A}$ is $(2 - 1)$ therefore this is the first column of $\mathbf{A}^T$. The second row of $\mathbf{A}$ is $(6\ 3)$ therefore this is the second column of $\mathbf{A}^T$. This gives $\mathbf{A}^T$ as follows.

$$\mathbf{A}^T = \begin{pmatrix} 2 & 6 \\ -1 & 3 \end{pmatrix}$$

Similarly

$$\mathbf{B}^T = \begin{pmatrix} 2 & -1 \\ 1 & 0 \\ 8 & 1 \end{pmatrix}$$

Some special types of matrices

A *square matrix* has the same number of rows as columns.

$$\begin{pmatrix} 2 & -1 \\ 6 & 3 \end{pmatrix}$$

is a square matrix of dimension 2.

$$\begin{pmatrix} 8 & 6 & 2 \\ -3 & 1 & 0 \\ 3 & 2 & 1 \end{pmatrix}$$

is a square matrix of dimension 3.

A square matrix has a leading diagonal, which comprises the elements lying along the diagonal from the top left-hand corner to the bottom right-hand corner as marked below. These elements have the same row number as they have column number.

$$\begin{pmatrix} 8 & 6 & 2 \\ -3 & 1 & 0 \\ 3 & 2 & 1 \end{pmatrix}$$

The leading diagonal is shown by the dotted line in the above matrix.

A *diagonal matrix* is a square matrix which has zero elements everywhere except, possibly, on its leading diagonal, for example

$$\begin{pmatrix} 4 & 0 & 0 \\ 0 & -2 & 0 \\ 0 & 0 & 3 \end{pmatrix}$$

An *upper triangular matrix* is a square matrix which has zeros below the leading diagonal, for example

$$\begin{pmatrix} 1 & 1 & 2 \\ 0 & 6 & 6 \\ 0 & 0 & 8 \end{pmatrix}$$

A *lower triangular matrix* has zeros above the leading diagonal, for example

$$\begin{pmatrix} 1 & 0 & 0 \\ 3 & -1 & 0 \\ 6 & 8 & 2 \end{pmatrix}$$

A *symmetric matrix* is such that $\mathbf{A}^T = \mathbf{A}$, that is, the elements are symmetric about the leading diagonal, for example

$$\mathbf{A} = \begin{pmatrix} 1 & 6 & -3 \\ 6 & 0 & -2 \\ -3 & -2 & 8 \end{pmatrix}, \quad \mathbf{B} = \begin{pmatrix} 1 & 6 \\ 6 & 1 \end{pmatrix}$$

are symmetric matrices. If you take the transpose of one of these matrices they result in the original matrix.

A *skew-symmetric matrix* is such that $\mathbf{A}^T = -\mathbf{A}$.

Example 13.8 Show that

$$\mathbf{A} = \begin{pmatrix} 0 & 6 \\ -6 & 0 \end{pmatrix}$$

is skew symmetric.

Solution

$$\mathbf{A}^T = \begin{pmatrix} 0 & -6 \\ 6 & 0 \end{pmatrix}$$

Multiplying $\mathbf{A}$ by -1, we get

$$-\mathbf{A} = \begin{pmatrix} 0 & -6 \\ 6 & 0 \end{pmatrix}$$

We can see that $\mathbf{A}^T = -\mathbf{A}$ and hence we have shown that $\mathbf{A}$ is skew symmetric.

Hermitian matrix

A Hermitian matrix is such that $\mathbf{A}^{*T} = \mathbf{A}$.

Example 13.9 Show that

$$\mathbf{A} = \begin{pmatrix} 3 & 7+j2 \\ 7-j2 & -2 \end{pmatrix} \quad \text{and} \quad \mathbf{B} = \begin{pmatrix} 2 & 3\,e^{-j2} \\ 3\,e^{j2} & 1 \end{pmatrix}$$

are Hermitian.

Solution Taking the complex conjugates of each of the elements in **A** and **B** gives

$$\mathbf{A}^* = \begin{pmatrix} 3 & 7-j2 \\ 7+j2 & -2 \end{pmatrix} \quad \text{and} \quad \mathbf{B}^* = \begin{pmatrix} 2 & 3\,e^{j2} \\ 3\,e^{-j2} & 1 \end{pmatrix}$$

Now taking the transposes of **A** and **B**, we get

$$\mathbf{A}^{*T} = \begin{pmatrix} 3 & 7+j2 \\ 7-j2 & -2 \end{pmatrix} \quad \text{and} \quad \mathbf{B}^{*T} = \begin{pmatrix} 2 & 3\,e^{-j2} \\ 3\,e^{j2} & 1 \end{pmatrix}$$

So we can see that

$$\mathbf{A}^{*T} = \mathbf{A} \quad \text{and} \quad \mathbf{B}^{*T} = \mathbf{B}$$

showing that they are Hermitian.

In the rest of this chapter we shall assume that our matrices are real. A *column vector* is a matrix with only one column, for example

$$\mathbf{v} = \begin{pmatrix} 1 \\ 2 \\ 3 \end{pmatrix}$$

A *row vector* is a matrix with only one row, for example

$$\mathbf{v} = (1 \quad 2 \quad 3).$$

The inverse of a matrix

The inverse of a matrix **A** is a matrix $\mathbf{A}^{-1}$ such that $\mathbf{A}\mathbf{A}^{-1} = \mathbf{A}^{-1}\mathbf{A} = \mathbf{I}$ (the unit matrix).

Example 13.10 Show that

$$\begin{pmatrix} \frac{1}{3} & \frac{1}{3} \\ \frac{1}{3} & -\frac{2}{3} \end{pmatrix}$$

is the inverse of

$$\begin{pmatrix} 2 & 1 \\ 1 & -1 \end{pmatrix}.$$

Solution Multiply:

$$\begin{pmatrix} \frac{1}{3} & \frac{1}{3} \\ \frac{1}{3} & -\frac{2}{3} \end{pmatrix} \begin{pmatrix} 2 & 1 \\ 1 & -1 \end{pmatrix}$$

$$= \begin{pmatrix} \frac{1}{3}(2) + \frac{1}{3}(1) & \frac{1}{3}(1) + \frac{1}{3}(-1) \\ \frac{1}{3}(2) + \left(-\frac{2}{3}\right)(1) & \frac{1}{3}(1) + \left(-\frac{2}{3}\right)(-1) \end{pmatrix}$$

$$= \begin{pmatrix} 1 & 0 \\ 0 & 1 \end{pmatrix}.$$

Also

$$\begin{pmatrix} 2 & 1 \\ 1 & -1 \end{pmatrix} \begin{pmatrix} \frac{1}{3} & \frac{1}{3} \\ \frac{1}{3} & -\frac{2}{3} \end{pmatrix}$$

$$= \begin{pmatrix} (2)\frac{1}{3} + (1)\frac{1}{3} & (2)\frac{1}{3} + (1)\left(-\frac{2}{3}\right) \\ (1)\frac{1}{3} + (-1)\frac{1}{3} & (1)\frac{1}{3} + (-1)\left(-\frac{2}{3}\right) \end{pmatrix}$$

$$= \begin{pmatrix} 1 & 0 \\ 0 & 1 \end{pmatrix}.$$

Not all matrices have inverses and only square matrices can possibly have inverses. A matrix does not have an inverse if its determinant is 0.

The determinant of

$$\begin{pmatrix} a & b \\ c & d \end{pmatrix}$$

is given by

$$\begin{vmatrix} a & b \\ c & d \end{vmatrix} = ad - cb$$

If the determinant of a matrix is 0 then it has no inverse and the matrix is said to be singular. If the determinant is non- zero then the inverse exists. The inverse of the 2×2 matrix

$$\begin{pmatrix} a & b \\ c & d \end{pmatrix}$$

is

$$\frac{1}{(ad - cb)} \begin{pmatrix} d & -b \\ -c & a \end{pmatrix}$$

That is, to find the inverse of a 2×2 matrix, we swap the diagonal elements, negate the off-diagonal elements, and divide the resulting matrix by the determinant.

Example 13.11 Find the determinants of the following matrices and state if the matrix has an inverse or is singular. Find the inverse in the cases where is exists and check that $\mathbf{AA}^{-1} = \mathbf{A}^{-1}\mathbf{A} = \mathbf{I}$

(a) $\begin{pmatrix} -1 & 3 \\ 2 & 1 \end{pmatrix}$,   (b) $\begin{pmatrix} 6 & -2 \\ -3 & 1 \end{pmatrix}$,   (c) $\begin{pmatrix} \frac{1}{\sqrt{2}} & -\frac{1}{\sqrt{2}} \\ \frac{1}{\sqrt{2}} & \frac{1}{\sqrt{2}} \end{pmatrix}$.

Solution

(a) $\begin{vmatrix} -1 & 3 \\ 2 & 1 \end{vmatrix} = (-1) \times 1 - 2 \times 3 = -7.$

As the determinant is not zero the matrix

$$\begin{pmatrix} -1 & 3 \\ 2 & 1 \end{pmatrix}$$

has an inverse found by swapping the diagonal elements and negating the off-diagonal elements, then dividing by the determinant. This gives

$$\frac{1}{-7} \begin{pmatrix} 1 & -3 \\ -2 & -1 \end{pmatrix} = \frac{1}{7} \begin{pmatrix} -1 & 3 \\ 2 & 1 \end{pmatrix}.$$

Check that $\mathbf{AA}^{-1} = \mathbf{I}$

$$\begin{pmatrix} -1 & 3 \\ 2 & 1 \end{pmatrix} \frac{1}{7} \begin{pmatrix} -1 & 3 \\ 2 & 1 \end{pmatrix}$$

$$= \frac{1}{7} \begin{pmatrix} (-1)(-1) + (3)(2) & (-1)3 + (3)(1) \\ (2)(-1) + (1)(2) & (2)(3) + (1)(1) \end{pmatrix} = \begin{pmatrix} 1 & 0 \\ 0 & 1 \end{pmatrix}$$

and that $\mathbf{A}^{-1}\mathbf{A} = \mathbf{I}$

$$\frac{1}{7} \begin{pmatrix} -1 & 3 \\ 2 & 1 \end{pmatrix} \begin{pmatrix} -1 & 3 \\ 2 & 1 \end{pmatrix}$$

$$= \frac{1}{7} \begin{pmatrix} (-1)(-1) + (3)(2) & (-1)3 + (3)(1) \\ (2)(-1) + (1)(2) & (2)(3) + (1)(1) \end{pmatrix} = \begin{pmatrix} 1 & 0 \\ 0 & 1 \end{pmatrix}.$$

(b) $\begin{vmatrix} 6 & -2 \\ -3 & 1 \end{vmatrix} = 6 \cdot 1 - (-3)(-2) = 0$

As the determinant is zero the matrix

$$\begin{pmatrix} 6 & -2 \\ -3 & 1 \end{pmatrix}$$

has no inverse. It is singular.

(c) $\begin{vmatrix} \frac{1}{\sqrt{2}} & -\frac{1}{\sqrt{2}} \\ \frac{1}{\sqrt{2}} & \frac{1}{\sqrt{2}} \end{vmatrix}$

$$= \frac{1}{\sqrt{2}} \frac{1}{\sqrt{2}} - \left(-\frac{1}{\sqrt{2}}\right) \frac{1}{\sqrt{2}} = 1.$$

Therefore, the matrix is invertible. Its inverse is given by swapping the diagonal elements, and negating the off-diagonal elements, and then dividing by the determinant. This gives

$$\begin{pmatrix} \frac{1}{\sqrt{2}} & \frac{1}{\sqrt{2}} \\ -\frac{1}{\sqrt{2}} & \frac{1}{\sqrt{2}} \end{pmatrix}$$

Check that $\mathbf{AA}^{-1} = \mathbf{I}$:

$$\mathbf{AA}^{-1} = \begin{pmatrix} \frac{1}{\sqrt{2}} & -\frac{1}{\sqrt{2}} \\ \frac{1}{\sqrt{2}} & \frac{1}{\sqrt{2}} \end{pmatrix} \begin{pmatrix} \frac{1}{\sqrt{2}} & \frac{1}{\sqrt{2}} \\ -\frac{1}{\sqrt{2}} & \frac{1}{\sqrt{2}} \end{pmatrix} = \begin{pmatrix} 1 & 0 \\ 0 & 1 \end{pmatrix}$$

Similarly, $\mathbf{A}^{-1}\mathbf{A} = \mathbf{I}$.

Solving matrix equations

To solve matrix equations, we use the same ideas about equivalent equations that we have used before. As in ordinary equations, we can 'do the same things to both sides' in order to find equivalent equations. It is important to remember that division by a matrix has not been defined. In order to 'undo' matrix multiplication we have to multiply by an inverse matrix, where it exists, and we need to specify whether we are pre- or post-multiplying. This is necessary because matrices do not obey the commutative law ($\mathbf{AB} \neq \mathbf{BA}$). If we pre- or post-multiply both sides of an equation by a matrix we must also be able to justify that the dimensions of the expressions are such that the multiplication is possible. Also if we add or subtract a matrix from both sides of the equation it must have exactly the same dimension as the current matrix expression.

Example 13.12   Given that **A**, **B**, and **C** are matrices and **AB** = **C** where **A** and **B** are non-singular, find expressions for **B** and **A**.

Solution   In this case, we are told that **A** and **B** are invertible, so they must be square and therefore **C** must also be square and of the same dimension. To find **B** we wish to 'get rid' of the **A** term on the left-hand side. We pre-multiply both sides of the equation by $\mathbf{A}^{-1}$

   AB = **C** and given **A** is invertible

$\Leftrightarrow$ $\mathbf{A}^{-1}\mathbf{AB} = \mathbf{A}^{-1}\mathbf{C}$.

Now using $\mathbf{A}^{-1}\mathbf{A} = \mathbf{I}$, the unit matrix, we have

$\mathbf{IB} = \mathbf{A}^{-1}\mathbf{C}$.

As the unit matrix multiplied by any matrix leaves it unchanged, we have

$\Leftrightarrow$ $\mathbf{B} = \mathbf{A}^{-1}\mathbf{C}$.

To find an expression for **A**, use

AB = **C**

given that **B** is invertible, we post-multiply by $\mathbf{B}^{-1}$

$\Leftrightarrow$ $\mathbf{ABB}^{-1} = \mathbf{CB}^{-1}$.

Now using $\mathbf{BB}^{-1} = \mathbf{I}$, the unit matrix, we have

$\mathbf{AI} = \mathbf{CB}^{-1}$.

As the unit matrix multiplied by any matrix leaves it unchanged, we have

$\Leftrightarrow$ $\mathbf{A} = \mathbf{CB}^{-1}$.

Remember that it is always important to specify whether you are pre- or post-multiplying when solving matrix equations. A term like $\mathbf{B}^{-1}\mathbf{AB}$ cannot be simplified because we cannot swap the order, as we would do with numbers.

13.3
Transformations

On a computer graphics screen an object is represented by a set of coordinates, either with reference to the screen origin or with reference to the origin of some window created by the graphical user interface (GUI). We may wish to move the object around inside its window. We shall consider in this section only two-dimensional objects as dealing with three-dimensional objects would add the complication of needing to represent a perspective view. Ideas about transformations are also important when considering movement of a robotic arm.

There are three ways of moving an object without affecting its overall size or shape: rotation, reflection and translation. We could also stretch it or compress it in some direction – the operation of scaling.

We shall look at how to perform these operations using matrices and vectors. We can check that the operations performed are those that we expected by looking at the effect on some simple shapes. In most of these examples, we look at the effect of a unit square at the origin, defined by the points A (0,0), B(1,0), C (1,1), D (0,1). The outcome of the transformation is called the image which we will represent by the points A′, B′, C′, D′. The transformation, T, is a function whose domain and codomain is the plane (which is referred to as $\mathbb{R}^2$). The term 'mapping' is also used in this context. It has exactly the same meaning as function, but is more often used when referring to geometrical problems.

Rotation

To perform a rotation through an angle θ, we multiply the position vector of the point

$$\begin{pmatrix} x \\ y \end{pmatrix}$$

by a matrix of the form

$$\begin{pmatrix} \cos(\theta) & -\sin(\theta) \\ \sin(\theta) & \cos(\theta) \end{pmatrix}.$$

Example 13.13   Find and draw the image of the unit square with vertices A(0,0), B(1,0), C(1,1), D(0,1) after rotation through 30° about the origin.

Solution   Rotation through 30° about the origin is found by multiplying the position vectors of the points by

$$\begin{pmatrix} \cos(30°) & -\sin(30°) \\ \sin(30°) & \cos(30°) \end{pmatrix} \approx \begin{pmatrix} 0.866 & -0.5 \\ 0.5 & 0.866 \end{pmatrix}$$

To find the image of the unit square, we multiply the position vectors of the vertices by this matrix

$$\begin{pmatrix} 0.866 & -0.5 \\ 0.5 & 0.866 \end{pmatrix} \begin{pmatrix} 0 \\ 0 \end{pmatrix} = \begin{pmatrix} 0 \\ 0 \end{pmatrix}$$

$$\begin{pmatrix} 0.866 & -0.5 \\ 0.5 & 0.866 \end{pmatrix} \begin{pmatrix} 1 \\ 0 \end{pmatrix} = \begin{pmatrix} 0.866 \\ 0.5 \end{pmatrix}$$

$$\begin{pmatrix} 0.866 & -0.5 \\ 0.5 & 0.866 \end{pmatrix} \begin{pmatrix} 1 \\ 1 \end{pmatrix} = \begin{pmatrix} 0.366 \\ 1.366 \end{pmatrix}$$

$$\begin{pmatrix} 0.866 & -0.5 \\ 0.5 & 0.866 \end{pmatrix} \begin{pmatrix} 0 \\ 1 \end{pmatrix} = \begin{pmatrix} -0.5 \\ 0.866 \end{pmatrix}$$

This transformation is shown in Figure 13.2.

Sometimes, it is useful to be able to rotate the axes rather than the object. For instance, the object may be held by a robotic arm and we want the arm to rotate but keep the orientation of the object the same. This is picture for the tea drinking robot in Figure 13.3.

In this case, if we rotate the axes Ox, Oy, by the position of the object remains the same but even so has new coordinates relative to the the transformed axes OX, OY. If the axes rotate through 30°, then the object moves relative to the axes by $-30°$. So to rotate the axes by we multiply the position vectors of the points

$$\begin{pmatrix} x \\ y \end{pmatrix}$$

by the matrix

$$\begin{pmatrix} \cos(-\theta) & -\sin(-\theta) \\ \sin(-\theta) & \cos(-\theta) \end{pmatrix} = \begin{pmatrix} \cos(\theta) & \sin(\theta) \\ -\sin(\theta) & \cos(\theta) \end{pmatrix}.$$

Example 13.14   A unit square has vertices A(0,0), B(1,0), C(1,1), D(0,1) relative to axes Ox, Oy. The axes are rotated through 30° to OX, OY, without moving the square. Find the coordinates of the vertices relative to the new axes OX, OY.

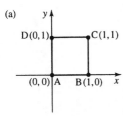

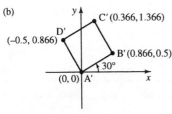

Figure 13.2   *(a) The unit square with vertices A(0,0), B(1,0), C (1,1) D(0,1). (b) The same unit square after rotation by 30°.*

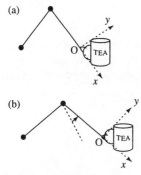

Figure 13.3   *In order not to spill the tea, the axes – defined with reference to the lower arm – rotate but the orientation of the tea cup must stay the same.*

(a)

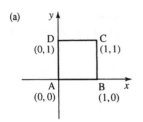

(b)

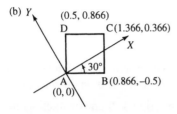

Figure 13.4   *(a) The unit square with vertices* A(0,0), B(1,0), C (1,1) D(0,1) *relative to axes Ox, Oy. (b) The same unit square shown relative to axes OX, OY found by rotating Ox, Oy through 30°.*

(a)

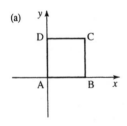

(b)

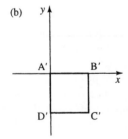

(c)

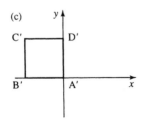

Figure 13.5   *(a) The unit square with vertices* A(0,0), B(1,0), C (1,1) D(0,1). *(b) The same unit square after reflection in the x axis. (c) After reflection in the y axis.*

Solution The effect of rotating the axes through $30°$ is found by multiplying the position vectors of the points by

$$\begin{pmatrix} \cos(30°) & \sin(30°) \\ \sin(-30°) & \cos(30°) \end{pmatrix} = \begin{pmatrix} 0.866 & 0.5 \\ -0.5 & 0.866 \end{pmatrix}.$$

To find the coordinates of the unit square relative to the new axes, we multiply the position vectors of the vertices by this matrix

$$\begin{pmatrix} 0.866 & 0.5 \\ -0.5 & 0.866 \end{pmatrix} \begin{pmatrix} 0 \\ 0 \end{pmatrix} = \begin{pmatrix} 0 \\ 0 \end{pmatrix}$$

$$\begin{pmatrix} 0.866 & 0.5 \\ -0.5 & 0.866 \end{pmatrix} \begin{pmatrix} 1 \\ 0 \end{pmatrix} = \begin{pmatrix} 0.866 \\ -0.5 \end{pmatrix}$$

$$\begin{pmatrix} 0.866 & 0.5 \\ -0.5 & 0.866 \end{pmatrix} \begin{pmatrix} 1 \\ 1 \end{pmatrix} = \begin{pmatrix} 1.366 \\ 0.366 \end{pmatrix}$$

$$\begin{pmatrix} 0.866 & 0.5 \\ -0.5 & 0.866 \end{pmatrix} \begin{pmatrix} 0 \\ 1 \end{pmatrix} = \begin{pmatrix} 0.5 \\ 0.866 \end{pmatrix}.$$

This is shown in Figure 13.4.

Reflection

To perform a reflection in the x-axis, we multiply the position vectors of the points

$$\begin{pmatrix} x \\ y \end{pmatrix}$$

by the matrix

$$\begin{pmatrix} 1 & 0 \\ 0 & -1 \end{pmatrix}$$

This has the effect of keeping the x-coordinate the same whilst changing the sign of the y-coordinate, hence turning the object upside down.

To perform a reflection in the y-axis, we multiply the position vectors of the points

$$\begin{pmatrix} x \\ y \end{pmatrix}$$

by the matrix

$$\begin{pmatrix} -1 & 0 \\ 0 & 1 \end{pmatrix}$$

which keeps the y-value constant while changing the sign of the x-coordinate. The effect on the unit square is shown in Figure 13.5.

Translation

Translation in the plane cannot be represented by multiplying by a 2×2 matrix. To perform a translation, we add the vector representing the translation to the original position vectors of the points.

Example 13.15 Find and draw the image of the unit square with vertices A(0,0), B(1,0), C(1,1), D(0,1) after translation through

$$\begin{pmatrix} 3 \\ 4 \end{pmatrix}.$$

Solution Add

$$\begin{pmatrix} 3 \\ 4 \end{pmatrix}$$

to the position vectors of the vertices, that is

$$\mathbf{v} + \begin{pmatrix} 3 \\ 4 \end{pmatrix}$$

which gives $\mathbf{A}'$ as (3,4), $\mathbf{B}'$ as (4,4), $\mathbf{C}'$ as (4,5), and $\mathbf{D}'$ as (3,5). This transformation is shown in Figure 13.6.

It is again often useful to consider what happens if the object stays where it is and the axes are translated. If the axes are translated through

$$\begin{pmatrix} 3 \\ 4 \end{pmatrix}$$

then the object appears to move relative to the axes by

$$\begin{pmatrix} -3 \\ -4 \end{pmatrix}.$$

Therefore, we subtract

$$\begin{pmatrix} 3 \\ 4 \end{pmatrix}$$

from the coordinates defining it. This is shown in Figure 13.7.

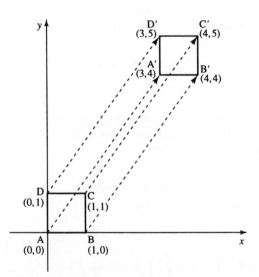

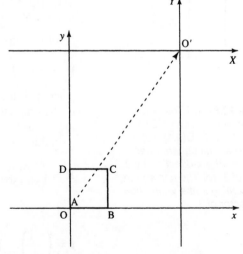

Figure 13.6 *(a) The unit square with vertices A(0,0), B(1,0), C (1,1) D(0,1). (b) The same unit square after translation through (3,4) becomes A′ (3,4), B′ (4,4), C′ (4,5), D′ (3,5).*

Figure 13.7 *(a) The unit square with vertices, relative to Ox, Oy A(0,0), B(1,0), C (1,1) D(0,1). (b) The unit square has co-ordinates* $(-3, -4), (-2, -4), (-2, -3), (-3, -3)$ *relative to the axes OX, OY which have been translated through (3,4).*

Scaling

To scale in the x-direction, we multiply the position vectors of the points

$$\begin{pmatrix} x \\ y \end{pmatrix}$$

by a matrix

$$\begin{pmatrix} S_x & 0 \\ 0 & 1 \end{pmatrix}$$

where S_x is the scale factor. Under this transformation, vectors that have no x-component will be unaffected. To scale in the y-direction, we multiply the position vectors of the points

$$\begin{pmatrix} x \\ y \end{pmatrix}$$

by a matrix

$$\begin{pmatrix} 1 & 0 \\ 0 & S_y \end{pmatrix}$$

where S_y is the scale factor. Under this transformation, vectors that have no y-component will be unaffected.

The effect on the unit square of scaling by 2 in the x-direction is shown in Figure 13.8(b) and of scaling by 3 in the y-direction is shown in Figure 13.8(c).

Combined transformations

Example 13.16   Find the coordinates of the vertices of the unit square after: (a) rotation about the origin through 50° followed by a translation of $(-1, 2)$; (b) translation of $(-1, 2)$ followed by rotation about the origin through 50°.

Solution   (a) We can write this combined transformation as

$$\mathbf{p}' = \mathbf{R}\mathbf{p} + \mathbf{t}$$

where $\mathbf{p}'$ is the position vector of the image point, $\mathbf{p}$ is the position vector of the original point, $\mathbf{R}$ is the matrix representing the rotation, and $\mathbf{t}$ is the vector representing the translation.

In this case

$$\mathbf{R} = \begin{pmatrix} \cos(50°) & -\sin(50°) \\ \sin(50°) & \cos(50°) \end{pmatrix} \approx \begin{pmatrix} 0.643 & -0.766 \\ 0.766 & 0.643 \end{pmatrix}$$

and

$$\mathbf{t} = \begin{pmatrix} -1 \\ 2 \end{pmatrix}, \quad \mathbf{p}' = \begin{pmatrix} x' \\ y' \end{pmatrix}, \quad \mathbf{p} = \begin{pmatrix} x \\ y \end{pmatrix}$$

So we have

$$\begin{pmatrix} x' \\ y' \end{pmatrix} = \begin{pmatrix} 0.643 & -0.766 \\ 0.766 & 0.643 \end{pmatrix} \begin{pmatrix} x \\ y \end{pmatrix} + \begin{pmatrix} -1 \\ 2 \end{pmatrix}$$

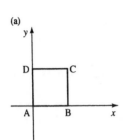

(a)

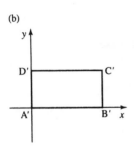

(b)

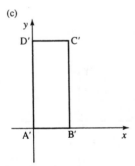

(c)

Figure 13.8   *(a) The unit square with vertices A(0,0), B(1,0), C (1,1), D(0,1). (b) The same unit square after scaling in the x-direction by a factor of 2. (c) The unit square after scaling in the y-direction by a factor of 3.*

For the coordinates of A′ substitute $x = 0$ and $y = 0$ giving

$$\begin{pmatrix} x' \\ y' \end{pmatrix} = \begin{pmatrix} 0.643 & -0.766 \\ 0.766 & 0.643 \end{pmatrix} \begin{pmatrix} 0 \\ 0 \end{pmatrix} + \begin{pmatrix} -1 \\ 2 \end{pmatrix} = \begin{pmatrix} 0 \\ 0 \end{pmatrix} + \begin{pmatrix} -1 \\ 2 \end{pmatrix} = \begin{pmatrix} -1 \\ 2 \end{pmatrix}$$

for B′

$$\begin{pmatrix} x' \\ y' \end{pmatrix} = \begin{pmatrix} 0.643 & -0.766 \\ 0.766 & 0.643 \end{pmatrix} \begin{pmatrix} 1 \\ 0 \end{pmatrix} + \begin{pmatrix} -1 \\ 2 \end{pmatrix} = \begin{pmatrix} 0.643 \\ 0.766 \end{pmatrix} + \begin{pmatrix} -1 \\ 2 \end{pmatrix}$$

$$= \begin{pmatrix} -0.357 \\ 2.766 \end{pmatrix}$$

for C′

$$\begin{pmatrix} x' \\ y' \end{pmatrix} = \begin{pmatrix} 0.643 & -0.766 \\ 0.766 & 0.643 \end{pmatrix} \begin{pmatrix} 1 \\ 1 \end{pmatrix} + \begin{pmatrix} -1 \\ 2 \end{pmatrix} = \begin{pmatrix} -0.123 \\ 1.409 \end{pmatrix} + \begin{pmatrix} -1 \\ 2 \end{pmatrix}$$

$$= \begin{pmatrix} -1.123 \\ 3.409 \end{pmatrix}$$

for D′

$$\begin{pmatrix} x' \\ y' \end{pmatrix} = \begin{pmatrix} 0.643 & -0.766 \\ 0.766 & 0.643 \end{pmatrix} \begin{pmatrix} 0 \\ 1 \end{pmatrix} + \begin{pmatrix} -1 \\ 2 \end{pmatrix} = \begin{pmatrix} -0.766 \\ 0.643 \end{pmatrix} + \begin{pmatrix} -1 \\ 2 \end{pmatrix}$$

$$= \begin{pmatrix} -1.766 \\ 2.643 \end{pmatrix}$$

The image of the unit square is pictured in Figure 13.9(b).
(b) We can write this combined transformation as

$$\mathbf{p}'' = \mathbf{R}(\mathbf{p} + \mathbf{t})$$

where $\mathbf{p}''$ is the position vector of the image point, $\mathbf{p}$ is the position vector of the original point, $\mathbf{R}$ is the matrix representing the rotation, and $\mathbf{t}$ is the vector representing the translation. We have put the brackets in to

Figure 13.9 *(a) The unit square with vertices* A(0,0), B(1,0), C (1,1) D(0,1). *(b) The same unit square after rotation through 50° about the origin and translation through* (−1, 2) *and the unit square after translation through* (−1, 2) *and then rotation of 50° about the origin.*

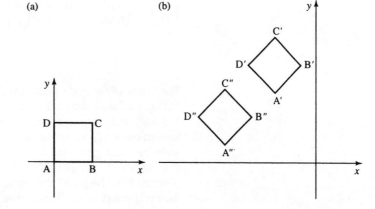

indicate that the translation is performed first. As before

$$\mathbf{R} = \begin{pmatrix} \cos(50°) & -\sin(50°) \\ \sin(50°) & \cos(50°) \end{pmatrix} \approx \begin{pmatrix} 0.643 & -0.766 \\ 0.766 & 0.643 \end{pmatrix}$$

and

$$\mathbf{t} = \begin{pmatrix} -1 \\ 2 \end{pmatrix}, \quad \mathbf{p}'' = \begin{pmatrix} x'' \\ y'' \end{pmatrix}, \quad \mathbf{p} = \begin{pmatrix} x \\ y \end{pmatrix}$$

So we have

$$\begin{pmatrix} x'' \\ y'' \end{pmatrix} = \begin{pmatrix} 0.643 & -0.766 \\ 0.766 & 0.643 \end{pmatrix} \left(\begin{pmatrix} x \\ y \end{pmatrix} + \begin{pmatrix} -1 \\ 2 \end{pmatrix} \right)$$

which is the same as

$$\begin{pmatrix} x'' \\ y'' \end{pmatrix} = \begin{pmatrix} 0.643 & -0.766 \\ 0.766 & 0.643 \end{pmatrix} \begin{pmatrix} x - 1 \\ y + 2 \end{pmatrix}$$

For the coordinates of A'', substitute $x = 0$ and $y = 0$ giving

$$\begin{pmatrix} x'' \\ y'' \end{pmatrix} = \begin{pmatrix} 0.643 & -0.766 \\ 0.766 & 0.643 \end{pmatrix} \begin{pmatrix} 0 - 1 \\ 0 + 2 \end{pmatrix} = \begin{pmatrix} 0.643 & -0.766 \\ 0.766 & 0.643 \end{pmatrix} \begin{pmatrix} -1 \\ 2 \end{pmatrix}$$

$$= \begin{pmatrix} -2.175 \\ 0.52 \end{pmatrix}$$

for B''

$$\begin{pmatrix} x'' \\ y'' \end{pmatrix} = \begin{pmatrix} 0.643 & -0.766 \\ 0.766 & 0.643 \end{pmatrix} \begin{pmatrix} 1 - 1 \\ 0 + 2 \end{pmatrix} = \begin{pmatrix} 0.643 & -0.766 \\ 0.766 & 0.643 \end{pmatrix} \begin{pmatrix} 0 \\ 2 \end{pmatrix}$$

$$= \begin{pmatrix} -1.532 \\ 1.286 \end{pmatrix}$$

for C''

$$\begin{pmatrix} x'' \\ y'' \end{pmatrix} = \begin{pmatrix} 0.643 & -0.766 \\ 0.766 & 0.643 \end{pmatrix} \begin{pmatrix} 1 - 1 \\ 1 + 2 \end{pmatrix} = \begin{pmatrix} 0.643 & -0.766 \\ 0.766 & 0.643 \end{pmatrix} \begin{pmatrix} 0 \\ 3 \end{pmatrix}$$

$$= \begin{pmatrix} -2.298 \\ 1.929 \end{pmatrix}$$

for D''

$$\begin{pmatrix} x'' \\ y'' \end{pmatrix} = \begin{pmatrix} 0.643 & -0.766 \\ 0.766 & 0.643 \end{pmatrix} \begin{pmatrix} 0 - 1 \\ 1 + 2 \end{pmatrix} = \begin{pmatrix} 0.643 & -0.766 \\ 0.766 & 0.643 \end{pmatrix} \begin{pmatrix} -1 \\ 3 \end{pmatrix}$$

$$= \begin{pmatrix} -2.941 \\ 1.163 \end{pmatrix}$$

The image of the unit square is pictured in Figure 13.9(b).

Note that the order of the transformations is important.

Sometimes, we might need to use a trick of temporarily moving the axes in order to perform certain transformations. Supposing we want to scale by 2 along the line $x = y$ we can rotate the axes temporarily so that the new X-axis lies along the line that was previously $x = y$, then perform X scaling, and then rotate back again, so the axes are back in their original position. This is done in the next example.

Example 13.17   Find a matrix that performs scaling by a factor of 2 along the direction $x = y$ and draw the image of the unit square defined by the vertices

$$A\left(-\tfrac{1}{2}, -\tfrac{1}{2}\right), B\left(\tfrac{1}{2}, -\tfrac{1}{2}\right), C\left(\tfrac{1}{2}, \tfrac{1}{2}\right), D\left(-\tfrac{1}{2}, \tfrac{1}{2}\right).$$

Solution   First, we rotate the axes by 45°, so that the OX-axis will lie along the line that was previously $x = y$. This is pictured in Figure 13.10.

The matrix that transforms the coordinates so they are relative to the new axes at an angle of 45° is given by:

$$\begin{pmatrix} \cos(45°) & \sin(45°) \\ -\sin(45°) & \cos(45°) \end{pmatrix}$$

A scaling of 2 in the X-direction is then performed by multiplying by

$$\begin{pmatrix} 2 & 0 \\ 0 & 1 \end{pmatrix}$$

We then need to rotate the axes back to their original position, that is, rotate the axes by −45°, this is done by multiplying by

$$\begin{pmatrix} \cos(-45°) & \sin(-45°) \\ -\sin(-45°) & \cos(-45°) \end{pmatrix} = \begin{pmatrix} \cos(45°) & -\sin(45°) \\ \sin(45°) & \cos(45°) \end{pmatrix}$$

Putting the three transformation matrices together we get

$$\begin{pmatrix} \cos(45°) & \sin(45°) \\ -\sin(45°) & \cos(45°) \end{pmatrix} \begin{pmatrix} 2 & 0 \\ 0 & 1 \end{pmatrix} \begin{pmatrix} \cos(45°) & -\sin(45°) \\ \sin(45°) & \cos(45°) \end{pmatrix}$$

which gives the matrix that represents a scaling along the line $x = y$.

Using $\cos(45°) = \tfrac{1}{\sqrt{2}} = \sin(45°)$, we get

$$\begin{pmatrix} \tfrac{1}{\sqrt{2}} & -\tfrac{1}{\sqrt{2}} \\ \tfrac{1}{\sqrt{2}} & \tfrac{1}{\sqrt{2}} \end{pmatrix} \begin{pmatrix} 2 & 0 \\ 0 & 1 \end{pmatrix} \begin{pmatrix} \tfrac{1}{\sqrt{2}} & \tfrac{1}{\sqrt{2}} \\ -\tfrac{1}{\sqrt{2}} & \tfrac{1}{\sqrt{2}} \end{pmatrix}$$

Taking out the two factors of $\tfrac{1}{\sqrt{2}}$ gives

$$\frac{1}{2}\begin{pmatrix} 1 & -1 \\ 1 & 1 \end{pmatrix} \begin{pmatrix} 2 & 0 \\ 0 & 1 \end{pmatrix} \begin{pmatrix} 1 & 1 \\ -1 & 1 \end{pmatrix}$$

Multiplying the second two matrices gives

$$\frac{1}{2}\begin{pmatrix} 1 & -1 \\ 1 & 1 \end{pmatrix} \begin{pmatrix} 2 & 2 \\ -1 & 1 \end{pmatrix}$$

and multiplying out the remaining two matrices gives

$$\frac{1}{2}\begin{pmatrix} 3 & 1 \\ 1 & 3 \end{pmatrix} = \begin{pmatrix} \tfrac{3}{2} & \tfrac{1}{2} \\ \tfrac{1}{2} & \tfrac{3}{2} \end{pmatrix}$$

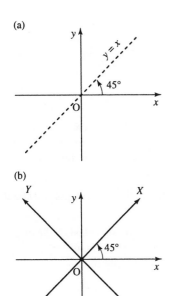

(a)

(b)

Figure 13.10   *(a) The line $x = y$ is at 45° to the Ox axis. If we rotate the axes by 45°, the new OX axis will lie in this direction. This is shown in (b).*

We can now multiply the position vectors representing the vertices of the square

$$\begin{pmatrix} \frac{3}{2} & \frac{1}{2} \\ \frac{1}{2} & \frac{3}{2} \end{pmatrix} \begin{pmatrix} -\frac{1}{2} \\ -\frac{1}{2} \end{pmatrix} = \begin{pmatrix} -1 \\ -1 \end{pmatrix}$$

$$\begin{pmatrix} \frac{3}{2} & \frac{1}{2} \\ \frac{1}{2} & \frac{3}{2} \end{pmatrix} \begin{pmatrix} \frac{1}{2} \\ -\frac{1}{2} \end{pmatrix} = \begin{pmatrix} \frac{1}{2} \\ -\frac{1}{2} \end{pmatrix}$$

$$\begin{pmatrix} \frac{3}{2} & \frac{1}{2} \\ \frac{1}{2} & \frac{3}{2} \end{pmatrix} \begin{pmatrix} \frac{1}{2} \\ \frac{1}{2} \end{pmatrix} = \begin{pmatrix} 1 \\ 1 \end{pmatrix}$$

$$\begin{pmatrix} \frac{3}{2} & \frac{1}{2} \\ \frac{1}{2} & \frac{3}{2} \end{pmatrix} \begin{pmatrix} -\frac{1}{2} \\ \frac{1}{2} \end{pmatrix} = \begin{pmatrix} -\frac{1}{2} \\ \frac{1}{2} \end{pmatrix}$$

The transformed figure is shown in Figure 13.11. We can see that has been stretched along the $x = y$ direction but has not been scaled along the other diagonal. The image is no longer a square but a rhombus.

Example 13.18   Find a transformation that will rotate any point **p** about $(1,1)$ through an angle of $90°$.

Solution   To rotate about a point not at the origin, we translate the origin temporarily, rotate, and then translate the origin back again.

Rotation through $90°$ is performed by multiplying by

$$\begin{pmatrix} \cos(90°) & -\sin(90°) \\ \sin(90°) & \cos(90°) \end{pmatrix} = \begin{pmatrix} 0 & -1 \\ 1 & 0 \end{pmatrix}$$

The combined transformation on a point **p** can be represented by

$$\mathbf{p}' = \begin{pmatrix} 0 & -1 \\ 1 & 0 \end{pmatrix} \left(\mathbf{p} - \begin{pmatrix} 1 \\ 1 \end{pmatrix} \right) + \begin{pmatrix} 1 \\ 1 \end{pmatrix}.$$

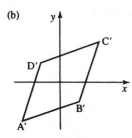

Figure 13.11   *(a) The unit square with vertices* $A(\frac{-1}{2},\frac{-1}{2})$, $B(\frac{1}{2},\frac{-1}{2})$, $C(\frac{1}{2},\frac{1}{2})$, $D(\frac{-1}{2},\frac{1}{2})$. *(b) The image after scaling by 2 along the line* $y = x$.

13.4 Systems of equations

Example 13.19   Using Ohm's law and Kirchoff's laws for the electrical network in Figure 13.12, show that

$$\begin{array}{rcrcrcl} I_1 & - & I_2 & - & I_3 & = & 0 \\ & & 3I_2 & - & 2I_3 & = & 0 \\ 7I_1 & & & + & 2I_3 & = & 8 \end{array}$$

Solution   Kirchoff's laws for an electrical network are as follows:

Kirchoff's voltage law (KVL): The sum of all the voltage drops around any closed loop is zero. This can also be expressed as: the voltage impressed on a closed loop is equal to the sum of the voltage drops in the rest of the loop.

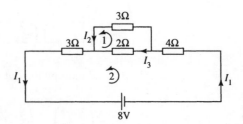

Figure 13.12   *The electrical network for Example 13.19.*

Kirchoff's current law (KCL): At any point of a circuit, the sum of the in-flowing currents is equal to the sum of the out-flowing currents.

By Ohm's law we know the voltage drop across a resistor is given by $V = IR$, where R is the resistance of the resistor. Two loops have been identified in Figure 13.12 and by using KVL and Ohm's law in loop 1 we get

$$3I_2 - 2I_3 = 0.$$

Now looking at loop 2, we get

$$3I_1 - 8 + 4I_1 + 2I_3 = 0$$
$$\Leftrightarrow \quad 7I_1 + 2I_3 = 8.$$

Finally, we use the current law at one of the nodes to give

$$I_1 = I_2 + I_3 \quad \Longleftrightarrow \quad I_1 - I_2 - I_3 = 0$$

Finally, we can list all the equations we have found

$$
\begin{array}{rcrcrcl}
I_1 & - & I_2 & - & I_3 & = & 0 \\
 & & 3I_2 & - & 2I_3 & = & 0 \\
7I_1 & & & + & 2I_3 & = & 8
\end{array}
$$

and the problem is now to find a solution which satisfies all of these equations simultaneously.

This is called a system of equations. In many electrical networks, there will be far more than three unknown currents. In such situations, it is impractical to solve the equations without the use of a computer. However, we can discover a number of important principles and problems involved in solving systems of linear equations by looking at some simple cases. The first problem we have is that it is possible to get more that these three equations from the network given in Figure 13.12. Using KVL in the outer loop would give

$$7I_1 + 3I_2 = 8$$

and KCL at the other node gives

$$I_2 + I_3 = I_1 \quad \Leftrightarrow \quad -I_1 + I_2 + I_3 = 0$$

We therefore have five equations and only three unknowns.

Luckily, it is possible to show that these equations are a consistent set, that is, it is possible to find a solution. We shall return to solve for I_1, I_2, and I_3 later. First, we shall examine all the possibilities when we have only two unknown quantities.

Systems of equations in two unknowns

The equation

$$ax + by = c$$

where a, b, c are constants is a linear equation in two unknowns (or variables) x and y. Because there are two unknowns we need two axes to represent it, and therefore the graph can be drawn in a plane.

Because the graph only involves terms in x, y and the constant term and no other powers of either x or y, we know that the graph of the equation

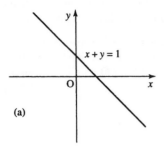

(a)

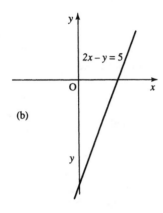

(b)

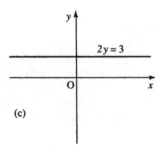

(c)

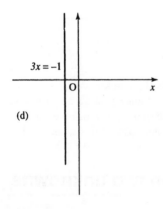

(d)

Figure 13.13 *An equation with two unknowns can be represented as a line in a plane: (a) $x + y = 1$; (b) $2x - y = 5$; (c) $2y = 3$; (d) $3x = -1$.*

is a straight line, as we saw in Chapter 2. Examples of graphs of linear equations in two unknowns are given in Figure 13.13.

We call a solution to the equation a pair of values for x and y which satisfy the equation; that is, when they are substituted they give a true expression. A solution to the equation $x + y = 1$ is $x = 0.5, y = 0.5$ because if we substitute these values we obtain a true expression:

$$0.5 + 0.5 = 1.$$

However, there are many other solutions to $x + y = 1$, for instance $x = 2, y = -1$ or $x = 2.5, y = -1.5$, etc. We say that the equation is indeterminate because there are any number of solutions to the equation $x + y = 1$. In fact, any point on the line $x + y = 1$ is a solution to the equation. We can express the solutions in terms of x or y (e.g. $x = 1 - y$) therefore the solutions are $(1 - y, y)$ where y can be any number. Alternatively $y = 1 - x$ gives solutions $(x, 1 - x)$ where x can be any number.

A system of two linear equations with two unknowns

We want to find values for x and y which solve both $a_1 x + b_1 y = c_1$ and $a_2 x + b_2 y = c_2$ simultaneously. The problem could be expressed as

$$(a_1 x + b_1 y = c_1) \wedge (a_2 x + b_2 y = c_2)$$

When we talk of systems of equations it is understood that we want all of the equations to hold simultaneously so they are usually just listed as

$$a_1 x + b_1 y = c_1$$
$$a_2 x + b_2 y = c_2$$

Each equation can be represented geometrically by a straight line. For example, the system

$$3x + 4y = 7$$
$$x + 2y = 2$$

can be represented by the pair of straight lines as in Figure 13.14.

We can find the point where the two straight lines cross by using substitution as follows.

Example 13.20 Solve the following system of equations using substitution:

$$3x + 4y = 7$$
$$x + 2y = 2.$$

Solution We begin by numbering the equations in order to identify them

$$3x + 4y = 7 \tag{13.1}$$
$$x + 2y = 2 \tag{13.2}$$

From Equation (13.2) we can express x in terms of y as

$$x + 2y = 2 \quad \Leftrightarrow \quad x = 2 - 2y \quad \text{(subtracting } 2y \text{ from both sides)}$$

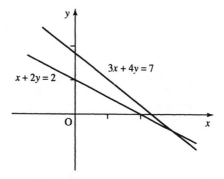

Figure 13.14 *A system of two equations in two unknowns:* $3x + 4y = 7$ *and* $x + 2y = 2$.

Now substitute $x = 2 - 2y$ into Equation (13.1) to give

$3(2 - 2y) + 4y = 7$

$\Leftrightarrow \quad 6 - 6y + 4y = 7$

$\Leftrightarrow \quad 6 - 2y = 7$

$\Leftrightarrow \quad -2y = 1 \quad$ (subtracting 6 from both sides)

$\Leftrightarrow \quad y = -\dfrac{1}{2} \quad$ (dividing both sides by -2)

$\Leftrightarrow \quad y = -0.5$

Now we can use $x = 2 - 2y$ to find x by substituting $y = -0.5$ to give

$x = 2 - 2(-0.5)$

$\Leftrightarrow \quad x = 2 + 1$

$\Leftrightarrow \quad x = 3$

The solution is given by $x = 3$ and $y = -0.5$, which can be represented by the pair of values for (x, y) of $(3, 0.5)$.

An alternative method of solution is to use elimination.

Example 13.21   Solve the following system of equations using elimination:

$3x + 4y = 7$

$x + 2y = 2.$

Solution   To solve the system of equations we look for a way of adding or subtracting multiples of one equation from the other in order to eliminate one of the variables. Multiply the second equation by 3 and leave the first the same. We choose these numbers in order to get the coefficients of x in both equations to be the same.

$3x + 4y = 7$

$3x + 6y = 6.$

Subtract the equations to give

$-2y = 1 \Leftrightarrow y = -\dfrac{1}{2} = -0.5.$

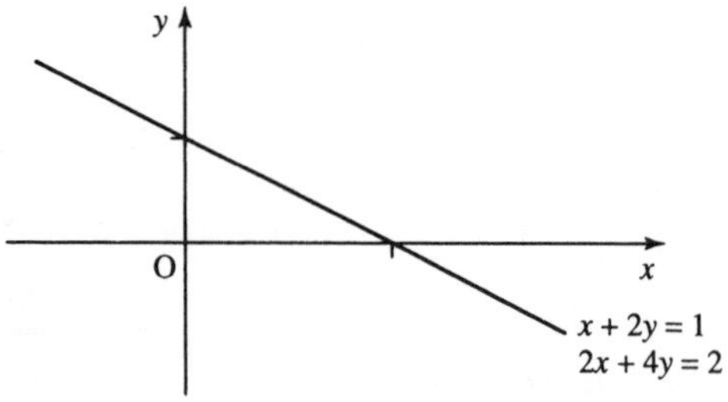

Figure 13.15 *The graphs of equations $x + 2y = 1$ and $2x + 4y = 2$ are coincident. The graph of one lies on top of the other.*

Substitute this into the first equation to give

$$3x + 4(-0.5) = 7$$
$$\Leftrightarrow \quad 3x - 2 = 7$$
$$\Leftrightarrow \quad 3x = 9$$
$$\Leftrightarrow \quad x = 3.$$

The solution is given by $x = 3$ and $y = -0.5$, which can be represented by the pair of values for (x, y) of $(3, -0.5)$.

The point $(3, -0.5)$ is the point on the graph where the two lines cross. This is the only point which lies both on the first graph and the second graph. It is the only point which satisfies both equations simultaneously. Hence, we say there is a unique solution to the system of equations. The system of equations is said to be determined because there is a single solution. The system of equations is also said to be consistent because it is possible to find a solution.

We could have the system of equations:

$$x + 2y = 1$$
$$2x + 4y = 2.$$

If we plot these lines we find that they are coincident, that is, one line lies on top of the other as in Figure 13.15.

In this case, the second equation, $2x + 4y = 2$, can be obtained by multiplying the first equation by 2. We say that the equations are dependent. Two equations are dependent if one can be obtained from the other by multiplying by a constant or by adding a constant to both sides. As any point that lies on $x + 2y = 1$ also lies on $2x + 4y = 2$, there is no unique solution to the system of equations.

We say that the system of equations is indeterminate as there exist any number of solutions to the system, that is, the system reduces to only one equation. However, the system of equations is said to be consistent, because at least one solution exists.

Example solutions are:

$$x = 2, y = 0.5 \quad \text{or} \quad (2, -0.5)$$
$$x = 3, y = -1 \quad \text{or} \quad (3, -1)$$
$$x = 4, y = -1.5 \quad \text{or} \quad (4, -1.5)$$

The solution set can be written as $(x, (1 - x)/2)$, where x can take any value or as $(1 - 2y, y)$, where y can take any value.

If we try to use elimination to solve the equations

$$x + 2y = 1$$
$$2x + 4y = 2$$

we find that one equation reduces to $0 = 0$, that is, a condition that is always true.

Example 13.22 Solve, using elimination:

$$x + 2y = 1$$
$$2x + 4y = 2.$$

Solution

$$x + 2y = 1$$
$$2x + 4y = 2$$

multiply the first equation by 2 to give

$$2x + 4y = 2$$
$$2x + 4y = 2$$

On subtraction we get

$$0 = 0$$

which is always true, thus indicating that the system of equations is indeterminate. The solutions are therefore any points lying on the line $x + 2y = 1$.

The third possibility for a system of equations is one that has no solutions at all. Such a system is as follows:

$$x + 2y = 1$$
$$2x + 4y = 5$$

If we plot these equations we find that they are parallel, as in Figure 13.16.

From the geometrical interpretation, it is therefore clear that no solution exists to this system of equations as no point on the line $x + 2y = 1$ lies on the line $2x + 4y = 5$. We say that the system of equations is inconsistent because no solutions exist.

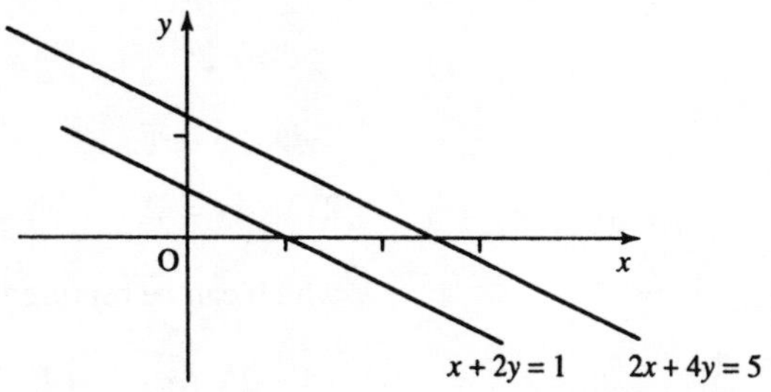

Figure 13.16 *The graphs of $x + 2y = 1$ and $2x + 4y = 5$ are parallel. There are no points in common between the two graphs.*

If we used elimination to attempt to solve an inconsistent system of equations like these, then we will find that we will get an impossible condition, such as

$$0 = 3$$

which is false. This situation indicates that there are no solutions and the equations are inconsistent.

Example 13.23   Solve, using elimination

$$x + 2y = 1$$
$$2x + 4y = 5.$$

Solution   Multiply the first equation by 2 to give

$$2x + 4y = 2$$
$$2x + 4y = 5.$$

Subtracting the equations, we get

$$0 = -3.$$

This condition is false. This indicates that the system of equations are inconsistent and there are no solutions.

We can express systems of equations in matrix form as

Av = b

where **A** is the matrix of coefficients,

$$\mathbf{v} = \begin{pmatrix} x \\ y \end{pmatrix}$$

and **b** is the vector of constants on the right-hand side of the equations. For the three cases we have looked at in Examples 13.21–13.23, we get the following:

Case 1:

$$3x + 4y = 7$$
$$x + 2y = 2$$

which can be represented in matrix form as

$$\begin{pmatrix} 3 & 4 \\ 1 & 2 \end{pmatrix} \begin{pmatrix} x \\ y \end{pmatrix} = \begin{pmatrix} 7 \\ 2 \end{pmatrix}.$$

Case 2:

$$x + 2y = 1$$
$$2x + 4y = 2$$

which can be represented in matrix form as

$$\begin{pmatrix} 1 & 2 \\ 2 & 4 \end{pmatrix} \begin{pmatrix} x \\ y \end{pmatrix} = \begin{pmatrix} 1 \\ 2 \end{pmatrix}.$$

Case 3:

$$x + 2y = 1$$
$$2x + 4y = 5$$

which can be represented in matrix form as

$$\begin{pmatrix} 1 & 2 \\ 2 & 4 \end{pmatrix} \begin{pmatrix} x \\ y \end{pmatrix} = \begin{pmatrix} 1 \\ 5 \end{pmatrix}.$$

We can look at the determinants of the coefficient matrices in order to help us analyse the system.

For Case 1 (the system with a unique solution) we find

$$\begin{vmatrix} 3 & 4 \\ 1 & 2 \end{vmatrix} = (3 \cdot 2) - (1 \cdot 4) = 6 - 4 = 2.$$

The fact that the determinant of the matrix of coefficients is non-zero shows that the system of equations has a unique solution.

For Case 2 (the system with many solutions) we find

$$\begin{vmatrix} 1 & 2 \\ 2 & 4 \end{vmatrix} = (1 \cdot 4) - (2 \cdot 2) = 0.$$

If we replace any column in this determinant by the constant terms

$$\begin{pmatrix} 1 \\ 2 \end{pmatrix}$$

we get the determinants

$$\begin{vmatrix} 1 & 1 \\ 2 & 2 \end{vmatrix} = (1 \cdot 2) - (2 \cdot 1) = 0$$

and

$$\begin{vmatrix} 1 & 2 \\ 2 & 4 \end{vmatrix} = 0.$$

This can be shown to hold in general. If all the determinants formed in this way are 0 then we have indeterminacy in the solutions. That is, there will be many solutions to the system.

For Case 3 (the system with no solutions) we find

$$\begin{vmatrix} 1 & 2 \\ 2 & 4 \end{vmatrix} = (1 \cdot 4) - (2 \cdot 2) = 0$$

If we replace any column in this determinant by the constant terms

$$\begin{pmatrix} 1 \\ 5 \end{pmatrix}$$

we get the determinants

$$\begin{vmatrix} 1 & 1 \\ 2 & 5 \end{vmatrix} = (1 \cdot 5) - (2 \cdot 1) = 3$$

and

$$\begin{vmatrix} 1 & 2 \\ 5 & 4 \end{vmatrix} = (1 \cdot 4) - (2 \cdot 5) = -6$$

This can be shown to hold in general. If the determinant of the matrix of coefficients is zero but any one of the determinants formed using the vector of constant terms are non-zero, then this shows that the system is inconsistent and there are no solutions.

We can summarize the results of this section as follows. For a system of equations (assuming we have as many equations as unknowns) there are three possibilities

Case 1: A determined system has a unique solution which can be found by using elimination. Geometrically, the solution is a single point which (in the case a a system in two unknowns) represents the intersection of the two lines. The determinant of the coefficients is non-zero. The system is both consistent and determined.

Case 2: An undetermined system has many solutions. If elimination is used to solve the system it will result in a condition like $0 = 0$, which is always true. Geometrically the solutions lie (for a system in two unknowns) anywhere along a line. The determinant of the coefficients is zero, as are any determinants found by replacing a column in the matrix of coefficients by the vector of constant terms. The system is undetermined but consistent (as there are solutions).

Case 3: An inconsistent system has no solutions. If elimination is used to solve the system it will result in a condition like $0 = 3$, which is always false. Geometrically, for a system in two unknowns, the system is represented by parallel lines which have no points in common, hence no solutions. The determinant of the coefficients is zero but at least one of the determinants found by replacing a column in the matrix of coefficients by the vector of constant terms is non-zero. The system is inconsistent.

For Case 1, the solution of the system can be found by using the inverse of the matrix of coefficients. We can represent the system by

$$\mathbf{Av} = \mathbf{b}.$$

As the determinant of $\mathbf{A}$ is non-zero, we know that $\mathbf{A}$ has an inverse $\mathbf{A}^{-1}$. We pre-multiply both sides of the matrix equation by $\mathbf{A}^{-1}$ giving $\mathbf{A}^{-1}\mathbf{Av} = \mathbf{A}^{-1}\mathbf{b}$

as $\mathbf{A}^{-1}\mathbf{A} = \mathbf{I}$, the unit matrix and $\mathbf{Iv} = \mathbf{v}$

we get

$$\mathbf{v} = \mathbf{A}^{-1}\mathbf{b}$$

$$\mathbf{Av} = \mathbf{b} \wedge |\mathbf{A}| \neq 0 \Leftrightarrow \mathbf{v} = \mathbf{A}^{-1}\mathbf{b}.$$

Example 13.24   Solve

$$3x + 4y = 7$$
$$x + 2y = 2$$

by finding the inverse of the matrix of coefficients.

Solution   The system can be expressed as

$$\begin{pmatrix} 3 & 4 \\ 1 & 2 \end{pmatrix} \begin{pmatrix} x \\ y \end{pmatrix} = \begin{pmatrix} 7 \\ 2 \end{pmatrix}.$$

As we know that if $\mathbf{Av} = \mathbf{b}$ and $\mathbf{A}$ is invertible then

$$\mathbf{v} = \mathbf{A}^{-1}\mathbf{b}$$

and in this case we have

$$\mathbf{A} = \begin{pmatrix} 3 & 4 \\ 1 & 2 \end{pmatrix} \quad \text{and} \quad \mathbf{b} = \begin{pmatrix} 7 \\ 2 \end{pmatrix}.$$

Then to find the solution we find the inverse of

$$\begin{pmatrix} 3 & 4 \\ 1 & 2 \end{pmatrix}.$$

We know that the inverse of

$$\begin{pmatrix} a & b \\ c & d \end{pmatrix}$$

is

$$\frac{1}{(ad - cb)} \begin{pmatrix} d & -b \\ -c & a \end{pmatrix}$$

this gives the inverse of

$$\begin{pmatrix} 3 & 4 \\ 1 & 2 \end{pmatrix}$$

as

$$\frac{1}{(3 \cdot 2) - (4 \cdot 1)} \begin{pmatrix} 2 & -4 \\ -1 & 3 \end{pmatrix} = \frac{1}{2} \begin{pmatrix} 2 & -4 \\ -1 & 3 \end{pmatrix}$$

Using $\mathbf{x} = \mathbf{A}^{-1}\mathbf{b}$ gives

$$\begin{pmatrix} x \\ y \end{pmatrix} = \frac{1}{2} \begin{pmatrix} 2 & -4 \\ -1 & 3 \end{pmatrix} \begin{pmatrix} 7 \\ 2 \end{pmatrix} = \frac{1}{2} \begin{pmatrix} 6 \\ -1 \end{pmatrix} = \begin{pmatrix} 3 \\ -0.5 \end{pmatrix}$$

So the solution of this system of equations is $x = 3$ and $y = -0.5$.

For a 2×2 system, this method of solving a system of equations is quite straightforward. However, for larger systems a solution by finding the inverse involves nearly twice as many operations as that by elimination of variables and therefore should not be used as a method of solving equations.

Equations with three unknowns

For three unknowns we need three axes to represent the equations. Each equation is represented by a plane, for example, Figure 13.17 shows the plane which represents the equation $x + y + z = 1$.

Two planes, if they intersect, will intersect along a line and if a third independent equation is given then the three planes will intersect at a point. More than three unknowns cannot be represented geometrically.

However many unknowns there are in a system of equations, the three types of systems which we identified as Cases 1–3 remain as do the methods to be used to distinguish between a determined, indeterminate, and inconsistent system.

We shall later look at finding the determinant and inverse of larger matrices, but first we look at a systematic way of doing elimination which is suitable for a computer solution of a system of equations.

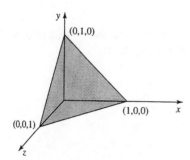

Figure 13.17   *The plane given by the equation $x + y + z = 1$.*

13.5 Gauss elimination

Gauss elimination is a structured process for the elimination of variables in one of the equations. It is easy to generalize to larger systems of equations and it is relatively numerically stable, making it suitable for use with a computer.

Gauss elimination is performed in stages. At Stage 1, we concentrate on the first column, the coefficients of x. The idea is to make the coefficient of the first equation 1 and eliminate the variable from the other equation(s) by using multiples of the first equation. Equation 1 is therefore the pivotal equation for Stage 1.

Example 13.25   Solve

$$3x + 4y = 7$$
$$5x - 8y = 8$$

using Gauss elimination.

Solution   We can either write out the equation each time we perform a step or we can abbreviate the solution by expressing the equations in short hand as an augmented matrix

$$\begin{pmatrix} 3 & 4 & 7 \\ 5 & -8 & 8 \end{pmatrix}.$$

We shall present both notations at the same time. We shall refer to the elements in the augmented matrix by

$$\begin{pmatrix} a_{11} & a_{12} & b_1 \\ a_{21} & a_{22} & b_2 \end{pmatrix}.$$

For the first step, the first equation is the pivotal equation:

$$\begin{array}{cc} 3x + 4y = 7 \\ 5x - 8y = 8 \end{array} \qquad \begin{pmatrix} 3 & 4 & 7 \\ 5 & -8 & 8 \end{pmatrix}.$$

Stage 1
Step 1: Divide the first equation by a_{11}:

$$\begin{array}{cc} x + \frac{4}{3}y = \frac{7}{3} \\ 5x - 8y = 8 \end{array} \qquad \begin{pmatrix} 1 & \frac{4}{3} & \frac{7}{3} \\ 5 & -8 & 8 \end{pmatrix}.$$

Step 2: Take 5 times the first equation away from the second equation in order to eliminate the term in x in the second equation:

$$x + \frac{4}{3}y = \frac{7}{3}$$
$$-8y - \left(\frac{4}{3}y \times 5\right) = 8 - \frac{7}{3} \times 5$$

which is the same as

$$\begin{array}{cc} x + \frac{4}{3}y = \frac{7}{3} \\ -\frac{44}{3}y = -\frac{11}{3} \end{array} \qquad \begin{pmatrix} 1 & \frac{4}{3} & \frac{7}{3} \\ 0 & -\frac{44}{3} & -\frac{11}{3} \end{pmatrix}.$$

Stage 2:

Divide the second equation through by the coefficient of $y(a_{22})$.

$$x + \tfrac{4}{3}y = \tfrac{7}{3} \qquad \begin{pmatrix} 1 & \tfrac{4}{3} & \tfrac{7}{3} \\ 0 & 1 & \tfrac{1}{4} \end{pmatrix}.$$
$$\qquad\quad y = \tfrac{1}{4}$$

We now already have the solution for y and we can obtain the solution for x by substitution into the first equation. This is called 'back-substitution'.

$$x + \left(\frac{4}{3}\right)\left(\frac{1}{4}\right) = \frac{7}{3}$$
$$\Leftrightarrow \quad x = \frac{7}{3} - \frac{1}{3}$$
$$\Leftrightarrow \quad x = 2.$$

Therefore, the solution is $(2, 0.25)$.

We see that the point of the exercise is to write the system of equations so that the final equation contains only one variable and the last but one equation has up to two variables, etc. The matrix of coefficients should be in upper triangular form like:

$$\begin{pmatrix} 1 & \tfrac{4}{3} \\ 0 & 1 \end{pmatrix}.$$

The augmented matrix is then like

$$\begin{pmatrix} 1 & \tfrac{4}{3} & \vdots & \tfrac{7}{3} \\ & & \vdots & \\ & & \vdots & \\ 0 & 1 & & \tfrac{1}{4} \end{pmatrix}$$

which is said to be in echelon form. Once this form has been achieved, then back-substitution can be performed to find the value of the variables.

Example 13.26   Solve the system of equations

$$\begin{aligned} 2x + y - 2z &= -1 \\ 2x - 3y + 2z &= 9 \\ -x + y - z &= -3.5. \end{aligned}$$

Solution

$$\begin{aligned} 2x + y - 2z &= -1 \\ 2x - 3y + 2z &= 9 \\ -x + y - z &= -3.5 \end{aligned} \qquad \begin{pmatrix} 2 & 1 & -2 & -1 \\ 2 & -3 & 2 & 9 \\ -1 & 1 & -1 & -3.5 \end{pmatrix}.$$

Stage 1

Stage 1 concerns the first column. We use the first row (the pivotal row) to eliminate the elements below a_{11}.

Step 1: Divide the first equation by a_{11}.

$$\begin{aligned} x + 0.5y - z &= -0.5 \\ 2x - 3y + 2z &= 9 \\ -x + y - z &= -3.5 \end{aligned} \qquad \begin{pmatrix} 1 & 0.5 & -1 & -0.5 \\ 2 & -3 & 2 & 9 \\ -1 & 1 & -1 & -3.5 \end{pmatrix}.$$

Step 2: Eliminate x from the second and third equations by taking away multiples of equation 1. To do this we take Row $2 - 2 \times$ (Row 1) and

Row $3 - (-1) \times$ (Row 1).

$$
\begin{array}{rcrcrcr}
x & + & 0.5y & - & z & = & -0.5 \\
 & & -4y & + & 4z & = & 10 \\
 & & 1.5y & - & 2z & = & -4
\end{array}
\qquad
\begin{pmatrix}
1 & 0.5 & -1 & -0.5 \\
0 & -4 & 0 & 10 \\
0 & 1.5 & -2 & -4
\end{pmatrix}
$$

The calculations can be done 'in the margin' and were:

Row $2 - 2 \times$ (Row 1)

$$
\begin{array}{r}
2x - 3y + 2z = 9 \\
-2 \times (x + 0.5y - z = -0.5) \\
\hline
-4y + 4z = 10
\end{array}
$$

Row$3 - (-1) \times$ (Row1)

$$
\begin{array}{r}
-x + y - z = -3.5 \\
-(-1) \times (x + 0.5y - z = -0.5) \\
\hline
1.5y - 2z = -4
\end{array}
$$

Stage 2

Stage 2 concerns the second column. We use the second row to eliminate the elements below a_{22}. Here Row 2 is the pivotal row.

Step 1: Divide the second equation by the coefficient a_{22}:

$$
\begin{array}{rcrcrcr}
x & + & 0.5y & - & z & = & -0.5 \\
 & & y & - & z & = & -2.5 \\
 & & 1.5y & - & 2z & = & -4
\end{array}
\qquad
\begin{pmatrix}
1 & 0.5 & -1 & -0.5 \\
0 & 1 & -1 & -2.5 \\
0 & 1.5 & -2 & -4
\end{pmatrix}.
$$

Step 2: Eliminate y from the third equation by taking away multiples of the second equation.

$$
\begin{array}{rcrcrcr}
x & + & 0.5y & - & z & = & -0.5 \\
 & & y & - & z & = & -2.5 \\
 & & & & -0.5z & = & -0.25
\end{array}
\qquad
\begin{pmatrix}
1 & 0.5 & -1 & -0.5 \\
0 & 1 & -1 & -2.5 \\
0 & 0 & -0.5 & -0.25
\end{pmatrix}.
$$

Here, the calculation was Row $3 - 1.5 \times$ (Row 2) and the calculation was as follows

$$
\begin{array}{r}
1.5y - 2z = -4 \\
-1.5 \times (y - z = -2.5) \\
\hline
-0.5z = -0.25
\end{array}
$$

Stage 3

Divide the third equation by the coefficient of z.

$$
\begin{array}{rcrcrcr}
x & + & 0.5y & - & z & = & -0.5 \\
 & & y & & - z & = & -2.5 \\
 & & & & z & = & 0.5
\end{array}
\qquad
\begin{pmatrix}
1 & 0.5 & -1 & -0.5 \\
0 & 1 & -1 & -2.5 \\
0 & 0 & 1 & 0.5
\end{pmatrix}
$$

Back-substitution: We have now finished the elimination stage and we can easily solve the equations using back-substitution.

From the third equation, $z = -0.5$

Find y from the second equation

$$y = -2.5 + z \Leftrightarrow y = -2.5 + 0.5$$
$$\Leftrightarrow \quad y = -2$$

Substitute into the first equation to find x

$$x + 0.5(-2) - 0.5 = -0.5$$
$$\Leftrightarrow \quad x - 1.5 = -0.5$$
$$\Leftrightarrow \quad x = 1$$

So the solution of the system of equations is $(1, -2, 0.5)$.

Check: To check, substitute $x = 1$, $y = -2$, and $z = 0.5$ into the original equations

$$2x + y - 2z = -1$$
$$2x - 3y + 2z = 9$$
$$-x + y - z = -3.5$$

giving

$2(1) + (-2) - 2(0.5) = -1$, which is true
$2(1) - 3(-2) + 2(0.5) = 9$, which is true
$-(1) + (-2) - 0.5 = -3.5$, which is true.

Now we can solve the system of equations for the electrical network, which was the introductory example of Section 13.4.

Example 13.27 Solve, using Gauss elimination, the system of equations

$$\begin{aligned} I_1 \; - \; I_2 \; - \; I_3 &= 0 \\ 3I_2 \; - \; 2I_3 &= 0 \\ 7I_1 \qquad\quad + \; 2I_3 &= 8 \end{aligned}$$

Solution We shall only show the augmented matrix in this example, so we begin with

$$\begin{pmatrix} 1 & -1 & -1 & 0 \\ 0 & 3 & -2 & 0 \\ 7 & 0 & 2 & 8 \end{pmatrix}$$

Stage 1
Stage 1 concerns the first column. We use the first row to eliminate the elements below a_{11}.
Step 1: Divide the first equation by a_{11}. As this is already 1 we do not need to divide by it.
Step 2: Eliminate elements in the first column below a_{11} by taking away multiples of Row 1 from Rows 2 and 3. Row 2 already has no entry in

the first column so we leave it alone. We take Row $3 - (7) \times$ (Row 1).

$$\begin{pmatrix} 1 & -1 & -1 & 0 \\ 0 & 3 & -2 & 0 \\ 0 & 7 & 9 & 8 \end{pmatrix}$$

The calculations performed here was: Row $2 - 7 \times$ (Row 1)

$$
\begin{array}{cccc}
 & 7 & 0 & 2 & 8 \\
-7 \times (& 1 & -1 & -1 & 0) \\
\hline
 & 0 & 7 & 9 & 8
\end{array}
$$

Stage 2

Stage 2 concerns the second column. We use the second row to eliminate the elements below a_{22}.

Step 1: Divide the second equation by the coefficient of a_{22}.

$$\begin{pmatrix} 1 & -1 & -1 & 0 \\ 0 & 1 & -\frac{2}{3} & 0 \\ 0 & 7 & 9 & 8 \end{pmatrix}$$

Step 2: Eliminate the element in the second column below a_{22} by taking away multiples of Row 2 from Row 3.

$$\begin{pmatrix} 1 & -1 & -1 & 0 \\ 0 & 1 & -\frac{2}{3} & 0 \\ 0 & 0 & \frac{41}{3} & 8 \end{pmatrix}$$

Here the calculation was Row $3 - (7) \times$ (Row 2), and the calculation was as follows:

$$
\begin{array}{cccc}
 & 0 & 7 & 9 & 8 \\
-7 \times (& 0 & 1 & -\frac{2}{3} & 0) \\
\hline
 & 0 & 0 & \frac{41}{3} & 8
\end{array}
$$

Stage 3

Divide the third equation by the coefficient of z:

$$\begin{pmatrix} 1 & -1 & -1 & 0 \\ 0 & 1 & -\frac{2}{3} & 0 \\ 0 & 0 & 1 & \frac{24}{41} \end{pmatrix}$$

Back-substitution: We have now finished the elimination stage and we can easily solve the equations using back substitution.

From the third equation, $I_3 = \frac{24}{41}$. Find I_2 from the second equation:

$$I_2 - \frac{2}{3} I_3 = 0$$

$$I_2 - \frac{2}{3} \times \frac{24}{41} = 0 \quad \Leftrightarrow \quad I_2 = \frac{16}{41}.$$

Substitute into the first equation to find I_1:

$$I_1 - \frac{16}{41} - \frac{24}{41} = 0 \quad \Leftrightarrow \quad I_1 = \frac{40}{41}$$

So the solution of the system of equations is $\left(\dfrac{40}{41}, \dfrac{16}{41}, \dfrac{24}{41} \right)$.

Check: To check, substitute $I_1 = \dfrac{40}{41}$, $I_2 = \dfrac{16}{41}$, and $I_3 = \dfrac{24}{41}$ into the original equations

$$I_1 - I_2 - I_3 = 0$$
$$3I_2 - 2I_3 = 0$$
$$7I_1 + 2I_3 = 8$$

giving

$$\frac{40}{41} - \frac{16}{41} - \frac{24}{41} = 0, \text{ which is true}$$

$$3\left(\frac{16}{41}\right) - 2\left(\frac{24}{41}\right) = 0, \text{ which is true}$$

$$7\left(\frac{40}{41}\right) + 2\left(\frac{24}{41}\right) = 8, \text{ which is true.}$$

Indeterminacy and inconsistency

When we analysed systems of equations in Section 13.5, we saw that one of the equations reducing to $0 = 0$, indicates that we have an indeterminate system; that is that there will be many solutions. If, in the process of performing Gauss elimination, we find a row of zeros then we know that we have an indeterminate system. We can use the remaining equations to eliminate as many of the unknowns as possible, giving a solution which will still involve one or more of the variables. This will give a whole line or possibly (in three dimensions) a plane of solutions.

If we come across a row that is zero everywhere in the matrix of coefficients but has a non-zero constant term we have found an equation $0 = c$, which is false. This is an inconsistent system and has no solutions.

Order of the equations

At the beginning of each stage in Gauss elimination, the order of the rows may be swapped (other than those already used in previous stages as the pivotal equation). We will have to swap the equations if the next 'pivotal' equation has a zero coefficient for the next variable to be eliminated. It is usual, although we have not illustrated this point, to always consider swapping the order of the equations, in order to choose the equation with the largest absolute value of the coefficient in the term to be used for eliminating, as the pivotal equation. This is called partial pivoting. This procedure as an attempt to avoid problems with equations that may become ill conditioned in the course if performing the elimination. The equations are ill conditioned when a small change in the coefficients of the equations causes a large change in the values of the solutions, and in such equations rounding errors can become large and cause significant inaccuracies in the solutions. We have not performed partial pivoting in these examples as they are only presented to give an idea of the method. It is assumed that for any real life problem a computer algorithm will be used to solve the system of equations, and such an algorithm will incorporate partial pivoting.

13.6 The inverse and determinant of a 3 × 3 matrix

Finding the inverse by elimination

To find the inverse using elimination, we write the matrix we need to invert on the left and the unit matrix on the right. We perform operations on both matrices at the same time. The method, called Gauss–Jordan elimination, begins in the same way as Gauss elimination. When we have upper triangular form for the matrix we metaphorically turn the problem upside down and eliminate the upper triangle also.

Example 13.28   Find the inverse of

$$\begin{pmatrix} 4 & 0 & -4 \\ 3 & 4 & 2 \\ -1 & -1 & 1 \end{pmatrix}.$$

Solution   We start by writing the matrix along with the unit matrix

$$\begin{pmatrix} 4 & 0 & -4 & 1 & 0 & 0 \\ 3 & 4 & 2 & 0 & 1 & 0 \\ -1 & -1 & 1 & 0 & 0 & 1 \end{pmatrix}.$$

Stage 1
Step 1: Divide the first row by a_{11}

$$\begin{pmatrix} 1 & 0 & -1 & 0.25 & 0 & 0 \\ 3 & 4 & 2 & 0 & 1 & 0 \\ -1 & -1 & 1 & 0 & 0 & 1 \end{pmatrix}.$$

Step 2: Eliminate the first column below a_{11} by subtracting multiples of the first row from the second and third rows

$$\begin{pmatrix} 1 & 0 & -1 & 0.25 & 0 & 0 \\ 0 & 4 & 5 & -0.75 & 1 & 0 \\ 0 & -1 & 0 & 0.25 & 0 & 1 \end{pmatrix}.$$

The calculations were as follows:
Row 2 − 3 × Row 1

$$\begin{array}{cccccc}
 & 3 & 4 & 2 & 0 & 1 & 0 \\
-3.(& 1 & 0 & -1 & 0.25 & 0 & 0) \\
\hline
 & 0 & 4 & 5 & -0.75 & 1 & 0
\end{array}$$

Row 3 − (−1) × Row 1

$$\begin{array}{cccccc}
 & -1 & -1 & 1 & 0 & 0 & 1 \\
-(-1) & (1 & 0 & -1 & 0.25 & 0 & 0) \\
\hline
 & 0 & -1 & 0 & 0.25 & 0 & 1
\end{array}$$

Stage 2
Step 1: Divide the second row by a_{22} (4)

$$\begin{pmatrix} 1 & 0 & -1 & 0.25 & 0 & 0 \\ 0 & 1 & 1.25 & -0.1875 & 0.25 & 0 \\ 0 & -1 & 0 & 0.25 & 0 & 1 \end{pmatrix}.$$

Step 2: Eliminate the elements in the second column below a_{22} by subtracting multiples of the second row from the third row

$$\begin{pmatrix} 1 & 0 & -1 & 0.25 & 0 & 0 \\ 0 & 1 & 1.25 & -0.1875 & 0.25 & 0 \\ 0 & 0 & 1.25 & 0.0625 & 0.25 & 1 \end{pmatrix}.$$

The calculations were as follows:

Row 3 − (−1) × Row 2

$$
\begin{array}{cccccc}
0 & -1 & 0 & 0.25 & 0 & 1 \\
-(-1)\,(0 & 1 & 1.25 & -0.1875 & 0.25 & 0) \\
\hline
0 & 0 & 1.25 & 0.0625 & 0.25 & 1
\end{array}
$$

Stage 3
Step 1: Divide the third row by a_{33} (1.25):

$$
\begin{pmatrix}
1 & 0 & -1 & 0.25 & 0 & 0 \\
0 & 1 & 1.25 & -0.1875 & 0.25 & 0 \\
0 & 0 & 1 & 0.05 & 0.2 & 0.8
\end{pmatrix}
$$

Step 2: Turn the problem metaphorically upside down and use the third row to eliminate elements in the third column above a_{33} by subtracting multiples of the third row from the first row and the second row

$$
\begin{pmatrix}
1 & 0 & 0 & 0.3 & 0.2 & 0.8 \\
0 & 1 & 0 & -0.25 & 0 & -1 \\
0 & 0 & 1 & 0.05 & 0.2 & 0.8
\end{pmatrix}
$$

The calculations were as follows:

Row 1 − (−1) × Row 3

$$
\begin{array}{cccccc}
1 & 0 & -1 & 0.25 & 0 & 0 \\
-(-1)\,(0 & 0 & 1 & 0.05 & 0.2 & 0.8) \\
\hline
1 & 0 & 0 & 0.3 & 0.2 & 0.8
\end{array}
$$

Row 2 − 1.25 × Row 3

$$
\begin{array}{cccccc}
0 & 1 & 1.25 & -0.1875 & 0.25 & 0 \\
-1.25\,(0 & 0 & 1 & 0.05 & 0.2 & 0.8) \\
\hline
0 & 1 & 0 & -0.25 & 0 & -1
\end{array}
$$

The matrix on the right-hand side is now the inverse of the original matrix.

The inverse is

$$
\begin{pmatrix}
0.3 & 0.2 & 0.8 \\
-0.25 & 0 & -1 \\
0.05 & 0.2 & 0.8
\end{pmatrix}
$$

Check: Multiply the original matrix by its inverse

$$
\begin{pmatrix}
4 & 0 & -4 \\
3 & 4 & 2 \\
-1 & -1 & 1
\end{pmatrix}
\begin{pmatrix}
0.3 & 0.2 & 0.8 \\
-0.25 & 0 & -1 \\
0.05 & 0.2 & 0.8
\end{pmatrix}
$$

$$
=\begin{pmatrix}
4(0.3)+0(-0.25)-4(0.05) & 4(0.2)+0(0)-4(0.2) & 4(0.8)+0(-1)-4(0.8) \\
3(0.3)+4(-0.25)+2(0.05) & 3(0.2)+4(0)+2(0.2) & 3(0.8)+4(-1)+2(0.8) \\
-1(0.3)-1(-0.25)+1(0.05) & -1(0.2)-1(0)+1(0.2) & -1(0.8)-1(-1)+1(0.8)
\end{pmatrix}
$$

$$
=\begin{pmatrix}
1 & 0 & 0 \\
0 & 1 & 0 \\
0 & 0 & 1
\end{pmatrix}
$$

Therefore we have correctly found the inverse of the matrix.

The determinant of a 3 × 3 matrix

The definition of the (2×2) determinant has been given as

$$\begin{vmatrix} a_1 & b_1 \\ a_2 & b_2 \end{vmatrix} = a_1 b_2 - a_2 b_1.$$

Each of the terms on the right-hand side of this definition is of the form $a_i b_j$ where i and j are different choices of the numbers 1 and 2. We can define higher order determinants by using ideas of permutations. We notice that the term $a_1 b_2$ above has a positive sign because the indices 1 and 2 appear in order, whereas the term $a_2 b_1$ has a negative sign because the indices 2,1 are reversed.

To define

$$\begin{vmatrix} a_1 & b_1 & c_1 \\ a_2 & b_2 & c_2 \\ a_3 & b_3 & c_3 \end{vmatrix}$$

we write down all terms of the form $a_i b_j c_k$ and give each term a $+$ sign or a $-$ sign depending on whether the permutation ijk is even or odd. A permutation of 123 is even if it can be achieved by an even number of swaps of the numbers, beginning with the order 123. If it can only be obtained by an odd number of swaps then the permutation is odd. For example, 231 is even because we can reach it by first swapping 1 and 2 giving 213 and then swapping 1 and 3. Alternatively, we could have interchanged 2 and 3 giving 132 and 1 and 3 giving 312, 2 and 1 giving 321, 3 and 2 giving 231. Whatever way we use to get to the order 231 involves an even number of steps. Similarly we say that a permutation of 123 is odd if it involves an odd number of adjacent interchanges.

This definition gives the determinant of a 3×3 array as

$$\begin{vmatrix} a_1 & b_1 & c_1 \\ a_2 & b_2 & c_2 \\ a_3 & b_3 & c_3 \end{vmatrix} = a_1 b_2 c_3 - a_1 b_3 c_2 - a_2 b_1 c_3 + a_2 b_3 c_1 + a_3 b_1 c_2 - a_3 b_2 c_1$$

This expression may be written in such a way that it involves 2×2 determinants as follows:

$$a_1 b_2 c_3 - a_1 b_3 c_2 - a_2 b_1 c_3 + a_3 b_1 c_2 + a_2 b_3 c_1 - a_3 b_2 c_1$$

$$= a_1 (b_2 c_3 - b_3 c_2) - b_1 (a_2 c_3 - a_3 c_1) + c_1 (a_2 b_3 - a_3 b_2)$$

$$= a_1 \begin{vmatrix} b_2 & c_2 \\ b_3 & c_3 \end{vmatrix} - b_1 \begin{vmatrix} a_2 & c_2 \\ a_3 & c_3 \end{vmatrix} + c_1 \begin{vmatrix} a_2 & b_2 \\ a_3 & b_3 \end{vmatrix}.$$

The 2×2 determinants that appear in this expression are called minors. This formula for the determinant is called the expansion by the first row, because the numbers a_1, b_1, c_1 which multiply the minors are from the first row of the matrix.

Note that the minor multiplying a_1 is the (2×2) determinant obtained from the original array by crossing out the row and column in which a_1 appears, as follows

$$\begin{vmatrix} \cancel{a_1} & \cancel{b_1} & \cancel{c_1} \\ \cancel{a_2} & b_2 & c_2 \\ \cancel{a_3} & b_3 & c_3 \end{vmatrix}$$

gives the minor of a_1 as

$$\begin{vmatrix} b_2 & c_2 \\ b_3 & c_3 \end{vmatrix}.$$

Similarly, the number multiplying b_1 is the determinant found by crossing out the row and the column in which b_1 appears.

We could also find the determinant by expanding about the first column

$$a_1 b_2 c_3 - a_1 b_3 c_2 - a_2 b_1 c_3 + a_2 b_3 c_1 + a_3 b_1 c_2 - a_3 b_2 c_1$$

$$= a_1(b_2 c_3 - b_3 c_2) - a_2(b_1 c_3 - b_3 c_1) + a_3(b_1 c_2 - b_2 c_1)$$

$$= a_1 \begin{vmatrix} b_2 & c_2 \\ b_3 & c_3 \end{vmatrix} - a_2 \begin{vmatrix} b_1 & c_1 \\ b_3 & c_3 \end{vmatrix} + a_3 \begin{vmatrix} b_1 & c_1 \\ b_2 & c_2 \end{vmatrix}.$$

Again we see that the minor of a_2, for instance, can be found by crossing out the row and column that a_2 appears in from the original array.

To find the sign multiplying each term in the expansion for the determinant we can remember the following pattern

```
+  -  +
-  +  -
+  -  +
```

To find the determinant we can expand about any row and column, multiplying each term a_{ij} by its respective minor and find the sign by multiplying by $(-1)^{i+j}$.

Example 13.29   Find the following determinant

$$\begin{vmatrix} -1 & 2 & 3 \\ 6 & -1 & 2 \\ 4 & 0 & -1 \end{vmatrix}$$

Solution   Expanding about the first row

$$\begin{vmatrix} -1 & 2 & 3 \\ 6 & -1 & 2 \\ 4 & 0 & -1 \end{vmatrix} = -1 \begin{vmatrix} -1 & 2 \\ 0 & -1 \end{vmatrix} - 2 \begin{vmatrix} 6 & 2 \\ 4 & -1 \end{vmatrix} + 3 \begin{vmatrix} 6 & -1 \\ 4 & 0 \end{vmatrix}$$

$$= -1(1 - 0) - 2(-6 - 8) + 3(0 + 4)$$

$$= -1 + 28 + 12 = 39.$$

Alternatively, expanding about the first column we get

$$-1 \begin{vmatrix} -1 & 2 \\ 0 & -1 \end{vmatrix} - 6 \begin{vmatrix} 2 & 3 \\ 0 & -1 \end{vmatrix} + 4 \begin{vmatrix} 2 & 3 \\ -1 & 2 \end{vmatrix}$$

$$= -1(1 - 0) - 6(-2 - 0) + 4(4 - (-3))$$

$$= -1 + 12 + 28 = 39.$$

The inverse of a matrix using (Adjoint(A))/|A|

We have already seen how to find the inverse of a matrix by using elimination. It is also possible to find the inverse by the following procedure:

(1)   Find the matrix of minors.
(2)   Multiply the minor for row i and column j by $(-1)^{i+j}$. This is then called the matrix of cofactors.

(3) Take the transpose of the matrix of cofactors to find the adjoint matrix.

(4) Divide by the determinant of the original matrix.

This procedure rarely needs to be used and only usually if we have a matrix which involves some unknown variables or expresses some formula and we would like to find the inverse formula. It would never be used as a numerical procedure, as it is both numerically unstable and also uses a very large number of operations (of the order of $n!$ operations, where n is the dimension of the matrix, whereas elimination is only of the order of n^3).

Example 13.30 Find the inverse of

$$\begin{pmatrix} 4 & 0 & -4 \\ 3 & 4 & 2 \\ -1 & -1 & 1 \end{pmatrix}$$

using $\mathbf{A}^{-1} = (\text{Adjoint}(\mathbf{A}))/|\mathbf{A}|$.

Solution Find the matrix of minors for each term in the matrix. The minor for the ith row and jth column is found by crossing out that row and column and finding the determinant of the remaining elements.

This gives the matrix of minors as

$$\begin{pmatrix} \begin{vmatrix} 4 & 2 \\ -1 & 1 \end{vmatrix} & \begin{vmatrix} 3 & 2 \\ -1 & 1 \end{vmatrix} & \begin{vmatrix} 3 & 4 \\ -1 & -1 \end{vmatrix} \\ \begin{vmatrix} 0 & -4 \\ -1 & 1 \end{vmatrix} & \begin{vmatrix} 4 & -4 \\ -1 & 1 \end{vmatrix} & \begin{vmatrix} 4 & 0 \\ -1 & -1 \end{vmatrix} \\ \begin{vmatrix} 0 & -4 \\ 4 & 2 \end{vmatrix} & \begin{vmatrix} 4 & -4 \\ 3 & 2 \end{vmatrix} & \begin{vmatrix} 4 & 0 \\ 3 & 4 \end{vmatrix} \end{pmatrix} = \begin{pmatrix} 6 & 5 & 1 \\ -4 & 0 & -4 \\ 16 & 20 & 16 \end{pmatrix}.$$

To find the matrix of cofactors we multiply by the pattern

$$\begin{array}{ccc} + & - & + \\ - & + & - \\ + & - & + \end{array}$$

giving

$$\begin{pmatrix} 6 & -5 & 1 \\ 4 & 0 & 4 \\ 16 & -20 & 16 \end{pmatrix}.$$

To find the adjoint, we take the transpose of the above, giving

$$\begin{pmatrix} 6 & 4 & 16 \\ -5 & 0 & -20 \\ 1 & 4 & 16 \end{pmatrix}.$$

Now we find the determinant – expanding about the first row, this gives

$$4(4 - (-2)) - 0(3(1) - 2(-1)) - 4(3(-1) - (-1)(4)) = 20$$

Finally, we divide the adjoint by the determinant to find the inverse giving

$$\frac{1}{20}\begin{pmatrix} 6 & 4 & 16 \\ -5 & 0 & -20 \\ 1 & 4 & 16 \end{pmatrix} = \begin{pmatrix} 0.3 & 0.2 & 0.8 \\ -0.25 & 0 & -1 \\ 0.05 & 0.2 & 0.8 \end{pmatrix}.$$

Check: To check that the calculation is correct, we multiply the original matrix by the inverse. If we get the unit matrix as the result we can conclude that we have indeed found the inverse.

$$\begin{pmatrix} 4 & 0 & -4 \\ 3 & 4 & 2 \\ -1 & -1 & 1 \end{pmatrix} \begin{pmatrix} 0.3 & 0.2 & 0.8 \\ -0.25 & 0 & -1 \\ 0.05 & 0.2 & 0.8 \end{pmatrix} = \begin{pmatrix} 1 & 0 & 0 \\ 0 & 1 & 0 \\ 0 & 0 & 1 \end{pmatrix}$$

which is correct.

13.7 Eigenvectors and eigenvalues

In Example 13.17, we looked at the problem of scaling along the line $x = y$ and we saw that the matrix

$$\begin{pmatrix} \frac{3}{2} & \frac{1}{2} \\ \frac{1}{2} & \frac{3}{2} \end{pmatrix}$$

represents a scaling along the line $y = x$ and it leaves points along the line $y = -x$ unchanged. This means that any vector in the direction $(1,1)$ will simply be multiplied by 2 and any vector in the direction $(-1,1)$ will remain unchanged after multiplication by this matrix. Other vectors will undergo a mixed effect.

Supposing we know that a matrix **A** represents a scaling but without knowing the direction of the scaling or by how much it scales. Is there any way we can find that direction and the scaling constant?

The problem then is to find a vector **v** which is simply scaled by some currently unknown amount λ when multiplied by **A**, and **v** must be such that

$$\mathbf{Av} = \lambda\mathbf{v}$$

If we manage to find values of λ and **v** we call these the eigenvalues and eigenvectors of the matrix **A**.

We shall solve this for

$$\begin{pmatrix} \frac{3}{2} & \frac{1}{2} \\ \frac{1}{2} & \frac{3}{2} \end{pmatrix}$$

as we know the result that we expect to get.

Example 13.31   Find λ and **v** such that $\mathbf{Av} = \lambda\mathbf{v}$ where

$$\mathbf{A} = \begin{pmatrix} \frac{3}{2} & \frac{1}{2} \\ \frac{1}{2} & \frac{3}{2} \end{pmatrix}.$$

Solution   Subtract $\lambda\mathbf{v}$ from both sides of the equation

$$\mathbf{Av} = \lambda\mathbf{v} \quad \Leftrightarrow \quad \mathbf{Av} - \lambda\mathbf{v} = \mathbf{0} \qquad \begin{pmatrix} \frac{3}{2} & \frac{1}{2} \\ \frac{1}{2} & \frac{3}{2} \end{pmatrix} \mathbf{v} - \lambda\mathbf{v} = \mathbf{0}$$

We put in the unit matrix as $\mathbf{v} = \mathbf{Iv}$ and combine the terms.

$$\begin{pmatrix} \frac{3}{2} & \frac{1}{2} \\ \frac{1}{2} & \frac{3}{2} \end{pmatrix} \mathbf{v} - \lambda \begin{pmatrix} 1 & 0 \\ 0 & 1 \end{pmatrix} \mathbf{v} = \begin{pmatrix} 0 \\ 0 \end{pmatrix}$$

$$\begin{pmatrix} \frac{3}{2} - \lambda & \frac{1}{2} \\ \frac{1}{2} & \frac{3}{2} - \lambda \end{pmatrix} \mathbf{v} = \begin{pmatrix} 0 \\ 0 \end{pmatrix}.$$

Now substitute

$$\mathbf{v} = \begin{pmatrix} x \\ y \end{pmatrix}$$

giving

$$\left(\frac{3}{2} - \lambda\right) x + \frac{1}{2} y = 0$$

$$\frac{1}{2} x + \left(\frac{3}{2} - \lambda\right) y = 0.$$

Unfortunately, the solution to this gives $x = 0$ and $y = 0$, which is not very enlightening (it is called the trivial solution).

However, we started by saying that we wanted to find the direction in which this matrix scaled any vector. That is, we want to find a whole line of solutions. We can use a result that we found from solving systems of equations. The equations may have a whole line of solutions if the determinant of the coefficients is 0.

Hence, we need to find λ such that

$$\begin{vmatrix} \frac{3}{2} - \lambda & \frac{1}{2} \\ \frac{1}{2} & \frac{3}{2} - \lambda \end{vmatrix} = 0.$$

Expanding the determinant gives

$$\left(\frac{3}{2} - \lambda\right)\left(\frac{3}{2} - \lambda\right) - \frac{1}{4} = 0$$

$$\frac{9}{4} - 3\lambda + \lambda^2 - \frac{1}{4} = 0 \Leftrightarrow \lambda^2 - 3\lambda + 2 = 0.$$

This factorizes to

$$(\lambda - 2)(\lambda - 1) = 0 \quad \Leftrightarrow \quad \lambda = 1 \vee \lambda = 1.$$

Hence, the eigenvalues of the matrix $\mathbf{A}$ are 1 and 2. To find the vectors which go with each of these eigenvalues we substitute into the equations

$$\left(\frac{3}{2} - \lambda\right) x + \frac{1}{2} y = 0$$

$$\frac{1}{2} x + \left(\frac{3}{2} - \lambda\right) y = 0.$$

For $\lambda = 2$

$$-\frac{1}{2} x + \frac{1}{2} y = 0$$

$$\frac{1}{2} x - \frac{1}{2} y = 0.$$

We notice that these equations are dependent, which we would have expected as by setting the determinant $= 0$ we were looking for an undetermined system.

We have

$$-\frac{1}{2} x + \frac{1}{2} y = 0 \quad \Leftrightarrow \quad x = y.$$

This means that any vector (x, y) where $y = x$ will be scaled by 2 if multiplied by the matrix $\mathbf{A}$. The eigenvector can be given as $(1,1)$ as it is only necessary to indicate the direction.

The other eigenvalue, $\lambda = 1$, gives

$$\frac{1}{2}x + \frac{1}{2}y = 0$$

$$\frac{1}{2}x + \frac{1}{2}y = 0.$$

Again, the equations are dependent and we have $x + y = 0$. The eigenvector is any vector (x, y) where $x = -y$ so this gives the direction $(-1, 1)$.

As not all matrices represent scaling, it is not always possible to find real eigenvalues. In particular, a rotation matrix has no real eigenvalues for angle of rotation, $\theta \neq 0$. Another point to note is that in this example the eigenvectors were at right angles to each other. This is only true for symmetric matrices (which we had in this case).

The method can be summarized as follows: To find the eigenvalues and eigenvectors of **A**

(1)   Solve $|\mathbf{A} - \lambda \mathbf{I}| = 0$ to find the eigenvalues. This is called the characteristic equation.

(2)   For each value of λ found, substitute into

$$(\mathbf{A} - \lambda \mathbf{I})\mathbf{v} = 0$$

and find **v**. This will be an undetermined system so we shall find at least a whole line of solutions. Choose any vector lying in the direction of the line.

Example 13.32   Find the eigenvectors and eigenvalues of

$$\begin{pmatrix} 1 & 3 \\ 2 & -4 \end{pmatrix}.$$

Solution   Solve $|\mathbf{A} - \lambda \mathbf{I}| = 0$, which gives

$$\begin{vmatrix} 1 - \lambda & 3 \\ 2 & -4 - \lambda \end{vmatrix} = 0$$

$$\Leftrightarrow (1 - \lambda)(-4 - \lambda) - 6 = 0$$

$$\Leftrightarrow \lambda^2 + 3\lambda - 10 = 0$$

$$\Leftrightarrow (\lambda + 5)(\lambda - 2) = 0 \Leftrightarrow \lambda = -5 \text{ or } \lambda = 2.$$

For each value of λ solve

$$\begin{pmatrix} 1 - \lambda & 3 \\ 2 & -4 - \lambda \end{pmatrix} \begin{pmatrix} x \\ y \end{pmatrix} = \begin{pmatrix} 0 \\ 0 \end{pmatrix}.$$

For $\lambda = -5$, this gives

$$\begin{pmatrix} 6 & 3 \\ 2 & 1 \end{pmatrix} \begin{pmatrix} x \\ y \end{pmatrix} = \begin{pmatrix} 0 \\ 0 \end{pmatrix} \quad \Rightarrow \quad \begin{matrix} 6x + 3y = 0 \\ 2x + y = 0 \end{matrix}$$

We see that these equations are dependent. Solving the first one

$$6x + 3y = 0$$

$$\Leftrightarrow \quad 2x + y = 0$$

$$y = -2x$$

The vector is therefore (x, y) where $y = -2x$, giving $(x, -2x)$. We only need the direction of the vector so choose $(1, -2)$ by substituting $x = 1$

For $\lambda = 2$ we get

$$\begin{pmatrix} -1 & 3 \\ 2 & -6 \end{pmatrix} \begin{pmatrix} x \\ y \end{pmatrix} = \begin{pmatrix} 0 \\ 0 \end{pmatrix} \Rightarrow \begin{array}{l} -x + 3y = 0 \\ 2x - 6y = 0 \end{array}$$

Solving

$$-x + 3y = 0 \quad \Leftrightarrow \quad x = 3y$$

Hence, we have (x, y) where $x = 3y$ giving $(3y, y)$. Substitute $y = 1$ giving the vector as (3,1).

We have shown that the matrix

$$\begin{pmatrix} 1 & 3 \\ 2 & -4 \end{pmatrix}$$

has eigenvalue -5 with eigenvector $(1, -2)$ and eigenvalue 2 with eigenvector $(3,1)$.

13.8 Least squares data fitting

Matrix methods can be employed to the problem of finding the 'best fit' line through a set of data. In Chapter 2 we performed a fit by eye to a set of data points. We drew a scatter diagram of the data and if the data appeared, more or less, to fit on a line then we would draw the line by hand and then use any two points lying on the line, to find the equation of the line. We do not expect experimental data to be exact and that is the reason that the data points do not lie exactly on a line. We shall now look at the method of least squares, which can be used to compute the 'best fit' line. The method is called 'least squares' because it minimizes the squared error between the data points and the equation of the line found. The method is justified in the following example.

We start with two sets of data which we suspect are related linearly.

Example 13.33 A student was late for a particularly interesting engineering maths lecture and was therefore walking briskly toward the lecture hall in a straight line at approximately constant speed b. The student's position x (metres) at time t (seconds) is given by

t	0	5	10	15	20	25
x	100	111	119	132	140	151

We can plot these points on a scatter diagram, as in Figure 13.18.

We can see that they lie on an approximate straight line. We need to decide which is the 'best' straight line to draw. One way is to fit the straight line, $y = a + bx$, to a set of data points (x_i, y_i) so that the sum of the squares of the vertical distances of the points from the straight line drawn is minimum. This is illustrated in Figure 13.19.

In order to fit the line $y = a + bx$, we can vary the values of a and b until it satisfies our condition for minimum squared error. If the data points are (x_i, y_i), then the sum of the squares of the errors is given by

$$E = \sum_i (y_i - a - bx_i)^2$$

We can vary a and b to make this a minimum. However, E is a function of two variables a and b. We know how to find the minima or maxima with

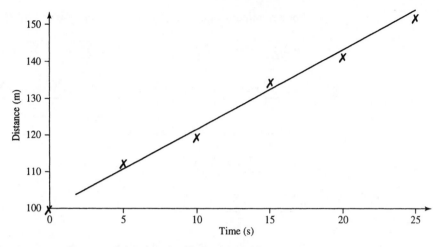

Figure 13.18   *A scatter diagram of the data for Example 13.33.*

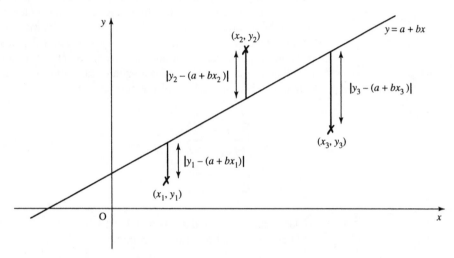

Figure 13.19   *Vertical distances between the data points and the line represent the error for each point. The method of least squares minimises the sum of the squares of these errors.*

respect to one variable, which we looked at in Chapter 11. We shall look in more detail at functions of two variables in Chapter 17. To minimize with respect to two variables we begin by differentiating with respect to each variable in turn keeping the other variable constant. This is called partial differentiation. The ideas used in finding stationary values are the same as that for one variable although the method of distinguishing between types of stationary values is slightly more involved because the function represents a two-dimensional surface drawn in three dimensions rather than a simple curve. A partial derivative is indicated by using a curly d, ∂, for the derivative and $\partial E / \partial a$ is read as 'partial dE by da'

$$\frac{\partial E}{\partial a} = -2 \sum_i (y_i - a - bx_i)$$

$$\frac{\partial E}{\partial b} = -2 \sum_i x_i (t_i - a - bx_i)$$

For a minimum value we must have $\partial E/\partial a = 0$ and $\partial E/\partial b = 0$ giving

$$-2\sum_i(y_i - a - bx_i) = 0 \Leftrightarrow \sum_i y_i - \sum_i a - b\sum_i x_i = 0$$

$$-2\sum_i x_i(y_i - a - bx_i) = 0 \Leftrightarrow \sum_i x_i y_i - a\sum_i x_i - b\sum_i x_i^2 = 0$$

Finally, we get the normal equations

$$an + b\sum_i x_i = \sum_i y_i$$

$$a\sum_i x_i + b\sum_i x_i^2 = \sum_i x_i y_i$$

where n is the number of data points.

Here, we have not attempted to justify that this is actually a minimum point (we have only shown it to give a stationary point). We can now illustrate the method for finding the values of a and b which minimize the sum of the squared errors. At the start of this example we had a set of data which we wish to fit to a function $x = a + bt$ where the dependent variable is x and the independent variable is t. We wish to find the values of a and b so that $x = a + bt$ gives a least squares fit to the data. The number of data points is 6, so we have as the normal equations:

$$6a + b\sum_{i=1}^{6} t_i = \sum_{i=1}^{6} x_i$$

$$a\sum_{i=1}^{6} t_i + b\sum_{i=1}^{6} t_i^2 = \sum_{i=1}^{6} t_i x_i$$

Make a table from the data, as in Table 13.1.

Then the normal equations become:

$$6a + 75b = 753$$

$$75a + 1375b = 10300$$

Solving

$$450a + 5625b = 56475$$

$$450a + 8250b = 61800$$

$$2625b = 5325$$

Table 13.1 *A table made from the data of Example 13.32*

t	x	t^2	tx
0	100	0	0
5	111	25	555
10	119	100	1190
15	132	225	1980
20	151	625	3775
75	753	1375	10 300

$\Rightarrow b \approx 2.03$

$6a + 75(2.03) = 754 \Rightarrow a \approx 100.14$

We have $a = 100.14$ and $b = 2.03$. Hence, the line of best fit is

$x = 100.14 + 2.03t$.

Curve fitting

The same method for fitting a straight line can be generalized to fit any polynomial. For example, it could appear that our data would be better fitted to a parabola.

$y = b_0 + b_1 x + b_2 x^2$

The normal equations in this case are

$$b_0 n + b_1 \sum_i x_i + b_2 \sum_i x_i^2 = \sum_i y_i$$

$$b_0 \sum_i x_i + b_1 \sum_i x_i^2 + b_2 \sum_i x_i^3 = \sum_i x_i y_i$$

$$b_0 \sum_i x_i^2 + b_1 \sum_i x_i^3 + b_2 \sum_i x_i^4 = \sum_i x_i^2 y_i$$

We can solve these system of equations using Gaussian elimination.

Example 13.34   Find the best fit parabola by the method of least squares for (0,3) (1,1) (2,0) (4,1) (6,4).

Solution   The data are given in Table 13.2. The normal equations are:

$5b_0 + 13b_1 + 57b_2 = 9$

$13b_0 + 57b_1 + 289b_2 = 29$

$57b_0 + 289b_1 + 1569b_2 = 161$

Solving these using Gauss elimination gives

$$\begin{pmatrix} 5 & 13 & 57 & 9 \\ 13 & 57 & 289 & 29 \\ 57 & 289 & 1569 & 161 \end{pmatrix}$$

Stage 1

$$\begin{pmatrix} 1 & 2.6 & 11.4 & 1.8 \\ 13 & 57 & 289 & 29 \\ 57 & 289 & 1569 & 161 \end{pmatrix}$$

Table 13.2 *Table made from the data of Example 13.34*

x	y	x^2	x^3	x^4	xy	x^2y
0	3	0	0	0	0	
1	1	1	1	1	1	1
2	0	4	8	16	0	0
4	1	16	64	256	4	16
6	4	36	216	1296	24	144
13	9	57	289	1569	29	161

$$\begin{pmatrix} 1 & 2.6 & 11.4 & 1.8 \\ 0 & 23.2 & 140.8 & 5.6 \\ 0 & 140.8 & 919.2 & 58.4 \end{pmatrix}$$

Stage 2

$$\begin{pmatrix} 1 & 2.6 & 11.4 & 1.8 \\ 0 & 1 & 6.069 & 0.241 \\ 0 & 140.8 & 919.2 & 58.4 \end{pmatrix}$$

$$\begin{pmatrix} 1 & 2.6 & 11.4 & 1.8 \\ 0 & 1 & 6.069 & 0.241 \\ 0 & 0 & 64.685 & 24.467 \end{pmatrix}$$

Stage 3

$$\begin{pmatrix} 1 & 2.6 & 11.4 & 1.8 \\ 0 & 1 & 6.069 & 0.241 \\ 0 & 0 & 1 & 0.378 \end{pmatrix}$$

Back-substitution

$$b_0 + 2.6b_1 + 11.4b_2 = 1.8$$
$$b_1 + 6.069b_2 = 0.241$$
$$b_2 = 0.378$$

gives $b_2 = 0.378$, $b_1 = -2.054$, and $b_0 = 2.831$.
Hence, the best fit parabola is $y = 2.831 - 2.054x + 0.378x^2$.

13.9 Summary

1. Matrices are used to represent information in a way suitable for use by a computer. They can represent, among other things, systems of linear equations, transformations, and networks. A matrix is a rectangular array of numbers of dimension $m \times n$ where m is the number of rows and n is the number of columns.

2. To add or subtract matrices add or subtract each corresponding element. The matrices must be of exactly the same dimension. To multiply two matrices, $\mathbf{C} = \mathbf{AB}$, the number of columns in matrix $\mathbf{A}$ must equal the number of rows in matrix $\mathbf{B}$. The i, jth element of $\mathbf{C}$ is found by multiplying the ith row of $\mathbf{A}$ by the jth column of $\mathbf{B}$.

3. The unit matrix, $\mathbf{I}$, leaves any matrix unchanged under multiplication.

$$I = \begin{pmatrix} 1 & 0 \\ 1 & 0 \end{pmatrix} \text{ (2 dimensions)}$$

$$I = \begin{pmatrix} 1 & 0 & 0 \\ 0 & 1 & 0 \\ 0 & 0 & 1 \end{pmatrix} \text{ (3 dimensions)}$$

$\mathbf{IA} = \mathbf{AI} = \mathbf{A}$ where $\mathbf{A}$ is any matrix.

4. The inverse of a matrix is represented by $\mathbf{A}^{-1}$ and can be found for square, non-singular matrices. A matrix is singular if its determinant is 0.

5. The 2×2 determinant is defined by

$$\begin{vmatrix} a & b \\ c & d \end{vmatrix} = ad - cb.$$

6. The inverse of a 2×2 non-singular matrix $\begin{pmatrix} a & b \\ c & d \end{pmatrix}$ is

$$\frac{1}{(ad - cb)} \begin{pmatrix} d & -b \\ -c & a \end{pmatrix}.$$

7. Transformations of the plane ($\mathbb{R}^2$) can be defined using vectors and matrices as in Section 13.13.

8. Systems of linear equations may be determined (a single solution), indeterminate (many solutions) or inconsistent (no solutions).

9. Systems of linear equations can be solved using Gaussian elimination.

10. The inverse of a matrix, if it exists, can be found using Gauss–Jordan elimination.

11. The determinant of a 3×3 matrix may be found by expanding about any row or column, where

$$\begin{vmatrix} a_1 & b_1 & c_1 \\ a_2 & b_2 & c_2 \\ a_3 & b_3 & c_3 \end{vmatrix} = a_1 \begin{vmatrix} b_2 & c_2 \\ b_3 & c_3 \end{vmatrix} - b_1 \begin{vmatrix} a_2 & c_2 \\ a_3 & c_3 \end{vmatrix} + c_1 \begin{vmatrix} a_2 & b_2 \\ a_3 & b_3 \end{vmatrix}$$

gives the expansion about the first row.

12. The inverse of a non-singular matrix can be found using

$$\mathbf{A}^{-1} = (1/|\mathbf{A}|)(\text{Adjoint}(\mathbf{A})).$$

13. The eigenvalues and eigenvectors of a matrix $\mathbf{A}$ are the values of and $\mathbf{v}$ such that $\mathbf{Av} = \lambda\mathbf{v}$.

14. The method of least squares is used to fit a line or a curve through experimental data in such a way as the sum of the square of the errors is a minimum.

13.10 Exercises

13.1
$$\mathbf{A} = \begin{pmatrix} 2 & 3 \\ 0 & -4 \\ 0 & 2 \end{pmatrix} \quad \mathbf{B} = \begin{pmatrix} 2 & -1 & 0 \\ 1 & 6 & 1 \\ 0 & 3 & 4 \end{pmatrix}$$

$$\mathbf{C} = (4 \quad 0 \quad -1) \quad \mathbf{D} = \begin{pmatrix} 0 & 3 \\ -1 & 4 \end{pmatrix}$$

$$\mathbf{E} = \begin{pmatrix} 3 & 2 \\ -1 & \frac{-2}{3} \end{pmatrix}$$

Find the following, where possible
(a) $\mathbf{A}$ (b) $\mathbf{A}^T\mathbf{B}$ (c) $\mathbf{AB}$ (d) $\mathbf{BA}$
(e) $\mathbf{C}^T$ (f) $\mathbf{AC}$ (g) $\mathbf{CA}$ (h) $\mathbf{CB}$
(i) $3\mathbf{CA}$ (j) $\mathbf{D} + \mathbf{E}$ (k) $3\mathbf{D} - \frac{1}{2}\mathbf{E}$ (l) $\frac{1}{2}\mathbf{B} + \mathbf{A}$
(m) $\mathbf{B}^2$ (n) $\mathbf{E}^3$ (o) $\mathbf{AC}^2$ (p) $\mathbf{A}^T\mathbf{BC}^T$.

13.2 Use

$$\mathbf{A} = \begin{pmatrix} a_{11} & a_{12} \\ a_{21} & a_{22} \end{pmatrix} \quad \mathbf{B} = \begin{pmatrix} b_{11} & b_{12} \\ b_{21} & b_{22} \end{pmatrix} \quad \mathbf{C} = \begin{pmatrix} c_{11} & c_{12} \\ c_{21} & c_{22} \end{pmatrix}$$

to justify the associative law for 2×2 matrices

$$\mathbf{A}(\mathbf{BC}) = (\mathbf{AB})\mathbf{C}$$

13.3 Find a real square matrix which is both symmetric and skew symmetric.

13.4 A network as in Figure 13.20(a), (b), or (c) can be used to represent an electrical network, as system of one-way streets, or a communication system. An incidence matrix can be defined for a network in the following way (the lines are called arcs and the dots are called vertices).

$a_{ij} = 1$ if the arc j is leaving the vertex i

$a_{ij} = -1$ if the arc j is entering the vertex i

$a_{ij} = 0$ if the arc j does not touch the vertex i

For instance, Figure 13.20(a) has incidence matrix

$$\begin{array}{c} \\ \text{Vertex} \end{array} \begin{array}{c} \\ 1 \\ 2 \\ 3 \end{array} \overset{\displaystyle \overset{\text{Arc}}{\begin{array}{ccc} a & b & c \end{array}}}{\begin{pmatrix} -1 & 0 & -1 \\ 1 & -1 & 0 \\ 0 & 1 & 1 \end{pmatrix}}$$

(i) Find incidence matrices for the networks in Figure 13.20(b) and (c).
(b) Draw networks that have the following incidence matrices

(i)

$$
\text{Vertex}\quad
\begin{array}{c}
 \\
1 \\
2 \\
3 \\
4
\end{array}
\begin{array}{c}
\text{Arc} \\
\begin{array}{ccc}
a & b & c
\end{array} \\
\left(
\begin{array}{ccc}
-1 & 1 & 1 \\
0 & -1 & 0 \\
1 & 0 & -1 \\
0 & 0 & 0
\end{array}
\right)
\end{array}
$$

(ii)

$$
\text{Vertex}\quad
\begin{array}{c}
 \\
1 \\
2 \\
3
\end{array}
\begin{array}{c}
\text{Arc} \\
\begin{array}{ccccc}
a & b & c & d & e
\end{array} \\
\left(
\begin{array}{ccccc}
-1 & 0 & -1 & 1 & 0 \\
1 & -1 & 0 & -1 & 1 \\
0 & 1 & 1 & 0 & -1
\end{array}
\right)
\end{array}
$$

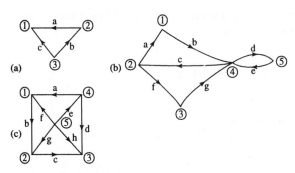

Figure 13.20 *Networks for Exercise 13.4.*

13.5 Represent the following transformations using matrices and vectors. In each case, apply the transformations to A(0,0), B(1,0), C(1,1), and D(0,1) to find their images A', B', C' and D' and draw your result:

(a) Rotation about the origin through $120°$.
(b) Translation by $(-4, 1)$.
(c) Reflection in the x-axis.
(d) Scaling in the y-direction by 5.
(e) Rotation through $120°$ about the origin followed by a translation by $(-2, 3)$.
(f) Translation by $(-1, 3)$ followed by rotation through $120°$ about the origin.
(g) Rotation through $120°$ about the origin followed by scaling by 5 in the y-direction.
(h) Reflection in the line $y = x$.
(i) Scaling along a line at a $30°$ angle to the x-axis, by a factor of 4.
(k) Find inverse transformations for those given in parts (a),(b),(c),(e),(g), and (i) and check in each case that the inverse transformation returns $A', B', C',$ and D' to A,B,C,D.

13.6 Sketch the following systems of equations and solve them using Gauss elimination. In each case, state whether the system is determined, indeterminate, or inconsistent.

(a) $4x + y = 3$ (b) $x + y = 3$
$-2x - y = -3$ $x - y = 7$
(c) $3x + 2y = 1$ (d) $-5x - y = 5$
$-2x - \frac{4}{3}y = 0$ $10x + 2y = -10$
(e) $3x + 2y = -17$ (f) $x + 6y = 4$
$10x + y = 0$ $3x + 18y = 10$
(g) $-3x + 2y = 1$ (h) $-7x + 8y = -10$
$1.5x - y = 0.5$ $-2x + y = -8.$

13.7 Solve the following using Gauss elimination. In each case state whether the system is determined, indeterminate or inconsistent.

(a) $2x - 3y - z = 4$ (b) $4x - y - 2z = 40$
$5x + 5y + 2z = -23$ $3x + y + 9z = 5$
$x + z = -1$ $x - y + z = -55$
(c) $x + y - 3z = 8$ (d) $x - y + z = 3$
$-10x + 6y = -14$ $x + z = 3$
$12x - 4y - 6z = 30$ $-4x - y - 4z = -10.$

13.8 Find inverses of the following matrices $\mathbf{A}$, if they exist, and check that $\mathbf{A}^{-1}\mathbf{A} = \mathbf{I}$

(a) $\begin{pmatrix} 1 & -2 \\ 0 & 1 \end{pmatrix}$ (b) $\begin{pmatrix} 2 & 3 \\ -1 & 6 \end{pmatrix}$

(c) $\begin{pmatrix} 2 & -1 \\ 10 & -5 \end{pmatrix}$ (d) $\begin{pmatrix} 1 & -1 & 0 \\ -2 & 4 & 2 \\ -3 & 1 & 2 \end{pmatrix}$

(e) $\begin{pmatrix} 0 & 5 & 1 \\ -1 & 5 & 3 \\ 2 & 0 & 2 \end{pmatrix}$ (f) $\begin{pmatrix} 1 & -1 & 2 \\ 6 & 1 & 3 \\ -5 & -2 & -1 \end{pmatrix}$

13.9 Find the following determinants:

(a) $\begin{vmatrix} 1 & 6 \\ 2 & 3 \end{vmatrix}$ (b) $\begin{vmatrix} 5 & 1 \\ -6 & 2 \end{vmatrix}$

(c) $\begin{vmatrix} 3 & -2 & 1 \\ 0 & 1 & 0 \\ 3 & 2 & 1 \end{vmatrix}$ (d) $\begin{vmatrix} 6 & -3 & -2 \\ 1 & -1 & 8 \\ 0 & -1 & 0 \end{vmatrix}.$

13.10 The vector product of two three-dimensional vectors can be defined using a determinant as follows:

$$(a_1, a_2, a_3) \times (b_1, b_2, b_3) = \begin{vmatrix} \mathbf{i} & \mathbf{j} & \mathbf{k} \\ a_1 & a_2 & a_3 \\ b_1 & b_2 & b_3 \end{vmatrix}$$

where $\mathbf{i}, \mathbf{j},$ and $\mathbf{k}$ are the unit vectors in the x, y, and z directions, respectively. Use this definition to find the following:

(a) $(1, 3, 6) \times (-1, 2, 2)$
(b) $(\frac{1}{2}, \frac{1}{2}, -1) \times (0, 0, 3)$.

13.11 The scalar triple produce of three vectors $\mathbf{a}, \mathbf{b}, \mathbf{c}$, given by $\mathbf{a} \cdot (\mathbf{b} \times \mathbf{c})$ can be found using a determinant as follows:

$$(a_1, a_2, a_3) \cdot ((b_1, b_2, b_3) \times (c_1, c_2, c_3))$$

$$= \begin{vmatrix} a_1 & a_2 & a_3 \\ b_1 & b_2 & b_3 \\ c_1 & c_2 & c_3 \end{vmatrix}$$

The absolute value of this can be interpreted as the volume of the parallelepiped which has $\mathbf{a}, \mathbf{b}, \mathbf{c}$ as its adjacent edges.

(a) Find the volume of the parallelepiped with adjacent edges given by the vectors

  (i) $(1, 0, -3)$, $(0,1,1)$, and $(3,0,1)$
  (ii) $(1, -2, 2)$, $(3, 2, -1)$, and $(2, 1, 1)$

(b) Explain why in case (a), (ii) you could conclude that the three vectors lie in the same plane.

13.12 In a homogeneous, isotropic and linearly elastic material it is found that the strains on a section of the material, represented by ε_x, ε_y, and ε_z for the x-, y-, and z-directions, respectively, can be related to the stresses, σ_x, σ_y, and σ_z by the following matrix equation.

$$\begin{pmatrix} \varepsilon_x \\ \varepsilon_y \\ \varepsilon_z \end{pmatrix} = \frac{1}{E} \begin{pmatrix} 1 & -\nu & -\nu \\ -\nu & 1 & -\nu \\ -\nu & -\nu & 1 \end{pmatrix} \begin{pmatrix} \sigma_x \\ \sigma_y \\ \sigma_z \end{pmatrix}$$

where E is the modulus of elasticity (also called Young's modulus) and ν is Poisson's ratio which relates the lateral and axial strains. Find σ_x, σ_y, σ_z, in terms of ε_x, ε_y, and ε_z and express the relationship in matrix form.

13.13 Find eigenvalues and eigenvectors of the following:

(a) $\begin{pmatrix} 6 & -1 \\ 0 & 2 \end{pmatrix}$ (b) $\begin{pmatrix} 1 & 5 \\ 1 & -3 \end{pmatrix}$ (c) $\begin{pmatrix} 3 & 2 \\ 1 & 2 \end{pmatrix}$.

13.14 A simple circuit comprising a variable voltage V_s, a diode and a resistor is shown in Figure 13.21. As V_s is varied, values of V_R and I are recorded. The values are given in Table 13.3.

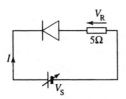

Figure 13.21 *A simple circuit with a variable voltage V_s (Exercise 13.14).*

Table 13.3 *Voltage (V_R) against current (I) for Exercise 13.14*

V_R (V)	I (A)
2	0.18
4	0.58
6	0.98
10	1.81

Using the method of least squares, determine Z and V_D in the equation $V_R = ZI + V_D$.

13.15 In an attempt to measure the stiffness of a spring the length of the spring under different loads was measured and the data is given in Table 13.4

Table 13.4 *Loads against spring length for Problem 13.15*

Load (kg)	Length (cm)
0	10
0.5	10.8
1	11.5
2	14
3	15.5
4	17.5

(a) Use the data to find an equation that could be used to find the length of the spring given the weight.

(b) From your equation estimate, if possible, the length of the spring when
  (i) the load is 2.5 kg,
  (ii) the load is 5 kg.

13.16 The power dissipation of a n–p–n silicon transistor is thought to vary linearly with temperature. The data given in Table 13.5 were recorded experimentally

Table 13.5 *Power dissipation of a n–p–n silicon transistor recorded against temperature*

Temperature (°C)	Power dissipation (W)
25	10
60	7.9
100	5.7
120	4.8
140	3.5

Plot these points on a scatter diagram and use the method of least squares to obtain a and b in the equation relating the power (P) to the temperature (T); $P = a + bT$.

13.17 Fit parabolas to the following sets of data:
(a) $(-1, 0)$, $(0, -1)$, $(1, 4)$, $(2, 14)$ $(3, 32)$
(b) $(-1, 5.5)$, $(0, 1.5)$, $(0.5, 0.5)$, $(2, 5)$.

14 Differential equations and difference equations

In order to model most physical situations we need to simplify their description. The idea of a system is central to this simplification. We identify the important elements that interact within the system. The only external interactions characterized as system inputs and the outputs are the quantities related to the system behaviour which we are interested in measuring. For instance, in examining the suspension system of a car we consider the system input to be forces exerted on the wheels of the car. The important elements in the system are the mass of the car and the springs and dampers used to connect the body of the car to the suspension links. We are interested in measuring the amount of vertical movement of the car and the velocity of that movement in response, mainly, to a bumpy road. We can examine the behaviour of the system for various external forces – building up a picture of how the system behaves in normal driving conditions. We do not need to concern ourselves with an exact model of the surface of the major motorways. In this way we have managed to separate the problem of modelling the system from modelling the environment in which it is operating. A system with a single input and output is pictured in Figure 14.1.

Dynamic physical processes involve variables which are interdependent and constantly changing. When modelling the system we will obtain relationships between variables, many of which will be related through derivatives. Even in a simple system with a mass, spring, and damper, there are many variables which can be identified, for example the extension of the spring, and its velocity and acceleration, the potential energy stored in the spring and the force exerted by it, and the force exerted by the damper. A choice of variables that completely describe the system is an important task in systems analysis. These variables are then called the state variables. These can be related to each other and the system inputs through a system of first-order differential equations which are found by using scientific laws and engineering principles to model the system. The system output, which is some quantity that we wish to compare with the response of a real system, can then be calculated from the values of the state variables.

A differential equation is one that involves some derivatives of the dependent variable, for example, $dy/dt = ky$ is a differential equation.

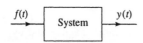

Figure 14.1 *A system has an input $f(t)$ which produces an output response $y(t)$.*

We have already looked at a number of examples of differential equations. The simplest case, when the solution can be found by integration, was looked at in Chapter 7 and then we found that exponential functions are solutions to differential equations of the form $dy/dt = ky$ in Chapter 8 and in Chapter 10 we looked at the problem of circular motion and found that exponential functions are also solutions to equations of the form $d^2y/dt^2 = -\omega^2 y$. In simple equations, we look for a solution which is a number. When solving a differential equation our idea of a solution is to find an explicit function for y in terms of t.

We will only be concerned with single input, single output (SISO) systems in which case there are two ways of proceeding in order to solve the system of differential equations. The system may be reduced to a single differential equation and then solved, or the system may be solved directly. These two methods correspond, in systems theory, to the transfer function description of the system or a state variable description.

The types of differential equations we shall look at are all linear; that is, they involve no powers of y, the independent variable or powers of the derivative or products of any of these. Any real system will, of course have some non-linear behaviour. However, we usually assume that either the non-linearities present are not very important or that it is possible to analyse a system as locally linear. Another important assumption used is that time invariance, that is, operating the system now or in an hours time with exactly the same input and initial conditions will yield the same result.

The method of solution we use for finding a particular solution of differential equations in this chapter is called the method of unknown coefficients. An alternative method, using Laplace transforms, is more widely applied, and we shall look at them in the next chapter.

Differential equations are used to analyse systems that are subject to continuous or piecewise continuous inputs. It may be necessary, or preferable, to develop a discrete model of the system. A discrete model may be used if the exact nature of the system is unknown and we attempt to predict its nature by recording its response to various inputs. The recording of the system inputs and response will generally be stored in digital form and the data processed in digital form. Another major application of digital systems is for digital filter design. Here we wish to design digital filters with certain desirable characteristics. We can find the characteristics of digital systems by solving difference equations. We give a method of solution of simple difference equations. More frequently, z-transforms are used and these are discussed in the next chapter.

14.2 Modelling simple systems

Damped forced motion of a spring

A spring of length l has a mass m attached to it and a dashpot damping the motion. It is subject to a force $f(t)$ forcing the motion (see Figure 14.2).

The input to this system is the forcing function $f(t)$ and the output is the displacement of the spring from its original length, x. In order to model this system we make a number of assumptions about its behaviour.

1. We assume Newton's second law, $F_T = ma$ where $a = m\,d^2x/dt^2$ and F_T is the total force operating on the mass: m is the mass on the spring.

2. The spring does not become distorted, that is, it is perfectly elastic. This means, from Hooke's Law, that $F_S = -kx$, where x is the displacement from the original spring length l and k is the spring constant.

3. The damping force due to the dashpot is $F_D = -r\,dx/dt$

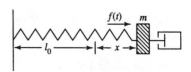

Figure 14.2 *Damped forced motion of a spring.*

To relate these quantities together we use the fact that the total force on the mass is made up of the spring force, the damping force, and the external force $F_T = F_D + F_S + f(t)$.

If we wish to combine these equations to express them using a single differential equation we simply substitute for F_T in terms of x and the derivatives of x, giving

$$m\frac{\mathrm{d}^2 x}{\mathrm{d}t^2} = -r\frac{\mathrm{d}x}{\mathrm{d}t} - kx + f(t)$$

$$\Leftrightarrow \quad m\frac{\mathrm{d}^2 x}{\mathrm{d}t^2} + r\frac{\mathrm{d}x}{\mathrm{d}t} + kx = f(t)$$

Should we wish to solve the system as a system of first order differential equations then we choose some state variables, usually the displacement and the velocity. Called these x_1 and x_2, and as x_1 is the displacement $x_1 = x$ and as x_2 is the velocity this is the derivative of the displacement giving $x_2 = \mathrm{d}x_1/\mathrm{d}t$.

As the acceleration is the derivative of the velocity, we can write the acceleration as the derivative of x_2

$$a = \frac{\mathrm{d}^2 x_1}{\mathrm{d}t^2} = \frac{\mathrm{d}}{\mathrm{d}t}\frac{\mathrm{d}x_1}{\mathrm{d}t} = \frac{\mathrm{d}x_2}{\mathrm{d}t}.$$

Now $F_T = ma$ can be written in terms of x_2 as $F_T = m\,\mathrm{d}x_2/\mathrm{d}t$ and the force due to the dashpot becomes $F_D = -rx_2$. So $F_T = F_D + F_S + f(t)$ gives

$$m\frac{\mathrm{d}x_2}{\mathrm{d}t} = -rx_2 - kx_1 + f(t)$$

$$\Leftrightarrow \quad \frac{\mathrm{d}x_2}{\mathrm{d}t} = -\frac{k}{m}x_1 - \frac{r}{m}x_2 + \frac{1}{m}f(t).$$

So the system of equations that represent the system is

$$\frac{\mathrm{d}x_1}{\mathrm{d}t} = x_2$$

$$\frac{\mathrm{d}x_2}{\mathrm{d}t} = -\frac{k}{m}x_1 - \frac{r}{m}x_2 + \frac{1}{m}f(t).$$

The first equation expresses the fact that x_2 is the velocity and the second equation was obtained by considering the total force on the mass and using Newton's second law.

This system of equations may be expressed in matrix form as

$$\frac{\mathrm{d}}{\mathrm{d}t}\begin{pmatrix} x_1 \\ x_2 \end{pmatrix} = \begin{pmatrix} 0 & 1 \\ -k/m & -r/m \end{pmatrix}\begin{pmatrix} x_1 \\ x_2 \end{pmatrix} + \begin{pmatrix} 0 \\ 1/m \end{pmatrix} f(t).$$

Writing

$$x = \begin{pmatrix} x_1 \\ x_2 \end{pmatrix}$$

$$A = \begin{pmatrix} 0 & 1 \\ -k/m & -r/m \end{pmatrix}$$

and

$$B = \begin{pmatrix} 0 \\ 1/m \end{pmatrix}$$

we can give the system of equations as a matrix differential equation

$$\frac{dx}{dt} = Ax + Bf.$$

The system input is f and the output $y = x_1$.

To write the derivative of a variable when it is clear that it is in respect to its independent variable we may write x' (read as 'x dashed') or x'' (read as 'x double dashed'). In addition, when the independent variable is time we often write $\dot{x}$ (read as 'x dot') and for the second derivative $\ddot{x}$ (read as 'x double dot'). Using this notation we have

$$\dot{x} = Ax + Bf$$

as the matrix differential equation.

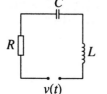

Figure 14.3 *An LRC circuit.*

Modelling electrical circuits

We can model an LRC circuit as shown in Figure 14.3 using the following assumptions:

1. Kirchoff's voltage law (KVL): the sum of all the voltage drops around any closed loop is zero.
2. The voltage drop v_R across a resistor is proportional to the current i i.e.

   $$v_R = Ri \quad \text{(Ohm's law)}$$

   where the constant of proportionality R is called the resistance of the resistor.
3. The voltage drop across a capacitor is proportional to the electric charge q on the capacitor

   $$v_c = \frac{1}{C}q$$

   where C is the capacitance and is measured in farads, the charge q is measured in coulombs.
4. The voltage drop across an inductor is proportional to the rate of change of the current i

   $$v_L = L\frac{di}{dt}$$

   where L is the inductance, measured in henries.

Then, from KVL:

$$L\frac{di}{dt} + Ri + \frac{q}{C} = v(t)$$

Since

$$i(t) = \frac{dq}{dt}$$

we have

$$\frac{di}{dt} = \frac{d}{dt}\left(\frac{dq}{dt}\right) = \frac{d^2q}{dt^2}$$

So we can write this differential equation in terms of q:

$$L\frac{d^2q}{dt^2} + R\frac{dq}{dt} + \frac{q}{C} = v(t).$$

Here, the system input is v and the output is the charge on the capacitor q.

To obtain a differential equation in terms of i we can differentiate the whole of this equation and use $i = dq/dt$, giving

$$L\frac{d^2i}{dt^2} + R\frac{di}{dt} + \frac{i}{C} = \frac{dv}{dt}.$$

Here the input is dv/dt and the output is i. Notice that L, R, and $1/C$ are constants.

Should we wish to solve the system as a system of first order differential equations then we choose some state variables. The general procedure is to choose capacitor voltages and inductor currents. Hence, we choose

$$x_1 = i$$

and

$$x_2 = v_c = q/C.$$

We can obtain one equation relating x_1 and x_2 by using

$$\frac{dq}{dt} = i \qquad \frac{1}{C}\frac{dq}{dt} = \frac{i}{C} \qquad \frac{dx_2}{dt} = \frac{x_1}{C}.$$

Now we use KVL to give the other equation

$$L\frac{di}{dt} + Ri + \frac{q}{C} = v(t)$$

substituting $i = x_1$ and $x_2 = q/C$ gives

$$\frac{dx_1}{dt} = -\frac{R}{L}x_1 - \frac{1}{L}x_2 + \frac{1}{L}v(t).$$

So the system of equations that give the state variables are

$$\frac{dx_1}{dt} = -\frac{R}{L}x_1 - \frac{1}{L}x_2 + \frac{1}{L}v(t)$$

$$\frac{dx_2}{dt} = \frac{1}{C}x_1.$$

The first equation expresses Kirchoff's voltage law the second equation expresses the relationship between the current and the voltage across the capacitor.

This system of equations may be expressed in matrix form as

$$\frac{d}{dt}\begin{pmatrix} x_1 \\ x_2 \end{pmatrix} = \begin{pmatrix} -R/L & -1/L \\ 1/C & 0 \end{pmatrix}\begin{pmatrix} x_1 \\ x_2 \end{pmatrix} + \begin{pmatrix} 1/L \\ 0 \end{pmatrix}v.$$

Writing

$$\mathbf{x} = \begin{pmatrix} x_1 \\ x_2 \end{pmatrix}$$

$$\mathbf{A} = \begin{pmatrix} -R/L & -1/L \\ 1/C & 0 \end{pmatrix}$$

and

$$\mathbf{B} = \begin{pmatrix} 1/L \\ 0 \end{pmatrix}$$

This can be written as a matrix differential equation

$$\frac{d\mathbf{x}}{dt} = \mathbf{A}\mathbf{x} + \mathbf{B}v.$$

A rotational mechanical system

A rotor of moment of inertia J is supported by a shaft of torsional stiffness k and the motion is damped by a rotational damper of torque c per unit angular velocity (see Figure 14.4). T is the external torque applied to the rotor.

The torque due to the rotational spring is proportional to the angle through which it has been twisted, giving $T_S = -k\theta$. The damper provides a torque of $-c\omega$ and the total rotational torque of the system is given by $J\,d\omega/dt$ where ω is the angular velocity and J is the moment of inertial of the rotating body.

$$J\frac{d\omega}{dt} = -c\omega - k\theta + T$$

$$J\frac{d\omega}{dt} + c\omega + k\theta = T.$$

Using $\omega = d\theta/dt$, we can write this equation in terms of the angle of rotation as

$$J\frac{d^2\theta}{dt^2} + c\frac{d\theta}{dt} + k\theta = T$$

Alternatively, we can choose state variables as $x_1 = \theta$ and $x_2 = \omega = dx_1/dt$. This gives the system of equations

$$\frac{d}{dt}\begin{pmatrix} x_1 \\ x_2 \end{pmatrix} = \begin{pmatrix} 0 & 1 \\ -k/J & -c/J \end{pmatrix}\begin{pmatrix} x_1 \\ x_2 \end{pmatrix} + \begin{pmatrix} 0 \\ 1/J \end{pmatrix}T.$$

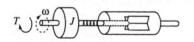

Figure 14.4 *A damped rotational system.*

Initial conditions and boundary values

To solve these equations we also need knowledge of the state of the system at some moment or moments in time. This is usually given in terms of the initial displacement and the initial velocity, that is, x and $\mathrm{d}x/\mathrm{d}t$ at $t = 0$. This is then called an initial value problem. If two values of the displacement are given at different times or if the velocity and the displacement are given at different times, then the problem is called a boundary value problem. Boundary value problems are more difficult to specify correctly to ensure that they determine a solution.

14.3 Ordinary differential equations

An ordinary differential equation is an equation which involves derivatives of y (the dependent variable) as well as functions of y and t.

$$\frac{\mathrm{d}y}{\mathrm{d}t} = \cos(t)$$

$$\frac{\mathrm{d}^2 y}{\mathrm{d}t^2} + 9t = 0$$

$$t^2 \frac{\mathrm{d}^3 y}{\mathrm{d}t^3} \frac{\mathrm{d}y}{\mathrm{d}t} + \mathrm{e}^t = t$$

are all ordinary differential equations.

The other sort of differential equations, which are introduced in Chapter 17, are partial differential equations which are used to describe a dependency on two independent variables and involve partial derivatives like $\partial y/\partial t$ (read as 'partial dy by dt').

The order of differential equations

The order of the differential equation is the order of the highest derivative of y (the dependent variable) in the equation.

$$R\frac{\mathrm{d}q}{\mathrm{d}t} + \frac{q}{C} = 3 \qquad \text{is first order in } q.$$

$$\frac{\mathrm{d}\theta}{\mathrm{d}t} = \sin(\theta) \qquad \text{is first order in } \theta.$$

$$x'' + 4t^2 = 0 \qquad \text{is second order in } x.$$

$$\frac{\mathrm{d}^3 u}{\mathrm{d}t^3} - \frac{\mathrm{d}u}{\mathrm{d}t} + u = 4t^4 \qquad \text{is third order in } u.$$

The solution of a differential equation

If a function, $y = g(t)$, is a solution to a differential equation in y then if, in the differential equation, we replace y and by $g(t)$ and all the derivatives of y by the corresponding derivatives of $g(t)$ then the resulting predicate should be an identity for t. That is, it should be true for all values of t where it is defined.

Example 14.1   Show that $x = t^3$ is a solution to the differential equation $\mathrm{d}x/\mathrm{d}t = 3t^2$.

Solution   Replace x by t^3 in the differential equation gives

$$\frac{\mathrm{d}}{\mathrm{d}t}(t^3) = 3t^2$$

$$\Leftrightarrow \quad 3t^2 = 3t^2$$

As this is true for all values of t, we have shown that $x = t^3$ is a solution to $\mathrm{d}x/\mathrm{d}t = 3t^2$.

Example 14.2   Show that $y = t^2 - 3t + 3.5$ is a solution to

$$y'' + 3y' + 2y = 2t^2.$$

Solution   Set $y = t^2 - 3t + 3.5$, then differentiating gives

$$y' = 2t - 3$$

and differentiating again

$$y'' = 2$$

We substitute these expressions for y and the derivatives of y into

$$y'' + 3y' + 2y = 2t^2$$

so we get

$$2 + 3(2t - 3) + 2(t^2 - 3t + 3.5) = 2t^2$$

$$\Leftrightarrow \quad 2 + 6t - 9 + 2t^2 - 6t + 7 = 2t^2$$

$$\Leftrightarrow \quad 2t^2 = 2t^2$$

which is true for all values of t showing that $y = t^2 - 3t + 3.5$ is a solution to $y'' + 3y' + 2y = 2t^2$.

A differential equation has many solutions. For instance, the equation $\mathrm{d}x/\mathrm{d}t = 3t^2$ has solutions $x = t^3$, $x = t^3 + 4$, $x = t^3 - 5$, etc. These are called particular solutions. A general solution is one which contains some arbitrary constants and encompasses all possible solutions to the differential equation. This means that by choosing values of these arbitrary constants we are able to find any of the particular solutions of the equation. For instance, $x = t^3 + C$ is a general solution to $\mathrm{d}x/\mathrm{d}t = 3t^2$.

Linear differential equations

A linear differential equation can be recognized by its form. It is linear if the coefficients of y (the dependent variable) and all order derivatives of y, are functions of t, or constant terms, only.

$$\mathrm{d}y/\mathrm{d}t = 4t$$

$$\mathrm{d}^2y/\mathrm{d}t^2 = 6t$$

$$t\,\mathrm{d}y/\mathrm{d}t = 6$$

$$ay'' + by' + cy = f(t)$$

$$3\,\mathrm{d}^2y/\mathrm{d}t^2 + t^2\,\mathrm{d}y/\mathrm{d}t + 6y = t^5$$

are all linear.

$$(\mathrm{d}y/\mathrm{d}t)^2 + 3 = 12t + y\,\mathrm{d}y/\mathrm{d}t - t^2$$

$$\theta'' + k\sin(\theta) = 0$$

$$u'' - (1 - u^2)u' + u = 0$$

are all non-linear.

Linearity and superposition of solutions

The property of linearity is very important. We have applied the term *linear* to equations whose graphs are straight *lines*. However, in an algebraic sense the term does not mean that everything represents a straight line! The property of linearity is defined for an operator, which we shall call O. A linear operator is one that, when operating on the sum of two terms gives the same result as operating on the terms and then taking the sum. This can be expressed as

$$\mathrm{O}(f_1 + f_2) = \mathrm{O}(f_1) + \mathrm{O}(f_2).$$

The other condition is that

$$\mathrm{O}(af) = a\mathrm{O}(f)$$

The two conditions can be combined to say that a linear operation on a linear combination of inputs produces a linear combination of the outputs. A linear combination of f_1 and f_2 is $af_1 + bf_2$, where a and b are constants and therefore if an operator, O, is linear then

$$\mathrm{O}(af_1 + bf_2) = a\mathrm{O}(f_1) + b\mathrm{O}(f_2)$$

Examples of linear operators are:

Matrices

$\mathbf{A}(\mathbf{x}_1 + \mathbf{x}_2) = \mathbf{A}\mathbf{x}_1 + \mathbf{A}\mathbf{x}_2$ and $\mathbf{A}(a\mathbf{x}) = a(\mathbf{A}\mathbf{x})$ where a is a number and $\mathbf{A}$ is a matrix and $\mathbf{x}_1, \mathbf{x}_2,$ and $\mathbf{x}$ are vectors

The differential operator

$$\frac{\mathrm{d}}{\mathrm{d}t}(3t + t^2) = 3\frac{\mathrm{d}}{\mathrm{d}t}(t) + \frac{\mathrm{d}}{\mathrm{d}t}(t^2).$$

The integral operator

$$\int (3t + t^2)\mathrm{d}t = \int 3t\,\mathrm{d}t + \int t^2\,\mathrm{d}t$$

An example of a non-linear operator is the sine function as

$$\sin(A + B) \neq \sin(A) + \sin(B)$$

For a differential equation the fact that it is linear leads to the fact that if a solution to the differential equation with input function $f_1(t)$ is y_1 and a solution to the differential equation with input $f_2(t)$ is y_2, then one solution to the equation with an input function $af_1 + bf_2$, is $ay_1 + by_2$, where a and b are constants. This result is central to the method of solving differential equations used in this chapter and also to the Laplace transform method of the next chapter.

Example 14.3   A differential equation

$$d^2x/dt^2 + 9x = f(t)$$

is found to have a particular solution $x = t/9$ when $f(t) = t$ and a particular solution

$$\tfrac{1}{5}\sin(2t) \quad \text{when} f(t) = \sin(2t).$$

Suggest a particular solution when $f(t) = t + 5\sin(2t)$ and check that your hypothesis is correct.

Solution   As the differential equation is linear, if we sum the input functions then we should be able to sum the solutions. Hence, a solution to

$$d^2x/dt^2 + 9x = t + 5\sin(2t)$$

should be given by

$$x = \frac{t}{9} + 5 \times \frac{1}{5}\sin(2t) = \frac{t}{9} + \sin(2t)$$

Check:

$$x = t/9 + \sin(2t)$$
$$dx/dt = (1/9) + 2\cos(2t)$$
$$d^2x/dt^2 = -4\sin(2t)$$

Then substituting these into

$$d^2x/dt^2 + 9x = t + 5\sin(2t)$$

gives

$$-4\sin(2t) + 9((t/9) + \sin(2t)) = t + 5\sin(2t)$$
$$-4\sin(2t) + t + 9\sin(2t) = t + 5\sin(2t)$$

which is true for all values of t, showing that

$$x = (t/9) + \sin(2t)$$

is a solution to

$$d^2x/dt^2 + 9x = t + 5\sin(2t).$$

Linear equations with constant coefficients

A linear equation is said to have constant coefficients if the coefficients multiplying y and the derivatives of y (the dependent variable) are all constants, that is, do not involve functions of t.

$$3\,\mathrm{d}^2y/\mathrm{d}t^2 + 2\,\mathrm{d}y/\mathrm{d}t + 4y = f(t)$$

has constant coefficients

$$t^2\,\mathrm{d}^2y/\mathrm{d}t^2 + 3t\,\mathrm{d}y/\mathrm{d}t + 4y = f(t)$$

does not have constant coefficients as t^2 and $3t$ are functions of time.

Time invariance

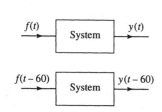

Figure 14.5   *If a system is time invariant, then a time-shifted input yields a time-shifted output.*

A linear differential equation with constant coefficients displays time invariance. If we use the same input and starting conditions for a system now or at some later time then the result relative to the initial starting time will be identical. Another way of expressing this is that if the input is time shifted then so is the output. This idea is represented in Figure 14.5.

Example 14.4   For the differential equation

$$\mathrm{d}^2y/\mathrm{d}t^2 + 4y = \sin(3t)$$

show that

$$y = \sin(2t) - \tfrac{1}{5}\sin(3t)$$

is a solution and find a solution for the equation with the same input function delayed by 1 s, that is, find a solution to

$$\mathrm{d}^2y/\mathrm{d}t^2 + 4y = \sin(3(t-1))$$

Solution   First, we check that

$$y = \sin(2t) - \tfrac{1}{5}\sin(3t)$$

is a solution to the differential equation.

To do this, we must find the first and second derivatives

$$\mathrm{d}y/\mathrm{d}t = 2\cos(2t) - (3/5)\cos(3t)$$

$$\mathrm{d}^2y/\mathrm{d}t^2 = -4\sin(2t) + (9/5)\sin(3t)$$

Substitute into

$$\mathrm{d}^2y/\mathrm{d}t^2 + 4y = \sin(3t)$$

giving

$$-4\sin(2t) + (9/5)\sin(3t) + 4(\sin(2t) - \tfrac{1}{5}\sin(3t)) = \sin(3t)$$

$$\Leftrightarrow \quad \sin(3t) = \sin(3t)$$

which is true for all t. Hence,

$$y = \sin(2t) - \tfrac{1}{5}\sin(3t)$$

is a solution to

$$\mathrm{d}^2y/\mathrm{d}t^2 + 4y = \sin(3t).$$

To find a solution to $d^2y/dt^2 + 4y = \sin(3(t-1))$, we use the property of time invariance, which means that a solution should be given by a time-shifted version of the solution to the first equation, that is

$$y = \sin(2(t-1)) - \tfrac{1}{5}\sin(3(t-1))$$

then

$$dy/dt = 2\cos(2(t-1)) - (3/5)\cos(3(t-1))$$
$$d^2y/dt^2 = -4\sin(2(t-1)) + (9/5)\sin(3(t-1)).$$

Substitute into

$$d^2y/dt^2 + 4y = \sin(3(t-1))$$

giving

$$-4\sin(2(t-1)) + (9/5)\sin(3(t-1))$$
$$+ 4(\sin(2(t-1)) - (1/5)\sin(3(t-1)))$$
$$= \sin(3(t-1))$$
$$\Leftrightarrow \quad \sin(3(t-1)) = \sin(3(t-1))$$

which is true for all t.

Example 14.5   For the differential equation

$$t\,dy/dt + y = 6t^2$$

(a)   show that $y = 2t^2$ is a solution
(b)   show that the equation $t\,dy/dt + y = 6t^2$ cannot represent a time invariant system.

Solution   (a) To show that $y = 2t^2$ is a solution to

$$t\,dy/dt + y = 6t^2$$

we need to find dy/dt

$$y = 2t^2$$
$$\Rightarrow dy/dt = 4t$$

substituting into

$$t\,dy/dt + y = 6t^2$$

gives

$$t(4t) + 2t^2 = 6t^2$$
$$\Leftrightarrow \quad 6t^2 = 6t^2, \text{ which is true for all } t.$$

(b) To show that this cannot represent a time-invariant system, we take an equation with a time-shifted input, for instance, shifted by 2 s to give

$$t\,dy/dt + y = 6(t-2)^2.$$

If it were to be time-invariant, then a solution to this equation would be a time-shifted solution of the solution to the equation in part (a), that is, $y = 2(t - 2)^2$. To show that the equation does not represent a time-invariant system we just need to show that $y = 2(t - 2)^2$ is not a solution

$$y = 2(t - 2)^2$$
$$\Rightarrow \mathrm{d}y/\mathrm{d}t = 4(t - 2)$$

Substitution into

$$t \, \mathrm{d}y/\mathrm{d}t + y = 6(t - 2)^2$$

gives

$$t(4(t - 2)) + 2(t - 2)^2 = 6(t - 2)^2$$
$$\Leftrightarrow \quad 4t(t - 2) + 2(t - 2)^2 - 6(t - 2)^2 = 0$$
$$\Leftrightarrow \quad (t - 2)(4t + 2t(t - 2) - 6(t - 2)) = 0$$
$$\Leftrightarrow \quad (t - 2)(4t + 2t - 4 - 6t + 12) = 0$$
$$\Leftrightarrow \quad 8(t - 2) = 0$$

which is not true for all values of t, showing that $y = 2(t - 2)^2$ is not a solution to $t \, \mathrm{d}y/\mathrm{d}t + y = 6(t - 2)^2$ and therefore we have shown that $t \, \mathrm{d}y/\mathrm{d}t + y = f(t)$ does not represent a time-invariant system.

We have seen that linear differential equations with constant coefficients represent linear time invariant (LTI) systems.

14.4 Solving first-order LTI systems

To solve $ay' + by = f(t)$, we begin by ignoring the forcing function setting it equal to 0. This is then called the homogeneous equation

$$ay' + by = 0.$$

The general solution to this equation is called the complementary function.

We then find any particular solution and the sum of the complementary function and the particular solution gives us the general solution to the differential equation. Summing the two solutions in this way we are making use of the linearity property of the differential equation.

It may seem strange that a system with no input, as represented by the homogeneous equation $ay' + by = 0$ may have a non-zero output! This is due to the fact that the initial conditions may be non-zero. For instance, a pendulum released from rest and performing small oscillations, approximately obeys the differential equation

$$\mathrm{d}^2 y/\mathrm{d}t^2 + \omega^2 y = 0$$

where y is the distance the tip of pendulum has moved from its rest position and ω is its angular velocity. The pendulum will only begin to move if either it is given a non-zero initial velocity of if it is lifted slightly and then released. This means that either the initial position or the initial velocity or both must be non-zero for y to be non-zero.

The method we shall use for solving the complementary function relies on the fact that we know that $y' = ky$ has the solution $y = C \, \mathrm{e}^{kt}$. That is, the derivative of an exponential function is a scaled version of the original function. This was discussed in Chapter 8. We can also show that any order derivative of an exponential function is a scaled version of the original function. Consider

$$\mathrm{d}y/\mathrm{d}t = ky.$$

Differentiating this equation gives

$$\mathrm{d}^2 y/\mathrm{d}t^2 = k\,\mathrm{d}y/\mathrm{d}t.$$

Substituting $\mathrm{d}y/\mathrm{d}t = ky$ we get

$$\mathrm{d}^2 y/\mathrm{d}t^2 = k^2 y$$

and hence the second derivative is a scaled version of the original exponential function with scaling factor k^2. Differentiating again, we get

$$\mathrm{d}^3 y/\mathrm{d}t^3 = k^2 \mathrm{d}y/\mathrm{d}t$$

and on substituting $\mathrm{d}y/\mathrm{d}t = ky$, we get

$$\mathrm{d}^3 y/\mathrm{d}t^3 = k^3 y$$

and hence the third derivative is a scaled version of the original function with scaling factor k^3. We use this property of exponential functions to solve the homogeneous equation by trying a solution of the form $y = C\,\mathrm{e}^{\lambda t}$. We then find value(s) for λ which are appropriate for the equation under consideration.

The equations we found in Section 14.2 were all second-order equations or systems of equations involving more than one state variable. However, in some circumstances, we get simple first order equations. For instance if their is no inductor present in an LRC circuit we can put $L = 0$ to give

$$R\,\mathrm{d}q/\mathrm{d}t + q/C = v(t),$$

which is first order in q.

The method of solution can be characterized by four steps. To find the solution to $ay' + by = f(t)$:

Step 1: Find the complementary function
The solution of the homogeneous equation

$$ay' + by = 0$$

is found by assuming a solution of the form $y = A\,\mathrm{e}^{\lambda t}$. Hence, $y' = A\lambda\,\mathrm{e}^{\lambda t}$ and substituting into $ay' + by = 0$ we get

$$aA\lambda\,\mathrm{e}^{\lambda t} + bA\,\mathrm{e}^{\lambda t} = 0$$

We are not interested in the trivial solution $A = 0$ and $\mathrm{e}^{\lambda t}$ is never zero, so we can divide through by $A\,\mathrm{e}^{\lambda t}$ giving

$$\lambda a + b = 0$$

which is called the auxiliary, or characteristic equation.

The auxiliary equation has the solution $\lambda = -b/a$. Hence, the complementary function is

$$y = A\,\mathrm{e}^{-(b/a)t}.$$

Table 14.1   *A table of trial solutions to be used to find a particular solution. The form of the trial solution is suggested by the form of the forcing function f(t). The coefficients $c_0, c_1, \ldots$ and c,d are to be determined*

Input function	Trial solution
Polynomial of order n	
$f(t) = a_0 + a_1 t + a_2 t^2 + \cdots + a_n t^n$	$y = c_0 + c_1 t + c_2 t^2 + \cdots + c_n t^n$
	(Use all terms up to n for the trial solution)
Exponential function	
$f(t) = a\,e^{\alpha t}$	$y = c\,e^{\alpha t}$
Sine or cosine function	
$f(t) = a\cos(\omega t)$ or $b\sin(\omega t)$ or $f(t) = a\cos(\omega t) + b\sin(\omega t)$	$y = c\cos(\omega t) + d\sin(\omega t)$

Step 2: Find a particular solution
The system output, after the effect of any initial conditions have died out, we would expect to mimic the system input. To find a particular solution we try a trial solution with some undetermined coefficients. Some good guesses to use for the trial solution, depending on the type of input function $f(t)$, are given in Table 14.1.

The coefficients in the trial solution are determined by substituting into

$$ay' + by = f(t).$$

These trial solutions will not always work. Laplace transforms may be used to find a particular solution in that case (see Chapter 15).

Step 3: Find the general solution
The general solution to the equation is given by the sum of the complementary function plus the particular solution.

Step 4: Find the particular solution
The final step is to use the initial condition to find the appropriate value of A, the arbitrary constant. This solution is then the particular solution which solves the differential equation with the given initial condition.

Example 14.6   The RC circuit in Figure 14.6 is initially relaxed and is closed at time $t = 0$. Hence, applying a DC voltage of v_0 the charge of the capacitor obeys the differential equation

$$R\,dq/dt + q/C = v_0.$$

Solve for q and find an expression for the voltage across the capacitor at a time t after the circuit was closed.

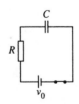

Figure 14.6   *An RC circuit with an applied DC voltage.*

Step 1
To find the complementary function we solve the homogeneous equation

$$R\,dq/dt + q/C = 0.$$

Substitute $q = A\,e^{\lambda t}$ and $dq/dt = A\lambda\,e^{\lambda t}$ to give

$$R\lambda A\,e^{\lambda t} + A\,e^{\lambda t}/C = 0.$$

Divide by $A\,e^{\lambda t}$

$$R\lambda + 1/C = 0 \quad \Leftrightarrow \quad \lambda = -1/(RC).$$

The complementary function is

$$q = A\,e^{-t/RC}$$

Step 2
Try a particular solution of the form $q = a$, where a is a constant. Then $q' = 0$, substituting into

$$R\,dq/dt + q/C = v_0$$

we get

$$a/C = v_0$$

$$\Leftrightarrow \quad a = v_0C$$

Step 3
The general solution is therefore the sum of the complementary function (found in Step 1) and a particular solution (found from Step 2), so the general solution for q is given by

$$q = A\,e^{-t/RC} + Cv_0$$

Step 4
Finally, we use the initial conditions to find the arbitrary constant A. We are told that initially the circuit was relaxed meaning that all voltages and currents are zero. In this case, this means that when $t = 0$, $q = 0$. Substituting this condition into

$$q = A\,e^{-t/RC} + Cv_0$$

we get

$$0 = A + Cv_0$$

$$\Leftrightarrow \quad A = -Cv_0$$

Therefore, the solution is

$$q = -Cv_0\,e^{-t/RC} + Cv_0$$

$$\Leftrightarrow \quad q = Cv_0(1 - e^{-t/RC}).$$

This gives the voltage across the capacitor $v_c = q/C$ as $v_c = v_0 \times (1 - e^{-t/RC})$ where v_0 is the applied DC voltage.

Check:
To check that $q = Cv_0(1 - e^{-t/RC})$ is in fact a solution to

$$R\,dq/dt + q/C = v_0$$

we carry out the following steps:

$$q = Cv_0(1 - e^{-t/RC})$$

$$\Rightarrow dq/dt = (Cv_0/RC)e^{-t/RC} \quad \Leftrightarrow \quad dq/dt = (v_0/R)e^{-t/RC}$$

Substituting into

$$R\,dq/dt + q/C = v_0$$

gives

$$(Rv_0/R)e^{-t/RC} + (v_0C/C)(1 - e^{-t/RC}) = v_0$$

$$\Leftrightarrow \quad v_0 = v_0$$

which is an identity for t, showing that $q = Cv_0(1 - e^{-t/RC})$ is a solution to the differential equation.

We also check that the solution satisfies the given initial condition by substituting $q = 0$ and $t = 0$ into $q = Cv_0(1 - e^{-t/RC})$. We find

$$0 = Cv_0(1 - 1) \quad \Leftrightarrow \quad 0 = 0,$$

which is true.

Example 14.7   Solve the differential equation

$$y' - 3y = 6\cos(2t),$$

given $y(0) = 2$.

Step 1
To find the complementary function we solve the homogeneous equation

$$y' - 3y = 0$$

substitute $y = A e^{\lambda t}$ so that $dy = A\lambda e^{\lambda t}$, giving

$$A\lambda e^{\lambda t} - 3A e^{\lambda t} = 0.$$

Therefore, we get the auxiliary equation

$$\lambda - 3 = 0 \quad \Leftrightarrow \quad \lambda = 3.$$

The complementary function is $y = A e^{3t}$

Step 2
Try a particular solution (chosen from Table 14.1) of the form $y = c\cos(2t) + d\sin(2t)$ then $y' = -2c\sin(2t) + 2d\cos(2t)$. Substituting into $y' - 3y = 6\cos(2t)$ we get

$$- 2c\sin(2t) + 2d\cos(2t) - 3(c\cos(2t) + d\sin(2t)) = 6\cos(2t)$$
$$\Leftrightarrow \quad (-2c - 3d)\sin(2t) + (2d - 3c)\cos(2t) = 6\cos(2t).$$

We want this to be true for all values of t, so we can equate the coefficients of the $\sin(2t)$ terms on both sides of the equation and also the $\cos(2t)$ terms.
Equating the coefficients of $\sin(2t)$ gives:

$$-2c - 3d = 0.$$

Equating the coefficients $\cos(2t)$ gives:

$$-2d - 3c = 6.$$

Therefore,

$$-2c - 3d = 0$$
$$-3c + 2d = 6.$$

Solving these equations simultaneously gives

$$c = -18/13 \quad d = 12/13.$$

Hence, a particular solution is

$$y = -(18/13)\cos(2t) + (12/13)\sin(2t).$$

Step 3
The general solution is given by the sum of the complementary function and a particular solution giving

$$y = A\,e^{3t} - (18/13)\cos(2t) + (12/13)\sin(2t).$$

Step 4
Use the initial condition to find A.
   When $t = 0$, $y = 2$. Substituting this into

$$y = A\,e^{3t} - (18/13)\cos(2t) + (12/13)\sin(2t)$$

gives

$$2 = A - (18/13)\cos(0) + (12/13)\sin(0)$$
$$\Leftrightarrow \quad 2 = A - 18/13 \quad \Leftrightarrow \quad A = 44/13.$$

Therefore, the solution is

$$y = (44/13)e^{3t} - (18/13)\cos(2t) + (12/13)\sin(2t).$$

Check: Substitute

$$y = (44/13)e^{3t} - (18/13)\cos(2t) + (12/13)\sin(2t)$$

and

$$y' = (132/13)e^{3t} + (36/13)\sin(2t) + (24/13)\cos(2t)$$

into

$$y' - 3y = 6\cos(2t) \text{ giving}$$
$$(132/13)e^{3t} + (36/13)\sin(2t) + (24/13)\cos(2t)$$
$$\quad - 3(44/13)e^{3t} - (18/13)\cos(2t) + (12/13)\sin(2t) = 6\cos(2t)$$
$$\Leftrightarrow \quad (78/13)\cos(2t) = 6\cos(2t)$$

which is correct.
   We also check the initial condition by substituting $t = 0$, $y = 2$ into

$$y = (44/13)e^{3t} - (18/13)\cos(2t) + (12/13)\sin(2t).$$

We find

$$2 = (44/13)e^{0} - (18/13)\cos(0) + (12/13)\sin(0)$$
$$\Leftrightarrow \quad 2 = (44/13) - (18/13)$$
$$\Leftrightarrow \quad 2 = 26/13, \text{ which is true.}$$

14.5 Solution of a second-order LTI systems

We can solve a second-order system in a manner similar to first-order systems. To solve $ay'' + by' + cy = f(t)$ we solve the homogeneous equation

$$ay'' + by' + cy = 0.$$

The solution to this equation is called the complementary function.

We then find any particular solution and the sum of the complementary function and a particular solution gives us the general solution to the differential equation. We can again list the steps involved.

To solve $ay'' + by' + cy = f(t)$

Step 1 – Find the complementary function
The solution of the homogeneous equation

$$ay'' + by' + cy = 0$$

is found by assuming solutions of the form $y = A\,e^{\lambda t}$.

Hence, $y' = A\lambda\,e^{\lambda t}$ and $y'' = A\lambda^2\,e^{\lambda t}$ and we find

$$aA\lambda^2\,e^t + bA\lambda\,e^{\lambda t} + cA\,e^{\lambda t} = 0$$

giving the auxiliary equation

$$a\lambda^2 + \lambda b + c = 0.$$

The auxiliary equation is a quadratic equation. It has solutions

$$\lambda_1, \lambda_2 = \frac{-b \pm \sqrt{b^2 - 4ac}}{2a}.$$

There are three possible types of solutions:

Case (1): λ_1 and λ_2 are real and distinct then the solution is:

$$y = A\,e^{\lambda_1 t} + B\,e^{\lambda_2 t}.$$

Case (2): λ_1 and λ_2 are complex. We set $\lambda_1 = k + j\omega_0$ and $\lambda_2 = k - j\omega_0$ so that $k = -b/2a$ and

$$\omega_0 = \frac{\sqrt{4ac - b^2}}{2a}.$$

Then the complex exponentials can be written in terms of cosines and sines and the solution becomes

$$y = e^{kt}(A\cos(\omega_0 t) + B\sin(\omega_0 t)).$$

Case (3): The roots are equal, that is, $\lambda_1 = \lambda_2 = k = -b/2a$, then the solution is $y = (At + B)e^{kt}$.

Step 2: Find a particular solution
The particular solution is any solution of the equation

$$ay'' + by' + cy = f(t).$$

As for the first-order system we expect the system output, after the effect of any initial conditions have died out, to mimic the system input. Again to find a particular solution we try a trial solution as given in Table 14.1. The 'guess' at the particular solution is substituted into the differential equation and the unknowns coefficients can be determined. If this guess does not produce a solution then Laplace transform methods can be used as in Chapter 15.

Step 3
The general solution to the equation is given by the sum of the complementary function plus a particular solution.

Step 4
The final stage is to use the initial conditions or boundary conditions to find the appropriate values of the arbitrary constants A and B.

Example 14.8   Solve the differential equation $5y'' + 6y' + 5y = 6\cos(t)$, where $y(0) = 0$ and $y'(0) = 0$.

Step 1

To find the complementary function we solve the homogeneous equation $5y'' + 6y' + 5y = 0$. Trying solutions of the form $y = A\,e^{\lambda t}$ leads to the auxiliary equation $5\lambda^2 + 6\lambda + 5 = 0$. Notice that a quick way to get the auxiliary equation is to 'replace' y'' by λ^2, y' by λ, and y by 1. The auxiliary equation has solutions

$$\lambda = \frac{-6 \pm \sqrt{36 - 100}}{10}$$

$$= \frac{-6 \pm \sqrt{-64}}{10}$$

$$= -0.6 \pm j0.8.$$

Comparing this with $\lambda = k \pm j\omega_0$ gives $k = -0.6$, $\omega_0 = 0.8$. This means that the complementary function has the form

$$y = e^{kt}(A\cos(\omega_0 t) + B\sin(\omega_0 t))$$

(given as Case (2) in the general method), with $k = -0.6$ and $\omega_0 = 0.8$. This gives

$$y = e^{-0.6t}(A\cos(0.8t) + B\sin(0.8t)).$$

Step 2

As $f(t) = 6\cos(t)$, from Table 14.1 we decide to try a particular solution of the form $y = c\cos(t) + d\sin(t)$. Then

$$y' = -c\sin(t) + d\cos(t)$$

and

$$y'' = -c\cos(t) - d\sin(t).$$

Substituting in

$$5y'' + 6y' + 5y = 6\cos(t)$$

we find

$$5(-c\cos(t) - d\sin(t)) + 6(-c\sin(t) + d\cos(t))$$
$$+ 5(c\cos(t) + d\sin(t)) = 6\cos(t)$$
$$\Leftrightarrow \quad (-5c + 6d + 5c)\cos(t) + (-5d - 6c + 5d)\sin(t) = 6\cos(t)$$
$$\Leftrightarrow \quad 6d\cos(t) - 6c\sin(t) = 6\cos(t).$$

As we want this to be an identity we equate the coefficients of $\cos(t)$ and the coefficients of $\sin(t)$ and get the two equations

$$6d = 6 \quad \Leftrightarrow \quad d = 1$$
$$6c = 0 \quad \Leftrightarrow \quad c = 0.$$

Hence, a particular solution is $y = \sin(t)$.

Step 3

The general solution is given by the sum of the complementary function and a particular solution, so

$$y = e^{-0.6t}(A\sin(0.8t) + B\cos(0.8t)) + \sin(t).$$

Step 4

Use the given initial conditions to find values for the constants A and B. Substituting $y = 0$ when $t = 0$ into

$$y = e^{-0.6t}(A\cos(0.8t) + B\sin(0.8t)) + \sin(t)$$

then

$$0 = e^0(A\cos(0) + B\sin(0)) + \sin(0)$$

$$\Leftrightarrow \quad 0 = A.$$

To use the other condition, that $y' = 0$ when $t = 0$, we need to differentiate the general solution to find an expression for y'. Differentiating (using the product rule):

$$y = e^{-0.6t}(A\cos(0.8t) + B\sin(0.8t)) + \sin(t)$$

$$\Rightarrow y' = -0.6\,e^{-0.6t}(A\cos(0.8t) + B\sin(0.8t))$$

$$+ e^{-0.6t}(-0.8A\sin(0.8t) + 0.8B\cos(0.8t)) + \cos(t)$$

and using $y' = 0$ when $t = 0$ gives

$$0 = -0.6A + 0.8B + 1.$$

We have already found $A = 0$ from the first condition, so

$$B = -1/(0.8) = -1.25.$$

Therefore, the solution is

$$y = -1.25\,e^{-0.6t}\sin(0.8t) + \sin(t).$$

Check

$$y = -1.25\,e^{-0.6t}\sin(0.8t) + \sin(t)$$

$$y' = 0.75\,e^{-0.6t}\sin(0.8t) - e^{-0.6t}\cos(0.8t) + \cos(t)$$

$$y'' = -0.45\,e^{-0.6t}\sin(0.8t) + 0.6\,e^{-0.6t}\cos(0.8t)$$

$$+ 0.6\,e^{-0.6t}\cos(0.8t) + 0.8\,e^{-0.6t}\sin(0.8t) - \sin(t).$$

Substitute into

$$5y'' + 6y' + 5y = 6\cos(t)$$

giving

$$5(-0.45\,e^{-0.6t}\sin(0.8t) + 0.6\,e^{-0.6t}\cos(0.8t)$$

$$+ 0.6\,e^{-0.6t}\cos(0.8t) + 0.8\,e^{-0.6t}\sin(0.8t) - \sin(t))$$

$$+ 6(0.75\,e^{-0.6t}\sin(0.8t) - e^{-0.6t}\cos(0.8t) + \cos(t))$$

$$+ 5(-1.25\,e^{-0.6t}\sin(0.8t) + \sin(t)) = 6\cos(t)$$

$$\Leftrightarrow \quad e^{-0.6t}\sin(0.8t)(1.75 + 4.5 - 6.25) + e^{-0.6t}\cos(0.8t)(6 - 6)$$

$$+ \sin(t)(-5 + 5) + 6\cos(t) = 6\cos(t)$$

which is true for all values of t.

We can also check that the solution satisfies the given initial conditions.

Example 14.9   Find the current in an initially relaxed LRC circuit with $R = 5\,\Omega, L = 4\,\mu\text{H}, C = 1\,\mu\text{F}$ where the AC voltage source is given by $v(t) = 20\sin(10t)$.

Solution   The differential equation is given in Section 14.2 as

$$L\frac{\mathrm{d}^2 i}{\mathrm{d}t^2} + R\frac{\mathrm{d}i}{\mathrm{d}t} + \frac{i}{C} = \frac{\mathrm{d}v}{\mathrm{d}t}$$

As $v = 20\sin(10t)$

$$\frac{\mathrm{d}v}{\mathrm{d}t} = 200\cos(10t)$$

and substituting in the given values of L, R, and C we get

$$(4 \times 10^{-6})\mathrm{d}^2 i/\mathrm{d}t^2 + 5\,\mathrm{d}i/\mathrm{d}t + \frac{i}{10^{-6}} = 200\cos(10t)$$

$$4 \times 10^{-6}\,\mathrm{d}^2 i/\mathrm{d}t^2 + 5\,\mathrm{d}i/\mathrm{d}t + 10^6 i = 200\cos(10t).$$

Step 1
Solve the homogeneous equation

$$(4 \times 10^{-6})\mathrm{d}^2 i/\mathrm{d}t^2 + 5\,\mathrm{d}i/\mathrm{d}t + 10^6 i = 0$$

This has auxiliary equation

$$4 \times 10^{-6}\lambda^2 + 5\lambda + 10^6 = 0$$

with solutions

$$\Leftrightarrow \quad \lambda = \frac{-5 \pm \sqrt{25 - 4(4 \times 10^{-6} \times 10^6)}}{2 \times 10^{-6}} \quad \Leftrightarrow \quad \lambda = \frac{-5 \pm \sqrt{9}}{2 \times 10^{-6}}$$

$$\Leftrightarrow \quad \lambda = -4 \times 10^6 \vee \lambda = -10^6.$$

The complementary function is

$$i = A\,\mathrm{e}^{-10^6 t} + B\,\mathrm{e}^{-4 \times 10^6 t}.$$

Step 2
Find a particular solution. As the forcing function is a cosine function we try

$$i = c\cos(10t) + d\sin(10t)$$

$$\mathrm{d}i/\mathrm{d}t = -10c\sin(10t) + 10d\cos(10t)$$

$$= -100c\cos(10t) - 100d\sin(10t).$$

Substituting into

$$(4 \times 10^{-6})\mathrm{d}^2 i/\mathrm{d}t^2 + 5\,\mathrm{d}i/\mathrm{d}t + 10^6 i = 200\cos(10t)$$

we get

$$4 \times 10^{-6}(-100c\cos(10t) - 100d\sin(10t))$$

$$+ 5(-10c\sin(10t) + 10d\cos(10t))$$

$$+ 10^6(c\cos(10t) + d\sin(10t)) = 200\cos(10t)$$

$$\Leftrightarrow \quad -0.0004c\cos(10t) - 0.0004d\sin(10t) - 50c\sin(10t) + 50d\cos(10t)$$

$$+ 10^6 c\cos(10t) + 10^6 d\sin(10t).$$

Equating coefficients of sine and cosine terms to make this an identity gives

$$-0.0004c + 50d + 10^6 c = 200$$

$$-0.0004d - 50c + 10^6 d = 0$$

and solving these equations we find $d \approx 10^{-8}$ and $c \approx 0.0002$, so

$$i = 0.0002\cos(10t) + 10^{-8}\sin(10t).$$

Step 3
Hence, the general solution is

$$i = A\,\mathrm{e}^{-10^6 t} + B\,\mathrm{e}^{-4 \times 10^6 t} + 0.0002\cos(10t) + 10^{-8}\sin(10t).$$

Step 4
Given than $i = 0$ and $\mathrm{d}i/\mathrm{d}t = 0$ when $t = 0$ initially, then we can substitute into

$$i = A\,\mathrm{e}^{-10^6 t} + B\,\mathrm{e}^{-4 \times 10^6 t} + 0.0002\cos(10t) + 10^{-8}\sin(10t)$$

and

$$\frac{\mathrm{d}i}{\mathrm{d}t} = -10^6 A\,\mathrm{e}^{-10^6 t} - 4B10^6\,\mathrm{e}^{-4 \times 10^6} - 0.002\sin(10t) + 10^{-7}\cos(10t)$$

to find equations for A and B

$$A + 4B = -10^{-13}$$

$$A + B = -0.0002$$

which we solve to find $A \approx -0.000267$ and $B = 0.0000667$.
So, the solution to the initial value problem is approximately

$$i = -0.000267\,\mathrm{e}^{-10^6 t} + 0.0000667\,\mathrm{e}^{-4 \times 10^6 t}$$

$$+ 0.0002\cos(10t) + 10^{-8}\sin(10t).$$

The transient and steady state solution – system stability

From the systems modelled in Section 14.2 we can see that the term by' in the equation:

$$ay'' + by' + cy = f(t)$$

comes from the damping factor in the system. For electrical systems this is provided by the resistance. b must be a positive quantity in any of those systems as are the other coefficients.

The auxiliary equation is $a\lambda^2 + b\lambda + c = 0$ with solutions

$$\lambda_1, \lambda_2 = \frac{-b \pm \sqrt{b^2 - 4ac}}{2a}$$

and this gives three possibilities for the complementary function.

Case (1): Real distinct roots $y = A e^{\lambda_1 t} + B e^{\lambda_2 t}$. Notice that $\lambda_1 \lambda_2$ must be negative when a, b, and c are positive and therefore the complementary function will die out as $t \to \infty$. In this case the system is said to be overdamped.

Case (2): Complex roots $y = e^{kt}(A \cos(\omega_0 t) + B \sin(\omega_0 t))$ where $k = -b/2a$ which is negative, so e^{kt} is a negative exponential term. This then represents dying oscillations. The system is said to be underdamped.

Case (3): The roots are equal, then $= k = -b/2a$ and $y = (At + B)e^{kt}$ Again this has a negative exponential part causing the complementary function to tend to zero as $t \to \infty$. This case is referred to as critical damping.

Because the contribution from the complementary function dies out as $t \to \infty$, it is referred to as the transient solution. Graphs of the form of the transients, for positive initial displacement and zero initial velocity, are shown in Figures 14.7(a)–(c) for Cases (1)–(3) respectively.

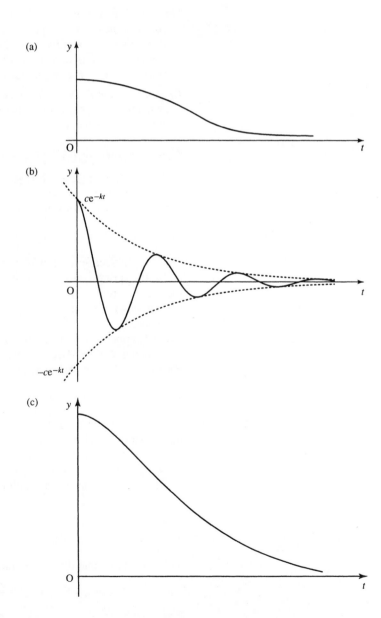

Figure 14.7 *The complementary function, found by solving the homogenous equation, provides the transient solution. The graphs shown are for positive initial displacement and zero velocity. (a) The graph of the transient for the overdamped case where $y = A e^{\lambda_1 t} + B e^{\lambda_2 t}$, λ_1, λ_2 negative. (b) Graph of the underdamped case where the transient consists of dying oscillations $y = e^{kt}(A \cos(\omega_0 t) + B \sin(\omega_0 t))$ (k is negative). (c) The critically damped case where $y = (At + B)e^{kt}$, k negative.*

The other part of the solution, where we consider the effect of the forcing function, is called the steady state solution. For the solution found in Example 14.8 where

$$y = -1.25\,e^{-0.6t}\,\sin(0.8t) + \sin(t)$$

we find that $-1.25\,e^{-0.6t}\,\sin(0.8t)$ is the transient and $\sin(t)$ is the steady state solution. If we consider the system after some time has passes then the transient will effectively be zero and we are left with $y = \sin(t)$, the steady state solution.

We have said that in any truly linear system, as represented, for instance, by our models in Section 14.2, the constants must be positive with positive damping. However, we may wish to analyse non-linear systems by using a locally linear approximation. In this situation they may exhibit unstable behaviour, where the 'transients', instead of dying out, display positive exponential behaviour and grow very large. This is then referred to as an unstable system.

We can analyse systems in the following way:

Stable system:A system is stable if all the solutions to the auxiliary equation have negative real parts. A system with some purely imaginary solutions to the auxiliary equation can also considered to be stable although the complementary function does not die out as $t \to \infty$ but represents sustained oscillations.

Unstable system:A system is unstable if any solutions to the auxiliary equation have positive real parts or, for systems of higher order, if there is a repeated, purely imaginary, solution.

Damped oscillations of a mechanical system – resonance

For damped oscillations of a spring we have the differential equation

$$\Leftrightarrow \quad m\,d^2x/dt^2 + r\,dx/dt + kx = f(t)$$

as shown in Section 14.2, where m is the mass, r the damping constant, and k the spring constant. We shall look at the steady state solution in the case where there is a single frequency input. We are interested in the magnitude of the response in this case in order to analyse the problem of resonance.

For simplicity we can regard the single frequency input as a complex exponential $f(t) = F_0\,e^{j\omega t}$ where F_0 is a constant. Having found the response to the complex exponential, then as

$$F_0\,e^{j\omega t} = F_0\cos(\omega t) + jF_0\sin(\omega t)$$

we can find the response to a cosine function by taking the real part of the output or to a sine function by taking the imaginary part of the output (this is an application of the linearity property).

We have the equation $m\ddot{x} + r\dot{x} + kx = F_0\,e^{j\omega t}$ and we want to find the steady state solution. Try a solution of the form $x = c\,e^{j\omega t}$ then

$$\dot{x} = cj\omega\,e^{j\omega t}$$

$$\ddot{x} = -c\omega^2\,e^{j\omega t}$$

and substituting into the differential equation we find

$$-mc\omega^2\,e^{j\omega t} + crj\omega\,e^{j\omega t} + kc\,e^{j\omega t} = F_0\,e^{j\omega t}.$$

On dividing both sides by $e^{j\omega t}$ (which is non-zero), we get

$$c(-m\omega^2 + jr\omega + k) = F_0.$$

Hence

$$c = F_0/(-m\omega^2 + jr\omega + k)$$

giving the steady state solution as

$$x = \frac{F_0\, e^{j\omega t}}{-m\omega^2 + jr\omega + k}$$

We are interested in the magnitude of the oscillations, which is given by

$$\left| \frac{F_0}{-m\omega^2 + jr\omega + k} \right|.$$

We can find this by taking the magnitude of the top and bottom lines. From Chapter 14 we know that the magnitude of a complex number, $z = x + jy$, is given by $r = \sqrt{x^2 + y^2}$. Hence, we find the amplitude of the oscillations in response to a single frequency of input of angular frequency is given by

$$\frac{F_0}{\sqrt{(m\omega^2 - k)^2 + r^2\omega^2}}.$$

This was plotted in Figure 11.10 for the case $m = k = 1$ and various values of r (there called c). If the damping, r, is quite small then we can set it to be effectively 0. In this case

$$|x| = \begin{cases} \dfrac{F_0}{k - \omega^2 m} & \text{for } \omega^2 < k/m \\[2mm] \dfrac{F_0}{\omega^2 m - k} & \text{for } \omega^2 > k/m \end{cases}$$

In which case there is an infinite discontinuity at

$$\omega = \sqrt{\frac{k}{m}}.$$

We see that $\sqrt{k/m}$ represents the natural angular frequencies of m the system in the case where r is 0. We can show this by examining the solution to the homogeneous in the case where $r = 0$.

$mx'' + kx = 0$ has auxiliary equation

$$m\lambda^2 + k = 0$$

$$\Leftrightarrow \quad \lambda^2 = -k/m$$

and setting $\lambda = j\omega_0$ we find that $\omega_0^2 = k/m$ and the complementary function $x = A\cos(\omega_0 t) + B\sin(\omega_0 t)$.

We can see from this limiting case that if the damping constant is small and the oscillations of the forcing function are near to the natural oscillations of the underdamped system then we have a situation of resonance where the magnitude of the response can become very large. This effect has led to the destruction of some physical systems before the phenomenon of resonance was well understood.

14.6 Solving systems of differential equations

To solve a system of differential equations we can combine the equations into a single differential equation using substitution. In this case we can use the method as outlined in Section 14.5. Alternatively, we can solve the system directly using matrices.

We follow the same pattern as in Section 14.4 for first-order systems. We solve the homogeneous equation, to find the complementary function and then find a particular solution. The sum of these two terms gives a general solution to the system.

Example 14.10   Find the displacement of the spring after time t described by the system of differential equations

$$\frac{d}{dt}\begin{pmatrix} x_1 \\ x_2 \end{pmatrix} = \begin{pmatrix} 0 & 1 \\ -k/m & -r/m \end{pmatrix}\begin{pmatrix} x_1 \\ x_2 \end{pmatrix}\begin{pmatrix} 0 \\ 1/m \end{pmatrix} f(t)$$

in the case where the mass on the spring is 1, the damping constant $r = 5$, the spring constant $k = 4$, the forcing function $f(t) = t$, and the initial extension of the spring is 0 with 0 initial velocity.

Solution   Substituting the given values for m, r, k, and f, we have the system

$$\frac{d}{dt}\begin{pmatrix} x_1 \\ x_2 \end{pmatrix} = \begin{pmatrix} 0 & 1 \\ -4 & -5 \end{pmatrix}\begin{pmatrix} x_1 \\ x_2 \end{pmatrix} + \begin{pmatrix} 0 \\ 1 \end{pmatrix} f(t).$$

Step 1

Solve the homogeneous system with $f(t) = 0$, giving

$$\frac{d\mathbf{x}}{dt} = \mathbf{A}\mathbf{x} \quad \text{where } \mathbf{A} = \begin{pmatrix} 0 & 1 \\ -4 & -5 \end{pmatrix}.$$

Try solutions of the form $\mathbf{x} = \mathbf{v}\,e^{\lambda t}$, where $\mathbf{v}$ is a constant vector. Then $d\mathbf{x}/dt = \lambda\mathbf{v}\,e^{\lambda t}$. Substitute for $\mathbf{x}$ and $d\mathbf{x}/dt$ into $d\mathbf{x}/dt = A\mathbf{x}$, giving

$$\lambda\mathbf{v}\,e^{\lambda t} = A\mathbf{v}\,e^{\lambda t}$$

as $e^{\lambda t} \neq 0$, we have

$$\lambda\mathbf{v} = A\mathbf{v}$$

$$A\mathbf{v} = \lambda\mathbf{v}$$

which we recognize as the eigenvalue problem of Chapter 13. In this case, however, we shall allow the possibility of complex eigenvalues.

The solutions for λ are found by solving $|\mathbf{A} - \lambda\mathbf{I}| = 0$ as

$$\mathbf{A} = \begin{pmatrix} 0 & 1 \\ -4 & -5 \end{pmatrix}$$

this gives

$$\begin{vmatrix} -\lambda & 1 \\ -4 & -5-\lambda \end{vmatrix} = 0$$

$$\Leftrightarrow \quad \lambda(5+\lambda) + 4 = 0$$

$$\Leftrightarrow \quad \lambda^2 + 5\lambda + 4 = 0$$

$$\Leftrightarrow \quad \lambda = \frac{-5 \pm \sqrt{25-16}}{2}$$

$$\Leftrightarrow \quad \lambda = \frac{-5 \pm 3}{2} \quad \Leftrightarrow \quad \lambda = -4 \vee \lambda = -1.$$

Now, we find the eigenvector $\mathbf{v}$ to go with each eigenvalue by solving $(\mathbf{A} - \lambda\mathbf{I})\mathbf{v} = 0$.

$$\begin{pmatrix} -\lambda & 1 \\ -4 & -5 - \lambda \end{pmatrix} \begin{pmatrix} v_1 \\ v_2 \end{pmatrix} = \begin{pmatrix} 0 \\ 0 \end{pmatrix}.$$

For $\lambda = -4$

$$\begin{pmatrix} 4 & 1 \\ -4 & -1 \end{pmatrix} \begin{pmatrix} v_1 \\ v_2 \end{pmatrix} = \begin{pmatrix} 0 \\ 0 \end{pmatrix}.$$

Multiplying out, this gives

$$4\,v_1 + v_2 = 0$$
$$-4\,v_1 - v_2 = 0.$$

These equations are dependent, solving either one of them we find

$$4\,v_1 = -v_2$$
$$v_2 = -4v_1.$$

The vector is given by $(v_1, -4v_1)$. As we are interested only in the direction of the vector we can set $v_1 = 1$ giving the eigenvector as $(1, -4)$.
For $\lambda = -1$ we get

$$\begin{pmatrix} 4 & 1 \\ -4 & -1 \end{pmatrix} \begin{pmatrix} v_1 \\ v_2 \end{pmatrix} = \begin{pmatrix} 0 \\ 0 \end{pmatrix}.$$

Multiplying out we get

$$v_1 + v_2 = 0$$
$$-4v_1 - 4v_2 = 0.$$

These equations are dependent, solving either one we get

$$v_2 = -v_1.$$

The vector is given by $(v_1, -v_1)$. As we are interested only in the direction of the vector we can set $v_1 = 1$ giving the eigenvector as $(1, -1)$.
Therefore, we have the complementary function for $\mathbf{x}$ as

$$\mathbf{x} = a \begin{pmatrix} 1 \\ -4 \end{pmatrix} \mathrm{e}^{-4t} + b \begin{pmatrix} 1 \\ -1 \end{pmatrix} \mathrm{e}^{-t}$$

where a and b are arbitrary constant scalars.

Step 2
Find a particular solution to the equation $\mathrm{d}\mathbf{x}/\mathrm{d}t = \mathbf{Ax} + \mathbf{b}t$. For this equation, the forcing function is $\mathbf{b}t$ so we try a particular solution of the form $\mathbf{x} = \mathbf{c}_1 t + \mathbf{c}_0$ where $\mathbf{c}_0$ and $\mathbf{c}_1$ are constant vectors. We see that this is similar to the choice of trial solution suggested in Table 14.1 but

the constant terms are now constant vectors.

$$\mathbf{x} = \mathbf{c}_1 t + \mathbf{c}_0$$

$$\Rightarrow \quad d\mathbf{x}/dt = \mathbf{c}_1.$$

Substitute this trial solution into

$$d\mathbf{x}/dt = \mathbf{A}\mathbf{x} + \mathbf{b}t$$

giving

$$\mathbf{c}_1 = \mathbf{A}(\mathbf{c}_1 t + \mathbf{c}_0) + \mathbf{b}t.$$

We can equate vector coefficients of t and the constant terms on both sides giving

$$0 = \mathbf{A}\mathbf{c}_1 + \mathbf{b}$$

$$\mathbf{c}_1 = \mathbf{A}\mathbf{c}_0.$$

We solve these two matrix equations

$$0 = \mathbf{A}\mathbf{c}_1 + \mathbf{b}$$

$$\Leftrightarrow \quad \mathbf{A}\mathbf{c}_1 = -\mathbf{b}.$$

Pre-multiply by $\mathbf{A}^{-1}$ (if it exists) giving

$$\mathbf{c}_1 = -\mathbf{A}^{-1}\mathbf{b}$$

As

$$\mathbf{A} = \begin{pmatrix} 0 & 1 \\ -4 & -5 \end{pmatrix}$$

we have

$$\mathbf{A}^{-1} = \frac{1}{4}\begin{pmatrix} -5 & -1 \\ 4 & 0 \end{pmatrix} = \begin{pmatrix} -1.25 & -0.25 \\ 1 & 0 \end{pmatrix}$$

and

$$\mathbf{b} = \begin{pmatrix} 0 \\ 1 \end{pmatrix}$$

so

$$\mathbf{c}_1 = -\begin{pmatrix} -1.25 & -0.25 \\ 1 & 0 \end{pmatrix}\begin{pmatrix} 0 \\ 1 \end{pmatrix} = \begin{pmatrix} 0.25 \\ 0 \end{pmatrix}.$$

To find $\mathbf{c}_0$ we can use

$$\mathbf{c}_1 = \mathbf{A}\mathbf{c}_0$$

we get

$$\mathbf{c}_0 = \mathbf{A}^{-1}\mathbf{c}_1$$

$$= \begin{pmatrix} -1.25 & 0.25 \\ 1 & 0 \end{pmatrix}\begin{pmatrix} 0.25 \\ 0 \end{pmatrix} = \begin{pmatrix} -0.3125 \\ 0.25 \end{pmatrix}.$$

Therefore, a particular solution $\mathbf{x} = \mathbf{c}_1 t + \mathbf{c}_0$ is given by

$$\mathbf{x} = \begin{pmatrix} 0.25 \\ 0 \end{pmatrix}t + \begin{pmatrix} -0.3125 \\ 0.25 \end{pmatrix}$$

Step 3

The general solution is the sum of the complementary function and a particular solution. This gives

$$\mathbf{x} = a \begin{pmatrix} 1 \\ -4 \end{pmatrix} e^{-4t} + b \begin{pmatrix} 1 \\ -1 \end{pmatrix} e^{-t} + \begin{pmatrix} 0.25 \\ 0 \end{pmatrix} t + \begin{pmatrix} -0.3125 \\ 0.25 \end{pmatrix}.$$

Step 4

We can apply the initial conditions. We are told that at $t = 0$ both the displacement and velocity is zero. These are the state variables x_1 and x_2. Therefore, we have

$$\mathbf{x} = \begin{pmatrix} 0 \\ 0 \end{pmatrix} \quad \text{when } t = 0$$

and substituting into the solution we find

$$\begin{pmatrix} 0 \\ 0 \end{pmatrix} = a \begin{pmatrix} 1 \\ -4 \end{pmatrix} + b \begin{pmatrix} 1 \\ -1 \end{pmatrix} + \begin{pmatrix} -0.3125 \\ 0.25 \end{pmatrix}$$

$$0 = a + b - 0.3125$$
$$0 = -4a - b + 0.25.$$

Solving these two equations gives

$$a \approx -0.02083$$
$$b \approx 0.3333$$

and therefore, the particular solution to this initial value problem is

$$\mathbf{x} \cong 0.02083 \begin{pmatrix} 1 \\ -4 \end{pmatrix} e^{-4t} + 0.3333 \begin{pmatrix} 1 \\ -1 \end{pmatrix} e^{-t} + \begin{pmatrix} 0.25 \\ 0 \end{pmatrix} t + \begin{pmatrix} -0.3125 \\ 0.25 \end{pmatrix}.$$

Check:

$$\frac{d\mathbf{x}}{dt} = -0.08332 \begin{pmatrix} 1 \\ -4 \end{pmatrix} e^{-4t} - 0.3333 \begin{pmatrix} 1 \\ -1 \end{pmatrix} e^{-t} + \begin{pmatrix} 0.25 \\ 0 \end{pmatrix}.$$

Substitute into

$$\frac{d\mathbf{x}}{dt} = \begin{pmatrix} 0 & 1 \\ -4 & -5 \end{pmatrix} \mathbf{x} + \begin{pmatrix} 0 \\ 1 \end{pmatrix} t$$

giving

$$-0.08332 \begin{pmatrix} 1 \\ -4 \end{pmatrix} e^{-4t} - 0.3333 \begin{pmatrix} 1 \\ -1 \end{pmatrix} e^{-t} + \begin{pmatrix} 0.25 \\ 0 \end{pmatrix}$$

$$= \begin{pmatrix} 0 & 1 \\ -4 & -5 \end{pmatrix} \begin{pmatrix} 0.02083\,e^{-4t} + 0.3333\,e^{-t} + 0.25t - 0.3125 \\ -0.08332\,e^{-4t} - 0.3333\,e^{-t} + 0.25 \end{pmatrix} + \begin{pmatrix} 0 \\ 1 \end{pmatrix} t.$$

Multiplying out and simplifying, we get

$$\begin{pmatrix} -0.08332\,e^{-4t} - 0.3333\,e^{-t} + 0.25 \\ 0.3333\,e^{-4t} + 0.3333\,e^{-t} \end{pmatrix}$$

$$= \begin{pmatrix} -0.08332\,e^{-4t} + 0.3333\,e^{-t} + 0.25 \\ 0.3333\,e^{-4t} + 0.3333\,e^{-t} \end{pmatrix}$$

which is true for all t.

14.7 Difference equations

Discrete systems are designed using digital adders, multipliers, and shift registers and are represented mathematically by difference equations. Difference equations are used in the design of digital filters and also to approximately model continuous systems.

The system has an input which is a sequence of values $f_0, f_1, f_2, \ldots$ and the output is a sequence $y_0, y_1, y_2, \ldots$ If y_n depends solely on the values of the input, then

$$y_n = b_0 f_n + b_1 f_{n-1} + b_2 f_{n-2} + \cdots.$$

This is a non-recursive system as we do not need previous knowledge of y to determine y_n. It is called a finite impulse response (FIR) system (the meaning of 'impulse response' will be explained in Chapter 15). If, however, the output y depends on the previous state of the system as well as the input then we have a recursive system, modelled by a difference equation (also called a recurrence relation). This is called a infinite impulse response (IIR) system.

We shall look at linear time-invariant discrete systems. We shall consider the solution of a second order difference equation

$$ay_n + by_{n-1} + cy_{n-2} = f_n.$$

The method of solution, as for differential equations, is to solve the homogeneous system and find a particular solution. The sum of these gives the general solution to the difference equation. We employ the initial conditions to find the value of the arbitrary constants.

Example 14.11 Solve the difference equation ($n \geqslant 2$):

$$6y_n - 5y_{n-1} + y_{n-2} = 1, \quad y_0 = 4.5 \text{ and } y_1 = 2.$$

Solution

Step 1
  Solve the homogeneous difference equation

$$6y_n - 5y_{n-1} + y_{n-2} = 0.$$

We know from Chapters 8 and 12 that the $y_{n+1} = ry_n$ defines a geometric series $y_n = ar^{n-1}$, where a is the first term of the sequence and r is the common ratio. If we start at the zeroth term, then we have $y_n = a_0 r^n$. We therefore try a solution of this form to the homogeneous equation and substitute $y_n = a\lambda^n$ into $6y_n - 5y_{n-1} + y_{n-2} = 0$. This gives

$$6a\lambda^n - 5a\lambda^{n-1} + a\lambda^{n-2} = 0.$$

Assuming that a and λ^{n-2} are non-zero, we can divide by $a\lambda^{n-2}$ to give

$$6\lambda^2 - 5\lambda + 1 = 0.$$

This is called the auxiliary equation and solving this we find

$$\lambda = \frac{5 \pm \sqrt{25 - 24}}{12} \quad \Leftrightarrow \quad \lambda = \frac{1}{3} \vee \lambda = \frac{1}{2}.$$

Hence, the general solution of the homogeneous equation is

$$y_n = a\left(\frac{1}{3}\right)^n + b\left(\frac{1}{2}\right)^n.$$

As in the case of differential equations, there are three possible types of solution, depending on whether the roots for λ are real and distinct, complex, or equal, where

$$\lambda_1, \lambda_2 = \frac{-b \pm \sqrt{b^2 - 4ac}}{2a}$$

Case (1): λ_1 and λ_2 are real and distinct

$$y_n = a(\lambda_1)^n + b(\lambda_2)^n.$$

Case (2): λ_1 and λ_2 are complex. Then we write in exponential form $\lambda_1 = r\, e^{j\theta}$ so $\lambda_2 = r\, e^{-j\theta}$ and the solution can be expressed as

$$y = r^n(a\cos(n\theta) + b\sin(n\theta)).$$

Case (3): λ_1 and λ_2 are equal real roots and the solution is $y_n = (a + bn)\lambda^n$.

Step 2
Find a particular solution. We use a trial solution which depends on the form of the input sequence f_n as suggested by Table 14.2.

In this case the input is a constant so we try a constant output. The trial solution is $y_n = c$. Substituting this into

$$6y_n - 5y_{n-1} + y_{n-2} = 1$$

gives the equation for c as

$$6c - 5c + c = 1$$
$$\Leftrightarrow \quad 2c = 1$$
$$c = 1/2$$

A particular solution for y is $y_n = 1/2$.

Step 3
The general solution is the sum of the solution of the homogeneous equation and a particular solution giving

$$y_n = a\left(\tfrac{1}{3}\right)^n + b\left(\tfrac{1}{2}\right)^n + \tfrac{1}{2}.$$

Table 14.2 *A table of trial solutions to be used to find a particular solution of a difference equation. The form of the trial solution is suggested by the form of the input sequence f_n. The coefficients c_0, $c_1, \ldots$ and c and d are to be determined*

Input function	Trial solution
Power series $f_n = n^k$ (k an integer)	$y = c_0 + c_1 n + c_2 n^2 + \cdots + c_k n^k$
An exponential function $f_n = a^n$	$y = ca^n$
Sine or cosine function $f_n = \cos(\omega n)$ or $\sin(\omega n)$	$y = c\cos(\omega n) + d\sin(\omega n)$

Step 4

Solve for the initial conditions $y_0 = 4.5$ and $y_1 = 2$. Substituting $n = 0$ and $y_0 = 4.5$ in the solution

$$y_n = a \left(\tfrac{1}{3}\right)^n + b \left(\tfrac{1}{2}\right)^n + \tfrac{1}{2}$$

gives

$$4.5 = a + b + \tfrac{1}{2} \quad \Leftrightarrow \quad a + b = 4.$$

Substituting $n = 1$ and $y_1 = 4.5$ gives

$$2 = \tfrac{1}{3}a + \tfrac{1}{2}b + \tfrac{1}{2} \quad \Leftrightarrow \quad 2a + 3b = 9.$$

Solving for the constants a and b gives

$$
\begin{array}{ccc}
a + b = 4 & & 2a + 2b = 8 \\
\quad\quad\quad \Leftrightarrow & & \quad\quad\quad \Rightarrow \quad -b = -1. \\
2a + 3b = 9 & & 2a + 3b = 9
\end{array}
$$

Substituting $b = 1$ into $a + b = 4$:

$$a + 1 = 4 \quad \Leftrightarrow \quad a = 3$$

$a = 3$ and $b = 1$, so the solution is

$$y_n = 3 \left(\tfrac{1}{3}\right)^n + \left(\tfrac{1}{2}\right)^n + \tfrac{1}{2}.$$

System stability

We can see that the solution, as in the continuous case, is made up of a transient, in this case, $3(1/3)^n + (1/2)^n$, and a steady state response, which in this case is $1/2$. The transient tends to zero as $n \to \infty$. This will be so for any system as long as the roots to the auxiliary equation are such that the modulus of λ is less than 1. If the roots have modulus of exactly 1, then the solution of the homogeneous equation does not die away as $n \to 0$ but neither does it tend to infinity. Therefore, we can identify a stable system as one with roots of the auxiliary equation such that $|\lambda| \leqslant 1$.

14.8 Summary

1. Dynamic physical processes involve variables which are interdependent and constantly changing. When modelling the system we will obtain relationships between variables, many of which will be related through derivatives. We can combine these to make a single differential equation relating the system input to the system output or we can choose state variables and express the relationship through a system of differential equations.

2. Linear time invariant (LTI) systems are represented by linear differential equations with constant coefficients. A second-order equation is of the form

$$a \, d^2 y / dt^2 + b \, dy/dt + cy = f(t)$$

where $f(t)$ represents the system input and y the system output.

3. If a function, $y = g(t)$, is a solution to a differential equation in y then if, in the differential equation, we replace y and by $g(t)$ and all the derivatives of y by the corresponding derivatives of $g(t)$ then the result should be an identity for t. That is, it should be true for all values of t.

4. To solve linear differential equations with constant coefficients, we solve the homogeneous equation (with $f(t) = 0$), thus finding the complementary function. This leads to the auxiliary equation which, for a second-order system is

$$a\lambda^2 + b\lambda + c = 0.$$

This gives three possibilities for the complementary function, depending on the roots for λ.

$$\lambda_1, \lambda_2 = \frac{-b \pm \sqrt{b^2 - 4ac}}{2a}$$

Case (1): Real distinct roots: $y = A\,e^{\lambda_1 t} + B\,e^{\lambda_2 t}$
Case (2): Complex roots: $y = e^{kt}(A\cos(\omega_0 t) + B\sin(\omega_0 t))$
Case (3): The roots are equal, then $\lambda = k = -b/2a$, $y = (At + B)e^{kt}$.

We then find a particular solution to the equation by using a trial solution as suggested in Table 14.1. The sum of the complementary function and a particular solution gives the general solution of the differential equation. The initial conditions are used to find the arbitrary constants A and B.

5. For a truly linear system, the complementary function dies out as $t \to \infty$ and is called the transient solution, The other part of the solution, in response to the input $f(t)$, is called the steady state solution. When analysing a system as locally linear, we say that the system is unstable if the roots of the complementary function have a positive real part. In these circumstances, the complementary function $\to \infty$ as $t \to \infty$. Higher order systems are also unstable if there is a repeated purely imaginary root.

6. Resonance occurs when, in an underdamped system, the forcing frequency approaches the natural frequency of the system.

7. Systems of differential equations may be solved using a matrix method.

8. Discrete systems may be represented by difference equations (also called recurrence relations). A second-order linear system is represented by $ay_n + by_{n-1} + cy_{n-2} = f_n$. To solve this equation we find the solution to the homogeneous system (setting $f_n = 0$), which leads to the auxiliary equation

$$a\lambda^2 + b\lambda + c = 0.$$

This leads to three possibilities for the complementary function depending on the roots for λ:

$$\lambda_1, \lambda_2 = \frac{-b \pm \sqrt{b^2 - 4ac}}{2a}$$

Case (1): λ_1 and λ_2 are real and distinct.

$$y_n = a(\lambda_1)^n + b(\lambda_2)^n.$$

Case (2): λ_1 and λ_2 are complex. Then we write in exponential form $\lambda_1 = r\,\mathrm{e}^{\mathrm{j}\theta}$ so $\lambda_2 = r\,\mathrm{e}^{-\mathrm{j}\theta}$ and the solution can be expressed as

$$y = r^n(a\cos(n\theta) + b\sin(n\theta)).$$

Case (3): λ_1 and λ_2 are equal real roots and the solution is $y_n = (a + bn)\lambda^n$. The system is stable if $|\lambda| \leqslant 1$.

We then find a particular solution by attempting a trial solution, as suggested in Table 14.2. The general solution of the difference equation is given by the sum of the complementary function and a particular solution.

14.9 Exercises

14.1. For the following differential equations:

(a) State whether the equation is linear
(b) Give the order of the equation
(c) Show that the given function represents a solution to the differential equation

    (i) $\dfrac{t}{4}\dfrac{\mathrm{d}y}{\mathrm{d}y} = y, \quad y = ct^4$

    (ii) $\dfrac{\mathrm{d}^2y}{\mathrm{d}t^2}\left(t - \dfrac{1}{2}\right) - \dfrac{\mathrm{d}y}{\mathrm{d}t} = 0, \quad y = t^2 - t$

    (iii) $t\dfrac{\mathrm{d}^2y}{\mathrm{d}t^2} + t\dfrac{\mathrm{d}y}{\mathrm{d}t} + y = 4, \quad y = 3t\,\mathrm{e}^{-t} + 4$

    (iv) $y\dfrac{\mathrm{d}y}{\mathrm{d}t} = t, y^2 = t^2 + c, \quad y \geqslant 0$

    (v) $3y^2\dfrac{\mathrm{d}^2y}{\mathrm{d}t^2} + 6y\left(\dfrac{\mathrm{d}y}{\mathrm{d}t}\right)^2 = 2, \quad y^3 = t^2$

14.2. A linear differential equation

$$\frac{\mathrm{d}y}{\mathrm{d}t} + 4y = f(t)$$

is found to have a particular solution $y = 2t^2 - t + 1$ when $f(t) = 8t^2 + 3$ and a particular solution $y = 0.8\cos(3t) + 0.6\sin(3t)$, when $f(t) = 5\cos(3t)$.

(a) Suggest a particular solution when $f(t) = 10\cos(3t) + 4t^2 + (3/2)$ and show by substitution that your solution is correct.
(b) Suggest a particular solution when $f(t) = 5\cos(3(t - 10))$ and show by substitution that your solution is correct.

14.3. Show that the differential equation

$$\cos(t)\mathrm{d}y/\mathrm{d}t - \sin(t)y = f(t)$$

has a solution $y = t^2/\cos(t)$, when $f(t) = 2t$ and a solution $y = \tan(t)$ when $f(t) = \cos(t)$

(a) Use the linearity property to find a solution when $f(t) = 2(t - \cos(t))$.
(b) By considering a time-shifted input of one of the functions for $f(t)$ given, show that

$$\cos(t)\mathrm{d}y/\mathrm{d}t - \sin(t)y = f(t)$$

cannot represent a time-invariant system.

14.4. Solve the following differential equations with the given initial conditions

(a) $\dfrac{\mathrm{d}y}{\mathrm{d}t} - 3y = 0, \quad y(0) = 1$

(b) $4\dfrac{\mathrm{d}y}{\mathrm{d}t} - y = 4, \quad y(0) = -2$

(c) $\dfrac{\mathrm{d}y}{\mathrm{d}t} + 5y = \sin(12t), \quad y(0) = 0$

(d) $3\dfrac{\mathrm{d}y}{\mathrm{d}t} + 2y = \mathrm{e}^{-t}, \quad y(0) = 3$

(e) $\dfrac{\mathrm{d}y}{\mathrm{d}t} - 6t = 0, \quad y(1) = 6$

(f) $\dfrac{1}{2}\dfrac{\mathrm{d}y}{\mathrm{d}t} + 6y - 3\sin(5t) = 2\cos(5t), \quad y(0) = 0.$

14.5. A spring of length l has a mass m attached to it and a dashpot damping the motion. It is subject to a force $f(t)$ forcing the motion, as in Figure 14.2. The extension x of the spring obeys the differential equation

$$m\,\mathrm{d}^2x/\mathrm{d}t^2 + r\,\mathrm{d}x/\mathrm{d}t + kx = f(t).$$

Solve for x given the following information
(a) $m = 2, \quad r = 7, \quad k = 5, \quad f(t) = t^2, \quad x(0) = 0, \dfrac{\mathrm{d}x(0)}{\mathrm{d}t} = 0.16$
(b) $m = 1, \quad r = 4, \quad k = 4 f(t) = 4\cos(6t), \quad x(0) = 0, \dfrac{\mathrm{d}x(0)}{\mathrm{d}t} = 0$
(c) $m = 1, \quad r = 4, \quad k = 5, \quad f(t) = \mathrm{e}^{-t}, \quad x(0) = 0, \dfrac{\mathrm{d}x(0)}{\mathrm{d}t} = 0.$

14.6. An LRC circuit, as in Figure 14.3 obeys the equation

$$L\mathrm{d}^2q/\mathrm{d}t^2 + R\,\mathrm{d}q/\mathrm{d}t + q/C = v(t)$$

where q is the charge on the capacitor, $v(t)$ the applied voltage, L the inductance, R the resistance, and C

the capacitance. Find a steady state solution for q and hence calculate the voltages across the capacitor, resistor, and inductor, given by $v_c = q/C$, $v_R = R\,dq/dt$ and $v_L = L\,d^2q/dt^2$ in each of the following cases:

(a) $R = 120\,\Omega$, $L = 0.06\,\text{H}$, $C = 0.0001\,\text{F}$, $v(t) = 130\cos(1000t)$

(b) $R = 1\,\text{k}\Omega$, $L = 0.001\,\text{H}$, $C = 1\,\mu\text{F}$, $v(t) = t$.

14.7. An RC circuit is subjected to a single frequency input of angular frequency ω and magnitude v_i.

(a) Find the steady state solution of the equation
$R\,dq/dt + (q/C) = v_i e^{j\omega t}$
and hence find the magnitude of
(i) The voltage across the capacitor $v_c = q/C$
(ii) The voltage across the resistor $v_R = R\,dq/dt$
(b) Using the impedance method of Chapter 10 confirm your results to part (a) by calculating
(i) The voltage across the capacitor v_c
(ii) The voltage across the resistor v_R
in response to a single frequency of angular frequency ω and magnitude v_i
(c) For the case where $R = 1\,k\Omega$, $C = 1\,\mu\text{F}$ find the ratio $|v_c|/|v_i|$ and fill in the table below:

ω	10	10^2	10^3	10^4	10^5	10^6
$|v_c|/|v_i|$						

Explain why the table results show that an RC circuit acts as a high-cut filter and find the value of the high-cut frequency, defined as $f_{hc} = \omega_{hc}/2\pi$, such that $|v_c|/|v_i| = 1/\sqrt{2}$.

14.8. A spring of length, l has a mass m attached to it and a dashpot damping the motion. It is subject to a force $f(t)$ forcing the motion as in Figure 14.2. The extension x_1 of the spring obeys the system of differential equations

$$\frac{d}{dt}\begin{pmatrix} x_1 \\ x_2 \end{pmatrix} = \begin{pmatrix} 0 & 1 \\ -k/m & -r/m \end{pmatrix}\begin{pmatrix} x_1 \\ x_2 \end{pmatrix} + \begin{pmatrix} 0 \\ 1/m \end{pmatrix} f(t)$$

By solving this system of equations, find the extension of the spring after time t in the following cases
(a) $m = 1$, $r = 12$, $k = 11$, $f(t) = e^{-2t}$, with initial extension 0 and 0 initial velocity.
(b) $m = 2$, $r = 13$, $k = 20$, $f(t) = 10t$, $x_1(0) = 0 = x_2(0)$.

14.9. Solve the following difference equations

(a) $2y_n - y_{n-1} = -3$, $y_0 = 1$

(b) $4y_n + y_{n-1} = n$, $y_0 = 2$

(c) $y_n - 1.2y_{n-1} + 0.36y_{n-2} = 4$, $y_0 = 3$, $y_1 = 2$

(d) $y_n + 0.6y_{n-1} = 0.16y_{n-2} + (0.5)^n$, $y_0 = 1$, $y_1 = 2$

(e) $2y_n + y_{n-1} = \cos\left(\frac{\pi n}{2}\right)$, $y_0 = 1$.

15 Laplace and z transforms

15.1 Introduction

In this chapter, we will present a quick review of Laplace transform methods for continuous and piecewise continuous systems and z transform methods for discrete systems. The widespread application of these methods means that engineers need increasingly to be able to apply these techniques in multitudes of situations.

Laplace transforms are used to reduce a differential equation to a simple equation in s-space and a system of differential equations to a system of linear equations. We discover that in the case of zero initial conditions, we can solve the system by multiplying the Laplace transform of the input function by the transfer function of the system. Furthermore, if we are only interested in the steady state response, and we have a periodic input, we can find the response by simply adding the response to each of the frequency components of the input signal. In Chapter 16, we therefore look at finding the frequency components of a periodic function by finding its Fourier series.

Similar ideas apply to z transform methods applied to difference equations representing discrete systems.

The system transfer function is the Laplace transform of the impulse response function of the system. The impulse response is its response to an impulse function also called a delta function. Inputting an impulse function can be though of as something like giving the system a short kick to see what happens next. The impulse function is an idealized kick, as it lasts for no time at all and has energy of exactly 1. Because of these requirements, we find that the delta function is not a function at all, in the sense that we defined functions in Chapter 1. We find that it is the most famous example of a generalized function.

15.2 The Laplace transform – definition

The Laplace transform $F(s)$ of the function $f(t)$ defined for $t > 0$ is

$$F(s) = \int_0^\infty e^{-st} f(t) \, dt.$$

The Laplace transform is a function of s where s is a complex variable. Because the integral definition of the Laplace transform involves an integral to ∞ it is usually necessary to limit possible values of s so that the integral converges (i.e. does not tend to ∞). Also, for many functions the Laplace transform does not exist at all.

$\mathcal{L}\{\ \}$ is the symbolic representation of the process of taking a Laplace transform (LT). Therefore,

$$\mathcal{L}\{f(t)\} = F(s).$$

Example 15.1   Find from the definition $\mathcal{L}\{e^{3t}\}$.

Solution

$$\mathcal{L}\{e^{3t}\} = \int_0^\infty e^{3t} e^{-st}\, dt = \int_0^\infty e^{(3-s)t}\, dt = \left[\frac{e^{(3-s)t}}{3-s}\right]_0^\infty.$$

Now $e^{(3-s)t} \to 0$ as $t \to \infty$ only if the real part of $(3-s)$ is negative, that is, $\mathrm{Re}(s) > 3$. Then the upper limit of the integration gives 0. Hence

$$\mathcal{L}\{e^{3t}\} = \frac{1}{s-3}$$

where $\mathrm{Re}\{s\} > 3$.

Example 15.2   Find, from the definition, $\mathcal{L}\{\cos(at)\}$.

Solution   We need to find the integral

$$I = \int_0^\infty e^{-st} \cos(at)\, dt.$$

We employ integration by parts twice until we find an expression that involves the original integral, I. We are then able to express I in terms of a and s.

Integrating by parts, using the formula $\int u\, dv = uv - \int v\, du$, where $u = \cos(at)$, $dv = e^{-st}\, dt$, $du = -a\sin(at)\, dt$ and $v = e^{-st}/(-s)$, gives

$$I = \left[\frac{-e^{-st}}{s}\cos(at)\right]_0^\infty - \int_0^\infty \frac{e^{-st}}{s} a\sin(at)\, dt.$$

The expression $[e^{-st}\cos(at)]_0^\infty$ only has a finite value if $e^{-st} \to 0$ as $t \to \infty$. This will only be so if $\mathrm{Re}(s) > 0$. With that proviso we get

$$I = \frac{1}{s} - \frac{a}{s}\int_0^\infty e^{-st}\sin(at)\, dt$$

integrating the remaining integral on the right-hand side, again by parts, where this time $u = \sin(at)$, $dv = e^{-st}dt$, $du = a\cos(at)\, dt$, $v = e^{-st}/(-s)$, gives

$$I = \frac{1}{s} - \left[\frac{-a}{s^2}e^{-st}\sin(at)\right]_0^\infty - \frac{a^2}{s^2}\int_0^\infty e^{-st}\cos(at)\, dt$$

$$= \frac{1}{s} - \frac{a^2}{s^2}\int_0^\infty e^{-st}\cos(at)\, dt.$$

We see that the remaining integral is the integral we first started with, which we called I. Hence

$$I = \frac{1}{s} - \frac{a^2}{s^2}I$$

and solving this for I gives

$$I\left(1 + \frac{a^2}{s^2}\right) = 1/s, \quad I = \frac{s}{s^2 + a^2}.$$

So we have that

$$\mathcal{L}\{\cos(at)\} = \frac{s}{s^2 + a^2}, \quad \text{where } \mathrm{Re}(s) > 0.$$

15.3 The unit step function and the (impulse) delta function

The unit step function and delta function are very useful in systems theory.

The unit step function

The unit step function is defined as follows:

$$u(t) = \begin{cases} 1 & t \geqslant 0 \\ 0 & t < 0 \end{cases}$$

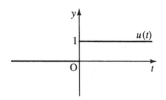

Figure 15.1 *The graph of the unit step function $u(t)$.*

and its graph is as shown in Figure 15.1.

The function is used to represent an idealized switch. It switches on at time $t = 0$. If it is multiplied by any other function then it acts to switch that function on at $t = 0$. That is, the function

$$f(t)u(t) = \begin{cases} 0 & t < 0 \\ f(t) & t \geqslant 0. \end{cases}$$

An example for $f(t) = t^2$ is given in Figure 15.2.

As the Laplace transform only involves the integral from $t = 0$, all functions can be thought of as multiplied by the unit step function because $f(t) = f(t)u(t)$ as long as $t > 0$.

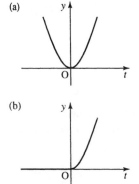

Figure 15.2 *Any function when multiplied by the unit step function is 'switched on' at $t = 0$. (a) the function $y = t^2$; (b) the function $y = t^2 u(t)$.*

Shifting the unit step function

As we saw in Chapter 2, the graph of $f(t - a)$ can be found by taking the graph of $f(t)$ and moving it a units to the right. There is an example of a shifted unit step function in Figure 15.3.

Notice that the unit step function 'switches on' where the argument of u is zero. Multiplying any function by the shifted unit step function changes where it is switched on. That is

$$f(t)u(t - a) = \begin{cases} 0 & t < a \\ f(t) & t \geqslant a. \end{cases}$$

The example of $\sin(t)u(t - 1)$ is shown in Figure 15.4.

The delta function

The impulse function, or delta function, is a mathematical representation of a kick. It is an idealized kick that lasts for no time at all and has energy of exactly 1.

Figure 15.3 *(a) The graph of $y = u(t)$. (b) The graph of $y = u(t - 2)$ is found by shifting the graph of $u(t)$ two units to the right.*

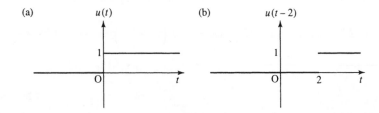

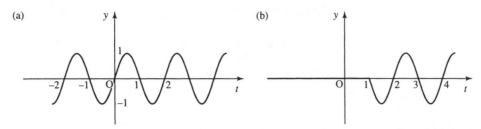

Figure 15.4   *(a) The graph of* $\sin(\omega t)$. *(b) The graph of* $u(t-1)\sin(\omega t)$. *Notice that this function is zero for* $t < 1$ *and equal to* $\sin(\omega t)$ *for* $t > 1$.

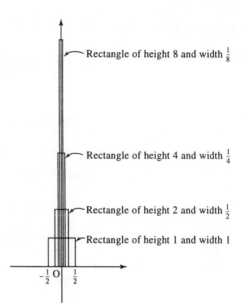

Figure 15.5   *The sequence of rectangular pulse functions of area 1. Starting from a pulse of height 1 and width 1 we double the height and half the width.*

The delta function, $\delta(t)$, is an example of a generalized function. A generalized function can be defined in terms of a sequence of functions. One way of defining it is as the limit of a rectangular pulse function, with area 1, as it halves in width and doubles in height. This sequence of functions is shown in Figure 15.5. Although the height of the pulse is tending to infinity, the area of the pulse remains 1.

Two important properties of the delta function are

1. $\delta(t-a) = 0$ for $t \neq a$,
2. $\int_{-\infty}^{\infty} \delta(t) = 1$.

The second property expresses the fact that the area enclosed by the delta function is 1.

The unit step function, $u(t)$, has no derivative at $t = 0$. Because of the sharp edges present in its graph and its jump discontinuity it is impossible to define a single tangent at that point. However, if we also consider the unit step function as a generalized function (by taking the limit of nice smooth, continuous curves as they approach the shape of the unit step function), we are able to define its derivative, which turns out to be the delta function. This gives the third definition

3. $\mathrm{d}u/\mathrm{d}t = \delta(t)$.

Symbolic representation of $\delta(t)$

$\delta(t)$ can be represented on a graph by an arrow of height 1. The height represents the weight of the delta function. A shifted delta function, $\delta(t-a)$, is represented by an arrow at $t = a$. Examples are given in Figure 15.6.

15.4 Laplace transforms of simple functions and properties of the transform

Rather than finding a Laplace transform from the definition we usually use tables of Laplace transforms. A list of some common Laplace transforms is given in Table 15.1. We can then use the properties of the transform to find Laplace transforms of many other functions. To find the inverse transform, represented by $\mathcal{L}^{-1}\{\}$ we use the same table backwards. That is, look for the function of s on the right-hand column; the left-hand column then gives its inverse:

$$F(s) = \int_0^\infty f(t)\,\mathrm{e}^{-st}\,\mathrm{d}t \qquad f(t) = \mathcal{L}^{-1}\{F(s)\}.$$

Properties of the Laplace transform

In all of the following, $F(s) = \mathcal{L}\{f(t)\}$

1. Linearity:

$$\mathcal{L}\{af_1(t) + bf_2(t)\} = aF_1(s) + F_2(s)$$

where a and b are constants.

2. First translation (or Shift rule):

$$\mathcal{L}\{\mathrm{e}^{at}f(t)\} = F(s - a)$$

3. Second translation:

$$\mathcal{L}\{f(t-a)u(t-a)\} = \mathrm{e}^{-as}F(s).$$

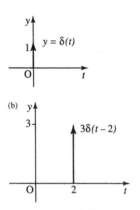

(a)

(b)

Figure 15.6 *Symbolic representations of delta functions: (a) $\delta(t)$; (b) $3\delta(t-a)$.*

Table 15.1 *Common Laplace transforms*

$f(t)$	$F(s) = \mathcal{L}\{f(t)\}$	
$u(t)$	$1/s$	$\mathrm{Re}(s) > 0$
$\delta(t)$	1	$\mathrm{Re}(s) > 0$
$\dfrac{t^{n-1}}{(n-1)!}$	$\dfrac{1}{s^n}$	$\mathrm{Re}(s) > 0$
e^{-at}	$\dfrac{1}{s+a}$	$\mathrm{Re}(s) > -a$
$\dfrac{1}{a}\sin(at)$	$\dfrac{1}{s^2 + a^2}$	$\mathrm{Re}(s) > 0$
$\cos(at)$	$\dfrac{s}{s^2 + a^2}$	$\mathrm{Re}(s) > 0$
$\dfrac{1}{a}\sinh(at)$	$\dfrac{1}{s^2 - a^2}$	$\mathrm{Re}(s) > |a|$
$\cosh(at)$	$\dfrac{s}{s^2 - a^2}$	$\mathrm{Re}(s) > |a|$

4. Change of scale:

$$\mathcal{L}\{f(at)\} = \frac{1}{a}F\frac{s}{a}.$$

5. Laplace transforms of derivatives:

$$\mathcal{L}\{f'(t)\} = sF(s) - f(0)$$

$$\mathcal{L}\{f''(t)\} = s^2 F(s) - sf(0) - f'(0)$$

and

$$\mathcal{L}\{f^{(n)}(t)\} = s^n F(s) - s^{n-1} f(0) - \cdots - sf^{(n-2)}(0) - f^{(n-1)}(0)$$

6. Integrals:

$$\mathcal{L}\left\{ \int_0^t f(\tau)\,d\tau \right\} = \frac{F(s)}{s}.$$

7. Convolution:

$$\mathcal{L}\{f * g\} = \mathcal{L}\left\{ \int_0^t f(\tau)g(t - \tau)\,d\tau \right\} = F(s)G(s).$$

8. Derivatives of the transform:

$$\mathcal{L}\{t^n f(t)\} = (-1)^n F^{(n)}(s)$$

where

$$F^n(s) = \frac{d^n F^{(n)}}{ds^n}(s).$$

Example 15.3 (*Linearity*)   Find $\mathcal{L}\{3t^2 + \sin(2t)\}$.

Solution   From Table 15.1

$$\mathcal{L}\left\{ \frac{t^2}{2} \right\} = \frac{1}{s^3}$$

and

$$\mathcal{L}\left\{ \frac{\sin(2t)}{2} \right\} = \frac{1}{(s^2 + 4)}.$$

Therefore

$$\mathcal{L}\{3t^2 + \sin(2t)\} = 6\mathcal{L}\left\{ \frac{t^2}{2} \right\} + 2\mathcal{L}\left\{ \frac{\sin(2t)}{2} \right\} = \frac{6}{s^3} + \frac{2}{s^2 + 4}$$

using

$$\mathcal{L}\{af_1(t) + bf_2(t)\} = aF_1(s) + bF_2(s).$$

Example 15.4 (*Linearity and the inverse transform*)   Find

$$\mathcal{L}^{-1}\left\{ \frac{2}{s + 4} + \frac{4s}{s^2 + 9} \right\}$$

Solution   From Table 15.1

$$\mathcal{L}^{-1}\left\{\frac{1}{s+4}\right\} = e^{-4t}$$

and

$$\mathcal{L}^{-1}\left\{\frac{s}{s^2+9}\right\} = \cos(3t)$$

Therefore

$$\mathcal{L}^{-1}\left\{\frac{2}{s+4} + \frac{4s}{s^2+9}\right\} = 2e^{-4t} + 4\cos(3t)$$

Example 15.5 (*First translation*)   Find $\mathcal{L}\{t^2 e^{-3t}\}$.

Solution   As

$$\mathcal{L}\{t^2\} = \frac{2}{s^3}$$

using

$$\mathcal{L}\{e^{-at}f(t)\} = F(s-a).$$

Then multiplying by e^{-3t} in the t domain will shift the function $F(s)$ by 3:

$$\mathcal{L}\{t^2 e^{-3t}\} = \frac{2}{(s+3)^2}$$

Example 15.6 (*First translation – inverse transform*)   Find

$$\mathcal{L}^{-1}\left\{\frac{s+2}{(s+2)^2+9}\right\}.$$

Solution   As

$$\mathcal{L}^{-1}\left\{\frac{s}{s^2+9}\right\} = \cos(3t)$$

and as

$$\frac{s+2}{(s+2)^2+9}$$

is $s/(s^2+9)$ translated by 2, using the first translation rule this will multiply in the t domain by e^{-2t}, so

$$\mathcal{L}^{-1}\left\{\frac{s+2}{(s+2)^2+9}\right\} = \cos(3t)e^{-2t}$$

Example 15.7 (*Second translation*)   Find the Laplace transform of

$$f(t) = \begin{cases} 0 & t < \frac{2}{3} \\ \sin(3t-2) & t \geq \frac{2}{3} \end{cases}$$

Solution $f(t)$ can be expressed using the unit step function

$$f(t) = \sin(3t - 2)u\left(t - \frac{2}{3}\right) = \sin\left(3\left(t - \frac{2}{3}\right)\right)u\left(t - \frac{2}{3}\right).$$

Using $\mathcal{L}\{f(t-a)u(t-a)\} = e^{-as}F(s)$ and as

$$\mathcal{L}\{\sin(3t)\} = \frac{3}{s^2 + 9}$$

we have

$$\sin\left(3\left(t - \frac{2}{3}\right)\right)u\left(t - \frac{2}{3}\right) = \frac{3e^{-2s/3}}{s^2 + 9}$$

Example 15.8 (*Second translation – inverse transform*)   Find

$$\mathcal{L}^{-1}\left\{\frac{e^{-s}}{(s+2)^2}\right\}.$$

Solution   As

$$\mathcal{L}^{-1}\left\{\frac{1}{(s+2)^2}\right\} = t\,e^{-2t}$$

(using first translation), the factor of e^{-s} will translate in the t domain, $t \to t - 1$. Then

$$\mathcal{L}^{-1}\left\{\frac{e^{-s}}{(s+2)^2}\right\} = (t-1)e^{-2(t-1)}u(t-1)$$

by second translation.

Example 15.9 (*Change of scale*)   Given

$$\mathcal{L}\{\cos(t)\} = \frac{s}{s^2 + 1}$$

find $\mathcal{L}\{\cos(3t)\}$ using the change of scale property of the Laplace transform

$$\mathcal{L}\{f(at)\} = \frac{1}{a}F\frac{s}{a}.$$

Solution   As

$$\mathcal{L}\{\cos(t)\} = \frac{s}{s^2 + 1} = F(s)$$

to find $\cos(3t)$ we put $a = 3$ into the change of scale property

$$\mathcal{L}\{f(at)\} = \frac{1}{a}F\frac{s}{a}$$

giving

$$\mathcal{L}\{\cos(3t)\} = \frac{1}{3}\frac{s/3}{(s/3)^2 + 1} = \frac{s}{s^2 + 9}$$

Example 15.10 (*Derivatives*)   Given

$$\mathcal{L}\{\cos(2t)\} = \frac{s}{s^2 + 4}$$

find $\mathcal{L}\{\sin(2t)\}$ using the derivative rule.

Solution   If $f(t) = \cos(2t)$ then $f'(t) = -2\sin(2t)$:

$$\mathcal{L}\{\cos(2t)\} = \frac{s}{s^2 + 4}$$

and $f(0) = \cos(0) = 1$. From the derivative rule $\mathcal{L}\{f'(t)\} = sF(s) - f(0)$,

$$\mathcal{L}\{-2\sin(2t)\} = s\left(\frac{s}{s^2 + 4}\right) - 1 = \frac{s^2}{s^2 + 4} - 1 = \frac{s^2 - s^2 - 4}{s^2 + 4}$$

$$= \frac{-4}{s^2 + 4}.$$

By linearity,

$$\mathcal{L}\{\sin(2t)\} = \frac{1}{-2}\left(\frac{-4}{s^2 + 4}\right) = \frac{2}{s^2 + 4}.$$

Example 15.11 (*Integrals*)   Using

$$\mathcal{L}\left\{\int_0^t f(\tau)\,d\tau\right\} = \frac{F(s)}{s}$$

and

$$\mathcal{L}\{t^2\} = \frac{2}{s^3}$$

find

$$\mathcal{L}\left\{\frac{t^3}{3}\right\}.$$

Solution   As

$$\int_0^t \tau^2\,d\tau = \left[\frac{\tau^3}{3}\right]_0^t = \frac{t^3}{3}$$

and

$$\mathcal{L}\{t^2\} = \frac{2}{s^3}$$

then

$$\mathcal{L}\left\{\int_0^t \tau^2\,d\tau\right\} = \frac{2/s^3}{s} = \frac{2}{s^4} \quad\Rightarrow\quad \mathcal{L}\left\{\frac{t^3}{3}\right\} = \frac{1}{s}\left(\frac{2}{s^3}\right) = \frac{2}{s^4}.$$

Example 15.12 (*Convolution*)   Find

$$\mathcal{L}^{-1}\left\{\frac{1}{(s-2)(s-3)}\right\}$$

using the convolution property:

$$\mathcal{L}\left\{\int_0^1 f(t)g(t-\tau)d\tau\right\} = F(s)G(s).$$

Solution

$$\frac{1}{(s-2)(s-3)} = \frac{1}{(s-2)}\frac{1}{(s-3)}.$$

Therefore, call

$$F(s) = \frac{1}{s-2}, \qquad G(s) = \frac{1}{s-3}$$

$$f(t) = \mathcal{L}^{-1}\left\{\frac{1}{s-2}\right\} = e^{2t}$$

$$g(t) = \mathcal{L}^{-1}\left\{\frac{1}{s-3}\right\} = e^{3t}.$$

Then by the convolution rule

$$\mathcal{L}^{-1}\left\{\frac{1}{(s-2)(s-3)}\right\} = \int_0^t e^{2\tau}e^{3(t-\tau)}\,d\tau.$$

This integral is an integral over the variable τ. t is a constant as far as the integration process is concerned. We can use the properties of powers to separate out the terms in τ and the terms in t, giving

$$\mathcal{L}^{-1}\left\{\frac{1}{(s-2)(s-3)}\right\} = e^{3t}\int_0^t e^{-\tau}\,d\tau = e^{3t}\left[\frac{e^{-\tau}}{-1}\right]_0^t$$

$$= e^{3t}(-e^{-t}+1)$$

$$= -e^{2t} + e^{3t}.$$

Example 15.13 (*Derivatives of the transform*)   Find $\mathcal{L}\{t\sin(3t)\}$.

Solution   Using the derivatives of the transform property

$$\mathcal{L}\{t^n f(t)\} = (-1)^n F^{(n)}(s)$$

we have $f(t) = \sin(3t)$ and

$$\mathcal{L}\{\sin(3t)\} = \frac{3}{s^2+9} = F(s)$$

(from Table 15.1).

As $\sin(3t)$ is multiplied by t in the expression to be transformed, we use the derivative of the transform property with $n = 1$:

$$\mathcal{L}\{t^1 f(t)\} = (-1)^1 F^{(1)}(s) = -F'(s)$$

to give

$$\mathcal{L}\{t\sin(3t)\} = -\frac{d}{ds}\left(\frac{3}{s^2+9}\right) = \frac{6s}{(s^2+9)^2}.$$

Example 15.14 (*Derivatives of the transform and the inverse transform*)
Find

$$\mathcal{L}^{-1}\left\{\frac{s}{(s^2+4)^2}\right\}$$

Solution   We notice that

$$\frac{s}{(s^2+4)^2} = -\frac{1}{2}\frac{\mathrm{d}}{\mathrm{d}s}\left(\frac{1}{s^2+4}\right).$$

Using the derivatives of the transform property

$$\mathcal{L}\{t^n f(t)\} = (-1)^n F^{(n)}(s)$$

and setting $n = 1$ we have

$$\mathcal{L}^{-1}\{(-1)\mathrm{d}F/\mathrm{d}s\} = tf(t)$$

and as

$$\mathcal{L}^{-1}\left\{\frac{1}{(s^2+4)}\right\} = \frac{1}{2}\sin(2t)$$

we have

$$\mathcal{L}^{-1}\left\{\frac{s}{(s^2+4)^2}\right\} = \frac{1}{2}t\left(\frac{1}{2}\sin(2t)\right) = \frac{t}{4}\sin(2t)$$

Using partial fractions to find the inverse transform

Partial fractions can be used to find the inverse transform of expressions like

$$\frac{11s+7}{s^2-1}$$

by expressing $F(s)$ as a sum of fractions with a simple factor in the denominator.

Example 15.15   Find the inverse Laplace transform of

$$\frac{11s+7}{s^2-1}$$

Solution   Factorize the denominator. This is equivalent to finding the values of s for which the denominator is 0, because if the denominator has factors s_1 and s_2 we know that we can write it as $c(s-s_1)(s-s_2)$ where c, is some number. In this case we could solve for $s^2 - 1 = 0$. However, it is not difficult to spot that $s^2 - 1 = (s-1)(s+1)$. The values for which the denominator is zero are called the poles of the function.

$$\frac{11s+7}{s^2-1} = \frac{11s+7}{(s-1)(s+1)}.$$

We assume that there is an identity such that

$$\frac{11s+7}{(s^2-1)} = \frac{A}{s-1} + \frac{B}{s+1}.$$

Multiply both sides of the equation by $(s-1)(s+1)$ to get

$$11s+7 = A(s+1) + B(s-1).$$

Substitute $s = 1$; then $11 + 7 = 2A \Rightarrow A = 9$.

Substitute $s = -1$; then $-11 + 7 = -2B \Rightarrow B = 2$.

So

$$\mathcal{L}^{-1}\left\{\frac{11s + 7}{s^2 - 1}\right\} = \mathcal{L}^{-1}\left\{\frac{9}{s - 1} + \frac{2}{s + 1}\right\} = 9e^t + 2e^{-t}.$$

A quick formula for partial fractions

There is a quick way of getting the partial fractions expansion, called the 'cover up' rule, which works in the case where all the roots of the denominator of $F(s)$ are distinct. If $F(s) = P(s)/Q(s)$ then write $Q(s)$ in terms of its factors $Q(s)$

$$Q(s) = c(s - s_1)(s - s_2)(s - s_3) \cdots (s - s_r) \cdots (s - s_n)$$

where $s_1 \ldots s_n$ are its distinct roots. Then we can find the constant A_r, etc., for the partial fraction expansion from covering up each of the factors of Q in term and substituting $s = s_r$ in the rest of the expression:

$$F(s) = \frac{P(s)}{c(s - s_1)(s - s_2)(s - s_3) \cdots (s - s_r) \cdots (s - s_n)}$$

then

$$F(s) = \frac{A_1}{(s - s_1)} + \frac{A_2}{(s - s_2)} + \cdots + \frac{A_r}{(s - s_r)} + \cdots + \frac{A_n}{(s - s_n)}$$

where

$$A_r = \frac{P(s_r)}{c(s_r - s_1)(s_r - s_2)(s_r - s_3) \cdots (s_r - s_r) \cdots (s_r - s_n)}$$

In this case

$$F(s) = \frac{11s + 7}{s^2 - 1} = \frac{11s + 7}{(s - 1)(s + 1)} = \frac{A}{s - 1} + \frac{B}{s + 1}$$

Then A is found by substituting $s = 1$ into

$$\frac{11s + 7}{(s - 1)(s + 1)} = \frac{11 + 7}{2} = 9$$

and B is found by substituting $s = -1$ into

$$\frac{11s + 7}{(s - 1)(s + 1)} = \frac{-11 + 7}{-2} = 2.$$

This gives the partial fraction expansion, as before, as

$$F(s) = \frac{9}{s + 1} + \frac{2}{s - 1}.$$

This method can also be used for complex poles; for instance,

$$\frac{1}{(s^2 + 4)(s + 3)} = \frac{1}{(s + 3)(s - j2)(s + j2)}.$$

The roots of the denominator are -3, j2 and $-$j2, so we get for the partial fraction expansion

$$\frac{1}{(s^2+4)(s+3)} = \frac{1}{13(s+3)} + \frac{1}{(-8+\text{j}12)(s-\text{j}2)}$$
$$+ \frac{1}{(-8-\text{j}12)(s+\text{j}2)}.$$

The inverse transform can be found directly or the last two terms can be combined to give the expansion as

$$\frac{1}{(s^2+4)(s+3)} = \frac{1}{13(s+3)} + \frac{-16s+48}{208(s^2+4)} = \frac{1}{13(s+3)} + \frac{3-s}{13(s^2+4)}.$$

Repeated poles

If there are repeated factors in the denominator, for example,

$$\frac{4}{(s+1)^2(s-2)}$$

then try a partial fraction expansion of the form

$$\frac{4}{(s+1)^2(s-2)} = \frac{A}{(s+1)^2} + \frac{B}{s+1} + \frac{C}{s-2}.$$

The 'cover up' rule cannot be used to find the coefficients A and B.

15.5 Solving linear differential equations with constant coefficients

The scheme for solving differential equations is as outlined below. The Laplace transform transforms the linear differential equation with constant coefficients to an algebraic equation in s. This can be solved and then the inverse transform of this solution gives the solution to the original differential equation.

Differential equation	Solution	t-domain
↓	↑	
Laplace transform	Inverse Laplace transform	
↓	↑	
Algebraic equation →	Solution	s-space

Example 15.16   A d.c. voltage of 3 V is applied to an RC circuit with $R = 2000\,\Omega$ and $C = 0.001\,\text{F}$, where $q(0) = 0$. Find the voltage across the capacitor as a function of t.

Solution   From Kirchoff's voltage law, we get the differential equation

$$2000\frac{\text{d}q}{\text{d}t} + \frac{q}{0.001} = 3$$

$$\frac{\text{d}q}{\text{d}t} + 0.5q = 0.0015$$

Taking Laplace transforms of both sides of the equation where $Q(s) = \mathcal{L}\{q(t)\}$,

$$sQ(s) - q(0) + 0.5Q(s) = \frac{0.0015}{s}.$$

Here, we have used the derivative property of the Laplace transform, $q'(t) = sQ(s) - q(0)$. Solve the algebraic equation

$$(s + 0.5)Q = \frac{0.0015}{s} \quad \Leftrightarrow \quad Q = \frac{0.0015}{s(s + 0.5)}.$$

We need to find the inverse Laplace transform of this function of s so that we expand using partial fractions, giving

$$Q = \frac{0.003}{s} - \frac{0.003}{s + 0.5}$$

$$q = \mathcal{L}^{-1}\left\{\frac{0.003}{s} - \frac{0.003}{s + 0.5}\right\} = 0.003 - 0.003e^{-0.5t}.$$

The voltage across the capacitor is

$$\frac{q(t)}{c} = \frac{0.003 - 0.003e^{-0.5t}}{0.001} = 3(1 - e^{-0.5t}).$$

Example 15.17   Solve the following differential equation using Laplace transforms:

$$\frac{d^2x}{dt^2} + 4\frac{dx}{dt} + 3x = e^{-3t}$$

given $x(0) = 0.5$ and $dx(0)/dt = -2$.

Solution   Transform the differential equation

$$\frac{d^2x}{dt^2} + 4\frac{dx}{dt} + 3x = e^{-3t}$$

to get

$$\left(s^2X(s) - sx(0) - \frac{dx}{dt}(0)\right) + 4(sX(s) - x(0)) + 3X(s) = \frac{1}{s + 3}.$$

Here we have used

$$\mathcal{L}\left\{\frac{d^2x}{dt^2}\right\} = s^2X(s) - sx(0) - \frac{dx}{dt}(0)$$

and

$$\mathcal{L}\left\{\frac{dx}{dt}\right\} = sX(s) - x(0).$$

Substitute $x(0) = 0.5$ and $dx(0)/dt = -2$

$$s^2X(s) - \frac{1}{2}s + 2 + 4sX(s) - 2 + 3X(s) = \frac{1}{s + 3}$$

$$\Leftrightarrow \quad X(s)(s^2 + 4s + 3) = \frac{1}{s + 3} + \frac{1}{2}s$$

$$\Leftrightarrow \quad X(s) = \frac{1}{(s + 3)(s^2 + 4s + 3)} + \frac{s}{2(s^2 + 4s + 3)}$$

Use partial fractions:

$$\frac{1}{(s+3)(s^2+4s+3)} = \frac{1}{(s+3)(s+3)(s+1)}$$

$$\frac{1}{(s+3)(s+3)(s+1)} = \frac{A}{(s+3)} + \frac{B}{(s+3)^2} + \frac{C}{s+1}$$

$$\Rightarrow \quad 1 = A(s+3)(s+1) + B(s+1) + C(s+3)^2.$$

Substituting $s = -3$

$$1 = -2B \quad \Leftrightarrow \quad B = -\tfrac{1}{2}.$$

Substituting $s = -1$

$$1 = 4C \quad \Leftrightarrow \quad C = \tfrac{1}{4}.$$

Substituting $s = 0$

$$1 = 3A + B + 9C.$$

Substituting $B = -\tfrac{1}{2}$ and $C = \tfrac{1}{4}$ gives

$$1 = 3A - \tfrac{1}{2} + \tfrac{9}{4} \quad \Leftrightarrow \quad A = -\tfrac{1}{4}.$$

Therefore

$$\frac{1}{(s+3)^2(s+1)} = -\frac{1}{4(s+3)} - \frac{1}{2(s+3)^2} + \frac{1}{4(s+1)}.$$

We can use the 'cover up' rule to find

$$\frac{s}{2(s+3)(s+1)} = \frac{3}{4(s+3)} - \frac{1}{4(s+1)}$$

so that

$$X(s) = -\frac{1}{4(s+3)} - \frac{1}{2(s+3)^2} + \frac{1}{4(s+1)} + \frac{3}{4(s+3)}$$

$$- \frac{1}{4(s+1)}$$

$$= \frac{-1}{2(s+3)^2} + \frac{1}{2(s+3)}.$$

Taking the inverse transform, we find

$$x(t) = -\frac{1}{2}te^{-3t} + \frac{1}{2}e^{-3t} = \frac{e^{-3t}}{2}(1-t)$$

15.6 Laplace transforms and systems theory

The transfer function and impulse response function

An important role is played in systems theory by the impulse response function, the Laplace transform of which is called the Transfer Function (or system function). We remember from Chapter 14 that a linear, time-invariant system is represented by a linear differential equation with constant coefficients. We will stick to second-order equations although the results can be generalized to any order.

An LTI system can be represented by

$$a\frac{d^2y}{dt^2} + b\frac{dy}{dt} + cy = f(t).$$

If we take $f(t)$ as the delta function $\delta(t)$, then we get

$$a\frac{d^2y}{dt^2} + b\frac{dy}{dt} + cy = \delta(t).$$

By definition of the impulse response function we consider all initial conditions to be 0.

Taking the Laplace transform we get

$$as^2Y(s) + bsY(s) + cY(s) = 1$$

$$Y(s)(as^2 + bs + c) = 1$$

$$Y(s) = \frac{1}{as^2 + bs + c}.$$

Notice that the poles of this function are found by solving $as^2+bs+c = 0$, which we recognize as the auxiliary equation or characteristic equation from Chapter 14.

This function, the Laplace transform of the impulse response function, is called the transfer function and is usually denoted by $H(s)$, so we have

$$H(s) = \frac{1}{as^2 + bs + c}$$

and $\mathcal{L}\{H(s)\} = h(t)$, where $h(t)$ is the impulse response function.

Note that the impulse response function describes the behaviour of the system after it has been given an idealized kick.

Example 15.18   Find the transfer function and impulse response of the system described by the following differential equation:

$$3\frac{dy}{dt} + 4y = f(t).$$

Solution   To find the transfer function replace $f(t)$ by $\delta(t)$ and take the Laplace transform of the resulting equation assuming zero initial conditions:

$$3\frac{dy}{dt} + 4y = \delta(t).$$

Taking the Laplace transform of both sides of the equation we get

$$3(sY - y(0)) + 4Y = 1$$

As $y(0) = 0$,

$$Y = \frac{1}{3s + 4} = H(s).$$

To find the impulse response function we take the inverse transform of the transfer function to find

$$h(t) = \mathcal{L}^{-1}\left\{\frac{1}{3s + 4}\right\} = e^{\frac{-4t}{3}}.$$

We now discover that we can find the system response to any input function $f(t)$, with zero initial conditions, by using the transfer function.

Response of a system with zero initial conditions to any input $f(t)$

Assuming all initial conditions are zero, that is, $y'(0) = 0$ and $y(0) = 0$, then the equation

$$a\frac{d^2y}{dt^2} + b\frac{dy}{dt} + cy = f(t)$$

transforms to become

$$(as^2 + bs + c)Y(s) = F(s) \quad \Leftrightarrow \quad Y(s) = \frac{F(s)}{as^2 + bs + c}.$$

We have just discovered that the transfer function, $H(s)$, is given by

$$H(s) = \frac{1}{as^2 + bs + c}.$$

Then we find that

$$Y(s) = F(s)H(s).$$

In order to find the response of the system to the function $f(t)$ we take the inverse transform of this expression. We are able to use the convolution property of Laplace transforms which states:

$$\mathcal{L}\left\{\int_0^t f(\tau)g(t - \tau)\,d\tau\right\} = F(s)G(s)$$

or equivalently

$$\mathcal{L}^{-1}\{F(s)G(s)\} = \int_0^t f(\tau)g(t - \tau)\,d\tau.$$

The integral $\int_0^t f(t)\,g(t - \tau)\,d\tau$ is called the convolution of f and g and can be expressed as $f(t) * g(t)$. So we find that the response of the system with zero initial conditions to any input function $f(t)$ is given by the convolution of $f(t)$ with the system's impulse response:

$$y(t) = \mathcal{L}^{-1}\{F(s)H(s)\} = f(t) * h(t).$$

We can use this result to solve two types of problems, given zero initial conditions. If we know the impulse response function of the system then we can find the system response to any input $f(t)$ either by convolving in the time domain $y(t) = f(t) * h(t)$ or by finding the Laplace transform of $f(t)$, $F(s)$, and finding the transfer function of the system $\mathcal{L}\{h(t)\} = H(s)$ and then finding $Y(s) = H(s)F(s)$ and taking the inverse transform of this to find $y(t)$. This type of problem is called a convolution problem. The other type of problem is that given any output, $y(t)$, and given the input, $f(t)$, we can deduce the impulse response of the system. This we do by finding $F(s) = \mathcal{L}\{f(t)\}, Y(s) = \mathcal{L}\{y(t)\}$ and then $H(s) = Y(s)/F(s)$, thus giving the transfer function. To find the impulse response we find $h(t) = \mathcal{L}^{-1}\{H(s)\}$

Example 15.19   The impulse response of a system is known to be $h(t) = e^{3t}$. Find the response of the system to an input of $f(t) = 6\cos(2t)$ given zero initial conditions.

Solution   Method 1. We can take Laplace transforms and use $Y(s) = H(s)F(s)$. In this case

$$h(t) = e^{3t} \quad \Leftrightarrow \quad H(s) = \mathcal{L}\{e^{3t}\} = \frac{1}{s-3}$$

$$f(t) = 6\cos(2t) \quad \Leftrightarrow \quad F(s) = \mathcal{L}\{6\cos(2t)\} = \frac{6s}{4+s^2}.$$

Hence

$$Y(s) = H(s)F(s) = \frac{6s}{(s-3)(4+s^2)}.$$

As we want to find $y(t)$, we use partial fractions:

$$\frac{6s}{(s-3)(4+s^2)} = \frac{6s}{(s-3)(s+j2)(s-j2)}$$

$$= \frac{18}{13(s-3)} + \frac{3}{(j2-3)(s-j2)}$$

$$\quad - \frac{3}{(j2+3)(s+j2)} \quad \text{(using the 'cover up' rule)}$$

$$= \frac{18}{13(s-3)} - \frac{3(6s-8)}{13(s^2+4)}$$

$$= \frac{18}{13(s-3)} - \frac{18s}{13(s^2+4)} + \frac{12}{13}\frac{2}{(s^2+4)}.$$

We can now take the inverse transform to find the system response:

$$y(t) = \mathcal{L}^{-1}\{Y(s)\}$$

$$= \mathcal{L}^{-1}\left\{\frac{18}{13(s-3)} - \frac{18s}{13(s^2+4)} + \frac{12}{13}\frac{2}{(s^2+4)}\right\}$$

$$= \frac{18}{13}e^{3t} - \frac{18}{13}\cos(2t) + \frac{12}{13}\sin(2t).$$

Alternative method. Find $y(t)$ by taking the convolution of $f(t)$ with the impulse response function

$$y(t) = f(t) * h(t) = (6\cos(2t)) * (e^{3t})$$

By definition of convolution

$$(6\cos(2t)) * (e^{3t}) = \int_0^t 6\cos(2\tau)\,e^{3(t-\tau)}\,d\tau.$$

As this is a real integral we can use the trick of writing $\cos(2\tau) = \text{Re}(e^{j2\tau})$ to make the integration easier. So we find

$$I = \int_0^t 6e^{j2\tau}e^{3(t-\tau)}\,d\tau$$

$$= 6e^{3t}\int_0^t e^{\tau(j2-3)}\,d\tau$$

$$= 6e^{3t}\left[\frac{e^{\tau(j2-3)}}{j2-3}\right]_0^t$$

$$= 6e^{3t}\left(\frac{e^{t(j2-3)}}{j2-3} - \frac{1}{j2-3}\right)$$

$$= \frac{6e^{3t}(-j2-3)(e^{-3t}(\cos(2t)+j\sin(2t))-1)}{4+9}.$$

Taking the real part of this result we get the system response as

$$\int_0^t 6\cos(2\tau)e^{3(t-\tau)}\,d\tau = \frac{6}{13}(-3\cos(2t)+2\sin(2t)) + \frac{18}{13}e^{3t}$$

$$= -\frac{18}{13}\cos(2t) + \frac{12}{13}\sin(2t) + \frac{18}{13}e^{3t}$$

which confirms the result of the first method.

Example 15.20   A system at rest has a constant input of $f(t) = 3$ applied at $t = 0$. The output is found to be $u(t)(\frac{3}{2} - \frac{3}{2}e^{-2t})$. Find the impulse response of the system.

Solution   Since

$$y(t) = u(t)\left(\frac{3}{2} - \frac{3}{2}e^{-2t}\right)$$

we have

$$y(s) = \frac{3}{2s} - \frac{3}{2(s+2)} = \frac{3s+6-3s}{2s(s+2)} = \frac{3}{s(s+2)}$$

and as $f(t) = 3$, we have

$$F(s) = \frac{3}{s}.$$

Hence

$$H(s) = \frac{Y(s)}{F(s)} = \frac{3/(s(s+2))}{3/s} = \frac{1}{s+2}$$

giving

$$h(t) = \mathcal{L}^{-1}\left\{\frac{1}{s+2}\right\} = e^{-2t}.$$

Hence the impulse response is $h(t) = e^{-2t}$.

The frequency response

In this section, we shall seek to establish a relationship between the transfer function and the steady state response to a single frequency input. Consider the response of the system

$$\ddot{y} + 3\dot{y} + 2y = f(t)$$

to a single sinusoidal input $e^{j\omega t} = \cos(\omega t) + j \sin(\omega t)$ with $y(0) = 2$ and $\dot{y}(0) = 1$. Taking Laplace transforms we find

$$s^2 Y - 2s - 1 + 3(sY - 2) + 2Y = \frac{1}{s - j\omega}$$

$$Y(s^2 + 3s + 2) = 2s + 7 + \frac{1}{s - j\omega}$$

$$Y = \frac{2s + 7}{s^2 + 3s + 2} + \frac{1}{(s - j\omega)(s^2 + 3s + 2)}.$$

Remember

$$H(s) = \frac{1}{s^2 + 3s + 2} = \frac{1}{(s + 2)(s + 1)}$$

$$Y = \frac{2s + 7}{(s + 2)(s + 1)} + \frac{1}{(s - j\omega)(s + 2)(s + 1)}.$$

The first term in this expression is due to non-zero initial conditions. We are particularly interested in the second term in the expression for $Y(s)$, which we notice may be written as $H(s)/(s - j\omega)$.

Using the 'cover up' rule to write $Y(s)$ in its partial fractions,

$$Y(s) = \frac{5}{s + 1} - \frac{3}{s + 2} + \frac{1}{(-1 - j\omega)(s + 1)} + \frac{1}{(2 + j\omega)(s + 2)} + \frac{H(j\omega)}{s - j\omega}.$$

Taking inverse transforms we find

$$y(t) = 5e^{-t} - 3e^{-2t} + \frac{1}{-1 - j\omega}e^{-t} + \frac{1}{2 + j\omega}e^{-2t} + H(j\omega)e^{j\omega t}.$$

The first two terms in this expression are caused by the non-zero initial values and decay exponentially. The next two terms also decay with increasing t and are as a result of the abrupt turn on of the input at $t = 0$. Thus for a stable system, all four terms are part of the transient solution that dies out as t increases. The fifth and final term is the sinusoidal input $f(t) = e^{j\omega t}$ multiplied by $H(j\omega)$. $H(j\omega)$ is a complex constant. This is the steady state response of the system to a single sinusoidal input, which we have shown in this case is given by

$$Y(t) = H(j\omega)e^{j\omega t}.$$

We can then see that for a single sinusoid input the steady state response is found by substituting $s = j\omega$ into the transfer function for the system and multiplying the resulting complex constant by the sinusoidal input. In other words, the steady state response is a scaled and phase shifted version of the input. We can find the response to a sine or cosine input by

$$y(t) = \text{Re}(H(j\omega)e^{j\omega t}) \quad \text{for } f(t) = \cos(\omega t)$$

and

$$y(t) = \text{Im}(H(j\omega)e^{j\omega t}) \quad \text{for } f(t) = \sin(\omega t)$$

Alternatively, we can find the response to $e^{j\omega t}$ and to $e^{-j\omega t}$ and use the fact that

$$\cos(\omega t) = \tfrac{1}{2}(e^{j\omega t} + e^{-j\omega t})$$
$$\sin(\omega t) = \tfrac{1}{2j}(e^{j\omega t} - e^{-j\omega t})$$

to write

for $f(t) = \cos(\omega t), \quad y(t) = \tfrac{1}{2}(H(j\omega)e^{j\omega t} + H(-j\omega)e^{-j\omega t})$

for $f(t) = \sin(\omega t), \quad y(t) = \dfrac{1}{2j}(H(j\omega)e^{j\omega t} - H(-j\omega)e^{-j\omega t}).$

All of these results make use of the principle of superposition for linear systems.

We have seen in this section that the response to a simple sinusoid can be characterized by multiplying the input by a complex constant $H(j\omega)$ where ω is the angular frequency of the input. The function $H(j\omega)$ is called the frequency response function. It is this result that motivates us towards the desirability of expressing all signals in terms of cosines and sines of single frequencies – a technique known as Fourier analysis. We shall look at Fourier Analysis for periodic functions in Chapter 16.

Example 15.21   A system transfer function is known to be

$$H(s) = \frac{1}{3s + 1}$$

then find the steady state response to the following:

(a) $f(t) = e^{j2t}$;
(b) $f(t) = 3\cos(2t)$.

Solution   (a) The steady state response to a single frequency $e^{j\omega t}$ is given $H(j\omega)e^{j\omega t}$. Here $f(t) = e^{j2t}$, so in this case $\omega = 2$ and $H(s)$ is given as $1/(3s + 1)$. Hence we get the steady state response as

$$H(j2)e^{j2t} = \frac{1}{3(j2) + 1}e^{j2t} = \frac{e^{j2t}}{1 + j6} = \frac{(1 - j6)e^{j2t}}{37}$$

(b) Using $(1/2)(H(j\omega)e^{j\omega t} + H(-j\omega)e^{-j\omega t})$ as the response to $\cos(\omega t)$ and substituting for H and $\omega = 2$ gives

$$\frac{1}{2}\left(\frac{(1 - 6j)e^{j2t}}{37} + \frac{(1 + j6)}{37}e^{-j2t}\right)$$

$$= \frac{1}{74}((1 - j6)(\cos(2t) + j\sin(2t))$$

$$+ (1 + j6)(\cos(2t) - j\sin(2t))$$

$$= \frac{1}{37}(\cos(2t) + 6\sin(2t)).$$

Hence, the steady state response to an input of $3\cos(2t)$ is

$\frac{3}{37}(\cos(2t) + 6\sin(2t))$.

15.7
z transforms

z transforms are used to solve problems in discrete systems in a manner similar to the use of Laplace transforms for piecewise continuous systems. We take z transforms of sequences. We shall assume that our sequences begin with the zeroth term and have terms for positive n. $f_0, f_1, f_2, \ldots, f_n, \ldots$ is an input sequence to the system. However, when considering the initial conditions for a difference equation it is convenient to assign them to $y_{-j}, \ldots, y_{-2}, y_{-1}$, etc., where j is the order of the difference equation. So, in that case, we shall allow some elements in the sequence with negative subscript. Our output sequence will be of the form $y_{-j}, \ldots, y_{-2}, y_{-1}, y_0, y_1, y_2, \ldots, y_n, \ldots$ where the difference equation describing the system only holds for $n \geq 0$.

z transform definition

The z transform of a sequence $f_0, f_1, f_2, \ldots, f_n, \ldots$ is given by

$$F(z) = \sum_{n=0}^{\infty} f_n z^{-n}.$$

As this is an infinite summation it will not always converge. The set of values of z for which it exists is called the region of convergence. The sequence, $f_0, f_1, f_2, \ldots, f_n, \ldots$ is a function of an integer, however, its z transform is a function of a complex variable z. The operation of taking the z transform of the sequence f_n is represented by $\mathcal{Z}\{f_n\} = F(z)$.

Example 15.22   Find the z transform of the finite sequence $1, 0, 0.5, 3$.

Solution   We multiply the terms in the sequence by z^{-n}, where $n = 0, 1, 2, \ldots$ and then sum the terms, giving

$$F(z) = 1 + 0z^{-1} + 0.5z^{-2} + 3z^{-3} = 1 + \frac{0.5}{z^2} + \frac{3}{z^3}.$$

Example 15.23   Find the z transform of the geometric sequence $a_0 r^n$ where $n = 0, 1, \ldots$

Solution

$$F(z) = \sum_{n=0}^{\infty} a_0 r^n z^{-n} = \sum_{n=0}^{\infty} a_0 \left(\frac{r}{z}\right)^n.$$

Writing this out we get

$$F(z) = a_0 + a_0 \left(\frac{r}{z}\right) + a_0 \left(\frac{r}{z}\right)^2 + a_0 \left(\frac{r}{z}\right)^3 + \cdots$$

From this we can see that we have another geometric progression with zeroth term a_0 and common ratio r/z; hence, we can sum to infinity provided $|r/z| < 1$, giving

$$F(z) = \frac{a_0}{1 - (r/z)} = \frac{a_0 z}{z - r}$$

where

$$\left| \frac{r}{z} \right| < 1 \quad \text{or} \quad |z| > |r|$$

We can see that in the case of infinite sequences there will be a region of convergence for the z transform.

The impulse function and the step function

The impulse function or delta function for a discrete system is the sequence

$$\delta_n = \begin{cases} 1 & n = 0 \\ 0 & n \neq 0. \end{cases}$$

The step function is the function

$$u_n = \begin{cases} 1 & n \geq 0 \\ 0 & n < 0 \end{cases}$$

and the shifted unit step function is

$$u_{n-j} = \begin{cases} 1 & n \geq j \\ 0 & n < j. \end{cases}$$

As we are mainly considering sequences defined for $n \geq 0$, we could consider that all of the sequences are multiplied by the step function u_n. That is, they are all 'switched on' at $n = 0$.

Rather than always using the definition to find the z transform, we will usually make use of a table of well-known transforms and properties of the z transform to discover the transform of various sequences. A list of z transforms is given in Table 15.2.

Table 15.2   *z transforms*

f_n	$F(z)$	
u_n	$\dfrac{z}{z-1}$	$|z| > 1$
δ_n	1	
n	$\dfrac{z}{(z-1)^2}$	$|z| > 1$
r^n	$\dfrac{z}{z-r}$	$|z| > |r|$
$\cos(\theta n)$	$\dfrac{z(z - \cos(\theta))}{z^2 - 2z\cos(\theta) + 1}$	$|z| > 1$
$\sin(\theta n)$	$\dfrac{z\sin(\theta)}{z^2 - 2z\cos(\theta) + 1}$	$|z| > 1$
$e^{j\theta n}$	$\dfrac{z}{z - e^{j\theta}}$	$|z| > 1$

Properties of the *z* transform

For the following

$$\mathcal{Z}\{f_n\} = \sum_{n=0}^{\infty} f_n z^{-n} = F(z), \qquad \mathcal{Z}\{g_n\} = \sum_{n=0}^{\infty} g_n z^{-n} = G(z).$$

1. Linearity:

$$\mathcal{Z}\{af_n + bg_n\} = aF(z) + bG(z).$$

2. Left shifting property:

$$\mathcal{Z}\{f_{n+k}\} = z^k F(z) - \sum_{i=0}^{k} z^{k-i} f_i.$$

3. Right shifting property (although usually we assume $f_n = 0$ for $n < 0$ we use f_{-1}, f_{-2} for the initial conditions when solving difference equations using z transforms):

$$\mathcal{Z}\{f_{n-1}\} = z^{-1}\mathcal{Z}\{f_n\} + f_{-1}$$

$$\mathcal{Z}\{f_{n-2}\} = z^{-2}\mathcal{Z}\{f_n\} + f_{-2} + z^{-1}f_{-1}$$

$$\mathcal{Z}\{f_{n-k}\} = z^{-k}\mathcal{L}\{f_n\} + \sum_{i=0}^{k-1} f_{i-k}z^{-i}.$$

4. Change of scale:

$$\mathcal{Z}\{a^n f_n\} = F\left(\frac{z}{a}\right)$$

where a is a constant.

5. Convolution:

$$\mathcal{Z}\left\{\sum_{k=0}^{n} g_k f_{n-k}\right\} = G(z)F(z).$$

The convolution of f and g can be written as

$$g * f = \sum_{k=0}^{n} g_k f_{n-k}.$$

where g_n and f_n are sequences defined for $n \geqslant 0$.

6. Derivatives of the transform:

$$\mathcal{Z}\{nf_n\} = -z\frac{dF}{dz}(z).$$

Example 15.24 (*Linearity*) Find the z transform of $3n + 2 \times 3^n$.

Solution From the linearity property

$$\mathcal{Z}\{3n + 2 \times 3^n\} = 3\mathcal{Z}\{n\} + 2\mathcal{Z}\{3^n\}$$

and from the Table 15.2

$$\mathcal{Z}\{n\} = \frac{z}{(z-1)^2} \quad \text{and} \quad \mathcal{Z}\{3^n\} = \frac{z}{z-3}$$

(r^n with $r = 3$). Therefore

$$\mathcal{Z}\{3n + 2 \times 3^n\} = \frac{3z}{(z-1)^2} + \frac{2z}{z-3}$$

Example 15.25 (*Linearity and the inverse transform*)   Find the inverse z transform of

$$\frac{2z}{z-1} + \frac{3z}{z-2}.$$

Solution   From Table 15.2

$$\mathcal{Z}^{-1}\left\{\frac{z}{z-1}\right\} = u_n$$

$$\mathcal{Z}^{-1}\left\{\frac{z}{z-2}\right\} = 2^n \quad (r = 2)$$

So

$$\mathcal{Z}^{-1}\left\{\frac{2z}{z-1} + \frac{3z}{z-2}\right\} = 2u_n + 3 \times 2^n$$

Example 15.26 (*Change of scale*)   Find the inverse z transform of

$$\frac{z}{(z-2)^2}$$

Solution

$$\frac{z}{(z-2)^2} = \frac{\frac{1}{2}(z/2)}{((z/2)-1)^2}.$$

From Table 15.2

$$\mathcal{Z}^{-1}\left\{\frac{z}{(z-1)^2}\right\} = n$$

Using the change of scale property and linearity:

$$\mathcal{Z}^{-1}\left\{\frac{\frac{1}{2}(z/2)}{((z/2)-1)^2}\right\} = \frac{1}{2}n(2)^n = n2^{n-1}.$$

Example 15.27 (*Convolution*)   Find the inverse z transform of

$$\frac{z}{z-1}\frac{z}{z-4}$$

Solution   Note that

$$\mathcal{Z}^{-1}\left\{\frac{z}{z-1}\right\} = u_n \quad \text{and} \quad \mathcal{L}^{-1}\left\{\frac{z}{z-4}\right\} = 4^n.$$

Hence, using convolution

$$\mathcal{Z}^{-1}\left\{\frac{z}{z-1}\frac{z}{z-4}\right\} = u_n * 4^n = \sum_{k=0}^{n} u_k 4^{n-k}.$$

Writing out this sequence for $n = 0, 1, 2, 3, \ldots$

$$\begin{array}{cccc} 1, & (1+4), & 1+4+16, & 1+4+16+64, \ldots \\ (n=0) & (n=1) & (n=2) & (n=3) \end{array}$$

We see that the nth term is a geometric series with $n + 1$ terms and first term 1 and common ratio 4. From the formula for the sum for n terms of a geometric progression, $S_n = a(r^n - 1)/(r - 1)$ where a is the first term, r is the common ratio and n is the number of terms. Therefore, for the n th term of the above sequence, we get:

$$\frac{4^{n+1} - 1}{4 - 1} = \frac{4^{n+1} - 1}{3}.$$

So we have found

$$\mathcal{Z}^{-1}\left\{\frac{z}{z - 1}\frac{z}{z - 4}\right\} = \frac{4^{n+1} - 1}{3}.$$

Example 15.28 (*Derivatives of the transform*)   Using

$$\mathcal{Z}\{n\} = \frac{z}{(z - 1)^2}$$

find $\mathcal{Z}\{n^2\}$.

Solution   Using the derivative of the transform property

$$\mathcal{Z}\{n^2\} = \mathcal{Z}\{nn\} = -z\frac{\mathrm{d}}{\mathrm{d}z}\mathcal{Z}\{n\}$$

$$= -z\frac{d}{dz}\left(\frac{z}{(z - 1)^2}\right).$$

As

$$\frac{d}{dz}\left(\frac{z}{(z - 1)^2}\right) = \frac{(z - 1)^2 - 2z(z - 1)}{(z - 1)^4} = \frac{z - 1 - 2z}{(z - 1)^3}$$

$$= \frac{-z - 1}{(z - 1)^3}$$

we obtain

$$\mathcal{Z}\{n^2\} = -z\left(\frac{-z - 1}{(z - 1)^3}\right) = \frac{z(z + 1)}{(z - 1)^3}.$$

Using partial fractions to find the inverse transform

Example 15.29   Find

$$\mathcal{Z}^{-1}\left\{\frac{z^2}{(z - 1)(z - 0.5)}\right\}.$$

Solution Notice that most of the values of the transform in Table 15.2 have a factor of z in the numerator. We write

$$\frac{z^2}{(z-1)(z-0.5)} = z\left(\frac{z}{(z-1)(z-0.5)}\right)$$

We use the 'cover up' rule to write

$$\frac{z}{(z-1)(z-0.5)} = \frac{1}{0.5(z-1)} - \frac{1}{z-0.5}$$

$$= \frac{2}{z-1} - \frac{1}{z-0.5}$$

So

$$\frac{z^2}{(z-1)(z-0.5)} = \frac{2z}{z-1} - \frac{z}{z-0.5}$$

and using Table 15.2 we find

$$\mathcal{Z}^{-1}\left\{\frac{2z}{z-1} - \frac{z}{z-0.5}\right\} = 2u_n - (0.5)^n$$

15.8 Solving linear difference equations with constant coefficients using *z* transforms

The scheme for solving difference equations is very similar to that for solving differential equations using Laplace transforms and is outlined below. The z transform transforms the linear difference equation with constant coefficients to an algebraic equation in z. This can be solved and then the inverse transform of this solution gives the solution to the original difference equation.

Difference equation	Solution	n-Domain
$\downarrow$	$\uparrow$	
z transform	Inverse z transform	
$\downarrow$	$\uparrow$	
Algebraic equation $\rightarrow$	Solution	z-space

Example 15.30 Solve the difference equation

$$y_n + 2y_{n-1} = 2u_n$$

for $n \geqslant 0$ given $y_{-1} = 1$.

Solution We take the z transform of both sides of the difference equation

$$y_n + 2y_{n-1} = 2u_n$$

and using the right shift property to find

$$\mathcal{Z}\{y_{n-1}\} = z^{-1}Y(z) + y_{-1}$$

we get

$$Y(z) + 2\left(z^{-1}Y(z) + y_{-1}\right) = \frac{2z}{z-1}.$$

As $y_{-1} = 1$,

$$Y(z)\left(1 + 2z^{-1}\right) = \frac{2z}{z - 1} - 2$$

$$Y(z) = \frac{2z^2}{(z - 1)(z + 2)} - \frac{2z}{(z + 2)}.$$

To take the inverse transform we need to express the first terms using partial fractions. Using the 'cover up' rule we get

$$\frac{2z}{(z - 1)(z + 2)} = \frac{2}{3(z - 1)} + \frac{4}{3(z + 2)}$$

So

$$Y(z) = \frac{2z}{3(z - 1)} + \frac{4z}{3(z + 2)} - \frac{2z}{(z + 2)} = \frac{2z}{3(z - 1)} - \frac{2z}{3(z + 2)}.$$

Taking inverse transforms we find

$$y_n = \tfrac{2}{3}u_n - \tfrac{2}{3}(-2)^n$$

Check: To check that we have the correct solution we can substitute in a couple of values for n and see that we get the same value from the difference equation as from the explicit formula found.

From the explicit formula and using $u_0 = 1$ (by definition of the unit step function), $n = 0$ gives

$$y_0 = \tfrac{2}{3}u_0 - \tfrac{2}{3}(-2)^0 = \tfrac{2}{3} - \tfrac{2}{3} = 0$$

From the difference equation, $y_n + 2y_{n-1} = 2u_n$, where $y_{-1} = 1, n = 0$ gives

$$y_0 + 2y_{-1} = 2u_0.$$

Substituting $y_{-1} = 1$ gives $y_0 = 0$ as before.

From the explicit formula, $n = 1$ gives

$$y_1 = \frac{2}{3}u_1 - \frac{2}{3}(-2)^1 = \frac{2}{3} + \frac{4}{3} = 2.$$

From the difference equation, $n = 1$ gives

$$y_1 + 2y_0 = 2u_1$$

substituting $y_0 = 0$ gives $y_1 = 2$, confirming the result of the explicit formula.

Example 15.31 Solve the difference equation

$$6y_n - 5y_{n-1} + y_{n-2} = (0.25)^n \quad n \geqslant 0$$

given $y_{-1} = 1, y_{-2} = 0$.

Solution   We take the z transform of both sides of the difference equation $6y_n - 5y_{n-1} + y_{n-2} = (0.25)^n$ and use the right shift property to find

$$\mathcal{Z}\{y_{n-1}\} = z^{-1}Y(z) + y_{-1}$$

$$\mathcal{Z}\{y_{n-2}\} = z^{-2}Y(z) + z^{-1}y_{-1} + y_{-2}$$

which gives

$$6Y(z) - 5(z^{-1}Y(z) + y_{-1}) + z^{-2}Y(z) + y_{-2} + z^{-1}y_{-1} = \frac{2}{z - 0.25}.$$

Substituting the initial conditions $y_{-1} = 1$ and $y_{-2} = 0$ and collecting the terms involving $Y(z)$ we get

$$Y(z)(6 - 5z^{-1}Y(z) + z^{-2}) - 5 + z^{-1} = \frac{z}{z - 0.25}$$

$$\Leftrightarrow Y(z) = \frac{z}{(z - 0.25)(6 - 5z^{-1} + z^{-2})} + \frac{5}{6 - 5z^{-1} + z^{-2}}$$

$$- \frac{z^{-1}}{6 - 5z^{-1} + z^{-2}}$$

$$= z\left(\frac{z^2}{(z - 0.25)(6z^2 - 5z + 1)}\right) + z\left(\frac{5z}{(3z - 1)(2z - 1)}\right)$$

$$- z\left(\frac{1}{(3z - 1)(2z - 1)}\right).$$

Using the 'cover up' rule we write each of these terms as partial fractions:

$$\frac{z^2}{(z - 0.25)(3z - 1)(2z - 1)}$$

$$= \frac{z^2}{6\left(z - \frac{1}{4}\right)\left(z - \frac{1}{3}\right)\left(z - \frac{1}{2}\right)}$$

$$= \frac{1}{2\left(z - \frac{1}{4}\right)} - \frac{4}{3}\frac{1}{\left(z - \frac{1}{3}\right)} + \frac{1}{\left(z - \frac{1}{2}\right)}$$

$$\frac{5z}{6\left(z - \frac{1}{3}\right)\left(z - \frac{1}{2}\right)} = \frac{-5}{3\left(z - \frac{1}{3}\right)} + \frac{5}{2\left(z - \frac{1}{2}\right)}$$

$$\frac{1}{3(z - 1)(2z - 1)} = \frac{1}{6\left(z - \frac{1}{3}\right)\left(z - \frac{1}{2}\right)}$$

$$= \frac{-1}{z - \frac{1}{3}} + \frac{1}{z - \frac{1}{2}}$$

giving

$$Y(z) = \frac{z}{2\left(z - \frac{1}{4}\right)} - \frac{4z}{3\left(z - \frac{1}{3}\right)} + \frac{z}{z - \frac{1}{2}} - \frac{5z}{3\left(z - \frac{1}{3}\right)}$$

$$+ \frac{5z}{2\left(z - \frac{1}{2}\right)} + \frac{z}{z - \frac{1}{3}} - \frac{z}{z - \frac{1}{2}}$$

$$\Leftrightarrow Y(z) = \frac{z}{2\left(z - \frac{1}{4}\right)} - \frac{2z}{z - \frac{1}{3}} + \frac{5}{2\left(z - \frac{1}{2}\right)}.$$

Taking the inverse transform:

$$y_n = \frac{1}{2}\left(\frac{1}{4}\right)^n - 2\left(\frac{1}{3}\right)^n + \frac{5}{2}\left(\frac{1}{2}\right)^n$$

15.9 z transforms and systems theory

The transfer function and impulse response function

As before, when considering Laplace transforms, we find that an important role is played by the impulse response function in systems theory of discrete systems. If this case is a sequence, h_n, its z transform of is called the transfer function (or system function). We remember from Chapter 14 that a linear, time-invariant system is represented by a linear difference equation with constant coefficients, that is, a second-order LTI system can be represented by

$$ay_n + by_{n-1} + cy_{n-2} = f_n.$$

If we take f_n as the delta function δ_n, we get

$$ay_n + by_{n-1} + cy_{n-2} = \delta_n.$$

By definition of the impulse response function we consider all initial conditions to be 0, that is, $y_{-1} = 0$, $y_{-2} = 0$; taking the z transform we get

$$aY(z) + bz^{-1}Y(z) + cz^{-2}Y(z) = 1$$

$$Y(z)(a + bz^{-1} + cz^{-2}) = 1$$

$$Y(z) = \frac{1}{a + bz^{-1} + cz^{-2}}$$

$$Y(z) = \frac{z^2}{az^2 + bz + c}.$$

Notice that the poles of this function are found by solving $az^2 + bz + c = 0$, which we recognize as the auxiliary equation or characteristic equation from Chapter 14.

This function, the z transform of the impulse response function, is called the transfer function and is usually denoted by $H(z)$, so we have

$$H(z) = z^2/(az^2 + bz + c)$$

and $\mathcal{Z}^{-1}\{H(z)\} = h_n$, where h_n is the impulse response function.

Example 15.32 Find the transfer function and impulse response of the system described by the following difference equation:

$$3y_n + 4y_{n-1} = f_n$$

Solution To find the transfer function replace f_n by δ_n and take the z transform of the resulting equation assuming zero initial conditions:

$$4y_n + 3y_{n-1} = \delta_n.$$

Taking the z transform of both sides of the equation we get

$$4Y + 3(z^{-1}Y + y_{-1}) = 1.$$

As $y_{-1} = 0$,

$$Y = \frac{z}{4z + 3} = H(z) = \frac{z/4}{z + (3/4)}.$$

To find the impulse response sequence we take the inverse transform of the transfer function to find

$$h_n = \mathcal{Z}^{-1}\left\{ \frac{z/4}{z + \frac{3}{4}} \right\} = \frac{1}{4}\left(-\frac{3}{4} \right)^n$$

We can now see why, in the design of digital filters, this is referred to as an infinite impulse response (IIR) filter. This impulse response has a non-zero value for all n . Hence it represents an IIR system.

We now discover that we can find the system response to any input sequence f_n, with zero initial conditions, by using the transfer function.

Response of a system with zero initial conditions to any input f_n

Assuming all initial conditions are zero, that is, $y_{-1} = 0$ and $y_{-2} = 0$, then the equation

$$ay_n + by_{n-1} + cy_{n-2} = f_n$$

transforms to become

$$aY(z) + bz^{-1}Y(z) + cz^{-2}Y(z) = F(z) \quad \Leftrightarrow \quad y(z) = \frac{F(z)z^2}{az^2 + bz + c}.$$

We have just discovered that the transfer function, $H(z)$, is given by

$$H(z) = \frac{z^2}{az^2 + bz + c}.$$

Then we find that

$$Y(z) = F(z)H(z).$$

In order to find the response of the system to the function f_n we take the inverse transform of this expression. We are able to use the convolution property of z transforms, which states:

$$\mathcal{Z}\left\{ \sum_{k=0}^{n} f_k g_{n-k} \right\} = F(z)G(z)$$

or equivalently

$$\mathcal{Z}^{-1}\{F(z)G(z)\} = \sum_{k=0}^{n} f_k g_{n-k}.$$

So we find that the response of the system with zero initial conditions to any input sequence f_n is given by the convolution of f_n with the system's impulse response:

$$y_n = \mathcal{Z}^{-1}\{F(z)H(z)\} = f_n * h_n.$$

We can use this result to solve two types of problems, given zero initial conditions. If we know the impulse response function of the system then we can find the system response to any input f_n either by convolving the two sequences

$$y_n = \sum_{k=0}^{n} f_k h_{n-k} = f_n * h_n$$

or by finding the z transform of f_n, $F(z)$, finding the transfer function of the system $\mathcal{Z}\{h_n\} = H(z)$, then finding $Y(z) = H(z)F(z)$ and taking the inverse transform of this to find y_n. This type of problem is called a convolution problem. The other type of problem is that given any output, y_n, and given the input, f_n, we can deduce the impulse response of the system. This we do by finding $F(z) = \mathcal{Z}\{f_n\}$, $Y(z) = \mathcal{Z}\{y_n\}$ and then $H(z) = Y(z)/F(z)$, thus giving the transfer function. To find the impulse response we find $h_n = \mathcal{Z}^{-1}\{H(z)\}$.

Example 15.33   The impulse response of a system is known to be $h_n = 3(0.5)^n$. Find the response of the system to an input of $f(t) = 2u_n$ given zero initial conditions.

Solution   Method 1. We can take z transforms and use

$$Y(z) = H(z)F(z)$$

In this case

$$h_n = 3(0.5)^n \quad \Leftrightarrow \quad H(z) = \mathcal{Z}\{3(0.5)^n\} = \frac{3z}{z - 0.5}$$

$$f(t) = 2u_n \quad \Leftrightarrow \quad F(z) = \mathcal{Z}\{2u_n\} = \frac{2z}{z - 1}$$

Hence,

$$Y(z) = H(z)F(z) = \frac{6z^2}{(z - 0.5)(z - 1)}.$$

As we want to find y_n, we use partial fractions

$$\frac{6z}{(z - 0.5)(z - 1)} = \frac{-6}{z - 0.5} + \frac{12}{z - 1}$$

(using the 'cover up' rule). We can know take the inverse transform to find the system response

$$y_n = \mathcal{Z}^{-1}\{Y(z)\} = \mathcal{Z}^{-1}\left\{\frac{-6z}{z - 0.5} + \frac{12z}{z - 1}\right\}$$

$$= -6(0.5)^n + 12u_n.$$

Alternative method. Find y_n by taking the convolution of f_n with the impulse response function

$$y_n = f_n * h_n = (2u_n) * (3(0.5)^n).$$

By definition of convolution

$$y_n = \sum_{k=0}^{n} 2u_k\big(3(0.5)^{n-k}\big) = \sum_{k=0}^{n} 6(0.5)^{n-k}.$$

Writing out some of these terms we get

$$\begin{array}{ccc}
6, & 6(0.5 + 1), & 6((0.5)^2 + 0.5 + 1), \\
(n = 0) & (n = 1) & (n = 2)
\end{array}$$

$$6((0.5)^3 + 0.5^2 + 0.5 + 1).$$
$$(n = 3)$$

We see that each term in the sequence is a geometric progression with first term 6 and common ratio 0.5 and $n + 1$ terms. Hence

$$y_n = \frac{6(1 - 0.5)^{n+1}}{1 - 0.5} = 12(1 - (0.5)^{n+1}) = 12 - 12(0.5)(0.5)^n$$

$$= 12 - 6(0.5)^n$$

which confirms the result of the first method.

The frequency response

As in the case of Laplace transforms and continuous systems, we find we are able to establish a relationship between the transfer function and the steady state response to a single frequency input.

The steady state response of the system to a sequence representing a single frequency input, $e^{j\omega n}$, is found to be

$$Y(z) = H(e^{j\omega})e^{j\omega n}.$$

We can then see that for a single sinusoidal input the steady state response is found by substituting $z = e^{j\omega}$ into the transfer function for the system and multiplying the resulting complex constant by the sinusoidal input. In other words, the steady state response is a scaled and phase shifted version of the input. The function $H(e^{j\omega})$ is called the frequency response function for a discrete system.

Example 15.34   A system transfer function is known to be

$$H(z) = \frac{z}{z + 0.2}.$$

Find the steady state response to the following:

(a) $f_n = e^{j2n}$;
(b) $f(t) = \cos(2n)$.

Solution   (a) The steady state response to a single frequency $e^{j\omega n}$ is given by $H(e^{j\omega})e^{j\omega n}$. $f_n = e^{j2n}$, so in this case $\omega = 2$ and $H(z)$ is given as $z/(z + 0.2)$. Hence, we get the steady state response as

$$\frac{e^{j2}}{e^{j^2} + 0.2}e^{j2n}.$$

(b) Using $\frac{1}{2}H(e^{j\omega})e^{j\omega t} + H(e^{-j\omega})e^{-i\omega t}$ as the response to $\cos(\omega n)$ and substituting for H and $\omega = 2$ gives

$$y_n = \frac{1}{2}\left(\frac{e^{j2}e^{j2n}}{e^{j^2} + 0.2} + \frac{e^{-j2}}{e^{-j^2} + 0.2}e^{-j2n}\right)$$

Expressing this over a real denominator and simplifying gives

$$\frac{\cos(2n)(1 + 0.2\cos(2)) + 0.2\sin(2)\sin(2n)}{1.4 + 0.4\cos(2)}$$

Hence, the steady state response to an input of $3\cos(2n)$ is

$$\frac{3\cos(2n)(1 + 0.2\cos(2)) + 0.6\sin(2)\sin(2n)}{1.4 + 0.4\cos(2)}.$$

15.10 Summary

1.   The Laplace transform $F(s)$ of the function $f(t)$ defined for $t \geqslant 0$ is:

$$F(s) = \int_0^\infty e^{-st} f(t)\, dt.$$

The Laplace transform is a function of s, where s is a complex variable. Because the integral definition of the Laplace transform involves an integral to ∞, it is usually necessary to limit possible values of s so that the integral above converges (i.e. does not tend to ∞).

2. The impulse function, $\delta(t)$, also called a delta function, is the most famous example of a generalized function. The impulse function represents an idealized kick as it lasts for no time at all and has energy of exactly 1.

3. Laplace transforms are usually found by using a table of transforms (as in Table 15.1) and also by using properties of the transform, some of which are listed in Section 15.4.

4. Laplace transforms are used to reduce a differential equation to a simple equation in s-space. This can then be solved and the inverse transform used to find the solution to the differential equation.

5. The transfer function of the system, $H(s)$, is the Laplace transform of its impulse response function with zero initial conditions, $h(t)$. The Laplace transform of the response to any input function, with zero initial conditions, can be found by multiply the Laplace transform of the input function by the transfer function of the system $Y(s) = H(s)F(s)$.

6. The steady state response to a single frequency input $e^{j\omega t}$ is $H(j\omega)e^{j\omega t}$. $H(j\omega)$ is called the frequency response function.

7. The z transform of a sequence $f_0, f_1, f_2, \ldots, f_n, \ldots$ is given by

$$F(z) = \sum_{n=0}^{\infty} f_n z^{-n}.$$

As this is an infinite summation it will not always converge. The set of values of z for which it exists is called the region of convergence.

8. The discrete impulse function or delta function is defined by

$$\delta_n = \begin{cases} 1 & n = 0 \\ 0 & n \neq 0. \end{cases}$$

9. z transforms are usually found by using a table of transforms (as in Table 15.2) and also by using properties of the transform, some of which are listed in Section 15.7

10. z transforms are used to reduce a difference equation to a simple equation in z-space. This can then be solved and the inverse transform used to find the solution to the difference equation.

11. The transfer function of a discrete system, $H(z)$, is the z transform of its impulse response function with zero initial conditions, h_n. The z transform of the response to any input function, with zero initial conditions, can be found by multiply the z transform of the input function by the transfer function of the system, $Y(z) = H(z)F(z)$.

12. The steady state response of a discrete system to a single frequency input $e^{j\omega n}$ is $H(e^{j\omega})e^{j\omega n} \cdot H(e^{j\omega n})$ is called the frequency response function.

15.11 Exercises

15.1. Using the definition of the Laplace transform

$$\mathcal{L}\{f(t)\} = \int_0^\infty e^{-st} f(t)\, dt$$

show the following. In each case specify the values of s for which the transform exists.

(a) $\mathcal{L}\{2e^{4t}\} = \dfrac{2}{s-4}$

(b) $\mathcal{L}\{3e^{-2t}\} = \dfrac{3}{s+2}$

(c) $\mathcal{L}\{5t - 3\} = \dfrac{5}{s^2} - \dfrac{3}{s}$

(d) $\mathcal{L}\{3\cos(5t)\} = \dfrac{3s}{s^2 + 25}$

15.2. Find Laplace transforms of the following using Table 15.1:

(a) $5\sin(3t)$ (b) $\cos\left(\dfrac{t}{2}\right)$ (c) $\dfrac{t^4}{3}$ (d) $\dfrac{1}{2}e^{-5t}$

15.3. Find inverse Laplace transforms of the following using tables:

(a) $\dfrac{1}{s-4}$ (b) $\dfrac{3}{s+1}$ (c) $\dfrac{4}{s}$ (d) $\dfrac{2}{s^2+3}$

(e) $\dfrac{4s}{9s^2+4}$ (f) $\dfrac{1}{s^3}$ (g) $\dfrac{5}{s^4}$

15.4. Find Laplace transforms of the following using the properties of the transform:

(a) te^{-3t} (b) $4\sin(t)e^{-2t}$ (c) $t\sinh(4t)$

(d) $u(t-1)$ (e) $f(t) = \begin{cases} \sin(t-\pi) & t \geqslant \pi \\ 0 & t < \pi \end{cases}$

15.5. Find inverse Laplace transforms of the following using the properties of the transform:

(a) $\dfrac{1}{(s-2)^2+9}$ (b) $\dfrac{s+4}{(s+4)^2+1}$

(c) $\dfrac{e^{-2s}}{s+1}$ (d) $\dfrac{s-1}{s^2-2s+10}$

15.6. Find inverse Laplace transforms using partial fractions:

(a) $\dfrac{1}{s(s+1)}$ (b) $\dfrac{1}{s(s^2+2)}$

(c) $\dfrac{s}{(s^2-4)(s+1)}$ (d) $\dfrac{4}{(s^2+1)(s-3)}$

15.7. In each case solve the given differential equation using Laplace transforms:

(a) $y'+5y=0$, where $y(0)=3$
(b) $y'+y=t$, where $y(0)=0$
(c) $y''+4y=9t$, where $y(0)=0$, $y'(0)=7$
(d) $y''-3y+2y=4e^{2t}$, where $y(0)=-3$, $y'(0)=5$
(e) $y''+y=t$, where $y(0)=1$, $y'(0)=-2$
(f) $y''+y'+2y=4$, $y(0)=0$, $y'(0)=0$

15.8. A capacitor of capacitance C in an RC circuit, as in Figure 15.7, is charged so that initially its potential is V_0. At $t=0$, it begins to discharge. Its charge q is then described by the differential equation

$$R\frac{dq}{dt} + \frac{q}{C} = 0$$

Using Laplace transforms, find the charge on the capacitor at time t after the switch was closed.

Figure 15.7   *An RC circuit for Exercise 15.8.*

15.9. Find the response of a system with zero initial conditions to an input of $f(t) = 2e^{-3t}$, given that the impulse response of the system is $h(t) = 1/2(e^{-t} - e^{-2t})$.

15.10. (a) The impulse response of a system is given by $h(t) = 3e^{-4t}$. Find the system's step response, that is, the response of the system to an input of the step function, $u(t)$.
(b) Use the result that $Y(s) = H(s)F(s)$, where Y is the Laplace transform of the system output, $F(s)$ the Laplace transform of the input and $H(s)$ the system transfer function to show that the step response can be found by

$$y_u(t) = \mathcal{L}^{-1}\left\{\frac{H(s)}{s}\right\}$$

(c) Given that the step response of a system is

$$-\frac{4}{3}u(t) - e^{-2t} - \frac{2}{3}e^{-3t}$$

then
    (i) find the system's transfer function; and
    (ii) find its response to an input of e^{-t}.

15.11. A system has a known impulse response of $h(t) = e^{-t}\sin(2t)$. Find the input function $f(t)$ that would produce an output of

$$y(t) = -0.16u(t) + 0.4t + 0.16e^{-t}\cos(2t)$$
$$- 0.04\sin(2t)e^{-t}$$

given zero initial conditions.

15.12. A system has transfer function

$$H(s) = \frac{1}{(s+2)^2+4}$$

(a) Find its steady state response to a single frequency input of e^{j5t};
(b) Find the steady-state response to an input of $\cos(5t)$ and $\sin(5t)$.

15.13. Using the definition of the z transform

$$\mathcal{Z}\{f_n\} = \sum_{n=0}^{\infty} f_n z^{-n} = F(z)$$

show the following. In each case specify the values of z for which the transform exists:

(a) $\mathcal{Z}\{3\delta_n\} = 3$ (b) $\mathcal{Z}\{6u_n\} = \dfrac{6z}{z-1}$

(c) $\mathcal{Z}\{3^n\} = \dfrac{z}{z-3}$

15.14. Find z transforms of the following using Table 15.2:

(a) $2u_n + \frac{1}{2}n$ (b) $\cos(3n) + 2\sin(3n)$

(c) $4(0.2)^n - 6(2)^n$ (d) $2e^{j4n}$

15.15. Find inverse z transforms of the following using Table 15.2:

(a) $\dfrac{2z}{z-1} + 2$ (b) $\dfrac{z}{2(z-1)^2} + \dfrac{1}{1+3/z}$

(c) $\dfrac{2z}{2z-3} + \dfrac{4z}{2z+3}$ (d) $\dfrac{z^2 + z(\sin(1) - \cos(1))}{z^2 - 2z\cos(1) + 1}$

15.16. Find z transforms of the following using the properties of the transform:

(a) $u_{n+2}3^{n+2}$ (b) $\left(\frac{1}{2}\right)^n n$ (c) ne^{jn} (d) n^3

15.17. Find inverse z transforms of the following using the properties of the transform:

(a) $\dfrac{z}{(z-4)^2}$ (b) $\dfrac{z}{z - 2e^{j4}}$ (c) $\dfrac{z}{(z-0.4)^2}$

(d) $\dfrac{z}{z-1}\dfrac{2z}{z-2}$

15.18. Find inverse z transforms using partial fractions

(a) $\dfrac{z^2}{(z-1)(z+1)}$ (b) $\dfrac{z^3}{(z-1)(z^2-2)}$

(c) $\dfrac{z^2}{(z-0.1)(5z-2)}$ (d) $\dfrac{z^2}{(z-1)^2(z+2)}$

(e) $\dfrac{z^2}{z^2+1}$

15.19. In each case solve the given difference equation using z transforms, $n \geqslant 0$

(a) $y_n + 5y_{n-1} = 0$, where $y_{-1} = 3$;

(b) $y_n + y_{n-1} = n$, where $y_{-1} = 0$;

(c) $y_n + 4y_{n-1} = 9$, where $y_{-1} = 1$;

(d) $y_n - 3y_{n-1} + 2y_{n-2} = 4 \times 2^n$, where $y_{-1} = -3, y_{-2} = 5$;

(e) $y_n + y_{n-1} = n$, where $y_{-1} = 0, y_{-2} = 0$

(f) $10y_n - 3y_{n-1} - y_{n-2} = 4, y_{-1} = -1, y_{-2} = 2$.

15.20. Find the response of a system with zero initial conditions, to an input of $f_n = 2(0.3)^n$, given that the impulse response of the system is $h_n = (0.1)^n + (-0.5)^n$.

15.21. (a) The impulse response of a discrete system is given by $h_n = (0.8)^n$. Find the system's step response, that is, the response of the system to an input of the step function, u_n.

(b) Use the result that $Y(z) = H(z)F(z)$, where Y is the z transform of the system output, $F(z)$ is the z transform of the input and $H(z)$ is the system transfer function to show that the step response can be found by

$$y_n = Z^{-1}\left\{\frac{H(z)z}{z-1}\right\}$$

(c) Given that the step response of a discrete system is

$$\frac{1}{24}(0.2)^n + \frac{10}{48}u_n$$

then
  (i) find the system's transfer functions; and
  (ii) find its response to an input of $6(0.5)^n$.

15.22. A system has a known impulse response of $h_n = (0.5)^n$. Find the input function f_n that would produce an output of $2(0.5)^n + 2n - 2u_n$ given zero initial conditions.

15.23. A system has transfer function $H(z) = z/(10z - 3)$
  (i) find its steady state response to a single frequency input of e^{j5n};
  (ii) find the steady state response to an input of $\cos(5n)$ and $\sin(5n)$.

16 Fourier series

16.1 Introduction

Fourier analysis is the theory behind frequency analysis of signals. This chapter is concerned with the Fourier analysis of periodic, piecewise continuous functions. A periodic function can be represented by a Fourier series. A non-periodic function can be represented by its Fourier transform which we shall not be concerned with here. Discrete functions may be represented by a discrete Fourier transform, which also we shall not look at in this book.

Any periodic signal is made up of the sum of single frequency components. These components consist of a fundamental frequency component, multiples of the fundamental frequency, called the harmonics and a bias term, which represents the average off-set from zero. There are three ways of representing this information which are equivalent. We can represent the frequency components as the sum of a sine and cosine terms, or by considering the amplitude and phase of each component, or we can represent them using a complex Fourier series. The use of the complex Fourier series simplifies the calculation.

Having found the Fourier components we can use the system's frequency response function, as found in the previous chapter, to find the steady state response to any periodic signal.

16.2 Periodic Functions

In Chapter 5 we discussed the property of periodicity of the trigonometric functions. A periodic function is one whose graph can be translated to the right or left by an amount, called the period, such that the new graph fits exactly on top of the original graph. The fundamental period, also called the cycle, is the minimum non-zero amount the graph needs to be shifted in order to fit over the original graph.

A periodic function, with period τ, satisfies $f(t + \tau) = f(t)$ for all values of t. Examples of periodic functions are given in Figure 16.1.

The fundamental frequency of a periodic function is the number of cycles in an interval of unit length, $f = 1/\tau$. The fundamental angular frequency is then given by $\omega_0 = 2\pi f = 2\pi/\tau$. A periodic function need only be defined in one cycle, as the periodicity property will then define it everywhere. For example, the graph of the periodic square wave

$$f(t) = \begin{cases} 1/2 & 0 < t < 1 \\ -1/2 & 1 < t < 2 \end{cases}$$

is drawn in Figure 16.2. First, we draw the section of the graph as given in the definition and then shift the section along by the period, in this case 2, and copy the section. By repeatedly shifting and copying in this way, both to the left and right, we get the graph as shown.

As we mentioned in the introduction there are three ways of expressing the Fourier series. As a sine and cosine series or in amplitude and phase form or in complex form.

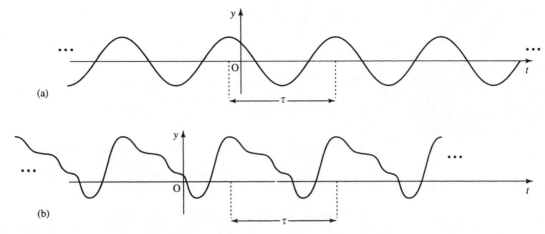

Figure 16.1 *Some periodic functions with their fundamental periods marked.*

Figure 16.2 *The periodic square wave defined by:*

$$f(t) = \begin{cases} 1/2 & 0 < t < 1 \\ -1/2 & 1 < t < 2 \end{cases}$$

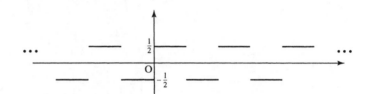

16.3 Sine and cosine series

If $f(t)$ is periodic with period $\tau = 2\pi/\omega_0$, then the Fourier series for f is given by:

$$f(t) = \frac{1}{2}a_0 + \sum_{n=1}^{\infty} a_n \cos(n\omega_0 t) + b_n \sin(n\omega_0 t)$$

where the coefficients are given by

$$a_0 = \frac{2}{\tau} \int_{-\tau/2}^{\tau/2} f(t)\, \mathrm{d}t$$

$$a_n = \frac{2}{\tau} \int_{-\tau/2}^{\tau/2} f(t) \cos(n\omega_0 t)\, \mathrm{d}t$$

$$b_n = \frac{2}{\tau} \int_{-\tau/2}^{\tau/2} f(t) \sin(n\omega_0 t)\, \mathrm{d}t$$

The steps for finding the Fourier series are:

Step 1: Plot the periodic function $f(t)$.
Step 2: Determine its fundamental period τ and its fundamental angular frequency $\omega_0 = 2\pi/\tau$.
Step 3: Evaluate a_0, a_n, and b_n as given above.
Step 4: Write down the resulting Fourier series.

Example 16.1 Find the Fourier series for

$$f(t) = \begin{cases} 1/2 & 0 < t < 1 \\ -1/2 & 1 < t < 2 \end{cases}$$

Solution

Step 1

We have already plotted the graph as shown in Figure 16.2.

Step 2

The fundamental period of this is $\tau = 2$, so that $\omega_0 = 2\pi/2 = \pi$.

Step 3

Calculate a_0, a_n, and b_n. We find

$$a_0 = \frac{2}{\tau} \int_{-\tau/2}^{\tau/2} f(t) \, dt = \frac{2}{2} \int_{-1}^{1} f(t) \, dt$$

$$= \int_{-1}^{0} \frac{1}{2} \, dt + \int_{0}^{1} \frac{1}{2} \, dt$$

$$= \left[-\frac{1}{2}t \right]_{-1}^{0} + \left[\frac{1}{2}t \right]_{0}^{-1} = -\frac{1}{2} + \frac{1}{2} = 0$$

$$a_n = \frac{2}{\tau} \int_{-\tau/2}^{\tau/2} f(t) \cos(n\omega_0 t) \, dt = \frac{2}{2} \int_{-1}^{1} f(t) \cos(n\pi t) \, dt$$

$$= \int_{-1}^{0} -\frac{1}{2} \cos(n\pi t) \, dt + \int_{0}^{1} \frac{1}{2} \cos(n\pi t) \, dt$$

$$= \left[-\frac{1}{2} \frac{\sin(n\pi t)}{n\pi} \right]_{-1}^{0} + \left[\frac{1}{2} \frac{\sin(n\pi t)}{n\pi} \right]_{0}^{1}$$

$$= -\frac{1}{2} \frac{\sin(0)}{n\pi} + \frac{\sin(-n\pi)}{n\pi} + \frac{1}{2} \frac{\sin(n\pi)}{n\pi} - \frac{1}{2} \frac{\sin(0)}{n\pi}$$

$$= -0 + 0 + 0 - 0 = 0$$

$$b_n = \frac{2}{\tau} \int_{-\tau/2}^{\tau/2} f(t) \sin(n\omega_0 t) \, dt$$

$$= \frac{2}{2} \int_{-1}^{1} f(t) \sin(n\pi t) \, dt$$

$$= \int_{-1}^{0} -\frac{1}{2} \sin(n\pi t) \, dt + \int_{0}^{1} \frac{1}{2} \sin(n\pi t) \, dt$$

$$= \left[\frac{1}{2} \frac{\cos(n\pi t)}{n\pi} \right]_{-1}^{0} + \left[-\frac{1}{2} \frac{\cos(n\pi t)}{n\pi} \right]_{0}^{1}$$

$$= \frac{1}{2n\pi} (\cos(0) - \cos(-n\pi)) + \frac{1}{2\pi n} (-\cos(n\pi) + \cos(0))$$

$$= \frac{1}{2n\pi} (1 - (-1)^n) + \frac{1}{2n\pi} (-(-1)^n + 1)$$

$$= \frac{1}{n\pi} (1 - (-1)^n)$$

since $\cos(n\pi) = (-1)^n$.

Step 4

The Fourier series for $f(t)$ is

$$f(t) = \frac{1}{2}a_0 + \sum_{n=1}^{\infty} a_n \cos(n\omega_0 t) + b_n \sin(n\omega_0 t)$$

in this case giving

$$f(t) = \frac{1}{\pi} \sum_{n=1}^{\infty} \frac{1-(-1)^n}{n} \sin(n\pi t)$$

Note that the even values of n all give zero coefficients as $1-(-1)^n = 0$ for n even. Odd values give $2/(n\pi)$. In this case, we can change the variable for the summation, using $n = 2m - 1$, which is always odd. This gives

$$f(t) = \frac{1}{\pi} \sum_{m=1}^{\infty} \frac{2}{2m-1} \sin((2m-1)\pi t)$$

It is interesting to plot graphs of the first few partial sums that we obtain from this series. In Figure 16.3 we have plotted the graph given by the terms up to $n = 3, n = 5$, and $n = 7$:

$$S_3 = \frac{2}{\pi} \sin(\pi t) + \frac{2}{3\pi} \sin(3\pi t)$$

$$S_5 = \frac{2}{\pi} \sin(\pi t) + \frac{2}{3\pi} \sin(3\pi t) + \frac{2}{5\pi} \sin(5\pi t)$$

$$S_7 = \frac{2}{\pi} \sin(\pi t) + \frac{2}{3\pi} \sin(3\pi t) + \frac{2}{5\pi} \sin(5\pi t) + \frac{2}{7\pi} \sin(7\pi t).$$

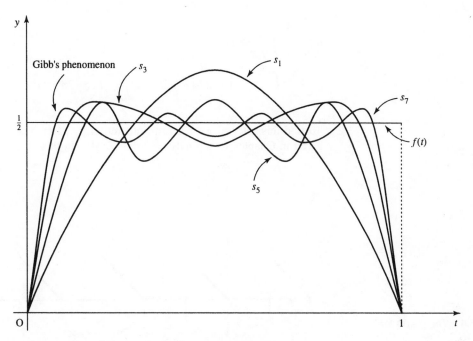

Figure 16.3 *Partial sums of the Fourier series for the square wave.*

You will notice that there is an overshoot of the value near the jump discontinuities. This is an example of Gibb's phenomenon. However many terms we take in the partial sum this overshoot remains significant, at about 10% of the function value. It is also interesting to see the value that the Fourier series takes at the discontinuous points, for example, $t = 1$. Substituting $t = 1$ into S_3, S_5, and S_7 gives 0. This is half way between the values of f at either side of the point $t = 1$.

Example 16.2   Find the Fourier series for the periodic function, defined in the interval $0 < t < 1$ by

$$f(t) = \begin{cases} t & 0 < t < 1/2 \\ 0 & 1/2 < t < 1 \end{cases}$$

Solution

Step 1

We plot the graph as shown in Figure 16.4.

Step 2

The fundamental period of this is $\tau = 1$ so that $\omega_0 = 2\pi/1 = 2\pi$.

Step 3

Calculate a_0, a_n, and b_n. We find

$$a_0 = \frac{2}{1} \int_{-1/2}^{1/2} f(t) \, dt = 2 \int_0^{1/2} t \, dt = [t^2]_0^{1/2} = (1/4) - 0 = 1/4$$

$$a_n = \frac{2}{\tau} \int_{-\tau/2}^{\tau/2} f(t) \cos(\omega_0 n t) \, dt$$

$$= 2 \int_{-\tau/2}^{\tau/2} f(t) \cos(2\pi n t) \, dt = 2 \int_0^{1/2} t \cos(2\pi n t) \, dt.$$

To find this integral we perform integration by parts. The formula, as given in Chapter 7, is $\int u \, dv = uv - \int v \, du$. In this case, we choose

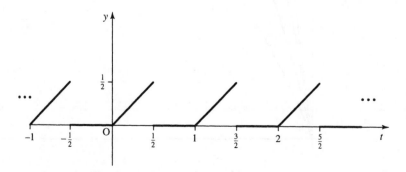

Figure 16.4 *The graph for Example 16.2.*

$u = t, dv = \cos(2\pi nt)\,dt, du = dt$ and $v = \sin(2\pi nt)/(2\pi n)$

$$a_n = 2\left(\left[t\frac{\sin(2\pi nt)}{2\pi n}\right]_0^{1/2} - \int_0^{1/2}\frac{\sin(2\pi nt)}{2\pi n}\,dt\right)$$

as $\sin(\pi n) = 0$ for all n, we get

$$a_n = 2\left(0 - \left[-\frac{\cos(2\pi nt)}{4\pi^2 n^2}\right]_0^{1/2}\right) = 2\left(\frac{\cos(\pi n)}{4\pi^2 n^2} - \frac{1}{4\pi^2 n^2}\right)$$

$\cos(\pi n)$ is -1 when n is odd and 1 when n is even. This means that $\cos(\pi n) = (-1)^n$. Then

$$a_n = 2\frac{((-1)^n - 1)}{4\pi^2 n^2} = \frac{(-1)^n - 1}{2\pi^2 n^2}$$

Next:

$$b_n = \frac{2}{\tau}\int_{-\tau/2}^{\tau/2} f(t)\sin(2\pi nt)\,dt = 2\int_0^{1/2} t\sin(2\pi nt)dt$$

Integrating by parts, using the formula $\int u\,dv = uv - \int v\,du$ and choosing $u = t, dv = \sin(2\pi nt)\,dt$
$\quad du = dt, v = -\cos(2\pi nt)/(2\pi n)$, we obtain

$$b_n = 2\left(\left[t\left(-\frac{\cos(2\pi nt)}{2\pi n}\right)\right]_0^{1/2} + \int_0^{1/2}\frac{\cos(2\pi nt)}{2\pi n}\,dt\right)$$

$$= 2\left(-\frac{\cos(\pi n)}{4\pi n} + \left[\frac{\sin(2\pi nt)}{4\pi^2 n^2}\right]_0^{1/2}\right)$$

$$= 2\left(-\frac{(-1)^n}{4\pi n}\right) = -\frac{(-1)^n}{2\pi n}$$

Step 4

The Fourier series for $f(t)$ is

$$f(t) = \frac{1}{2}a_0 + \sum_{n=1}^{\infty} a_n\cos(n\omega_0 t) + b_n\sin(n\omega_0 t)$$

in this case, giving

$$f(t) = \frac{1}{8} + \sum_{n=1}^{\infty}\frac{(-1)^n - 1}{2\pi^2 n^2}\cos(2\pi nt) + \sum_{n=1}^{\infty}\frac{(-1)^{n+1}}{2\pi n}\sin(2\pi nt)$$

The signal bias: the direct current (DC) component

The term $\frac{1}{2}a_0$ is called the bias term, or the DC component (a name adopted from electronic signals), as it corresponds to the average value of the function over a single period.

16.4 Fourier series of symmetric periodic functions

We looked at even functions and odd functions in Chapter 2.

Even functions

We find that even functions, which have the property that $f(-t) = f(t)$, have all $b_n = 0$ in their Fourier series. They are represented by cosine terms only. This is not surprising as the cosine is an even function and the sine function is odd. We would expect that an even function would be expressed in terms of other even functions. Another simplification in this case is

$$a_n = \frac{2}{\tau} \int_{-\tau/2}^{\tau/2} f(t) \cos(n\omega_0 t)\, dt$$

$$= \frac{2}{\tau} \int_{-\tau/2}^{0} f(t) \cos(n\omega_0 t)\, dt + \frac{2}{\tau} \int_{0}^{\tau/2} f(t) \cos(n\omega_0 t)\, dt$$

$$= \frac{4}{\tau} \int_{0}^{\tau/2} f(t) \cos(n\omega_0 t)\, dt$$

and it is therefore only necessary to integrate over a half cycle. To summarize, for an even function

$$a_0 = \frac{4}{\tau} \int_{0}^{\tau/2} f(t)\, dt$$

$$a_n = \frac{4}{\tau} \int_{0}^{\tau/2} f(t) \cos(n\omega_0 t)\, dt$$

$$b_n = 0 \quad \text{all } n$$

Odd functions

Odd functions, where $f(-t) = -f(t)$ have all $a_n = 0$ and only have sine terms in their Fourier series. We only need to consider the half cycle, because

$$b_n = \frac{2}{\tau} \int_{-\tau/2}^{\tau/2} f(t) \sin(n\omega_0 t)$$

$$= \frac{2}{\tau} \int_{-\tau/2}^{0} f(t) \sin(n\omega_0 t)\, dt + \frac{2}{\tau} \int_{0}^{\tau/2} f(t) \sin(n\omega_0 t)\, dt$$

$$= \frac{4}{\tau} \int_{0}^{1/2} f(t) \sin(n\omega_0 t)\, dt$$

To summarize, for an odd function

$$a_n = 0 \quad \text{all } n$$

$$b_n = \frac{4}{\tau} \int_{0}^{\tau/2} f(t) \sin(n\omega_0 t)\, dt.$$

Half-wave symmetry

There is another sort of symmetry that has an important effect on the Fourier series representation. This is called half-wave symmetry. A function with half-wave symmetry obeys $f(t + \frac{1}{2}\tau) = -f(t)$, that is, the graph of the function in the second half of the period is the same as the graph of the function in the first half turned upside down. A function with half-wave symmetry has no even harmonics. This can be shown by considering one of the even terms where $n = 2m$. Then

$$b_{2m} = \frac{2}{\tau} \int_{-\tau/2}^{\tau/2} f(t) \sin(2m\omega_0 t)\, dt$$

$$= \frac{2}{\tau} \int_{-\tau/2}^{0} f(t) \sin(2m\omega_0 t)\, dt + \frac{2}{\tau} \int_{0}^{\tau/2} f(t) \sin(2m\omega_0 t)\, dt$$

Substitute $t' = t - (1/2)\tau$ in the second term, so that $dt' = dt$, giving

$$\frac{2}{\tau} \int_{-\tau/2}^{0} f\left(t' + \frac{\tau}{2}\right) \sin\left(2m\omega_0 \left(t' + \frac{\tau}{2}\right)\right) dt'$$

As $\tau = 2\pi/\omega_0$

$$\sin\left(2m\omega_0 \left(t' + \frac{\tau}{2}\right)\right) = \sin\left(2m\omega_0 t' + 2m\frac{\omega_0 2\pi}{\omega_0 2}\right)$$

$$= \sin\left(2m\omega_0 t' + 2m\pi\right) = \sin\left(2m\omega_0 t'\right)$$

As $f(t' + \frac{1}{2}\tau) = -f(t')$, the second term in b_{2m} becomes

$$\frac{2}{\tau} \int_{-\tau/2}^{0} -f(t') \sin(2m\omega_0 t')\, dt'$$

which cancels the first term, giving $b_{2m} = 0$.

A similar argument shows that the coefficients of the cosine terms for even n are also zero. In this case also it is only necessary to consider the half-cycle, as

$$\frac{2}{\tau} \int_{-\tau/2}^{\tau/2} f(t) \sin(n\omega_0 t)\, dt = \frac{4}{\tau} \int_{0}^{\tau/2} f(t) \sin(n\omega_0 t)\, dt \quad n \text{ odd}$$

To summarize, for a function with half wave symmetry

$$a_n = \frac{4}{\tau} \int_{0}^{\tau/2} f(t) \cos(n\omega_0 t)\, dt \quad n \text{ odd}$$

$$b_n = \frac{4}{\tau} \int_{0}^{\tau/2} f(t) \sin(n\omega_0 t)\, dt \quad n \text{ odd}$$

$$a_n = b_n = 0 \quad n \text{ even}$$

An even function, an odd function, and a function with half-wave symmetry are shown in Figure 16.5.

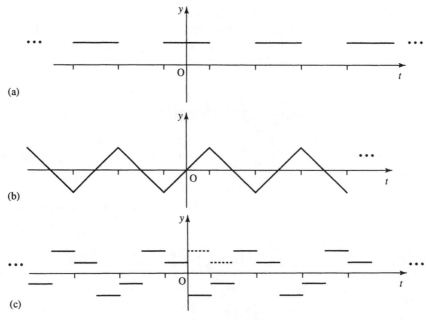

Figure 16.5   *(a) An even function satisfies $f(-t) = f(t)$, that is reflecting the graph in the y-axis results in the same graph. This function has only cosine terms in its Fourier series. (b) An odd function satisfies $f(-t) = -f(t)$, that is reflecting the graph in the y-axis results in an upside down version of the same graph. This function has only sine terms in its Fourier series. (c) A function with half-wave symmetry satisfies $f(t + \tau/2) = -f(t)$ that is the graph of the function in the second half of the period is the same as the graph of the function in the first half reflected in the x-axis. This function has no even harmonics.*

16.5 Amplitude and phase representation of a Fourier series

In Chapter 9, when considering using vectors to represent single frequency waves, we saw that terms like $c \cos(\omega t) - d \sin(\omega t)$ can be represented by a single cosine term $A \cos(\omega t + \phi)$ where A is the amplitude and ϕ is the phase. The terms A and ϕ can be found by expressing the vector (c, d) in polar form. We can employ this idea to represent the Fourier series in amplitude and phase form. This can be very useful because, for instance, a filter may be designed to attenuate frequencies outside of the desired pass band. this requirement specifies its amplitude characteristics. The phase characteristics may then be considered separately.

The Fourier series becomes

$$f(t) = \frac{1}{2}a_0 + \sum_{n=1}^{\infty} c_n \cos(n\omega_0 t + \phi_n)$$

where

$$c_n \cos(n\omega_0 t + \phi_n) = a_n \cos(n\omega_0 t) + b_n \sin(n\omega_0 t)$$

From the trigonometric identity for $\cos(A + B)$, we can expand the left-hand side of the above expression to get

$$c_n \cos(n\omega_0) \cos(\phi_n) - c_n \sin(n\omega_0) \sin(\phi_n) = a_n \cos(n\omega_0 t) + b_n \sin(n\omega_0 t)$$

Equating terms in $\cos(n\omega_0 t)$ and $\sin(n\omega_0 t)$, we get

$$c_n \cos(\phi_n) = a_n$$
$$-c_n \sin(\phi_n) = b_n$$

giving

$$c_n = \sqrt{a_n^2 + b_n^2} \quad \text{and} \quad \phi_n = -\tan^{-1}\left(\frac{b_n}{a_n}\right) \quad (\pm\pi \text{ if } a_n \text{ is negative})$$

We see that $c_n \angle \phi_n$ can be found by expressing $(a_n, -b_n)$ in polar form.

The amplitudes can be plotted against the frequency f (or angular frequency ω) giving the amplitude spectrum and the phases can be plotted, giving the phase spectrum. The amplitude gives information about the distribution of energy among the different frequencies.

For the unbiased square wave which we considered in Example 16.1 we found

$$f(t) = \frac{1}{\pi} \sum_{n=1}^{\infty} \frac{1-(-1)^n}{n} \sin(n\pi t) = \frac{1}{\pi} \sum_{m=1}^{\infty} \frac{2}{2m-1} \sin((2m-1)\pi t)$$

$$a_m = 0$$

$$b_m = \frac{2}{(2m-1)\pi}$$

So

$$c_m = \sqrt{a_m^2 + b_m^2} = \frac{2}{\pi(2m-1)}$$

as there is only a sine term then the phase is given by $-\pi/2$. So we get

$$f(t) = \sum_{m=1}^{\infty} \frac{2}{\pi(2m-1)} \cos\left((2m-1)\pi t - \frac{\pi}{2}\right)$$

where $n = 2m - 1$.

On plotting these amplitudes and phases we get Figure 16.6. The phases are not defined for frequencies with zero amplitude but we can consider them as 0.

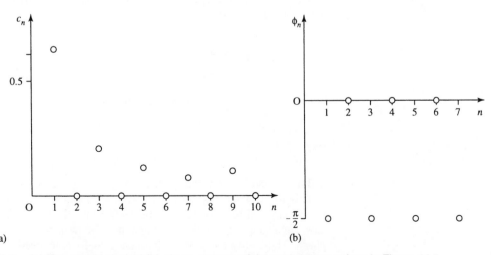

Figure 16.6 *(a) The amplitude and (b) phase spectra of the square wave given in Figure 16.1.*

16.6 Fourier series in complex form

The most useful form of the Fourier series is the amplitude and phase form of the previous section. c_n and ϕ_n can often be calculated more simply by considering the complex form of the Fourier series. We can find this by taking the expression from the previous section:

$$f(t) = \frac{1}{2}a_0 + \sum_{n=1}^{\infty} c_n \cos(n\omega_0 t + \phi_n)$$

and expressing the cosine terms as the sums of complex exponentials:

$$c_n \cos(n\omega_0 t + \phi_n) = \frac{c_n}{2}(e^{j(n\omega_0 t + \phi_n)} + e^{-j(n\omega_0 t + \phi_n)})$$

$$= \frac{c_n}{2}e^{j\phi_n}\,e^{jn\omega_0 t} + \frac{c_n}{2}e^{-j\phi_n}\,e^{-jn\omega_0 t}$$

Then setting

$$\alpha_n = \frac{c_n}{2}e^{j\phi_n} \quad \text{and} \quad \alpha_{-n} = \frac{c_n}{2}e^{-j\phi_n}$$

we get

$$f(t) = \frac{1}{2}a_0 + \sum_{n=1}^{\infty} \alpha_n\,e^{jn\omega_0 t} + \alpha_{-n}\,e^{-jn\omega_0 t}$$

$$\Rightarrow f(t) = \sum_{n=-\infty}^{n=\infty} \alpha_n\,e^{jn\omega_0 t}$$

where $\alpha_0 = \frac{1}{2}a_0$

$$\alpha_n = \frac{c_n}{2}e^{j\phi n} \quad \text{and} \quad \alpha_{-n} = \frac{c_n}{2}e^{-j\phi_n}$$

as c_n is real $|\alpha_n| = |\alpha_{-n}| = \dfrac{c_n}{2}$, and $\alpha_n^* = \alpha_{-n}$.

Therefore, $\alpha_n\,(n > 0)$ is a complex coefficient of the Fourier series with amplitude $c_n/2$ and phase ϕ_n.

The complex form is generally the most convenient form of the Fourier series because in the expression

$$f(t) = \sum_{n=-\infty}^{\infty} \alpha_n\,e^{jn\omega_0 t}$$

the complex Fourier components can be found from

$$\alpha_n = \frac{1}{\tau}\int_{-\tau/2}^{\tau/2} f(t)\,e^{-jn\omega_0 t}\,\mathrm{d}t$$

and hence involve performing only a single integration.

This form of the Fourier series gives apparent negative frequencies but for any real function of time the coefficients of negative frequencies have equal amplitude to the equivalent positive frequencies and negative phase. Thus, only the positive frequency coefficients need be given to totally specify the function.

From the complex form, we can easily find the amplitude and phase spectra. If we write the coefficients in exponential form, $|\alpha_n|e^{j\theta}$, we find $c_n = 2|\alpha_n|$ and $\phi_n = \theta_n = \arg(\alpha_n), n \geqslant 0$.

We can also easily find the sine and cosine form of the Fourier series by using

$$a_n = 2|\alpha_n|\cos(\phi_n) = 2\,\mathrm{Re}(\alpha_n)$$
$$b_n = -2|\alpha_n|\sin(\phi_n) = -2\,\mathrm{Im}(\alpha_n).$$

Example 16.3 Find the complex Fourier series for a function defined in the interval $0 < t < 1$ by

$$f(t) = \begin{cases} t & 0 < t < 1/2 \\ 0 & 1/2 < t < 1 \end{cases}$$

Solution

Step 1

We have already plotted the graph as shown in Figure 16.4.

Step 2

The fundamental period of this is $\tau = 1$ so that $\omega_0 = 2\pi$.

Step 3

Calculate α_n. We find

$$\alpha_n = \frac{1}{\tau}\int_{-\tau/2}^{\tau/2} f(t)\,e^{-jn\omega_0 t}\,\mathrm{d}t = \int_0^{1/2} t\,e^{-j2\pi nt}\,\mathrm{d}t$$

To find this integral, we perform integration by parts using $\int u\,\mathrm{d}v = uv - \int v\,\mathrm{d}u$. In this case we choose $u = t$, $\mathrm{d}v = e^{-j2\pi nt}\,\mathrm{d}t$, $\mathrm{d}u = \mathrm{d}t$ and

$$v = \frac{e^{-j2\pi nt}}{-j2\pi n} = j\frac{e^{-j2\pi nt}}{2\pi n}$$

giving

$$\alpha_n = \int_0^{1/2} t\,e^{-j2\pi nt} = \left[j\frac{t\,e^{-j2\pi nt}}{2\pi n}\right]_0^{1/2} - \int_0^{1/2}\frac{j\,e^{-j2\pi nt}}{(2\pi n)}\,\mathrm{d}t$$

$$= j\frac{e^{-j\pi n}}{4\pi n} - \left[j\frac{e^{-j2\pi nt}}{(2\pi n)(-j2\pi n)}\right]_0^{1/2} = \frac{j\,e^{-j\pi n}}{4\pi n} + \frac{e^{-j\pi n} - 1}{4\pi^2 n^2}$$

Since $e^{-j\pi n} = (-1)^n$, this means

$$\alpha_n = \frac{(-1)^n - 1}{4\pi^2 n^2} + j\frac{(-1)^n}{4\pi n}$$

Step 4

The complex Fourier series for $f(t)$ is

$$f(t) = \sum_{n=-\infty}^{n=\infty} \alpha_n \, e^{jn\omega_0 t}$$

in this case giving

$$f(t) = \sum_{n=-\infty}^{\infty} \left(\frac{(-1)^n - 1}{4\pi^2 n^2} + j\frac{(-1)^n}{4\pi n} \right) e^{-jn\omega_0 t}$$

To find the trigonometric form from this we can use

$$a_n = 2\,\text{Re}(\alpha_n) = \frac{(-1)^n - 1}{2\pi^2 n^2}$$

$$b_n = -2\,\text{Im}(\alpha_n) = \frac{-(-1)^n}{2\pi_n}$$

which agrees with our previous result found in Example 16.2.

16.7 Summary

1. If $f(t)$ is periodic, with period τ, then $\omega_0 = 2\pi/\tau$ and the Fourier series for f is given by:

$$f(t) = \frac{1}{2}a_0 + \sum_{n=1}^{\infty} a_n \cos(n\omega_0 t) + b_n \sin(n\omega_0 t)$$

where the coefficients are given by

$$a_0 = \frac{2}{\tau} \int_{-\tau/2}^{\tau/2} f(t)\mathrm{d}t$$

$$a_n = \frac{2}{\tau} \int_{-\tau/2}^{\tau/2} f(t) \cos(n\omega_0 t)\mathrm{d}t$$

$$b_n = \frac{2}{\tau} \int_{-\tau/2}^{\tau/2} f(t) \sin(n\omega_0 t)\mathrm{d}t$$

This is the sine and cosine form (trigonometric form) of the Fourier series.

2. An even function, such that $f(-t) = f(t)$, has $b_n = 0$ and a_n can be found by integrating over the half cycle. An odd function, such that $f(-t) = -f(t)$, has $a_n = 0$ and the terms b_n can be found by integrating over the half cycle. A function with half-wave symmetry, $f(t + \frac{1}{2}\tau) = f(t)$ has no even terms in the expansion. The terms a_n and b_n can be found by integrating over a half-cycle only.

3. The amplitude and phase form of the Fourier series is

$$f(t) = \frac{1}{2}a_0 + \sum_{n=1}^{\infty} c_n \cos(n\omega_0 t + \phi_n)$$

$c_n \angle \phi_n$ can be found by expressing $(a_n, -b_n)$ (as defined above) in polar form: $a_n = c_n \cos(\phi_n), b_n = -c_n \sin(\phi_n)$.

4. The complex form of the Fourier series is given by

$$f(t) = \sum_{n=-\infty}^{n=\infty} \alpha_n \, e^{jn\omega_0 t}$$

where

$$\alpha_n = \frac{1}{\tau} \int_{-\tau/2}^{\tau/2} f(t) \, e^{-jn\omega_0 t} \, dt$$

The coefficients are related to a_n and b_n by

$$a_n = 2\,\mathrm{Re}(\alpha_n), \quad b_n = -2\,\mathrm{Im}(\alpha_n), \quad \alpha_n = \frac{1}{2}(a_n - b_n)$$

and to the amplitude and phase form by

$$c_n = 2|\alpha_n| \quad \text{and} \quad \phi_n = \arg(\alpha_n), \quad n \geqslant 0.$$

16.8 Exercises

For the following periodic functions in Exercises 16.1–16.7, find the Fourier series in trigonometric and complex form

$$f(t) = \frac{1}{2}a_0 + \sum_{n=1}^{\infty} a_n \cos(n\omega_0 t) + b_n \sin(n\omega_0 t)$$

$$f(t) = \sum_{n=-\infty}^{\infty} \alpha_n \, e^{jn\omega_0 t}$$

and check that the results are equivalent in each case.

16.1 $f(t) = \begin{cases} 1 & 0 < t < 1 \\ 0 & 1 < t < 4 \end{cases}$

16.2 $f(t) = \begin{cases} t^2 & 0 < t < 1 \\ 0 & 1 < t < 2 \end{cases}$

16.3 $f(t) = \cos^2(t)$

16.4 $f(t) = \begin{cases} t & 0 < t < 1 \\ 0 & 1 < t < 2 \end{cases}$

16.5 $f(t) = \begin{cases} t^2 & 0 < t < 1 \\ 1 - (t-1)^2 & 1 < t < 2 \end{cases}$

16.6 $f(t) = \begin{cases} \cos(2\pi t) & 0 < t < 1 \\ 0 & 1 < t < 2 \end{cases}$

16.7 $f(t) = \begin{cases} t - 0.5 & 0 < t < 1 \\ 1.5 - t & 1 < t < 2 \end{cases}$

16.8 Which of the periodic functions whose graphs are shown in Figure 16.7 are even, odd, or have half-wave symmetry. What consequences will such properties have for the Fourier series in trigonometric form?

16.9 Find the fundamental period and give the amplitude and phase spectra of the Fourier series representation of the following periodic functions

(a) $f(t) = \sin(3t) - \sin(t)$

(b) $g(t) = \frac{1}{2}\cos(2t) + 2 - \frac{1}{2}\sin(2t)$

(c) $f(t) = \sum_{n=-\infty}^{\infty} \frac{1}{n^2} e^{jnt} \quad n \neq 0$

16.10 Find and plot the amplitude and phase spectra of your results from Exercises 16.1, 16.3, and 16.7.

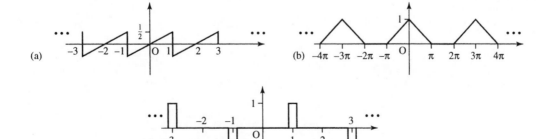

Figure 16.7 *Graphs for Exercise 16.8.*

16.11 Considering the functions in Exercises 16.1, 16.2, and 16.4 show that at the position of the jump discontinuities, $t = t_d$, the Fourier series converges to the value

$$\frac{f(t_d^+) + f(t_d^-)}{2}$$

where $f(t_d^+) = \lim_{t \to t_d^+} f(t)$ and $f(t_d^-) = \lim_{t \to t_d^-} f(t)$

You may assume that:

$$\sum_{n=1}^{\infty} \frac{(-1)^{n-1}}{2n-1} = \frac{\pi}{4}$$

$$\sum_{n=1}^{\infty} \frac{1}{n^2} = \frac{\pi^2}{6} \quad \text{and} \quad \sum_{n=1}^{\infty} \frac{(-1)^{n-1}}{n^2} = \frac{\pi^2}{12}$$

16.12 The steady state response to a signal $e^{j\omega t}$ is given by $H(j\omega)e^{j\omega t}$. Find the steady state response to the function $f(t) = \cos^2(t)$ from a system with transfer function $1/(s+2)$.

Part 3　Functions of more than one variable

17 Functions of more than one variable

17.1 Introduction

We concentrated so far in this book on functions of time. In Chapter 5 we looked at waves that were functions of distance, x, and time, t. For many applications, we need to analyse functions of the three possible spatial coordinates, x, y, and z and also of time, t. For instance, consider making a mug of tea using a tea bag. When first brewed, the temperature will be nearly at boiling point 100°C. The temperature of the water will start off being much the same everywhere but it will be cooler near the tea bag and at the sides of the mug. As time progresses the surface of the tea and those parts in touch with the mug, will cool quicker than the centre of the tea. We can see that the temperature of a mug of tea is a function both of where in the mug we measure it and also the length of time since the tea was first made. This is a function of more than one variable.

Some of the ideas that we have learnt for analysing a function of a single variable we revisit here for functions of more than one variable. In this chapter we consider graphs of functions and taking their partial derivatives, changing variables, using the chain rule and the derivative in any given direction. To begin we need to think about visualizing these functions and that means considering how to visualize graphs of functions that would require three or more dimensions when armed only with a two-dimensional sheet of paper.

17.2 Functions of two variables – surfaces

When drawing a graph we need one dimension to represent the actual function value, the dependent variable, for which we have used the vertical axis. When drawing a graph of distance, x or of time t, we are able to use the horizontal axis of the graph to represent the independent variable. Supposing now we have a function of two spatial dimensions – for instance, the height above sea level on a landscape or the temperature in a cup of tea at the surface of the tea. We would like to be able to picture these functions. However, we need more than one horizontal axis in order to be able to represent the spatial dimensions x and y when a picture on a page of this book has only one horizontal direction. We could build three-dimensional models only this would be time consuming and make book production even more costly. One of the ways of representing these graphs on a page of paper is to use a perspective representation of a surface. The other way of doing it is to draw a plan of the function from above and use contours to represent the function value at each point. You are familiar with this from maps that mark contours representing the height of the ground at each point. Another example is meteorological forecasts that use contours superimposed on maps of the country, to represent the

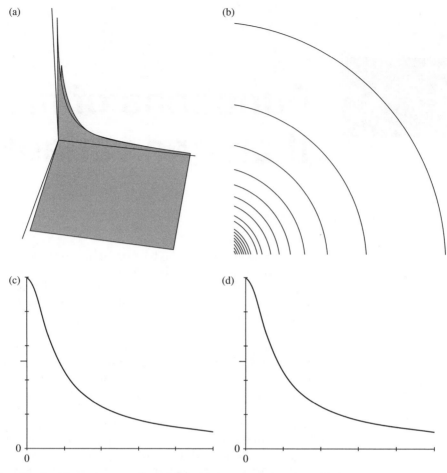

Figure 17.1 *Graphs of $\phi = 4\pi\varepsilon_0 q/(x^2 + y^2)^{1/2}$. (a) A perspective view. (b) Contour representation. (c) A cross-section parallel to the x-axis for a fixed value of y. (d) A cross-section parallel to the y-axis for a fixed value of x.*

air pressure. A third way to picture the functions is to take a cross section parallel to either one of the x- or y-axes.

Graphs of functions of two variables

The function:

$$\phi = \frac{4\pi\varepsilon_0 q}{(x^2 + y^2)^{1/2}}$$

represents the electrical potential field of a point source (where $z = 0$) positioned at the origin. Various graphs of this function are shown in Figure 17.1: (a) has a perspective view; (b) has a contour representation; and (c) and (d) show x and y cross-sections.

Graphs of the function $h = 4 - 2x^2 + y$ are shown in Figure 17.2.

17.3 Partial differentiation

From the graphs of functions that we looked at in the last section we can see that we are able to take a cross-section either parallel to the x-axis or parallel to the y-axis. Each cross-section can be represented by a function of one variable only. If we take a cross-section parallel to the x-axis, then

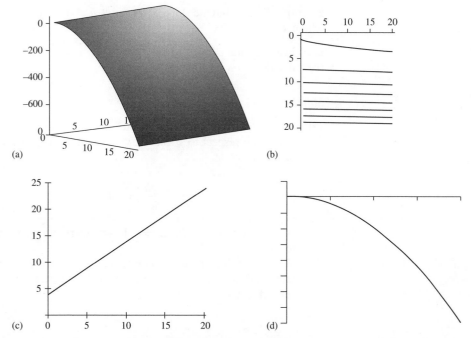

Figure 17.2 *Graphs of $h = 4 - 2x^2 + y$. (a) A perspective view. (b) Contour representation. (c) A cross-section parallel to the x-axis for a fixed value of y. (d) A cross-section parallel to the y-axis for a fixed value of x.*

we have fixed y and only x is varying. In this case, we can differentiate the function as a function of one variable.

It is this idea that is used for partial differentiation. We can find the derivative along a line parallel to the x-axis by 'freezing' y and differentiation with respect to x only. We can find the derivative along a line parallel to the y-axis by 'freezing' x and differentiation with respect to y only. Partial derivatives of the function $u = f(x, y)$ are defined by:

$$\frac{\partial u}{\partial x} = \lim_{\delta x \to 0} \frac{u(x + \delta x, y) - u(x, y)}{\delta x}$$

$$\frac{\partial u}{\partial y} = \lim_{\delta y \to 0} \frac{u(x, y + \delta y) - u(x, y)}{\delta y}$$

$\partial u / \partial x$ reads as 'partial du by dx' and $\partial u / \partial y$ reads as 'partial du by dy'.

Example 17.1

$u = x^3 - 3x^2 y + y$.

Find its partial derivatives $\partial u / \partial x$ and $\partial u / \partial y$.

Solution   To find $\partial u / \partial x$, we differentiate, u with respect to x and treat y as though it were a constant

$$\frac{\partial u}{\partial x} = 3x^2 - 3(2)xy = 3x^2 - 6xy.$$

To find $\partial u / \partial y$, we differentiate u with respect to y and treat x as though it were a constant

$$\frac{\partial u}{\partial y} = -3x^2 + 1.$$

We can substitute particular values for x and y in order to find the slope of the graph parallel to the x- and parallel to the y-axes.

Example 17.2 Find the slope of the curve $h = 4 - 2x^2 + y$ at the point where $x = 3$, $y = 2$

(a) in a direction parallel to the x-axis,
(b) in a direction parallel to the y-axis.

Solution

(a) To find the slope parallel to the x-axis, we freeze y and differentiating with respect to x, that is, we need the partial derivative with respect to x

$$h = 4 - 2x^2 + y \quad \Rightarrow \quad \frac{\partial h}{\partial x} = -4x.$$

At the point $(3, 2)$, $\partial h / \partial x = -12$ and therefore the slope parallel to the x-axis is -12.

(b) To find the slope parallel to the y-axis, we freeze x and differentiating with respect to y, that is, we need the partial derivative with respect to y.

$$h = 4 - 2x^2 + y \Rightarrow \frac{\partial u}{\partial y} = 1.$$

At the point $(3, 2)$, $\partial u / \partial y = 1$ and therefore the slope parallel to the y-axis is 1.

17.4 Changing variables – the chain rule

Supposing we have a function, $u = f(x, y)$, and x and y are functions of s and t so that $x = v(s, t)$, $y = w(s, t)$, and $u = F(s, t)$. We want to find a relationship between the partial derivatives of u with respect to s and t and those with respect to x and y. We assume in the following discussion that we have well-behaved functions, that is, the functions and their partial derivatives do not have discontinuities in the region of x, y.

We want to find the partial derivative of u with respect to s, that is, we would like to find

$$\frac{\partial u}{\partial s} = \lim_{\delta s \to 0} \frac{u(s + \delta s, t) - u(s, t)}{\delta s}.$$

As we are considering a change of variable so that the point (x, y) corresponds to the point (s, t) and therefore $u(x, y) = u(s, t)$. We consider a small change, δs, in the variable s so that the point $(s + \delta s, t)$ will correspond to some point $(x + \delta x, y + \delta y)$ in the (x, y) plane, where δx and δy are small. Again $u(s + \delta s, t)$ will have the same value as $u(x + \delta x, y + \delta y)$. Using $u(s + \delta s, t) = u(x + \delta y, y + \delta y) = u(x + \delta x, y + \delta y) - u(x, x + \delta y) + u(x, y + \delta y)$ and $u(x, y) = u(s, t)$ we find:

$$\frac{\partial u}{\partial s} = \lim_{\delta s \to 0} \frac{u(x + \delta x, y + \delta y) - u(x, y + \delta y) + u(x, y + \delta y) - u(x, y)}{\delta s}$$

$$= \lim_{\delta s \to 0} \frac{u(x + \delta x, y + \delta y) - u(x, y + \delta y)}{\delta s}$$

$$+ \lim_{\delta s \to 0} \frac{u(x, y + \delta y) - u(x, y)}{\delta s}.$$

As long as $\delta x, \delta y$ are not actually equal to 0 we can multiply the first term by $\delta x / \delta x$ and the second term by $\delta y / \delta y$. As $\delta s \to 0$, so also $\delta x \to 0$ and $\delta y \to 0$, hence we get

$$\frac{\partial u}{\partial s} = \lim_{\delta x \to 0, \delta y \to 0} \frac{u(x + \delta x, y + \delta y) - u(x, y + \delta y)}{\delta x} \lim_{\delta s \to 0} \frac{\delta x}{\delta s}$$

$$+ \lim_{\delta x \to 0} \lim_{\delta y \to 0} \frac{u(x, y + \delta y) - u(x, y)}{\delta y} \lim_{\delta s \to 0} \frac{\delta y}{\delta s}.$$

In the first term, we can take the limit as $\delta y \to 0$ to get:

$$\frac{\partial u}{\partial s} = \lim_{\delta x \to 0} \frac{u(x + \delta x, y) - u(x, y)}{\delta x} \lim_{\delta s \to 0} \frac{\delta x}{\delta s}$$

$$+ \lim_{\delta y \to 0} \frac{u(x, y + \delta y) - u(x, y)}{\delta y} \lim_{\delta s \to 0} \frac{\delta y}{\delta s}.$$

Using our previous definitions of the partial derivative we take the limits to find:

$$\frac{\partial u}{\partial s} = \frac{\partial u}{\partial x} \frac{\partial x}{\partial s} + \frac{\partial u}{\partial y} \frac{\partial y}{\partial s}.$$

Similarly, we can show that

$$\frac{\partial u}{\partial t} = \frac{\partial u}{\partial x} \frac{\partial x}{\partial t} + \frac{\partial u}{\partial y} \frac{\partial y}{\partial t}.$$

Example 17.3   Given $h = 4 - 2x^2 + y$ and $x = r \cos(\theta), y = r \sin(\theta)$, find $\partial h / \partial r$ and $\partial h / \partial \theta$.

Solution   We can find the following partial derivatives:

$$\frac{\partial h}{\partial x} = -4x \quad \frac{\partial h}{\partial y} = 1$$

$$\frac{\partial x}{\partial r} = \cos(\theta) \quad \frac{\partial y}{\partial r} = \sin(\theta)$$

$$\frac{\partial x}{\partial \theta} = -r \sin(\theta) \quad \frac{\partial y}{\partial \theta} = r \cos(\theta).$$

To find $\partial h / \partial r$ we use the chain rule:

$$\frac{\partial h}{\partial r} = \frac{\partial h}{\partial x} \frac{\partial x}{\partial r} + \frac{\partial h}{\partial y} \frac{\partial y}{\partial r}$$

$$\frac{\partial h}{\partial r} = -4x \cos(\theta) + \sin(\theta)$$

and substituting for x and y gives

$$\frac{\partial h}{\partial r} = -4r \cos^2(\theta) + \sin(\theta)$$

Similarly, to find $\partial h / \partial \theta$ we use the chain rule:

$$\frac{\partial h}{\partial \theta} = \frac{\partial h}{\partial x}\frac{\partial x}{\partial \theta} + \frac{\partial h}{\partial y}\frac{\partial y}{\partial \theta}$$

$$\frac{\partial h}{\partial \theta} = 4xr\sin(\theta) + r\cos(\theta)$$

and substituting for x and y gives

$$\frac{\partial h}{\partial \theta} = 4r^2\cos(\theta)\sin(\theta) + r\cos(\theta).$$

Check: We should get the same result by substituting for x and y in h and then directly finding $\partial h / \partial r$ and $\partial h / \partial \theta$.

Substituting $x = r\cos(\theta)$ and $y = r\sin(\theta)$ into $h = 4 - 2x^2 + y$ gives

$$h = 4 - 2r^2\cos^2(\theta) + r\sin(\theta).$$

Differentiating h with respect to r and keeping θ constant gives:

$$\frac{\partial h}{\partial r} = -4r\cos^2(\theta) + \sin(\theta).$$

Differentiating h with respect to θ and keeping r constant gives:

$$\frac{\partial h}{\partial \theta} = 2r^2 2\cos(\theta)\sin(\theta) + r\cos(\theta) = 4r^2\cos(\theta)\sin(\theta) + r\cos(\theta)$$

and we see that these are the same results as before.

17.5 The total derivative along a path

We have found the derivatives in the x- and y-directions for some function $u = f(x, y)$. We would now like to consider the derivative along any given path, where for instance $x = g(t)$ and $y = h(t)$, that is, both x and y are a function of t. Consider a small change in u, δu, in the direction of the path, with a change in t of δt. Then we can define the total derivative along the path as:

$$\frac{du}{dt} = \lim_{\delta t \to 0} \frac{u(x + \delta x, y + \delta y) - u(x, y)}{\delta t}.$$

Substituting $u(x + \delta y, y + \delta y) = u(x + \delta x, y + \delta y) - u(x, x + \delta y) + u(x, y + \delta y)$ and multiplying the first term by $\delta x / \delta x$ and the second term by $\delta y / \delta y$ gives:

$$\frac{du}{dt} = \lim_{\delta x \to 0, \delta y \to 0} \frac{u(x + \delta x, y + \delta y) - u(x, y + \delta y)}{\delta x} \lim_{\delta t \to 0} \frac{\delta x}{\delta t}$$
$$+ \lim_{\delta x \to 0, \delta y \to 0} \frac{u(x, y + \delta y) - u(x, y)}{\delta y} \lim_{\delta t \to 0} \frac{\delta y}{\delta t}.$$

As x and y are functions of t only,

$$\lim_{\delta t \to 0} \frac{\delta x}{\delta t} = \frac{dx}{dt} \quad \text{and} \quad \lim_{\delta t \to 0} \frac{\delta y}{\delta t} = \frac{dy}{dt}$$

so taking the limits in the expression above we get:

$$\frac{du}{dt} = \frac{\partial u}{\partial x}\frac{dx}{dt} + \frac{\partial u}{\partial y}\frac{dy}{dt}.$$

This is the second version of the chain rule, where we are not just making a change of variables but now we are limiting the direction of the derivative to be along a particular path, such that $x = g(t)$ and $y = h(t)$. This is the total derivative function along the path and du/dt is the gradient of the tangent along the path defined by $(x(t), y(t))$.

Approximations

As we now have a way of determining the gradient along any given path as

$$\frac{du}{dt} = \frac{\partial u}{\partial x}\frac{dx}{dt} + \frac{\partial u}{\partial y}\frac{dy}{dt}.$$

We can use this to estimate the value of the function near some known value (x, y). In Chapter 7, we used the approximation that for small $\delta x, \delta y/\delta x \approx dy/dx$. Therefore, along some known path we can use $\delta x/\delta t \approx dx/dt, \delta y/\delta t \approx dy/dt$, and $\delta u/\delta t \approx du/dt$. Substitute these into the above chain rule above to get: $\delta u/\delta t \approx (\partial u/\partial x)(\delta x/\delta t) + (\partial u/\partial y)(\delta y/\delta t)$. As long as δt is not actually equal to 0, we can multiply both sides of this equation by δt to get

$$\delta u \approx \frac{\partial u}{\partial x}\delta x + \frac{\partial u}{\partial y}\delta y.$$

This expression can be used to approximate the value of a function near a known value.

Example 17.4   The focal length of a convex lens is given by

$$\frac{1}{f} = \frac{1}{p} + \frac{1}{q}$$

where p is the distance of some object and q the distance of its image, measured in both case from the position of the lens. The maximum error in measurement of distances is known to be 5%. If p and q are measured as 10 and 2.5 cm, respectively, estimate the focal length of the lens and the maximum error in your estimate.

Solution   We have $p = 0.1$ m and $q = 0.025$ m. Substituting for f gives

$$\frac{1}{f} = \frac{1}{0.1} + \frac{1}{0.025}$$

$$\Rightarrow \quad \frac{1}{f} = 10 + 40 = 50$$

$$\Leftrightarrow \quad f = 0.02$$

We can approximate the error for f using:

$$\delta f \approx \frac{\partial f}{\partial p}\delta p + \frac{\partial f}{\partial q}\delta q$$

to estimate the maximum error in this value for f. If the error in measurement of p and q could be up to 5%, then the maximum errors for p and q are given by:

$$\delta p = p \times 5/100, \quad \text{substitute } p = 0.1, \text{giving } \delta p = 0.005$$
$$\delta q = q \times 5/100, \quad \text{substitute } q = 0.025, \text{giving } \delta q = 0.00125$$

We also find the first-order partial derivatives for f from the formula for f in terms of p and q, which we rewrite as

$$\frac{1}{f} = \frac{q+p}{pq} \Leftrightarrow f = \frac{pq}{p+q}$$

$$\frac{\partial f}{\partial p} = -\frac{pq}{(p+q)^2} + \frac{q}{p+q} = \frac{qp + q^2 - pq}{(p+q)^2} = \frac{q^2}{(p+q)^2}$$

$$\frac{\partial f}{\partial q} = -\frac{pq}{(p+q)^2} + \frac{p}{p+q} = \frac{p^2 + pq - pq}{(p+q)^2} = \frac{p^2}{(p+q)^2}.$$

Substituting $p = 0.1$ and $q = 0.025$ gives

$$\frac{\partial f}{\partial p}(0.1, 0.025) = 0.04 \quad \text{and} \quad \frac{\partial f}{\partial q}(0.1, 0.025) = 0.64.$$

Finally, we can estimate the error for f from

$$\delta f \approx \frac{\partial f}{\partial p}\delta p + \frac{\partial f}{\partial q}\delta q.$$

At $p = 0.1, q = 0.025$, we have $\delta p = 0.005, \delta q = 0.00125, \partial f/\partial p = 0.04$, and $\partial f/\partial q = 0.64$, which gives $\delta f \approx 0.04 \times 0.005 + 0.64 \times 0.00125 = 0.001$.

This means that the focal length of the lens is 0.02 m with a maximum error of approximately 0.001 m.

The chain rule holds whatever variable name we use in place of t, the parameter to describe the path in the x, y plane. It also holds whatever path we choose, that is, whatever the functions used for $x = g(t)$ and $y = h(t)$ (provided all the derivatives exist and are continuous in the region of the path). Therefore, we can use mathematical shorthand to give an expression for the total derivative, du:

$$du = \frac{\partial u}{\partial x}dx + \frac{\partial u}{\partial y}dy.$$

Example 17.5   Given $u = x^3 - 3y^2x + y^2$ and $x = 2t, y = 1 - 2t$, find du/dt:

(a)   using the chain rule,
(b)   by direct substitution.

Solution   (a) $u = x^3 - 3y^2x + y^2$ and $x = 2t$, $y = 1 - 2t$, so:

$$\frac{\partial u}{\partial x} = 3x^2 - 3y^2 \quad \text{and} \quad \frac{\partial u}{\partial y} = -6yx + 2y$$

$$\frac{\mathrm{d}x}{\mathrm{d}t} = 2 \quad \text{and} \quad \frac{\mathrm{d}y}{\mathrm{d}t} = -2$$

The chain rule

$$\frac{\mathrm{d}u}{\mathrm{d}t} = \frac{\partial u}{\partial x}\frac{\mathrm{d}x}{\mathrm{d}t} + \frac{\partial u}{\partial y}\frac{\mathrm{d}y}{\mathrm{d}t}$$

gives

$$\frac{\mathrm{d}u}{\mathrm{d}t} = (3x^2 - 3y^2)2 + (-6yx + 2y)(-2)$$

$$= 6x^2 - 6y^2 + 12yx - 4y.$$

Substituting for x and y as functions of t, we find:

$$\frac{\mathrm{d}u}{\mathrm{d}t} = 6(2t)^2 - 6(1 - 2t)^2 + 12(1 - 2t)2t - 4(1 - 2t)$$

$$= 24t^2 - 6 + 24t - 24t^2 + 24t - 48t^2 - 4 + 8t$$

$$= -48t^2 + 56t - 10.$$

(b) Substitute $x = 2t$, $y = 1 - 2t$ into $u = x^3 - -3y^2x + y^2$ giving

$$u = 8t^3 - 3(1 - 2t)^22t + (1 - 2t)^2$$

$$\Leftrightarrow u = 8t^3 - 3(1 - 4t + 4t^2)2t + (1 - 4t + 4t^2)$$

$$\Leftrightarrow u = 8t^3 - 6t + 24t^2 - 24t^3 + 1 - 4t + 4t^2$$

$$\Leftrightarrow u = -16t^3 + 28t^2 - 10t + 1$$

Therefore, $\mathrm{d}u/\mathrm{d}t = -48t^2 + 56t - 10$.

17.6 Higher-order partial derivatives

We can carry on differentiating partial derivatives to find higher-order partial derivatives such as $\partial^2 u/\partial x^2$, $\partial^2 u/\partial y^2$, $\partial^2 u/\partial x\partial y$, and $\partial^2 u/\partial y\partial x$, where

$$\frac{\partial^2 u}{\partial x^2} = \frac{\partial}{\partial x}\frac{\partial u}{\partial x}$$

$$\frac{\partial^2 u}{\partial y^2} = \frac{\partial}{\partial y}\frac{\partial u}{\partial y}$$

$$\frac{\partial^2 u}{\partial x\partial y} = \frac{\partial}{\partial x}\frac{\partial u}{\partial y}$$

$$\frac{\partial^2 u}{\partial y\partial x} = \frac{\partial}{\partial y}\frac{\partial u}{\partial x}.$$

For functions with continuous first-order derivatives

$$\frac{\partial^2 u}{\partial x\partial y} = \frac{\partial^2 u}{\partial y\partial x}.$$

The partial derivatives $\partial^2 u/\partial x^2$, $\partial^2 u/\partial y^2$, $\partial^2 u/\partial x\partial y$, and $\partial^2 u/\partial y\partial x$ are called the second-order partial derivatives of u because we have differentiated twice to find them. We can continue differentiating to find third- and higher-order partial derivatives.

Example 17.6

$u = x^4 - 2x^3 y + xy^2$.

Find its first- and second-order partial derivatives with respect to x and y

Solution

$$u = x^4 - 2x^3 y + xy^2$$

$$\frac{\partial u}{\partial x} = 4x^3 - 6x^2 y + y^2$$

$$\frac{\partial u}{\partial y} = -2x^3 + 2xy.$$

We differentiate $\partial u/\partial x$ with respect to x, keeping y constant, to find

$$\frac{\partial^2 u}{\partial x^2} = \frac{\partial}{\partial x}\frac{\partial u}{\partial x} = 12x^2 - 12xy.$$

We differentiate $\partial u/\partial y$ with respect to y, keeping x constant, to find

$$\frac{\partial^2 u}{\partial y^2} = \frac{\partial}{\partial y}\frac{\partial u}{\partial y} = 2x.$$

We differentiate $\partial u/\partial y$ with respect to x, keeping y constant, to find

$$\frac{\partial^2 u}{\partial x\partial y} = \frac{\partial}{\partial x}\frac{\partial u}{\partial y} = -6x^2 + 2y.$$

We differentiate $\partial u/\partial x$ with respect to y, keeping x constant, to find

$$\frac{\partial^2 u}{\partial y\partial x} = \frac{\partial}{\partial y}\frac{\partial u}{\partial x} = -6x^2 + 2y.$$

Notice that $\partial^2 u/\partial x\partial y = \partial^2 u/\partial y\partial x$ as expected.

17.7 Summary

1. A function $u(x, y)$ is a function of two variables x and y; u is the dependent variable and x and y are the independent variables. The graph of this function can be represented on the page of a book or a computer screen by using a perspective representation of a surface or by using contours to represent the lines where the value of the function is constant.

2. A function of two variables $f(x, y)$ has two first-order partial derivatives. The partial derivative with respect to x is found by considering y to be constant and then differentiating with respect to x only. The partial derivative with respect to y is found by considering x to be a constant and differentiating with respect to y only.

3. The chain rule can be used to change the variables of differentiation. Supposing we have a function, $u = f(x, y)$, and x and y are functions of s and t so that $x = v(s, t)$, $y = w(s, t)$, and $u = F(s, t)$, then:

$$\frac{\partial u}{\partial s} = \frac{\partial u}{\partial x}\frac{\partial x}{\partial s} + \frac{\partial u}{\partial y}\frac{\partial y}{\partial s} \quad \text{and} \quad \frac{\partial u}{\partial t} = \frac{\partial u}{\partial x}\frac{\partial x}{\partial t} + \frac{\partial u}{\partial y}\frac{\partial y}{\partial t}$$

4. If we know the path on which we are travelling across the surface $f(x, y)$, so that x and y are defined in terms of a parameter t, then it is possible to find the total derivative of f with respect to t because the gradient along the path will be unique at each point. It this case we can use:

$$\frac{du}{dt} = \frac{\partial u}{\partial x}\frac{dx}{dt} + \frac{\partial u}{\partial y}\frac{dy}{dt}.$$

5. Partial derivatives can be used for approximating the value of a function near a known value. For a function of two variables we have:

$$\delta u \approx \frac{\partial u}{\partial x}\delta x + \frac{\partial u}{\partial y}\delta y.$$

This can also be used to estimate the maximum error in a calculated value.

17.8 Exercises

17.1 Find the first-order partial derivatives of the following:

(a) $f(x, y) = x \ln(y^2)$

(b) $g(x, y) = x + xy - \dfrac{4x}{y}$

(c) $h(x, y) = \tan^{-1}\left(\dfrac{x}{y}\right)$

(d) $f(x, y) = \dfrac{4x}{y}e^{-x^2}.$

17.2 Use the chain rule to find partial derivatives of the following in terms of x and y where $r = \sqrt{x^2 + y^2}$ and $\theta = \tan^{-1}(\frac{y}{x})$.

(a) $u(r, \theta) = \sin(\theta)/r$

(b) $u(r, \theta) = r^2 + \tan(\theta).$

17.3 Given $u = (x^2 + y^2)/\sqrt{y}$, find the value of the following partial derivatives at the point $(0,4)$

(a) $\dfrac{\partial u}{\partial y}$ (b) $\dfrac{\partial^2 u}{\partial x \partial y}$ (c) $\dfrac{\partial^2 u}{\partial y \partial x}$

(d) $\dfrac{\partial^2 u}{\partial x^2}$ (e) $\dfrac{\partial^2 u}{\partial y^2}.$

17.4 Given $z = x^2 + y^2$ and $x = r \cos(\theta), y = r \cos(\theta)$, find $\partial z/\partial r$ and $\partial z/\partial \theta$ by substitution and by using the relationships between the partial derivatives. Show that the results obtained are equal.

17.5 Given that $z = 2x + 3y^2$ and $x = 3t + 5, y = -2t$, find $\partial z/\partial x$, $\partial z/\partial y$, dx/dt, dy/dt, and dz/dt.

17.6 Given $u = e^{-8t}(A\cos(2x) + B\sin(2x))$, evaluate $\partial u/\partial t$ and $\partial^2 u/\partial x^2$ and hence show that u is a solution to the partial differential equation $\partial u/\partial t = 2\partial^2 u/\partial x^2$.

17.7 Use the approximation

$$\delta u \approx \frac{\partial u}{\partial x}\delta x + \frac{\partial u}{\partial y}\delta y$$

with $u = \sqrt{x^2 + y^3}$ near the point $(6,4)$ to estimate the value of $\sqrt{5.96^2 + 4.03^3}$.

17.8 The power delivered to the load resistance R_L for the circuit is given by

$$P = 25R_L/(R + R_L)^2$$

If $R = 2000$ and $R_L = 1000$ with a possible error of 5% in either, find P and estimate the maximum error in P.

17.9 The height, width, and length of an open box are subject to errors of 1, 2, and 3%, respectively. Estimate to the nearest percentage the maximum relative error in calculating the surface area of the box for a desired height of 4 m, width of 3 m, and length of 2 m.

18 Vector calculus

18.1 Introduction

In the previous chapter, we looked at functions of more than one variable. For a function of two variables (x, y) we define a function $u = f(x, y)$ which can be represented by a surface. For each pair of values (x, y) we can find a single value for u, showing that u is a scalar quantity. For this reason, a function of spatial coordinates is called a scalar field. We also saw how to calculate a gradient of a function of two variables and that the gradient depends on the direction of the path that we choose across the surface. This means that the gradient must be described by both a magnitude and a direction. From Chapter 9, we know that vectors are used to represent quantities that have both magnitude and direction and we shall show in this chapter that we represent the gradient of a scalar field as a vector field.

A vector field is a vector function, which means that at each point in space the function has both magnitude and direction and can be expressed by a vector with x, y, and z components. In Chapter 6, we quoted many relationships between physical quantities that involve derivatives. There we considered only movement in a single spatial direction. Many of these equations should properly be described by vector field equations in space. In order to express these equations, we need to define the operations of divergence and curl of a vector field. Vector field equations are particularly important in electromagnetic field theory.

In this chapter we give an introduction to vector fields and operations on vector fields with applications to evaluating line and surface integrals.

18.2 The gradient of a scalar field

We define the gradient of a scalar field as follows

$$\nabla\phi = \frac{\partial\phi}{\partial x}\mathbf{i} + \frac{\partial\phi}{\partial y}\mathbf{j} + \frac{\partial\phi}{\partial z}\mathbf{k} = \left(\frac{\partial\phi}{\partial x}, \frac{\partial\phi}{\partial y}, \frac{\partial\phi}{\partial z}\right).$$

$\nabla\phi$ is the gradient of ϕ, or simply 'grad ϕ', ∇ is called the 'del' operator. The meaning of this definition is that we take the partial derivative of the scalar field with respect to x, y, and z, and these are the components of our vector field.

This definition is equivalent to

$$\nabla\phi = \left(\mathbf{i}\frac{\partial}{\partial x} + \mathbf{j}\frac{\partial}{\partial x} + \mathbf{k}\frac{\partial}{\partial x}\right)\phi$$

and also to

$$\nabla\phi = \left(\frac{\partial}{\partial x}, \frac{\partial}{\partial x}, \frac{\partial}{\partial x}\right)\phi.$$

If ϕ is defined for two spatial dimensions (x, y) only then we would have

$$\nabla\phi = \frac{\partial\phi}{\partial x}\mathbf{i} + \frac{\partial\phi}{\partial y}\mathbf{j} = \left(\frac{\partial\phi}{\partial x}, \frac{\partial\phi}{\partial y}\right).$$

Example 18.1 Given $\phi = x^2 + xyz$, find the vector field that describes its gradient.

Solution We use

$$\nabla\phi = \frac{\partial\phi}{\partial x}\mathbf{i} + \frac{\partial\phi}{\partial y}\mathbf{j} + \frac{\partial\phi}{\partial z}\mathbf{k}.$$

Differentiating ϕ partially with respect to x, y, and z, we find:

$$\frac{\partial\phi}{\partial x} = 2x + yz$$

$$\frac{\partial\phi}{\partial y} = xz$$

$$\frac{\partial\phi}{\partial z} = xy.$$

Therefore, we find that

$$\nabla\phi = (2x + yz)\mathbf{i} + xz\mathbf{j} + xy\mathbf{k} = (2x + yz, xz, xy).$$

Direction of maximum slope

We would like to know why this vector function is called *the gradient*. We saw in the last chapter that a function of more than one variable has many derivatives associated with it, depending on the direction that we choose to measure it. We found that for a function of ϕ of two variables, along a path defined by $(x(t), y(t))$, the derivative of ϕ is given by:

$$\frac{\mathrm{d}\phi}{\mathrm{d}t} = \frac{\partial\phi}{\partial x}\frac{\mathrm{d}x}{\mathrm{d}t} + \frac{\partial\phi}{\partial y}\frac{\mathrm{d}y}{\mathrm{d}t}.$$

Using the definition of the scalar product, we can write the above as

$$\frac{\mathrm{d}\phi}{\mathrm{d}t} = \left(\frac{\partial\phi}{\partial x}, \frac{\partial\phi}{\partial y}\right) \cdot \left(\frac{\mathrm{d}x}{\mathrm{d}t}, \frac{\mathrm{d}y}{\mathrm{d}t}\right) = \nabla\phi \cdot \left(\frac{\mathrm{d}x}{\mathrm{d}t}, \frac{\mathrm{d}y}{\mathrm{d}t}\right)$$

where $(\mathrm{d}x/\mathrm{d}t, \mathrm{d}y/\mathrm{d}t)$ has components representing the rate of change of x and y and therefore gives the direction of the path in the (x, y) plane.

We also know from Chapter 9 that for two vectors $\mathbf{a}$ and $\mathbf{b}$:

$$\mathbf{a} \cdot \mathbf{b} = ab\cos(\theta)$$

where $\mathbf{a} \cdot \mathbf{b}$ is the scalar product of vectors $\mathbf{a}$ and $\mathbf{b}$, θ is the angle between them and a, b are their magnitudes. If the direction of $\mathbf{b}$ could be chosen in order to maximize this scalar product we would choose $\theta = 0$, because $|\cos(\theta)| \leqslant 1$ with the maximum value occurring at $\theta = 0$, where $\cos(\theta) = 1$. So to maximize the scalar product we choose the direction of $\mathbf{b}$ to be along $\mathbf{a}$. So the direction of the path which maximizes the scalar product

$$\frac{\mathrm{d}\phi}{\mathrm{d}t} = \left(\frac{\partial\phi}{\partial x}, \frac{\partial\phi}{\partial y}\right)\left(\frac{\mathrm{d}x}{\mathrm{d}t}, \frac{\mathrm{d}y}{\mathrm{d}t}\right)$$

will be when the direction of $(\mathrm{d}x/\mathrm{d}t, \mathrm{d}y/\mathrm{d}t)$ is the same as that of $\nabla\phi$. So $\nabla\phi$ gives the magnitude and direction of maximum gradient of ϕ and the slope in any general direction can be found by taking the component of $\nabla\phi$ in the direction of interest.

Example 18.2 Given $\phi = z - x^{1/2} - y^{3/2}$, find the vector field that describes its gradient, where $x, y > 0$ and find the maximum slope at the point $(4, 9, 1)$.

Solution We use

$$\nabla\phi = \frac{\partial\phi}{\partial x}\mathbf{i} + \frac{\partial\phi}{\partial y}\mathbf{j} + \frac{\partial\phi}{\partial z}\mathbf{k}.$$

Differentiating ϕ partially with respect to x, y, and z, we find:

$$\frac{\partial\phi}{\partial x} = -\frac{1}{2}x^{-1/2}$$

$$\frac{\partial\phi}{\partial y} = -\frac{1}{2}y^{-1/2}$$

$$\frac{\partial\phi}{\partial z} = 1.$$

Therefore, we find that

$$\nabla\phi = -\tfrac{1}{2}x^{-1/2}\mathbf{i} - \tfrac{1}{2}y^{-1/2}\mathbf{j} + \mathbf{k} = \left(-\tfrac{1}{2}x^{-1/2}, -\tfrac{1}{2}y^{-1/2}, 1\right)$$

Therefore, the maximum slope at the point $(4, 9, 1)$ is given by substituting $x = 4$, $y = 9$, and $z = 1$ into the above expression for grad ϕ giving $(-1/4, -1/6, 1)$. This has a magnitude of

$$\sqrt{\left(-\tfrac{1}{4}\right)^2 + \left(-\tfrac{1}{6}\right)^2 + 1^2} \approx 1.044$$

in a direction given by the unit vector

$$\frac{1}{1.044}\left(-\frac{1}{4}, -\frac{1}{6}, 1\right) = (-0.261, -0.174, 1).$$

The ∇ (del) operator

We defined

$$\nabla\phi = \left(\frac{\partial}{\partial x}, \frac{\partial}{\partial x}, \frac{\partial}{\partial x}\right)\phi$$

or equivalently

$$\nabla\phi = \left(\mathbf{i}\frac{\partial}{\partial x} + \mathbf{j}\frac{\partial}{\partial x} + \mathbf{k}\frac{\partial}{\partial x}\right)\phi.$$

We can see that ∇(del) can be considered as an operator defined by

$$\nabla = \left(\frac{\partial}{\partial x}, \frac{\partial}{\partial x}, \frac{\partial}{\partial x}\right)$$

or equivalently

$$\nabla = \left(\mathbf{i}\frac{\partial}{\partial x} + \mathbf{j}\frac{\partial}{\partial x} + \mathbf{k}\frac{\partial}{\partial x}\right).$$

This same operator can be used when differentiating vector fields.

18.3
Differentiating vector fields

We now consider vector fields, which represent a vector at each point in space. Therefore, we have the field $\mathbf{F}$ which has components in the x, y, and z directions, all of which are also functions of x, y, and z, that is

$$\mathbf{F}(x, y, z) = \mathbf{i}F_x + \mathbf{j}F_y + \mathbf{k}F_z = (F_x, F_y, F_z)$$

where F_x, F_y, F_z are all functions of x, y, and z.

There are two important fields that can be found by differentiating vector fields. The divergence of a vector field produces a scalar field and the curl of a vector field produces a vector field. These are defined as follows

$$\nabla \cdot \mathbf{F} = \text{div } \mathbf{F} = \frac{\partial F_x}{\partial x} + \frac{\partial F_y}{\partial y} + \frac{\partial F_z}{\partial z}$$

$$\nabla \times \mathbf{F} = \text{curl } \mathbf{F} = \left(\frac{\partial F_z}{\partial y} - \frac{\partial F_y}{\partial z} \right) \mathbf{i} + \left(\frac{\partial F_x}{\partial z} - \frac{\partial F_z}{\partial x} \right) \mathbf{j}$$
$$+ \left(\frac{\partial F_y}{\partial x} - \frac{\partial F_x}{\partial y} \right) \mathbf{k}.$$

The last expression can also be represented by using a determinant to define the cross-product of a vector. We have that

$$\nabla \times \mathbf{F} = \left(\frac{\partial}{\partial x}, \frac{\partial}{\partial x}, \frac{\partial}{\partial x} \right) \times (F_x, F_y, F_z) = \begin{vmatrix} \mathbf{i} & \mathbf{j} & \mathbf{k} \\ \partial/\partial x & \partial/\partial y & \partial/\partial z \\ F_x & F_y & F_z \end{vmatrix}.$$

There is one more operator that is used in many equations of electromagnetic fields. This is the 'del squared' operator ∇^2 which operates on a scalar field. This is defined by

$$\nabla^2 = \nabla \cdot \nabla = \left(\frac{\partial}{\partial x}, \frac{\partial}{\partial y}, \frac{\partial}{\partial z} \right) \cdot \left(\frac{\partial}{\partial x}, \frac{\partial}{\partial y}, \frac{\partial}{\partial z} \right) = \frac{\partial^2}{\partial x} + \frac{\partial^2}{\partial y} + \frac{\partial^2}{\partial z}.$$

All of these operators are important in describing vector field relationships. For instance

$$\nabla^2 \phi = k^2 \frac{\partial^2 \phi}{\partial t^2}$$

represents the three-dimensional wave equation.

Example 18.3   For $\mathbf{F} = (2x^2 y, 4y^2 z, 8z^2)$ find $\nabla \cdot \mathbf{F}$ and $\nabla \times \mathbf{F}$.

Solution   From the definition

$$\nabla \cdot \mathbf{F} = \frac{\partial F_x}{\partial x} + \frac{\partial F_y}{\partial y} + \frac{\partial F_z}{\partial z}$$

we see that we find the scalar field by partially differentiating the first component by x, the second by y, and the third by z, and summing the result. Hence

$$\nabla \cdot \mathbf{F} = 4xy + 8yz + 16z.$$

To find $\nabla \times \mathbf{F}$ we use the definition in terms of the determinant and expand about the first row, which in this case gives:

$$\nabla \times \mathbf{F} = \begin{vmatrix} \mathbf{i} & \mathbf{j} & \mathbf{k} \\ \partial/\partial x & \partial/\partial y & \partial/\partial z \\ 2x^2 y & 4y^2 z & 8z^2 \end{vmatrix}$$

$$= \left(\frac{\partial}{\partial y}(8z^2) - \frac{\partial}{\partial z}(4y^2 z) \right) \mathbf{i} - \left(\frac{\partial}{\partial x}(8z^2) + \frac{\partial}{\partial z}(2x^2 y) \right) \mathbf{j}$$

$$+ \left(\frac{\partial}{\partial x}(4y^2 z) - \frac{\partial}{\partial y}(2x^2 y) \right) \mathbf{k}$$

$$= 4y^2 \mathbf{i} - 2x^2 y \mathbf{k} = (4y^2, 0, 2x^2 y).$$

Example 18.4   Show the vector identity

$$\nabla \cdot \nabla \times \mathbf{F} = 0.$$

Solution   Taking $\mathbf{F} = (F_x, F_y, F_z)$, we get from the definition of curl:

$$\nabla \times \mathbf{F} = \left(\frac{\partial F_z}{\partial y} - \frac{\partial F_y}{\partial z} \right) i + \left(\frac{\partial F_x}{\partial z} - \frac{\partial F_z}{\partial x} \right) j + \left(\frac{\partial F_y}{\partial x} - \frac{\partial F_x}{\partial y} \right) k.$$

Now taking the divergence of the resulting vector field we take the dot product of

$$\nabla = \left(\frac{\partial}{\partial x}, \frac{\partial}{\partial y}, \frac{\partial}{\partial z} \right)$$

with the above giving

$$\left(\frac{\partial}{\partial x}, \frac{\partial}{\partial y}, \frac{\partial}{\partial z} \right) \cdot \left(\left(\frac{\partial F_z}{\partial y} - \frac{\partial F_y}{\partial z} \right), \left(\frac{\partial F_x}{\partial z} - \frac{\partial F_z}{\partial x} \right), \left(\frac{\partial F_y}{\partial x} - \frac{\partial F_x}{\partial y} \right) \right)$$

$$= \left(\frac{\partial^2 F_z}{\partial x \partial y} - \frac{\partial^2 F_y}{\partial x \partial z} \right) + \left(\frac{\partial^2 F_x}{\partial y \partial z} - \frac{\partial^2 F_z}{\partial y \partial x} \right) + \left(\frac{\partial^2 F_y}{\partial z \partial x} - \frac{\partial^2 F_x}{\partial z \partial y} \right).$$

We use the fact that for functions with continuous partial derivatives

$$\frac{\partial^2 F_z}{\partial y \partial x} = \frac{\partial^2 F_z}{\partial x \partial y}$$

that is, the order of differentiation used to calculate higher-order partial derivatives is not important. Then all the terms in the above cancel out giving

$$\nabla \cdot \nabla \times \mathbf{F} = 0.$$

18.4 The scalar line integral

The result of integrating a scalar field along a given curve is important for calculating many physical quantities. We know that in one dimension we relate the work done by a force in moving from one location to another as $W = \int \mathbf{F}\,\mathrm{d}\mathbf{x}$, where $\mathbf{F}$ is the force, W is the work done or energy used and $\mathbf{x}$ is the distance moved in the direction of the force. In three dimensions, an object can move along a path and the position of the object will vary such that $\mathbf{r} = (x(t), y(t), z(t))$ where t is some parameter used to describe the path taken. The work done in any given direction will be given by the component of the force in that direction multiplied by the distance moved. Hence we find:

$$W = \int_C \mathbf{F} \cdot \mathrm{d}\mathbf{r}$$

where C is the path along which the object moves and $\mathbf{r}$ describes its position vector. To calculate this value we need to be able to integrate fields along a path, where the path is described in terms of a position vector, $\mathbf{r} = (x(t), y(t), z(t))$.

We use

$$\frac{\mathrm{d}\mathbf{r}}{\mathrm{d}t} = \left(\frac{\mathrm{d}x}{\mathrm{d}t}, \frac{\mathrm{d}y}{\mathrm{d}t}, \frac{\mathrm{d}z}{\mathrm{d}t} \right)$$

and the symbolic relationship

$$\mathbf{r} = \frac{\mathrm{d}\mathbf{r}}{\mathrm{d}t}\mathrm{d}t$$

to give

$$\mathrm{d}\mathbf{r} = \left(\frac{\mathrm{d}x}{\mathrm{d}t}, \frac{\mathrm{d}y}{\mathrm{d}t}, \frac{\mathrm{d}z}{\mathrm{d}t} \right) \mathrm{d}t$$

and

$$\int_C \mathbf{F} \cdot \mathrm{d}\mathbf{r} = \int_{t_1}^{t_2} (F_x, F_y, F_z) \cdot \left(\frac{\mathrm{d}x}{\mathrm{d}t}, \frac{\mathrm{d}y}{\mathrm{d}t}, \frac{\mathrm{d}z}{\mathrm{d}t} \right) \mathrm{d}t$$

$$= \int_{t_1}^{t_2} F_x \frac{\mathrm{d}x}{\mathrm{d}t} + F_y \frac{\mathrm{d}y}{\mathrm{d}t} + F_z \frac{\mathrm{d}z}{\mathrm{d}t} \mathrm{d}t$$

where t_1 and t_2 are the values of the parameter at the start and end points of the curve C.

Example 18.5 Given $\mathbf{F} = 2xyz\mathbf{i} - x^2 y\mathbf{j} + z^2 x\mathbf{k}$, Find the integral of $\mathbf{F}$ along a path defined by $2t\mathbf{i} - t\mathbf{j} + \mathbf{k}$ from $t = 1$ to $t = 4$.

Solution We have $\mathbf{r} = 2t\mathbf{i} - 3t\mathbf{j} + \mathbf{k}$, and therefore, $x = 2t$, $y = -3t$, and $z = 1$ giving

$$\mathrm{d}\mathbf{r} = \left(\frac{\mathrm{d}x}{\mathrm{d}t}, \frac{\mathrm{d}y}{\mathrm{d}t}, \frac{\mathrm{d}z}{\mathrm{d}t} \right) \mathrm{d}t = (2, -3, 0)\mathrm{d}t \quad \text{and} \quad \mathbf{F} = (2xyz, -x^2 y, z^2 x)$$

Therefore, we want to find

$$\int_1^4 (2xyz, -x^2 y, z^2 x) \cdot (2, -3, 0) \, \mathrm{d}t = \int_1^4 (4xyz + 3x^2 y) \, \mathrm{d}t$$

Substituting for x, y, z in terms of t, as above, we get

$$\int_1^4 -24t^2 - 24t^3 \, dt = \left[\frac{24t^3}{3} - \frac{24t^4}{4} \right]_1^4 = [-8t^3 - 6t^4]_1^4$$

$$= -(4 \times 64) - (6 \times 256) - (-8 - 6)$$

$$= -1792 + 14 = -1778.$$

Example 18.6   Find

$$\int_C \mathbf{F} \cdot d\mathbf{r}$$

where $\mathbf{F} = (2xy, 3z, 12xyz)$, and C is a path clockwise around a triangle ABC with vertices A(1,0,−1), B (1,1,1), C (0,1,1).

Solution   We need to integrate along each of the three sides of the triangle. We need, in each case to find the equation of the line along the side of the triangle. The vector equation of a line between two points $\mathbf{a}$ and $\mathbf{b}$ was found in Chapter 9 to be $\mathbf{r} = \mathbf{a}(1 - t) + \mathbf{b}t$ where t is some parameter and for points between A and B then $0 \leqslant t \leqslant 1$.

From A to B: $\mathbf{r} = (1, 0, -1)(1 - t) + (1, 1, 1)t = (1, t, -1 + 2t)$. So $x = 1, y = t$, and $z = -1 + 2t$, and

$$\frac{dx}{dt} = 0, \quad \frac{dy}{dt} = 1, \quad \frac{dz}{dt} = 2$$

$$d\mathbf{r} = \left(\frac{dx}{dt}, \frac{dy}{dt}, \frac{dz}{dt} \right) dt \quad \Rightarrow \quad d\mathbf{r} = (0, 1, 2)dt$$

$$\int_{A \text{ to } B} \mathbf{F} \cdot d\mathbf{r} = \int_{t_1}^{t_2} (F_x, F_y, F_z) \cdot \left(\frac{dx}{dt}, \frac{dy}{dt}, \frac{dz}{dt} \right) dt$$

$$= \int_0^1 (2xy, 3z, 12xyz) \cdot (0, 1, 2) \, dt$$

$$= \int_0^1 3z + 24xyz \, dt$$

$$= \int_0^1 3(-1 + 2t) + 24t(-1 + 2t)dt$$

$$= \int_0^1 -3 - 18t + 48t^2 \, dt$$

$$= [-3t - 9t^2 + 16t^3]_0^1 = 4.$$

Similarly, we find that from B to C: $\mathbf{r} = (1, 1, 1)(1 - t) + (0, 1, 1)t = (1 - t, 1, 1)$. So $x = 1 - t, y = 1$, and $z = 1$, and

$$\frac{dx}{dt} = -1, \quad \frac{dy}{dt} = 0, \quad \frac{dz}{dt} = 0$$

$$\int_{B \text{ to } C} \mathbf{F} \cdot d\mathbf{r} = \int_{t_1}^{t_2} (F_x, F_y, F_z) \cdot \left(\frac{dx}{dt}, \frac{dy}{dt}, \frac{dz}{dt} \right) dt$$

$$= \int_0^1 (2xy, 3z, 12xyz) \cdot (-1, 0, 0) \, dt$$

$$= \int_0^1 -2xy \, dt = \int_0^1 -2(1 - t) \, dt = [(1 - t)^2]_0^1 = -1$$

From C to A: $\mathbf{r} = (0, 1, 1)(1 - t) + (1, 0, -1)t = (t, 1 - t, 1 - 2t)$. So $x = t, y = 1 - t$, and $z = 1 - 2t$, and

$$\frac{dx}{dt} = 1, \quad \frac{dy}{dt} = -1, \quad \frac{dz}{dt} = -2$$

$$\begin{aligned}
\int_{C \text{ to } A} \mathbf{F} \cdot d\mathbf{r} &= \int_{t_1}^{t_2} (F_x, F_y, F_z) \cdot \left(\frac{dx}{dt}, \frac{dy}{dt}, \frac{dz}{dt} \right) dt \\
&= \int_0^1 (2xy, 3z, 12xyz) \cdot (1, -1, -2) \, dt \\
&= \int_0^1 2xy - 3z - 24xyz \, dt \\
&= \int_0^1 2t - 2t^2 - 3(1 - 2t) - 24t(1 - t)(1 - 2t) \, dt \\
&= \int_0^1 -48t^3 + 70t^2 - 16t - 3t \, dt \\
&= \left[-12t^4 + \frac{70t^3}{3} - 8t - 3t \right]_0^1 \\
&= -12 + \frac{70}{3} - 8 - 3 = \frac{1}{3}.
\end{aligned}$$

Therefore, the total integral around the path is given by the sum of the integral along the three sections – that is

$$4 - 1 + \tfrac{1}{3} = 3\tfrac{1}{3}.$$

Integrals round a closed curve

If we are calculating the line integral round a closed curve in a plane (where the field is a function of x and y only) we can use Green's theorem in a plane to convert the integral into a double integral over the enclosed surface. This theorem is as follows

$$\oint_C F_x \, dx + F_y \, dy = \int_s \frac{\partial F_y}{\partial x} - \frac{\partial F_x}{\partial y} \, dx \, dy.$$

The left-hand side of this expression represent the integral of $\mathbf{F} \cdot d\mathbf{r}$, as before, however now we are limited to considering a plane so that $\mathbf{r} = (x, y)$ and also the path of integration must be closed. The fact that C is a closed path is indicated by the small circle on the integral sign.

Example 18.7   Using Green's theorem find the integral

$$\oint_C \mathbf{F} \cdot d\mathbf{r}$$

where $\mathbf{F} = (3x^2, -4xy)$, and C a path clockwise along the perimeter of the rectangle $0 \leqslant x \leqslant 4, 0 \leqslant y \leqslant 1$.

Solution   We want to find

$$\oint_C \mathbf{F} \cdot d\mathbf{r} = \oint_C F_x \, dx + F_y \, dy$$

which, by Green's theorem in the plane, is equal to

$$\int_s \frac{\partial F_y}{\partial x} - \frac{\partial F_x}{\partial y} \, dx \, dy.$$

We have $\mathbf{F} = (3x^2, -4xy) = (F_x, F_y)$ and therefore

$$\frac{\partial F_y}{\partial x} = -4y \quad \text{and} \quad \frac{\partial F_x}{\partial y} = 0.$$

The surface is a rectangle so the limits for the integration are easy to establish as being for x from 0 to 4 and for y from 0 to 1. Hence, we get the surface integral.

$$\oint_C \mathbf{F} \cdot d\mathbf{r} = \int_0^1 \left(\int_0^4 -4y \, dx \right) dy.$$

To perform a multiple integral we simply do one integral and then integrate the result. The order of performing the integrations will not matter in this case because the limits of integration are independent of the other variable and the integrals exists and are continuous in the relevant region. We do the inner integration with respect to x. Here we are integrating $-4y$, which does not contain a term in x, so we treat it as a constant for the purposes of the first integration giving the integral as $-4yx$.

$$\int_0^1 \left[\int_0^4 -4y \, dx \right] dy = \int_0^1 [-4yx]_{x=0}^{x=4} \, dy = \int_0^1 -16y \, dy.$$

Now, we do the integration in y to give

$$[-8y^2]_0^1 = -8.$$

18.5 Surface integrals

Many problems in field theory involve the calculation of flux of a vector field out of some enclosed surface. This requires us to integrate a vector field over the surface. Such problems are simplified by using the divergence theorem, which relates the integral of a vector field over a bounding surface to the integral of the divergence of the field over the enclosed volume.

$$\oint_S \mathbf{F} \cdot d\mathbf{S} = \int_V (\nabla \cdot \mathbf{F}) \, dV.$$

S is a surface enclosing the volume V. This expresses the relationship between the amount of source material in a volume and the flux out of the enclosing surface. An example is the relationship between electric charge within a volume and the flux of the resulting electric field.

Example 18.8   Given $\mathbf{F} = (x - y + z, 2x^2y, 1)$ find the integral of $\mathbf{F}$ over the closed surface consisting of the edges of the cube $0 \leqslant x \leqslant 3$, $0 \leqslant y \leqslant 3$ and $0 \leqslant z \leqslant 3$.

Solution   We want to use the divergence theorem

$$\oint_S \mathbf{F} \cdot d\mathbf{S} = \int_V (\mathbf{\nabla \cdot F})\, dV$$

so we need to find the divergence of the vector field $\mathbf{F} = (x - y + z, 2x^2y, 1)$ giving $F_x = x - y + z$, $F_y = 2x^2y$, and $F_z = 1$ and

$$\mathbf{\nabla \cdot F} = \frac{\partial F_x}{\partial x} + \frac{\partial F_y}{\partial y} + \frac{\partial F_z}{\partial z} = 1 + 2x^2$$

The integral is

$$\oint_S \mathbf{F} \cdot d\mathbf{S} = \int_V (\mathbf{\nabla \cdot F})\, dV = \int_0^3 \int_0^3 \int_0^3 1 + 2x^2\, dx\, dy\, dz.$$

We can now perform each of the integrations one after the other. We begin with, the integration over x, which gives

$$\int_0^3 \int_0^3 \int_0^3 1 + 2x^2\, dx\, dy\, dz = \int_0^3 \int_0^3 \left[x + \frac{x^3}{3} \right]_0^3 dy\, dz$$

and evaluating the limits for x, this gives:

$$\int_0^3 \int_0^3 21\, dy\, dz = \int_0^3 [21y]_0^3\, dz,$$

now evaluating the limits for y, we get:

$$\int_0^3 63\, dz = [63z]_0^3 = 189.$$

Example 18.9   Use the divergence therem to evaluate the surface integral of $\mathbf{F}$ where $R = e^{-(x+y+z)}$ and S is the surface of a tetrahedron defined by the vertices $(0,0,0)$, $(1,0,0)$, $(0,1,0)$, and $(0,0,1)$.

Solution   We use

$$\oint_S \mathbf{F} \cdot d\mathbf{S} = \int_V (\mathbf{\nabla \cdot F})\, dV$$

but in order to express the limits of the integration we need to consider the geometry of the given tetrahedron. This is shown in Figure 18.1.

We can see from the figure that the sides of the tetrahedron lie along the x, y- and z-axes and the fourth side is the plane ABC, given by the equation $x + y + z = 1$. In this simple case, we can guess the equation of this plane and check that it is correct by substituting the values for the points A $(1,0,0)$, B $(0,1,0)$, and C $(0,0,1)$. We need to choose the limits of integration so that we integrate correctly over this tetrahedron. x must start from 0 and go up to values lying on the plane ABC. This means that x is from 0 to $1 - y - z$. If we integrate for x first then when considering the integration for y, we will have eliminated the x variable and will be left with the y, z plane. So y goes between $y = 0$ and the line BC, given

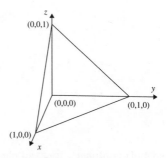

Figure 18.1   *Tetrahedron for Example 18.9.*

by $y = 1 - z$. Finally, in considering z we have the values of z from 0 to 1. This gives:

$$\int_0^1 \int_0^{1-z} \int_0^{1-y-z} e^{-(x+y+z)} \, dx \, dy \, dz.$$

We perform the integration with respect to x:

$$= \int_0^1 \int_0^{1-z} [e^{-(x+y+z)}]_0^{1-y-z} \, dy \, dz$$

$$= \int_0^1 \int_0^{1-z} -e^{-(1-y-z+y+z)} - e^{-(y+z)} \, dy \, dz$$

$$= \int_0^1 \int_0^{1-z} -e^{-1} - e^{-(y+z)} \, dy \, dz.$$

Now we integrate with respect to y:

$$\int_0^1 [-e^{-1}y + e^{-(y+z)}]_0^{1-z} \, dz = \int_0^1 -e^{-1}(1-z) + e^{-1} - e^{-z} \, dz$$

Finally, integrating with respect to z we get:

$$\left[e^{-1}\frac{(z-1)^2}{2} + e^{-1}z - e^{-z} \right]_0^1 = e^{-1} + e^{-1} - \frac{1}{2}e^{-1} - 1$$

$$= \frac{3}{2}e^{-1} - 1.$$

So the integral gives $\frac{3}{2}e^{-1} - 1$.

18.6 Summary

1. A scalar field is a function of spatial coordinates giving a single, scalar value at every point (x, y, z).
2. The gradient of a scalar field ϕ, grad ϕ, is defined by:

$$\nabla \phi = \frac{\partial \phi}{\partial x}\mathbf{i} + \frac{\partial \phi}{\partial y}\mathbf{j} + \frac{\partial \phi}{\partial z}\mathbf{k} = \left(\frac{\partial \phi}{\partial x}, \frac{\partial \phi}{\partial y}, \frac{\partial \phi}{\partial z} \right).$$

3. The gradient of a scalar field gives the magnitude and direction of the maximum slope at any point $\mathbf{r} = (x, y, z)$ on ϕ.
4. ∇ is the 'del' operator where

$$\nabla = \left(\frac{\partial}{\partial x}, \frac{\partial}{\partial x}, \frac{\partial}{\partial x} \right)$$

or equivalently

$$\nabla = \left(\mathbf{i}\frac{\partial}{\partial x} + \mathbf{j}\frac{\partial}{\partial x} + \mathbf{k}\frac{\partial}{\partial x} \right)$$

5. A vector field is a vector function of spatial coordinates, for example, $\mathbf{F}(x, y, z) = \mathbf{i}F_x + \mathbf{j}F_y + \mathbf{k}F_z = (F_x, F_y, F_z)$ where F_x, F_y, F_z are all functions of $x, y,$ and z.

6. The divergence and curl of a vector field are defined by

$$\nabla \cdot \mathbf{F} = \operatorname{div} \mathbf{F} = \frac{\partial F_x}{\partial x} + \frac{\partial F_y}{\partial y} + \frac{\partial F_z}{\partial z}$$

$$\nabla \times \mathbf{F} = \operatorname{curl} \mathbf{F} = \left(\frac{\partial F_z}{\partial y} - \frac{\partial F_y}{\partial z} \right) \mathbf{i} + \left(\frac{\partial F_x}{\partial z} - \frac{\partial F_z}{\partial x} \right) \mathbf{j}$$

$$+ \left(\frac{\partial F_y}{\partial x} - \frac{\partial F_x}{\partial y} \right) \mathbf{k}.$$

The expression for curl $\mathbf{F}$ can also be represented using a determinant, to define the cross product of two vectors. We have that:

$$\nabla \times \mathbf{F} = \left(\frac{\partial}{\partial x}, \frac{\partial}{\partial x}, \frac{\partial}{\partial x} \right) \times (F_x, F_y, F_z)$$

$$= \begin{vmatrix} \mathbf{i} & \mathbf{j} & \mathbf{k} \\ \partial/\partial x & \partial/\partial y & \partial/\partial z \\ F_x & F_y & F_z \end{vmatrix}$$

7. A line integral of a vector field, F, along some path C can be calculated by

$$\int_C \mathbf{F} \cdot d\mathbf{r} = \int_{t_1}^{t_2} F_x \frac{dx}{dt} + F_y \frac{dy}{dt} + F_z \frac{dz}{dt} dt$$

where $\mathbf{r} = (x(t), y(t), z(t))$ is the position vector of a position on the curve C represented using a parameter t and values t_1 and t_2 correspond to the end points of the path of C.

8. The line integral around a closed path in two dimensions can be converted to a surface integral by using Green's theorem in two dimensions:

$$\oint_C F_x \, dx + F_y \, dy = \int_s \frac{\partial F_y}{\partial x} - \frac{\partial F_x}{\partial y} \, dx \, dy.$$

This gives a double integral evaluated over the given surface in the plane.

9. A surface integral around a closed surface in three dimensions can be converted to a volume integral over the volume enclosed within the surface using the divergence theorem

$$\int_S \mathbf{F} \cdot d\mathbf{S} = \int_V (\nabla \cdot \mathbf{F}) dV.$$

This then gives a triple integral to be evaluated over the given volume.

18.7 Exercises

18.1. Given that $\phi = x^2 - yz + 5z^3$, $\psi = x^2 y^2 z$, $\mathbf{F} = (xy, 3xyz, x - z)$, and $\mathbf{G} = (3z, 4x, 2)$, find

(a) $\nabla \phi$ at $(1, 1, -2)$
(b) $\nabla \psi$ at $(0.5, 0.1, 0)$
(c) $\nabla \cdot \mathbf{F}$

(d) $\nabla \times \mathbf{G}$
(e) curl $(\phi \mathbf{i} + \psi \mathbf{j})$
(f) div$(\psi \mathbf{G})$
(g) The magnitude of the maximum slope and the unit vector in the direction of the maximum slope of ψ at the point $(1, 2, 0)$.

18.2. ϕ and ψ are scalar field functions of (x, y, z). Show the following vector identities:

(a) $\nabla(\phi\psi) = \phi\nabla\psi + \psi\nabla\phi$
(b) $\nabla \times (\nabla\phi) = 0$.

18.3. Calculate

$$\int_C \mathbf{F} \cdot d\mathbf{r}$$

where

(a) $\mathbf{F} = (y^2, x, zy)$ and C is along the line joining the points A $(2,0,0)$ and B $(1,1,0)$,
(b) $\mathbf{F} = x^2 y\mathbf{i} + 2xy\mathbf{j} + 3xyz\mathbf{k}$ and C is along the path given by $\mathbf{r} = t\mathbf{i} + (1-t)\mathbf{j} + t\mathbf{k}$ for t from 1 to 3.

18.4. Find the work done by the force $\mathbf{F} = (3x, -2y, z)$ in the displacement along the curve $y = x, z = 2x^2$ as x goes from 1 to 2.

18.5. Use Green's theorem in the plane to evaluate the following line integrals clockwise around the given closed curve

(a) $\mathbf{F} = x\mathbf{i} + y^2\mathbf{j}$, where C is the perimeter of a rectangle ABCD where A $= (0,0)$, B $= (0,2)$, C $= (1,2)$, and D $= (1,0)$;
(b) $\mathbf{F} = (x-y)\mathbf{i} + (-x-y)\mathbf{j}$ where C is the perimeter of the square given by $1 \leqslant x \leqslant 3$ and $0 \leqslant y \leqslant 2$;
(c) $\mathbf{F} = (\cos(y), \sin(x))$ where C is the region bounded by the lines $x = 0, x = 1$, $y = -\pi/2$, and $y = \pi/2$.

18.6. For the following vector fields, $\mathbf{F}$, use the divergence theorem to evaluate the surface integrals over the surface, S, indicated:

(a) $\mathbf{F} = xy\mathbf{i} + yz\mathbf{j} + xz\mathbf{k}$, where S is the surface of the cube given by $0 \leqslant x \leqslant a$, $-a/2 \leqslant y \leqslant a/2$, and $-a \leqslant z \leqslant 0$;
(b) $\mathbf{F} = z\mathbf{i} - y^2\mathbf{j}$ and S is the surface of $0 \leqslant x \leqslant 4$, $0 \leqslant y \leqslant 1$ and $0 \leqslant z \leqslant y$;
(c) $\mathbf{F} = (x+y, x-z, x^2+z^2)$ where S is the surface of a tetrahedron defined by the vertices $(0,0,0)$, $(1,0,0)$, $(0,1,0)$, and $(0,0,1)$.

Part 4 Graph and language theory

Graph theory

19.1 Introduction

A graph consists of points, called vertices, and lines connecting them, called edges. They can be used to represent many diverse situations, for example five cities and five roads connecting them as in Figure 19.1. The same graph could represent some people and the edge connections could represent those people who do business with each other.

Sometimes the relationship is 'one-way', for instance, in the case of a network of one-way streets, or a graph representing a circuit where the arrows on the edges represent the current flow. In this case, the graph is called a directed graph, shortened to 'digraph'.

We will look at a few definitions and applications of different types of graph and their matrix representations (which we also looked at in Chapter 13). We also look applications to the solving of routing problems through networks.

19.2 Definitions

Graph

A graph G consists of a finite set of vertices $V(G)$, which cannot be empty and a finite set of edges, $E(G)$, which connect pairs of vertices. The number of vertices of G is called the order of G. The number of edges in G is represented by $|E|$ and the number of vertices by $|V|$.

The graph in Figure 19.2 has $|V| = 6$ and $|E| = 9$.

Incidence, adjacency, and neighbours

Two vertices are adjacent if they are joined by an edge. In Figure 19.2, v_3 is adjacent to v_4 and is also said to be a neighbour of v_4. The edge which joins the vertices is said to be incident to them. That is e_3 is incident to v_3 and v_4.

Multiple edges and loops, simple graphs

Two or more edges joining the same pair of vertices are multiple edges. An edge joining a vertex to itself is called a loop. In Figure 19.2

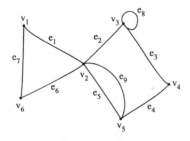

Figure 19.2 *A graph with six vertices and nine edges.*

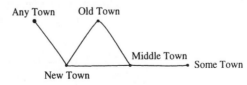

Figure 19.1 *Five cities with five roads connecting them represented by a graph.*

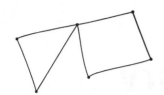

Figure 19.3 *A simple graph.*

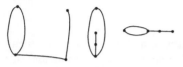

Figure 19.4 *Some isomorphic graphs.*

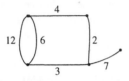

Figure 19.5 *A weighted graph.*

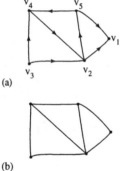

(a)

(b)

Figure 19.6 *(a) A digraph; and (b) its underlying graph.*

edges e_5 and e_9 are multiple edges and e_8 is a loop. A graph containing no multiple edges or loops is called a simple graph. There is an example of a simple graph in Figure 19.3.

Isomorphism

Graphs are isomorphic to each other if one can be obtained from a redrawing of the other one. Some isomorphic graphs are shown in Figure 19.4.

Weighted graph

A weighted graph has a number assigned to each of its edges, called its weight. The weight can be used to represent distances or capacities. A weighted graph is shown in Figure 19.5.

Digraphs

A digraph is a directed graph, that is, instead of edges in the definition of graphs we have arcs join pairs of vertices in a specified order. A digraph is shown in Figure 19.6(a).

Underlying graph

The underlying graph of a digraph D is the graph obtained by replacing each arc by an (undirected) edge as in Figure 19.6(b).

Degrees

The number of times edges are incident to a vertex, v, is called its degree, denoted by $d(v)$. In Figure 19.2 vertex v_5 has degree 3 and vertex v_3 has degree 4.

A vertex of a digraph has an in-degree $d_-(v)$ and an out-degree, $d_+(v)$. In Figure 19.6(a) v_5 has in-degree of 1 and out-degree of 2. Vertex v_1 has in-degree 1 and out-degree 1.

The sum of the values of the degree, $d(v)$, over all the vertices of a graph totals to twice the number of edges:

$$\sum_i d(v_i) = 2|E|$$

where $|E|$ is the number of edges.

Checking this result for Figure 19.2, we find that the degrees are $d(v_1) = 2, d(v_2) = 5, d(v_3) = 4, d(v_4) = 2, d(v_5) = 3$, and $d(v_6) = 2$. Summing these gives 18, which is the same as twice the number of edges.

For a digraph we get

$$\sum_i d_-(v_i) = |A| \quad \text{and} \quad \sum_i d_+(v_i) = |A|$$

where $|A|$ is the number of arcs.

Subgraph

A subgraph of G is a graph, H, whose vertex set is a subset of G's vertex set, and similarly its edge set is a subset of the edge set of G.

$$V(H) \subseteq V(G)$$
$$E(H) \subseteq E(G)$$

If it spans all of the vertices of G, that is, $V(H) = V(G)$ then H is called a spanning subgraph of G. A graph G, a subgraph, and a spanning subgraph are shown in Figure 19.7.

Complete graph

A simple graph in which every pair of two vertices is adjacent is called a complete graph. K_n is the complete graph with n vertices. K_n has $\frac{1}{2}n(n-1)$ edges. Some examples of complete graphs are shown in Figure 19.8.

Bipartite graph

This is a graph whose vertex set can be partitioned into two sets in such a way that each edge joins a vertex of the first set to a vertex of the second set. Some examples are given in Figure 19.9.

A complete bipartite graph is a bipartite simple graph in which every vertex in the first set is adjacent to every vertex in the second set. $K_{m,n}$ is the complete bipartite graph with m vertices in the first set and n vertices

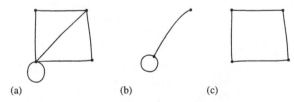

(a)   (b)   (c)

Figure 19.7   (a) A graph G. (b) A subgraph of G. (c) A spanning subgraph of G.

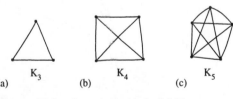

(a) K_3 (b) K_4 (c) K_5

Figure 19.8   Complete graphs (a) K_3; (b) K_4; (c) K_5.

Figure 19.9   Bipartite graphs. (b) and (c) are the complete bipartite graphs $K_{2,4}$ and $K_{3,1}$.

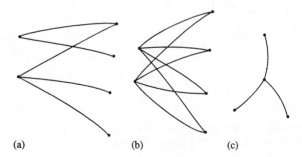

(a)   (b)   (c)

in the second set. Some examples of complete bipartite graphs are given in Figure 19.9(b) and (c).

Walks, paths, and circuits

A sequence of edges of the form

$$v_s v_i,\ v_i v_j,\ v_j v_k,\ v_k v_l,\ v_l v_t$$

is a walk of from v_s to v_t. If these edges are all distinct then the walk is called a trail and if the vertices are also distinct then the walk is called a path. A walk or trail is closed if $v_s = v_t$. A closed walk in which all the vertices are distinct except v_s and v_t is called a *cycle* or *circuit*. Examples are given in Figure 19.10.

Connected graph

A graph G is connected if there is a path from any one of its vertices to any other vertex. A disconnected graph is said to be made up of components. All the graphs we have drawn so far have been connected. In Figure 19.11(a), there is a disconnected graph with two components and in Figure 19.11(b) a disconnected graph with three components.

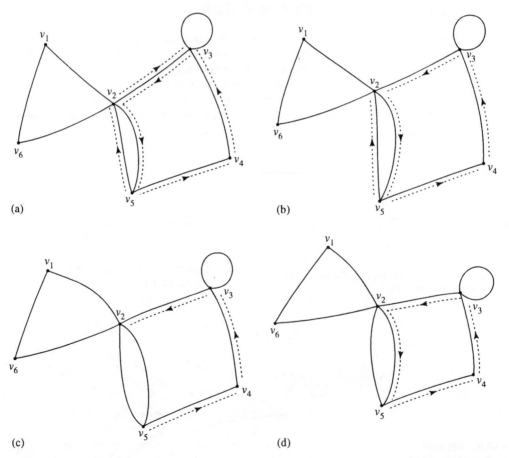

Figure 19.10   (a) A graph G with a walk marked. (b) A graph G with a trail marked. (c) A graph G with a path marked. (d) A graph G with a circuit marked.

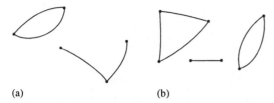

Figure 19.11   *(a) A disconnected graph with two components. (b) A disconnected graph with three components.*

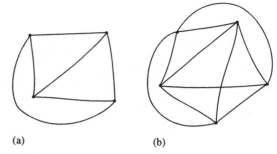

(a)   (b)

Figure 19.12   *(a) A planar graph. (b) A non-planar graph.*

Planar graphs

A planar graph is one which can be drawn in the plane without any edges intersecting except at vertices to which they are both incident. There is an example of a planar graph in Figure 19.12(a) and a non-planar graph in Figure 19.12(b).

19.3 Matrix representation of a graph

The incidence matrix and adjacency matrix

The incidence matrix of a graph G is a $|V| \times |E|$ matrix. The element a_{ij} = the number of times that vertex v_i is incident with the edge e_j.

The adjacency matrix of G is the $|V| \times |V|$ matrix. a_{ij} = the number of edges joining v_i and v_j. The incidence matrix for the graph in Figure 19.2 is given by

$$
\begin{array}{c}
 & \begin{array}{ccccccccc} e_1 & e_2 & e_3 & e_4 & e_5 & e_6 & e_7 & e_8 & e_9 \end{array} \\
\begin{array}{c} v_1 \\ v_2 \\ v_3 \\ v_4 \\ v_5 \\ v_6 \end{array} &
\left(\begin{array}{ccccccccc}
1 & 0 & 0 & 0 & 0 & 0 & 1 & 0 & 0 \\
1 & 1 & 0 & 0 & 1 & 1 & 0 & 0 & 1 \\
0 & 1 & 1 & 0 & 0 & 0 & 0 & 2 & 0 \\
0 & 0 & 1 & 1 & 0 & 0 & 0 & 0 & 0 \\
0 & 0 & 0 & 1 & 1 & 0 & 0 & 0 & 1 \\
0 & 0 & 0 & 0 & 0 & 1 & 1 & 0 & 0
\end{array} \right)
\end{array}
$$

and the adjacency matrix by

$$
\begin{array}{c}
 & \begin{array}{cccccc} v_1 & v_2 & v_3 & v_4 & v_5 & v_6 \end{array} \\
\begin{array}{c} v_1 \\ v_2 \\ v_3 \\ v_4 \\ v_5 \\ v_6 \end{array} &
\left(\begin{array}{cccccc}
0 & 1 & 0 & 0 & 0 & 1 \\
1 & 0 & 1 & 0 & 2 & 1 \\
0 & 1 & 1 & 1 & 0 & 0 \\
0 & 0 & 1 & 0 & 1 & 0 \\
0 & 2 & 0 & 1 & 0 & 0 \\
1 & 1 & 0 & 0 & 0 & 0
\end{array} \right)
\end{array}.
$$

19.4 Trees

A tree is a connected graph with no cycles. A forest is a graph with no cycles but may or may not be connected (i.e. a forest is a graph whose components are trees). Figure 19.13(a) shows a tree, while Figure 19.12(b) shows a forest.

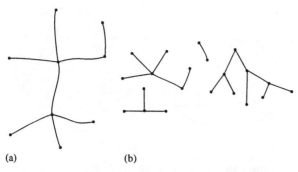

(a)   (b)

Figure 19.13 *(a) A tree is a connected graph with no cycles. (b) A forest has no cycles, but may or may not be connected.*

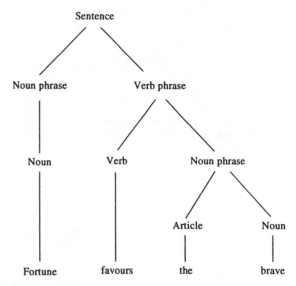

Figure 19.14 *A parsing tree for the sentence 'Fortune favours the brave'.*

If T is a tree with at least two vertices, then it has the three properties (T1), (T2), and (T3):

(T1) There is exactly one path from any vertex v_i in T to any other vertex v_j.

(T2) The graph obtained from T by removing any edge has two components, each of which is a tree.

(T3) $|E| = |V| - 1$.

Trees have many applications, particularly rooted trees. Decision trees are used to represent the possible decisions at each stage of a problem or algorithm. Probability trees can be used to analyse conditional probabilities, which we shall see in Chapter 21. Another application is to parsing of a sentence. The tree in Figure 19.14 represents the sentence 'Fortune favours the brave'. The vertices, other than the terminal vertices, represent grammatical categories and the terminal vertices represent the words of the sentence. The same sort of idea can be used to analyse allowed constructs of statements in programming languages and the syntax of arithmetic expressions. We look at this application of trees in the next chapter on language theory.

Spanning trees

A spanning tree of a graph G is a tree T which is a spanning subgraph of G. That is, T has the same vertex set as G. Examples of graphs with spanning trees marked are given in Figure 19.15.

How to grow a spanning tree

Take any vertex v of G as an initial partial tree. Add edges one by one so each new edge joins a new vertex to the partial tree. If there are n vertices in the graph G then the spanning tree will have n vertices and $n - 1$ edges.

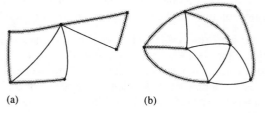

(a)   (b)

Figure 19.15   *Graphs with spanning trees shaded.*

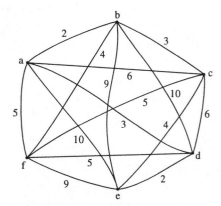

Figure 19.16   *A weighted complete graph. The vertices represent offices and the edges represent possible communication links. The weights on the edges represent the cost of construction of the link.*

Minimum spanning tree

Supposing we have a group of offices which need to be connected by a network of communication lines. The offices may communicate with each other directly or through another office. In order to decide on which offices to build links between we firstly work out the cost of all possible connections. This will then give us a weighted complete graph as shown in Figure 19.16.

The minimum spanning tree is then the spanning tree that represents the minimum cost.

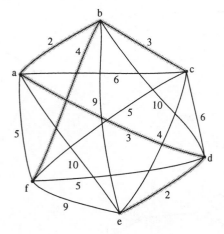

Figure 19.17   *The graph of Figure 19.16 with its minimum spanning tree marked.*

The greedy algorithm for the minimum spanning tree

1.   Choose any start vertex to form the initial partial tree (v_i).
2.   Add the cheapest edge, e_i, to a new vertex to form a new partial tree.
3.   Repeat Step 2 until all vertices have been included in the tree.

Example 19.1 Find the minimum spanning tree for the graph representing communication links between offices as shown in Figure 19.16.

Solution We start with any vertex and choose the one marked a. Add the edge ab which is the cheapest edge of those incident to a.

Add a new edge in order to form a partial tree and choose bc, which is one of the cheapest remaining edges incident either with a or b. Now we add edge ad which is the cheapest remaining edge of those incident with a or b or c. Continuing in this manner we find the minimum spanning tree, as shown in Figure 19.17. The total cost of the communication links in our solution is found to be $2 + 3 + 3 + 2 + 4 = 14$.

19.5 The shortest path problem

The weights on a graph may represent delays on a communication network or travel times along roads. A practical problem that we may wish to solve is to find the shortest path from any two vertices. The algorithm for solving this problem is illustrated in Example 19.2.

Example 19.2 The weighted graph in Figure 19.18 represents a communication network with weights indicating the delays associated with each edge. Find the minimum delay path from s to t.

Solution

Stage 1
To find the path we begin at the start vertex s. s is the reference vertex for stage 1. Label all the adjacent vertices with the lengths of the paths using only one edge. Mark all the other vertices with a very high number (bigger than the sum of all the weights in the graph) in this case we choose 100. This is shown in Figure 19.19.

At the same time start to fill in a table as in Table 19.1.

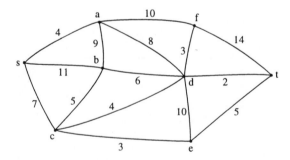

Figure 19.18 *The graph representing a communications network for Example 19.2.*

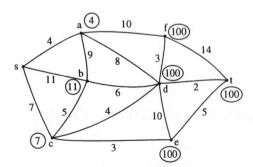

Figure 19.19 *Stage 1 of solving the shortest path problem of Example 19.2.*

Table 19.1 *The lengths of paths using one edge from s*

Reference vertex	a	b	c	d	e	f	t
s	4	11	7	100	100	100	100

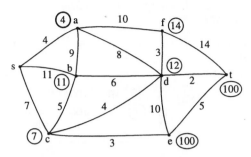

Figure 19.20 *Stage 2 of solving the shortest path problem.*

Table 19.2 *The lengths of paths using up to two edges from s*

Reference vertex	a	b	c	d	e	f	t
s	④	11	7	100	100	100	100
a		11	7	12	100	14	100

Stage 2

Choose the vertex with the smallest label, that has not already been a reference vertex, so that this becomes the new reference vertex. In this case we choose vertex a. Once a vertex has been used as a reference vertex it gains a permanent label and for this reason we circle the label in bold, as in Figure 19.20 and circle the label in the table as in Table 19.2. This label represents the length of the shortest path from s to the new reference vertex. Consider any vertex adjacent to the new reference vertex and calculate the length of the path via a to this vertex. If this is less than the current temporary label on the vertex then change the label on the vertex. For instance, b is adjacent to vertex a. The length of the path to b via a is $4 + 9 = 13$. This is not less than the current label of 11, so we keep the current label. However, looking at vertex d, which is adjacent to a, the length of the path via a is $4 + 8 = 12$, which is shorter than the current label of d, hence we change the current label. Considering also the vertices b and f, we end up at the end of Stage 2, with the labels as given in Figure 19.20 and in Table 19.2.

Stage 3

Choose the vertex with the smallest label that has not already been a reference vertex; this becomes the new reference vertex. In this case we choose vertex c, so this gains a permanent label which we circle in bold as in Figure 19.21; we also circle the label as in Table 19.3. This label represents the length of the shortest path from s to c. Consider any vertex adjacent to c which does not have a permanent label, and calculate the length of the path via c to this vertex. If this is less than the current temporary label on the vertex then change the label on the vertex. For instance, d is adjacent to vertex c. The length of the path to d via c is $7 + 4 = 11$. This is less than the current label of 12 so we change the label. However, looking at vertex b, which is adjacent to c, the length of the path via c is $7 + 5 = 12$ which is longer than the current label of 11, so we keep the current label. Considering also the vertex e we end up, at the end of Stage 3, with the labels as given in Figure 19.21 and Table 19.3.

Stage 4

Choose e as the new reference vertex and mark its label in bold. Compare paths via e to the labels on any adjacent vertices and re-label the vertices if the paths via e are found to be shorter. The result of this stage is shown in Figure 19.22 and Table 19.4.

Stage 5

Choose b as the new reference vertex (we could choose d instead but this would make no difference to the final result) and circle its label in bold. Compare paths via b to the labels on any adjacent vertices and relabel the

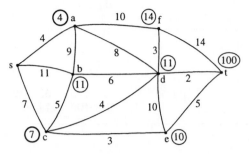

Figure 19.21   *Stage 3 of solving the shortest path problem.*

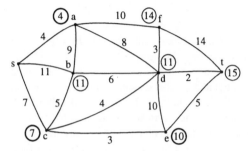

Figure 19.22   *Stage 4 of solving the shortest path problem.*

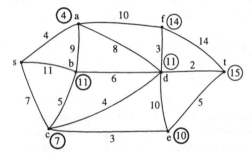

Figure 19.23   *Stage 5 of solving the shortest path problem.*

Table 19.3   *The lengths of paths using up to three edges from s*

Reference vertex	a	b	c	d	e	f	t
s	④	11	7	100	100	100	100
a		11	⑦	12	100	14	100
c		11		11	10	14	100

Table 19.4   *The lengths of paths using up to four edges from s*

Reference vertex	a	b	c	d	e	f	t
s	④	11	7	100	100	100	100
a		11	⑦	12	100	14	100
c		11		11	⑩	14	100
e		11		11		14	15

Table 19.5   *The lengths of paths using up to five edges from s*

Reference vertex	a	b	c	d	e	f	t
s	④	11	7	100	100	100	100
a		11	⑦	12	100	14	100
c		11		11	⑩	14	100
e		⑪		11		14	15
b				11		14	15

vertices if the paths via b are found to be shorter. The result of this stage is shown in Figure 19.23 and Table 19.5.

Stage 6

Choose d as the new reference vertex and mark its label in bold. The only vertices left without permanent labels are t and f. We find that the path via d to t gives a smaller value than the current label of 15 and hence we change the label to $11 + 2 = 13$. The result of this stage is shown in Figure 19.24 and Table 19.6.

Stage 7

The remaining vertex with the smallest label is t. We therefore give t the permanent label of 13. As soon as t receives a permanent label we can stop the algorithm as this must now represent the length of the shortest path from s to t which is the result we set out to find. To find the actual path that has this length we move backwards from t looking for consistent labels. To get to t we must have gone through d as adding the label on d to the length of edge dt gives the correct value of 13. Continuing this

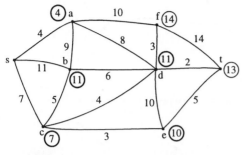

Figure 19.24 *Stage 6 of solving the shortest path problem.*

Table 19.6 *The lengths of paths using up to six edges from s*

Reference vertex	a	b	c	d	e	f	t
s	④	11	7	100	100	100	100
a		11	⑦	12	100	14	100
c		11		11	⑩	14	100
e		⑪		11		14	15
b				11		14	15
d						14	13

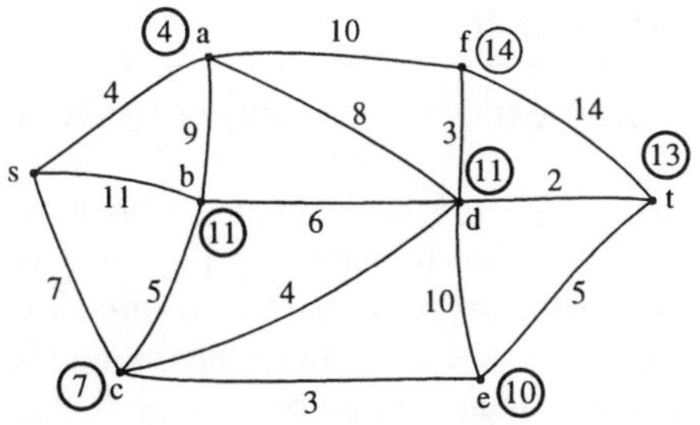

Figure 19.25 *The solution to the shortest path problem. Vertex t now has a permanent label, showing that the length of the shortest path from s to t is 13. By working backwards from t to s we find that the actual path must be scdt.*

Table 19.7 *The algorithm to find the shortest path stops as soon as t receives a permanent label*

Reference vertex	a	b	c	d	e	f	t
s	④	11	7	100	100	100	100
a		11	⑦	12	100	14	100
c		11		11	⑩	14	100
e		⑪		11		14	15
b			⑪			14	15
d						14	⑬

process, we find that the shortest path is scdt and is of length 13. The final solution is shown in Figure 19.25 and Table 19.7.

19.6 Networks and maximum flow

A network is a digraph with a weight function. The weight function represents the number of links between the vertices, for example the units of traffic (channels) in a communication network, or the maximum capacity of a network of one way streets in a busy town. This weight function value is called the capacity of the arc. An example of a network is shown in Figure 19.26. s is the source and t is the sink. The capacity of an arc, a_k will be referred to as $c(a_k)$. We wish to find the maximum possible flow from s the source to the sink t. That is we want to find a flow function $f(a_k)$ assigning a flow to each of the arcs which gives this maximum flow.

Flow function

A valid flow function must be such that:

1. $f(a_k) \leqslant c(a_k)$.
2. The out-flow at any vertex equals the in-flow except at the source or the sink.

An arc is saturated if $f(a_k) = c(a_k)$. The total flow is then the flow out of s (which equals the flow into t).

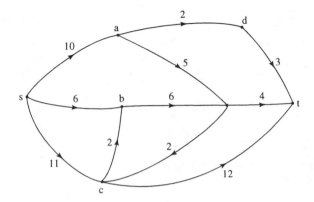

Figure 19.26 *A network. The weights on the arcs represent the capacity of the arc. s is the source and t is the sink.*

Finding a maximum flow

1. Assign an initial flow to each edge (this will probably be 0 flow on each arc).
2. Find a flow-augmenting path from s to t. Any arc along this flow-augmenting path may be either in the forward direction, that is, in the same direction as the path, or in the backward direction, that is, in the opposite direction to the path. If the arc along the path is in the forward direction, then it must have some 'extra' capacity $c(a_k) \geqslant f(a_k)$ then the extra capacity is represented by $\Delta_k = c(a_k) - f(a_k)$. If the arc along the path is in the backward direction, then $f(a_k)$ must be greater than 0 and we assign the possible increase in flow $\Delta_k = f(a_k)$. We can then calculate the amount that this path can augment the flow by as $\Delta = \text{minimum} (\Delta_k)$, where we consider all the values of along the path.
3. We can now change the value of the flow function along the flow-augmenting path to $f(a_k) + \Delta$ (for arcs in the forward direction along the path) and $f(a_k) - \Delta$ (for arcs in the backward direction along the path). We then repeat the second stage and look for another flow-augmenting path. If no more flow-augmenting paths exist, then we have found the maximum flow. This algorithm is due to Ford and Fulkerson. It can take a long while to terminate on complicated networks and better algorithms have been developed, for example, the Dinic algorithm which is not given here. We can check that we have in fact found the maximum flow in the network by finding the minimum cut. The max-flow, min-cut theorem states that the maximum value of a flow from s to t in a network is equal to the minimum capacity of a cut separating s and t.

Example 19.3 Find the maximum flow from s to t in the network shown in Figure 19.27(a). The weights on the arcs represent their capacities.

Solution We begin by identifying a flow-augmenting path s, e, t. As initially we consider the flow function to be zero then the spare capacities along the arcs are 12 and 14, the minimum of which is 12. We therefore assign a flow of 12 along the path to get Figure 19.27(b).

We spot another flow-augmenting path, s, c, d, t with a possible flow of 10 (the minimum of 14, 19, and 10). We add 10 to the flow function along this path to give Figure 19.27(c).

We identify the flow-augmenting path s, a, b, t with a possible flow of 8 (the minimum of the values of the spare capacities of 14, 8 and 15). We add 8 to the flow function along this path to give Figure 19.27(d).

Another flow-augmenting path is s, a, c, d, b, t with a possible flow of 6 (the minimum of the values of the spare capacities of 6, 14, 9, 6, 7). We add 6 to the flow function along this path to give Figure 19.27(e).

There are no more entirely forward paths but we can see one that involves an arc in the backward direction, that has a non-zero flow function. This path is s, c, a, e, t where ca is in the backward direction. We take the minimum of the spare capacity in the forward direction along with the value of the flow in the backward direction. This gives 2 as the minimum of 4, 6, 3, and 2. We add 2 to the flows on the arcs in the forward direction and subtract 2 from the flow on ca which is in the backward direction. Note that this process is equivalent to redirecting the flow. We have now reached Figure 19.27(f). We are unable to identify any more flow-augmenting paths. We can confirm that this is in fact a solution to the maximum flow problem by using the max-flow, min-cut theorem. If we look at the cut marked we see that this has a capacity of 38. To find this we add the capacities of all the arcs which cross the cut from the 's end' to the 't end', that is, db, with a capacity of 6, dt with a capacity of 10 and et with a capacity of 14. This cut has a capacity of 38 in total which is the value of the flow we have found.

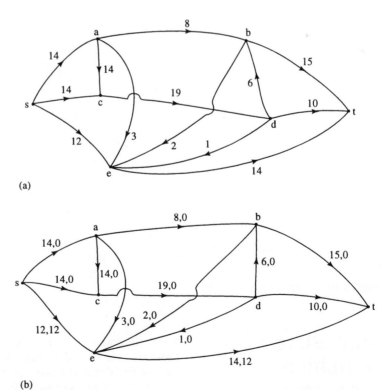

Figure 19.27 (a) The network for Example 19.3. The solution is presented in (b) − (f). (f) also shows the minimum cut which confirms that the maximum flow of 38 has been found.

(a)

(b)

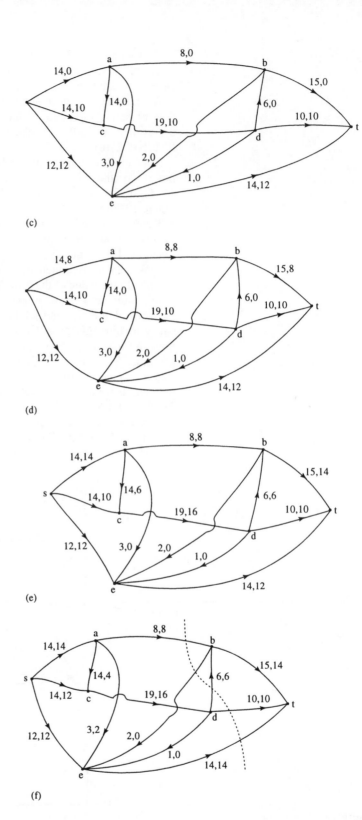

Figure 19.27 *continued.*

(c)

(d)

(e)

(f)

19.7 State transition diagrams

A state transition diagram is used to represent a finite state machine. These are used to model objects which have a finite number of possible states and whose interaction with the outside world can be described by its state changes in response to a finite number of events. A state transition diagram is a digraph whose nodes are states and whose directed arcs are

Table 19.8 *The states of a telephone*

State (off-hook, connected, dialing, tone, bell)	State name
(TRUE, TRUE, FALSE, none, OFF)	Connected
(TRUE, FALSE, TRUE, none, OFF)	In process of dialing
(TRUE, FALSE, FALSE, dial tone, OFF)	Prepared for dialing
(TRUE, FALSE, FALSE, engaged, OFF)	Engaged
(TRUE, FLASE, FALSE, ringing, OFF)	Waiting to connect
(TRUE, FALSE, FALSE, error, OFF)	Incorrect number or timeout
(TRUE, FALSE, FALSE, dead, OFF)	Connection fault
(FALSE, FALSE, FALSE, none, OFF)	Idle
(FALSE, FALSE, FALSE, none, ON)	Incoming call

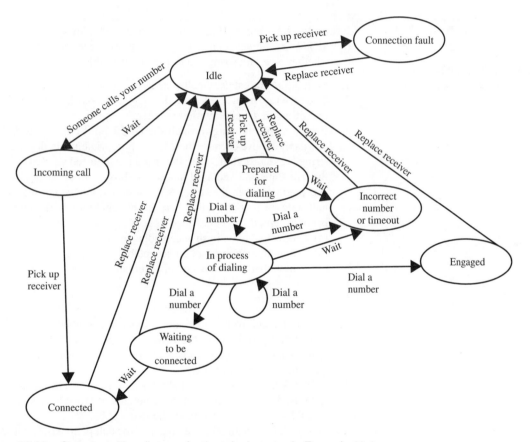

Figure 19.28 *State transition diagram for the telephone as in Example 19.4.*

transitions labelled by event names. A state is drawn as a rounded box containing an optional name. A transition is drawn as an arc with the arrow from the receiving state to the target state. The label on the arrow is the name of the event causing the transition. State transition diagrams have a number of applications. They are used in object-oriented modelling techniques to represent the life cycle of an object. They can be used as a finite state recognizer for a regular language, for instance, to describe regular expressions used as variables in computer languages.

Example 19.4 A telephone has the attributes of off-hook (TRUE, FALSE), connected (TRUE, FALSE), dialing (TRUE, FALSE), tone (dial tone , engaged, ringing, error, dead or none), and bell (ON, OFF). These can be represented as a list (off-hook, connected, dialing, tone, bell) which can take a selection of values to indicate the current state of the telephone, for example, TRUE, FALSE, FALSE, engaged, OFF indicates that the phone is off-hook, it is not connected, it is not dialing it has the engaged tone, and the bell is OFF. Use your knowledge of the operation of a telephone to give a full list of the possible states of the phone. Draw a state transition diagram to represent the life-cycle of the phone.

Solution   The possible states are given in Table 19.8.

The possible events are: Pick up receiver, replace receiver, wait, dial a number, someone calls your number. The events that are possible and the transitions that they cause depend on the current state of the phone. The state transition diagram is shown in Figure 19.28.

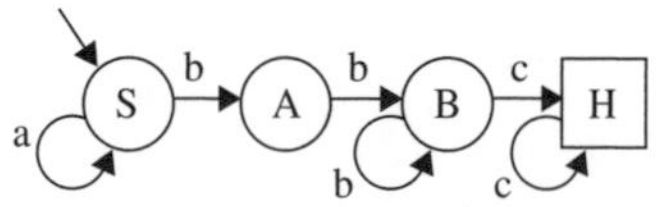

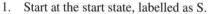

Figure 19.29   *A DFSR for Example 19.5.*

Example 19.5   Figure 19.29 shows a deterministic finite state recognizer (DFSR) for a language consisting of strings of the letters a, b, and c. To test whether a certain string is accepted by the DFSR:

1. Start at the start state, labelled as S.
2. Take a letter from the left of the string. If possible, move along a transition arc labelled with that letter.
3. Repeat step 2 moving through the DFSR until either you find that you cannot move because there is no appropriately labelled arc or the string is exhausted. If, when the string is exhausted you have reached the halt state (marked as H in a square box) then the string has been accepted. If, when the string is exhausted you have not reached the halt state, or if at any point there are no legal moves, then the string is not accepted.
   (a) show that abbc is an acceptable string,
   (b) show that abbb is not an acceptable string,
   (c) show that ababab is not an acceptable string.

Solution

(a) We find a path from S to H along arcs labelled abbc. To do this, we travel along arcs in the following order SS, SA, AB, BH. As this path begins with S and ends with H we have shown that the DFSR recognizes the string abbc.

(b) We look at the path labelled abbb beginning at S. The only possible path is: SSAB. As this path does not end at H we have shown that the DFSR does not recognize the string abbb.

(c) We look for a path labelled ababab. No such path exists as it would begin with the arcs SS, SA and then there is no arc out of A labelled 'b'. As there are no more legal moves we have shown that the string is not accepted.

19.8 Summary

1. A graph consists of a set of vertices, V and a set of edges, E. A digraph is a directed graph, with arcs instead of edges. Various definitions for different types of graphs are given in Section 19.2.
2. Graphs can be represented by an incidence matrix or an adjacency matrix.
3. Various optimization algorithms have been demonstrated in this chapter, including the greedy algorithm for the minimum spanning tree and the solution to the minimum path problem. For these two examples, the weights on the graph represent lengths, costs, or delays. We have also looked at one solution to the maximum flow problem.

For this problem the weights on the arcs of a digraph represent the maximum capacity of the arc and the problem is to maximise the flow through the network.

4. A state transition diagram is used to represent a finite state machine.

19.9 Exercises

19.1. From the graphs in Figure 19.30 identify any that are isomorphic.

19.2. The sequence of complete graphs $K_1, K_2, K_3, K_4 \ldots$ can be drawn by adding a vertex to the previous member of the sequence and drawing in all the edges necessary to make the new graph complete. Using this method, show that the number of edges in the complete graph K_n can be found from the series

$$1 + 2 + 3 + \cdots + n - 1$$

19.3. For what values of n is the complete graph K_n planar? For what values of m, n is the complete bipartite graph $K_{m,n}$ planar?

19.4. (a) Find the incidence and adjacency matrices for the graphs in Figure 19.31.
   (b) Show that in each case the column sum of the incidence matrix is 2 and explain this property.
   (c) What are the column sums of A and what property of the graph do they represent?

19.5. Find a minimum spanning tree for the weighted graph in Figure 19.32. Is there more than one possible minimum spanning tree for this graph?

19.6. Find the shortest path from s to t in the weighted graph shown in Figure 19.33.

19.7. Find the maximum flow in the network shown in Figure 19.34. The number next to each arc is its capacity. Use the max-flow, min-cut theorem to check you have found the maximum flow.

19.8. In the network in Figure 19.35, the arcs represent communication channels which have the maximum capacity indicated. The vertices represent switching centres, which also have a maximum capacity as indicated on the vertex. By replacing each of the vertices by two vertices with a connecting arc with a capacity equal to that of the original vertex, find the maximum

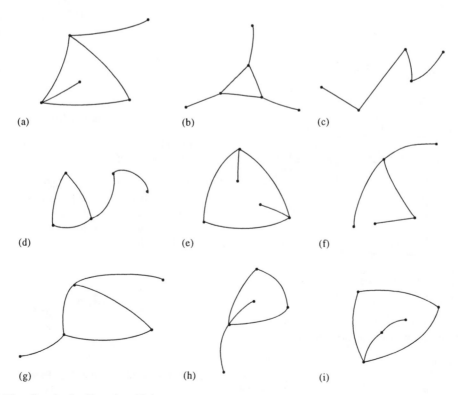

(a)   (b)   (c)

(d)   (e)   (f)

(g)   (h)   (i)

Figure 19.30 *Graphs for Exercise 19.1.*

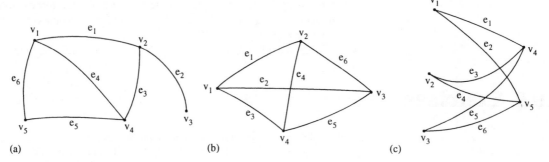

Figure 19.31   *Graphs for Exercise 19.4.*

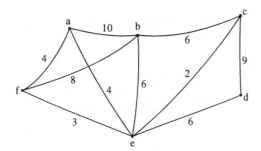

Figure 19.32   *Graph for Exercise 19.5.*

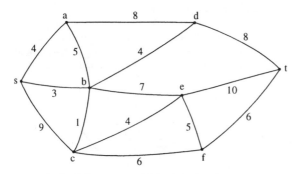

Figure 19.33   *Graph for Exercise 19.6.*

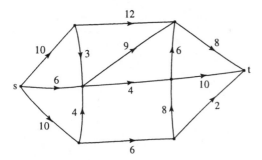

Figure 19.34   *The network for Exercise 19.7.*

flow from s to t. Use the max-flow, min-cut theorem to check you have found the maximum flow.

19.9. A bank's automatic telling machine (ATM) is characterized by the following attributes {money available, card, PIN number, authorization, display, money issued}.

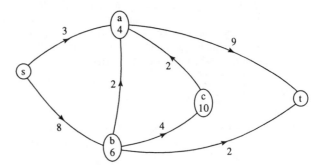

Figure 19.35   *The network for Exercise 19.8.*

'money available' is (TRUE, FALSE) depending on whether the cash available in the machine is sufficient for the current request.

'card' is TRUE or FALSE.

'PIN number' is TRUE or FALSE indicating whether any given pin number matches for the current card

'authorization' is TRUE or FALSE.

'display' is 'Insert card', 'Enter PIN', 'Enter amount', 'Take card', 'Take Money', 'Sorry – unable to complete request'.

'money issued' is TRUE or FALSE.

List all the possible states of the ATM and draw a state transition diagram.

19.10. Figure 19.36 shows a deterministic finite state recogniser (DFSR) for a language consisting of strings of the letters a, b, and c. Use the method of Example 19.5 to determine whether the following strings are accepted by the DFSR

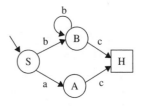

Figure 19.36   *A DFSR for Exercise 19.10.*

(a) bbbc, (b) abc, (c) acb.

20 Language theory

20.1 Introduction

We all use natural language to communicate. In a natural language, there is no strictly formal way of deciding exactly what is or is not an acceptable sentence. This is because natural language is constantly evolving and subject to cultural diversities. The meaning of a sentence in a natural language is yet another problem. A sentence may have different meanings in different parts of the world or because of ambiguities arising from its structure. If I say 'I locked my car keys in my boot' in England I have accidentally left my car keys in the boot (trunk) of my car. In America I have cleverly forced them into a lockable shoe. If I say 'Green fields like a wet blanket' I could be making a poetic reference to the covering of dew or rain being helpful for keeping a field green or I could be making disparaging comments about a cricket fielder called Green. Meanings associated with language are called semantics whereas its structure forms the syntax of the language.

There are reasons for requiring a simpler definition of language to use in formal situations. For instance, there are programming languages used for writing computer instructions. We would hope that the decision as to whether a particular sequence of symbols forms a program could be exactly determined. Also we would like to reduce any ambiguities as to its meaning so that any computer faced with the same program might take the same actions. Even in natural language it can be desirable to have rules that can be applied to the language to produce a generally accepted version of the language. This can be used for cross-cultural exchange and such a formalized version of a language can also simplify translation from one natural language to another.

A formal language consists of a set of sequences of symbols called sentences. The membership of the language set is usually defined by using grammar rules. Using the grammar rules to determine whether a particular string of symbols constitutes a sentence is called parsing. One way to do this would be to repeatedly use the grammar rules to list all the sentences in the language and stop if our sentence appears. This is not very practical because the number of sentences in a languages can be very large or even infinite. A refinement of this approach would be to use the grammar rules to find all sentences at least as long as the string of symbols we are interested in and check whether our sentence appears. These are called top-down approaches because we start with the rules and derive sentences. Alternatively, we could try and start with a string of symbols and test whether it obeys the grammar rules provided. This is a bottom-up approach which involves applying the grammar rules in reverse. We find that only certain sorts of grammars are easy, or possible, to use in reverse in this manner. For these grammars, which are easier to parse, we can define a procedure to test whether any given string is a sentence. The procedure that is used is called a parser. Identifying grammars with simple parsing procedures is an important consideration when designing a new computer language.

We look at eXtensible Markup Language (XML) as a method of defining a markup language. There are many easily accessible tools for checking whether an XML document is well formed and it is therefore ideal as a way of communicating data.

20.2 Languages and grammars

Definitions

A *language* is a set of sequences of symbols. A sequence of symbols is called a *string*.

An element of the language set is called a *sentence*. Some texts refer to the element of the language as a word or sometimes a well-formed formula or well-formed document depending on the type of language being discussed. For instance, logical expressions involving propositions or predicates, which we looked at in Chapter 3, can be regarded as sentences in the language of logic. In this case an expression which makes sense is called a well-formed formula or wff (pronounced 'woof'). An example of a logical sentence or wff is 'p∧q⇒p' but 'p⇔∧q' is not a sentence.

A symbol which forms part of a sentence is called a *terminal symbol*. For instance, the terminal symbols in the sentence 'The cat sat on the mat' are 'The', 'cat', 'sat', 'on', 'the' and 'mat'. The terminal symbols in 'p ∧ q⇒p' are 'p', '∧', 'q' and '⇒'.

Other symbols are used to express the grammar rules for the language. These symbols are called *non-terminal symbols*. In a natural language we could say that a simple sentence is made up of a noun phrase followed by a verb phrase possibly followed by a noun phrase. Then symbols representing 'simple sentence', 'noun phrase' and 'verb phrase' would be amongst our non-terminal symbols. In logic, we could say that a binary operation consists of a proposition followed by a binary operator followed by a proposition. Then symbols representing 'proposition' and 'binary operator' would be amongst our non-terminal symbols.

The grammar rules for the language give rules for replacing symbols by other symbols. For instance, in a natural language we might replace 'noun phrase' by 'article', 'adjective', and 'noun'. These grammar rules are called *rewriting rules* or *productions*.

When we write the rewriting rules or productions we will write one symbol followed by another. This indicates the operation of *concatenation*. Concatenation is the process of writing two or more symbols as one.

We carry on using rewriting rules or productions to replace symbols until we find a sequence of symbols that only contains terminal symbols. Then we have found a sentence.

In order to use the rewriting rules we must have a starting point. The starting point of the grammar is called the initial or start symbol. This is a symbol which represents an entire sentence. For a computer program it would represent the whole program.

We can now state that a grammar consists of four things:

1. A set of terminal symbols.
2. A set of non-terminal symbols.
3. A set of rewriting rules or productions. Every rule has a left-hand side and right-hand side. The meaning of a rule is that you can substitute its right-hand side for its left-hand side.
4. A designated start symbol.

Terminal symbols are often represented by lowercase letters and non-terminal symbols by uppercase letters.

A sentence is grammatical if it can be produced by a series of rewritings such that

1. It starts with the start symbol.
2. It uses grammar rules to get from one rewriting to the next.
3. It does not contain any non-terminals.

A grammar rule or production is a pair of strings of terminal and non-terminal symbols (P, Q). P is called the left-hand side of the rule and Q the right-hand side. This means that P may be replaced by Q. This is often represented as

$$P \rightarrow Q$$

A *derivation,* of one string of symbols, Y, from another string of symbols, X, is a sequence of application of productions which derives one string from another. If Y can be derived from X then we can write $X \Rightarrow^* Y$. The notation '$\Rightarrow^*$' indicates zero or more steps and '$\Rightarrow$' indicates a single step.

A string of terminal symbols is a sentence if it can be derived from the start symbol.

Example 20.1   A language, L, over the symbols a and b, is defined by the following grammar rules, where S is the start symbol.

$$S \rightarrow aSb$$

$$S \rightarrow ab$$

Find a definition of the set L.

Solution   We start with the symbol S and use either one of the grammar rules to make a replacement. We choose the rule

$$S \rightarrow aSb$$

to get

$$S \Rightarrow aSb$$

Using the rule $S \rightarrow aSb$ again to give

$$S \Rightarrow aSb \Rightarrow aaSbb$$

continuing using the same rule and we can see that we get

$$S \Rightarrow^* a^{n-1} S\, b^{n-1}$$

after $n - 1$ applications of the rule. Here, a^{n-1} has been used as shorthand for the repetition of 'a' n-1 times. There is still at least one non-terminal symbol in the right-hand side. We would need to use the second rule

$$S \rightarrow ab$$

to obtain

$$S \Rightarrow^* a^n b^n$$

We can see that the language consists of a number of 'a's followed by the same number of 'b's and so

$$L = \{a^n b^n \text{ where } n > 0\}.$$

Example 20.2   The language of the Boolean expressions over the set $\{0,1\}$ with the binary operations of $+$ and $\cdot$ with brackets, (,), can be defined with the following grammar rules

$S \rightarrow (S)B(S)$

$S \rightarrow (S)$

$S \rightarrow NBN$

$S \rightarrow (S)BN$

$S \rightarrow NB(S)$

$S \rightarrow N$

$N \rightarrow 0$

$N \rightarrow 1$

$B \rightarrow +$

$B \rightarrow \cdot$

(a) Show that $1 \cdot 10$ is not a sentence.
(b) Show that $(1 \cdot 1)$ is a sentence.

Solution   (a) We will find all the sentences of four terminal symbols. To do this, we repeatedly use the given grammar rules. For convenience, we can summarize the grammar rules by using '|' to represent 'or'. Then we can combine rules with the same left-hand side as follows:

$S \rightarrow (S)B(S) \mid (S)BN \mid NB(S) \mid (S) \mid NBN \mid N$

$N \rightarrow 0 \mid 1$

$B \rightarrow + \mid \cdot$

As none of the production rules have more symbols on the left than on the right and as we are looking for sentences of four terminal symbols, we only need to consider production rules with up to four symbols on the right-hand side. For this reason we only need to consider:

$S \rightarrow (S) \mid NBN \mid N$

$N \rightarrow 0 \mid 1$

$B \rightarrow + \mid \cdot$

Substituting $+$ or $\cdot$ for B and 0 or 1 for N in $S \rightarrow NBN$ gives the following derivations:

$S \Rightarrow^* 0 + 0$	$S \Rightarrow^* 0 + 1$	$S \Rightarrow^* 1 + 0$	$S \Rightarrow^* 1 + 1$
$S \Rightarrow^* 0 \cdot 0$	$S \Rightarrow^* 0 \cdot 1$	$S \Rightarrow^* 1 \cdot 0$	$S \Rightarrow^* 1 \cdot 1$

Substituting for B and N in $S \rightarrow N$ gives the following derivations:

$S \Rightarrow^* 0 \quad S \Rightarrow^* 1$

If we start with the production $S \rightarrow (S)$ and then apply $S \rightarrow NBN$ to give $S \Rightarrow^* (NBN)$ we can substitute for NBN as found above which gives:

$S \Rightarrow^* (0 + 0)$	$S \Rightarrow^* (0 + 1)$	$S \Rightarrow^* (1 + 0)$	$S \Rightarrow^* (1 + 1)$
$S \Rightarrow^* (0 \cdot 0)$	$S \Rightarrow^* (0 \cdot 1)$	$S \Rightarrow^* (1 \cdot 0)$	$S \Rightarrow^* (1 \cdot 1)$

Similarly, we can start with S → (S) and then substitute S → N followed by N → 0 or N → 1, to give:

$$S \Rightarrow^* (0) \quad S \Rightarrow^* (1)$$

So the language consists of only the following sentences of length less than or equal to 4

| $0+0$ | $0+1$ | $1+0$ | $1+1$ | $0 \cdot 0$ | $0 \cdot 1$ | $1 \cdot 0$ | $1 \cdot 1$ |

$0 \qquad 1 \qquad (0) \qquad (1)$

We can see that the string $1 \cdot 10$ does not appear in this list of sentences and so $1 \cdot 10$ is not a sentence.

(b) To show that (1.1) is a sentence we need to look at sentences of five symbols. In our solution to (a) we already found (1.1) to be a sentence as we derived $S \Rightarrow^* (1.1)$ by using S → (S) followed by S → NBN with N replaced by '1' on both occurrences and B replaced by '·' So we have the derivation $S \Rightarrow (S) \Rightarrow (NBN) \Rightarrow (1.1)$. We have shown that $S \Rightarrow^* (1.1)$. This string contains only terminal symbols and therefore (1.1) is a sentence.

Context-free languages

Formal grammars were studied by a linguist, Noam Chomsky, in the 1950s. Chomsky proposed a classification of grammars of which the most important classification, for our current interest, is that of context-free languages. Grammars that allow only a single non-terminal and nothing else on the left-hand side of the rule are called context free.

Chomsky proposed a transformation procedure to transform grammars of context-free languages to Chomsky Normal Form. His work influenced John Backus, who realized that programming languages are mostly context free. He used context-free grammars to describe the syntax of programs, and developed what is now called 'Backus Normal Form' or 'Backus Naur Form'. It has since been further revised and extended so nowadays programming language specifications are written in EBNF, Extended Backus-Naur Form. Before looking at EBNF we look at derivation graphs and derivation trees.

20.3 Derivations and derivation trees

A derivation can be represented using a digraph. We draw a vertex for the start symbol, a vertex for each production at each step of the derivation and a vertex for each symbol on the right-hand side of each step of the derivation. The following example is for a context-sensitive language.

Example 20.3   A language is defined over the symbols {a,b} with the following productions (labelled r_1–r_5):

$r_1 : S \rightarrow SBba$

$r_2 : bBa \rightarrow aAb$

$r_3 : bA \rightarrow AB$

$r_4 : Bb \rightarrow bBa$

$r_5 : S \rightarrow bB$

$r_6 : B \rightarrow a$

Represent the derivation

$$S \Rightarrow SBba \Rightarrow SbBaa \Rightarrow SaAba \Rightarrow bBaAba$$

as a digraph.

Solution We draw a digraph with a vertex for the start symbol, for each production used and for each symbol on the right-hand side of each step of the derivation. The productions used in this case are

$S \Rightarrow SBba$	Using r_1
$SBba \Rightarrow SbBaa$	Using r_4
$SbBaa \Rightarrow SaAba$	Using r_2
$SaAba \Rightarrow bBaAba$	Using r_5

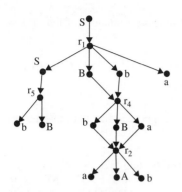

Figure 20.1 *A digraph representing the derivation of Example 20.3.*

This is represented as a digraph in Figure 20.1.

Context-free grammars have a single symbol on the left-hand side of each production. In this case the derivation graph is always a derivation tree. We do not need to include a vertex for each rule, but just for the start symbol and for each symbol on the right-hand side of each step of the derivation.

Example 20.4 Using the grammar of Example 20.1:

$$S \rightarrow (S)B(S) \mid (S)BN \mid NB(S) \mid (S) \mid NBN \mid N$$

$$N \rightarrow 0 \mid 1$$

$$B \rightarrow + \mid \cdot$$

draw the derivation tree for

$$S \Rightarrow^* (1 \cdot 1).$$

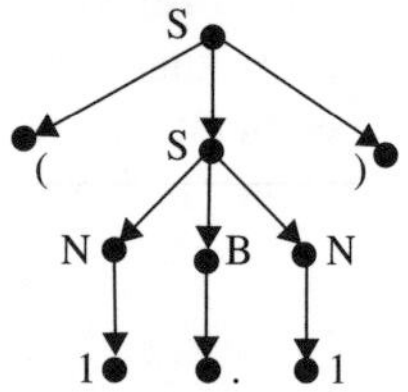

Solution The derivation is as follows:

$$S \Rightarrow (S)$$

$$(S) \Rightarrow (NBN)$$

$$(NBN) \Rightarrow (1 \cdot 1)$$

The derivation tree is shown in Figure 20.2.

Figure 20.2 *The derivation tree for Example 20.4.*

We are able to show that the derivation of a context-free grammar can be represented as a derivation tree. However, we want to efficiently solve the problem of deciding whether a particular string is a sentence in a context-free language. At the end of Chapter 19, we looked at DFSRs as an example of the use of state transition diagrams. These are able to decide whether a string is a sentence in some regular language. Regular languages are a sub-class of context-free languages where the right-hand side of all productions are of the form of either a single terminal or a terminal followed by a non-terminal. These languages are useful for defining the syntax of regular expressions but are not powerful enough to be used to define the syntax of a programming language. To build a parser for a context-free language we use the idea of a pushdown recognizer. This is a machine with a finite number of states, as before, but it also has the use of a stack. A stack can be compared to a stack of plates. You can only place plates on the top of the stack, when the operation is called pushing, and remove them again from the top, called popping. The bottom of the pile is inaccessible. In this way, any number of symbols can be pushed onto the top of the stack. When they are popped off again the last one pushed will be the first one popped. Stacks are often referred to as LIFOs (last in, first out).

Efficient parsers can be written if the grammar of the context-free language is converted to Extended Backus-Naur Form.

20.4 Extended Backus-Naur Form (EBNF)

Any grammar in EBNF is context-free. Each rule of the grammar defines one symbol of the form.

symbol ::= expression

In the above symbol and expression are non-terminals and represent syntactic categories. Terminal symbols, or string literals, are given inside quotes, for example "string" or 'string' is a literal string as given inside the quotes. Brackets (. . .) can be used to group constituents.

Symbols may be combined to represent more complex patterns as in Table 20.1.

Not all grammars can be converted to EBNF and a full description of methods of conversion, when possible, is outside the scope of this book. The following procedure works in simple cases.

To convert a grammar to EBNF:

1. Remove left recursion, that is

   N ::= E | NF

   replace by N ::= E | EF*.
   We can see that applying the rule N ::= NF twice gives N := NFF. As this is a recursive rule we can continue applying it any number of times, giving N ::= NFFF, N ::= NFFFF, N ::= NFFFFF, and so on. The only way to have a right-hand side which does not include N, that is, to remove the left recursion is to replace the N by E and finally we get N ::= E | EF*.

2. Left factor the grammar

   N ::= EFG | EF'G

   replace by N ::= E(F | F')G.

3. If N ::= E is not recursive remove it and replace all occurrences of N in the grammar with E.

Table 20.1 *Symbol patterns and their meaning in EBNF, where A and B are simple expressions*

Symbols	Represents
A?	A or nothing and is referred to as 'optional A'
A B	A followed by B
A|B	A or B but not both. This operator has lower precedence than followed by thus A B | C D is identical to (A B) | (C D)
A − B	any string that matches A but does not match B
A+	one or more repetitions of A
A*	zero or more repetitions of A

4. Reduce the grammar until there is a single production rule for each non-terminal symbol, using optional expressions where necessary.

Example 20.5 Express the grammar of Example 20.1, that is

$S \rightarrow (S)B(S) \mid (S)BN \mid NB(S) \mid (S) \mid NBN \mid N$

$N \rightarrow 0 \mid 1$

$B \rightarrow + \mid \cdot$

in EBNF.

Solution First, we change the notation, using quotes for literals and the ::= instead of $\rightarrow$

$S ::= \text{"("S")"}B\text{"("S")"} \mid \text{"("S")"}BN \mid NB\text{"("S")"} \mid \text{"("S")"} \mid NBN \mid N$

$N ::= \text{"0"} \mid \text{"1"}$

$B ::= \text{" + "} \mid \text{" } \cdot \text{ "}$

There is no left recursion. We left factor the grammar for the first expression and use optional expressions to reduce it to a single production rule. To do left factoring notice that three productions for S begin "("S")" and another three begin with N so we have:

$S ::= \text{"("S")"}(B(\text{"("S")"} \mid N))? \mid N(B(\text{"("S")"} \mid N))?$

$N ::= \text{"0"} \mid \text{"1"}$

$B ::= \text{" + "} \mid \text{" } \cdot \text{ "}$

where the productions for N and B were already in EBNF.

We can see why this form is easy for a parser to use to develop an algorithm for checking strings for validity by representing each of the productions in EBNF as a directed tree so that the grammar forms a forest. We use a different node for each of the non-terminals in the expression and join following on non-terminal symbols with arcs. If the production has one or more terminal symbols then label the arc with the terminal symbols. The final symbol in an expression of the production has a halt node attached. This is pictured in Figure 20.3.

Strings are checked for validity in the following manner. Start at the start symbol with an ingoing arc and take a symbol from the left of the string, which we call 't'. If there is an arc marked with 't' then follow that arc. If not then examine each of the possible nodes at the other end of an unmarked arc, which we will call N, and check whether 't' would be accepted by N. To do this use the sub-tree which has a start symbol N. Continue within that sub-tree until the halt state is reached and then return to the main sub-tree at our previous position at the node with label N. Continue in this manner until the string is exhausted or there are no further legal moves. If, when the string is exhausted, there is an unmarked arc leading to a halt state or we have already reached a halt state, then the string is accepted.

Example 20.6 Use Figure 20.3 to show that $(1 \cdot 1)$ is a valid sentence in the language.

Solution We begin at the start symbol marked S with an incoming arc. Take the first symbol from "$(1 \cdot 1)$" and this is "(". There is an arc labelled

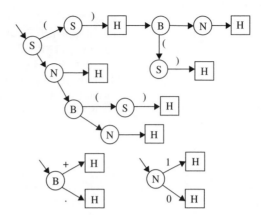

Figure 20.3 *Forest used for parsing a context-free grammar as in Example 20.6.*

"(" so we follow that to the node marked S. We enter the sub-graph for S, back at the original incoming arc, and take the next symbol from the string "1". The only possibility is to follow the arc to N. We check that N will accept "1" by going into its sub-graph. The sub-graph for N has an arc labelled "1" leading to a halt state and therefore N accepts "1". We take the next symbol from the string "·" and move on from N to B. We check that B accepts "·" which it does because the sub-graph for B has an arc labelled "·" leading to a halt symbol. We take the next symbol from the string "1" and move on from B to N. We check that N will accept "1" by going into its sub-graph. We have now reached a halt state in S so we have shown that the sub-string "1 · 1" has been accepted by S. We are back into the main graph at the S node following the "(". We take the next symbol from the string ")" and this leads to a halt state. As we have reached the end of the string we have shown that the string "(1 · 1)" is accepted.

This parsing algorithm is called LR parsing where LR stands for '*L*eft to Right *R*ightmost'. This is a bottom-up parsing technique which derives the parsing tree for a string in reverse, reducing the string by replacing right-hand sides of a production by its left-hand side until the string has been reduced to the start symbol of the grammar. This particular example is LR(1) because it is only ever necessary to lookahead by one symbol in order to take the decision as to which production should be used for the reduction.

There are programs that will take a context-free grammar as input and output the code for the corresponding parser. Such parser generators are called compiler compilers. Some examples of these are YACC (Yet Another Compiler Compiler) developed by Bell labs. The GNU version is called Bison. The Java Compiler Compiler, JavaCC, originally developed by Sun Microsystems, can be downloaded free from http://www.webgain.com/products/java_cc/ and runs on most systems.

20.5 Extensible markup language (XML)

Embedding mark-up in a document is not a new idea. Proofreaders refer to marking up a document in order to indicate the changes that are to be made. Using a precise set of symbols makes the interpretation of the proofreaders instructions much easier. Early word processing languages made extensive use of mark-up to define how the document was to be laid out. Markup became an expression that entered into almost everyone's (non-formal) vocabulary with the introduction of Hypertext Markup Language (HTML) which allowed easy exchange of document information over the Internet which could be displayed by browsers specially built to interpret HTML.

eXtensible Markup Language (XML) is a meta-language, a tool for defining markup languages using tags enclosed in angle brackets $< \cdots >$ to start an element of the language and $< / \cdots >$ to close an element. The elements may have properties, called attributes, and child elements. The main reasons for the importance of the development of XML is:

1. XML is a tool for defining markup languages.
2. XML languages can be used to develop standards for the interchange of data.
3. There are many XML parsers available which will check an XML document for well-formedness and validity.
4. The structure of XML lends itself easily to a tree representation – showing the derivation tree of the document.

An XML document has an associated definition of the allowed elements, element embeddings, that is, the possible parent-child relationships, and attributes. This is called a document type definition (DTD). The parser can check any XML document in two possible ways, either simply for well-formedness or it can also validate it against its DTD. A *well-formed document* must follow the general rules for any XML document whereas a *valid document* must also agree with the constraints defined in its DTD. We give an example here of a well-formed document and not concern ourselves at present with the structure of a DTD.

We might want to represent the following information about books: Publisher, Title, Author, ISBN, Category. As all books have a publisher then we could say that a book is a child of the publisher. In this case, we could represent some information in the following XML fragment.

```
<publisher name="The Book Company">
    <book ISBN="0–07–707975–2" category="Engineering
    Mathematics">
        <author>
            <forename>Mary</forename>
            <surname>Attenborough</surname>
        </author>
        <title>
            Engineering Mathematics for All
        </title>
    </book>
    <book ISBN="1–873432–07–0" category="Food Hygiene">
        <author>
            <forename>John</forename>
            <surname>O'Brien</surname>
        </author>
        <title>
            Understanding Food Hygiene
        </title>
    </book>
</publisher>
```

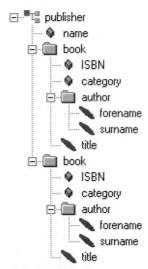

Figure 20.4 *Tree representation of a well-formed XML document.*

This fragment would constitute a well-formed XML document. We have indented it so it is clear the structure of document. We can see it as a tree displayed with the root of the tree on the left-hand side – this is pictured in Figure 20.4.

20.6 Summary

1. A language is a set of sequences of symbols. A sequence of symbols is called a string. An element of the language set is called a sentence. A symbol which forms part of a sentence is called a terminal symbol.

2. Other symbols are used to express the grammar rules for the language. These symbols are called non-terminal symbols.

3. A grammar consists of four things:
    (a) A set of terminal symbols.
    (b) A set of non-terminal symbols.
    (c) A set of rewriting rules or productions. Every rule has a left-hand side and right-hand side. The meaning of a rule is that you can substitute its right-hand side for its left-hand side.
    (d) A designated start symbol.

4. A sentence is grammatical if it can be produced by a series of rewritings such that
    (a) It starts with the start symbol.
    (b) It uses grammar rules to get from one rewriting to the next.
    (c) It does not contain any non-terminals.

5. A derivation, of one string of symbols, Y, from another string of symbols, X, is a sequence of applications of productions which derives one string from another. If Y can be derived from X then we can write $X \Rightarrow^* Y$. The notation '$\Rightarrow^*$' indicates zero or more steps and '$\Rightarrow$' indicates a single step.

6. Grammars that allow only a single non-terminal and nothing else on the left-hand side of the rule are called context-free.

7. A derivation can be represented using a digraph. A derivation of a context-free grammar can be represented as a directed tree called the derivation tree.

8. Extended Backus-Naur Form (EBNF) is used to represent the syntax of programming languages. Any grammar in EBNF is context-free. This form of a grammar lends itself to the easy development of a parser to check strings for validity.

9. To convert a grammar to EBNF then remove left recursion, left factor the grammar, if a rule N ::= E is not recursive remove it and replace all occurrences of N in the grammar with E. Reduce the grammar until there is a single production rule for each non-terminal symbol, using optional expressions where necessary.

10. eXtensible Markup Language (XML) is a tool for defining markup languages. XML languages can be used to develop standards for the interchange of data and there are many XML parsers available which will check an XML document for well-formedness and validity. The structure of XML lends itself easily to a tree representation – showing the derivation tree of the document.

20.7 Exercises

20.1. A language, L, over the symbols a, b, and c, is defined by the following grammar rules, where S is the start symbol.

$S \rightarrow aS$

$S \rightarrow aB$

$B \rightarrow bc$

Find a definition of the set L.

20.2. The language of the Boolean expressions over the set {0,1} with the binary operations of + and · with brackets, (,), can be defined with the following grammar rules

$S \rightarrow (S)B(S) \,|\, (S)BN \,|\, NB(S) \,|\, (S) \,|\, NBN \,|\, N$

$N \rightarrow 0 \,|\, 1$

$B \rightarrow + \,|\, \cdot$

Use a top-down method to show that
(a) $(1 \cdot 0) + 1$ is a sentence
(b) $1 \cdot 0 + 1$ is not a sentence.

20.3. For the grammar

$$S \rightarrow a \mid b \mid c \mid aSa \mid bSb \mid cSc$$

find derivation trees for the following:
(a) aaa
(b) abbcbba.

20.4. Write the following grammar in EBNF

$$S \rightarrow Ab \quad C \rightarrow c$$

$$A \rightarrow AC \quad C \rightarrow a$$

$$A \rightarrow a$$

20.5. For the grammar in Exercise 20.4 use a bottom-up parsing technique to show that
(a) accb is a sentence
(c) abc is not a sentence.

20.6. Represent the data given in the table below as a well-formed XML document. (Note: there is no need to define the DTD for your XML.) There are several possible ways of presenting this data in XML – presenting one is sufficient.

Subject	Student ID	Marks	
		First semester	Second semester
Electronics 1	121	40	58
	122	35	38
	123	75	65
Engineering Maths	121	66	54
	122	54	45
	123	72	22
Communications	121	58	60
	122	44	55
	123	70	68

Part 5 Probability and statistics

21 Probability and statistics

21.1 Introduction

Consider a system of computers and communication networks supporting a bank's cash dispensing machines. The machines provide instant money on the street on production of a cash card and a personal identification code with the only prerequisite being that of a positive bank balance.

The communication network must be reliable – that is, not subject too often to break downs. The reliability of the system, or each part of the system, will depend on each component part.

A very simple system, with only two components, can be configured in series or in parallel. If the components are in series then the system will fail if one component fails (Figure 21.1).

If the components are in parallel then only one component need function (Figure 21.2) and we have built in redundancy just in case one of the components fails.

A network could be made up of tens, hundreds, or even thousands of components. It is important to be able to estimate the reliability of a complex system and therefore the rate of replacement of components necessary and the overall annoyance level of our customers should the system fail.

The reliability will depend on the reliability of each one of the components but it is impossible to say exactly how long even one single component will last.

We would like to be able to answer questions like:

1. What is the likely lifetime of any one element of the system?
2. How can we estimate the reliability of the whole system by considering the interaction of its component parts?

Figure 21.1 *Component A and B must function for the system to function.*

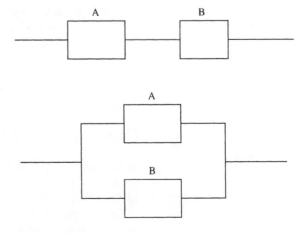

Figure 21.2 *Either component A or component B must function for the system to function.*

This is just one example of the type of problem for which we need the ideas of probability and statistics. Others are, a factory produces 1000 electronic components each day where we wish to be able to estimate how many defective components are produced each day or a company with eight telephone lines where we wish to say whether that is a sufficient number, in general, for the number of calls likely to be received at any one time.

All of these questions are those that need the ideas of probability and statistics. Statistics is used to analyse data and produce predictions about it . Probability is the theory upon which statistical models are built.

For the engineering situations in previous chapters we have considered that we know all the factors that determine how a system will behave, that is, we have modelled the system deterministically. This is not very realistic. Any real situation will have some random element. Some problems, as the one we have just introduced, contain a large random element so that it is difficult to predict the behaviour of each part of the system. However, it is possible to say, for instance, what the average overall behaviour would be.

21.2 Population and sample, representation of data, mean, variance and standard deviation

In statistics, we are generally interested in a 'population' that is too large for us to measure completely. For instance, we could be interested in all the light bulbs that are produced by a certain manufacturer. The manufacturer may claim that her light bulbs have a lifetime of 1500 h. Clearly this cannot be exact but we might be satisfied to agree with the manufacturer if most of the bulbs have a lifetime greater than 1500 h. If the factory produces half a million light bulbs per year then we do not have the time to test them all. In this case we test a 'sample'. The larger the sample, the more accurate will be the comparison with the whole population.

We have recorded the lifetimes of samples of light bulbs in Table 21.1. We can build up a table to give information about our sample. Columns 8 and 9 are given for comparison with statistical modelling and columns 4, 6, and 7 are to help with the calculation. The following describes each column of Table 21.1, beginning with a list of the 'raw' data.

Column 1: the class intervals

Find the minimum and the maximum of the data. The difference between these gives the range of the data. Choose a number of class intervals (usually up to about 20) so that the class range can be chosen as some multiple of 10, 100, 1000, etc. (like the class range of 100 above). The data range divided by the class range and then rounded up to the integer above gives the number of classes. The class intervals are chosen so that the lowest class minimum is less than the minimum data value. In our example, the lowest class minimum has been chosen as 900. Add the class range of 100 to find the class interval, for example, 900–1000. Carry on adding the class range to find the next class interval until you pass the maximum value.

Column 2: the class midpoint

The class midpoint is found by the maximum value in the class interval – the minimum value in the class interval divided by 2. For the interval 1300–1400, the class midpoint is $(1400 - 1300)/2 = 1350$.

The class midpoint is taken as a representative value for the class. That is, for the sake of simplicity we treat the data in the class 1300–1400 as

Table 21.1 *Frequency distribution of the lifetimes of a sample of light bulbs showing method of calculating the mean and standard deviation. N represents the number of classes, n the number in the sample*

(1) Lifetime (h)	(2) Class mid-point x_i	(3) Class frequency f_i	(4) $f_i(x)$	(5) Cumulative frequency F_i	(6) $(x_i - \bar{x})^2$	(7) $f_i(x_i - \bar{x})^2$	(8) Relative frequency f_i/n	(9) Relative cumulative frequency F_i/n
900–1000	950	2	1900	2	536 849	1 073 698	0.002	0.002
1000–1100	1050	0	0	2	400 309	0	0.002	0.002
1100–1200	1150	5	5750	7	283 769	1 418 845	0.005	0.007
1200–1300	1250	23	28 750	30	187 229	4 306 267	0.023	0.03
1300–1400	1350	47	63 450	77	110 689	5 202 383	0.047	0.077
1400–1500	1450	103	149 350	180	54 149	5 577 347	0.103	0.18
1500–1600	1550	160	248 000	340	17 609	2 817 440	0.16	0.34
1600–1700	1650	190	313 500	530	1069	203 110	0.19	0.53
1700–1800	1750	165	288 750	695	4529	747 285	0.165	0.695
1800–1900	1850	164	303 400	859	27 989	4 590 196	0.164	0.859
1900–2000	1950	92	179 400	951	71 449	6 573 308	0.092	0.951
2000–2100	2050	49	100 450	1000	134 909	6 610 541	0.049	1
		1000	1 682 700	$F_N = n$		39 120 420	1	
		$\sum_i f_i = n$					$\sum_i \dfrac{f_i}{n} = 1$	$\dfrac{F_N}{n} = 1$

$$\bar{x} = \frac{1\,682\,700}{1000}$$

$$\bar{x} = \frac{1}{n}\sum_i f_i x_i$$

$$\sigma^2 = \frac{1}{n}\sum_i f_i(x_i - \bar{x})^2$$

$$\sigma^2 = \frac{39\,120\,420}{1000}$$

$$\sigma \simeq 198$$

though there were 47 values at 1350. This is an approximation that is not too serious if the classes are not too wide.

Column 3: class frequency and the total number in the sample

The class frequency is found by counting the number of data values that fall within that class range. For instance, there are 47 values between 1300 and 1400. You have to decide whether to include values that fall exactly on the class boundary in the class above or the class below. It does not usually matter which you choose as long as you are consistent within the whole data set.

The total number of values in the sample is given by the sum of the frequencies for each class interval.

Column 4: $f_i x_i$ and the mean of the sample, $\bar{x}$

This is the product of the frequency and the class midpoint. If the class midpoint is taken as a representative value for the class then $f_i x_i$ gives the sum of all the values in that class. For instance, in the eighth class, with class midpoint 1650, there are 190 values. If we say that each of the values is approximately 1650 then the total of all the values in that class is $1650 \times 190 = 313\,500$. Summing this column gives the approximate total value for the whole sample $= 1\,682\,700$ h. If we only used one light bulb at a time and changed it when it failed then the 1000 light bulbs would last approximately $1\,682\,700$ h. The mean of the sample is this total divided by the number in the sample, 1000. Giving $1\,682\,700/1000 = 1682.7$ h. The mean is a measure of the central tendency. That is, if you wanted a simple number to sum up the lifetime of these light bulbs you would say that the average life was 1683 h.

Column 5: the cumulative frequency

The cumulative frequency gives the number that falls into the current class interval or any class interval that comes before it. You could think of it as 'the number so far' function. To find the cumulative frequency for a class, take the number in the current class and add on the previous cumulative frequency for the class below, for example, for 1900–2000 we have a frequency of 92. The cumulative frequency for 1800–1900 is 859. Add $859 + 92$ to get the cumulative frequency of 951. That is, 951 light bulbs in the sample have a lifetime below 2000 h. Notice that the cumulative frequency of the final class must equal the total number in the sample. This is because the final class must include the maximum value in the sample and all the others have lifetimes less than this.

Column 6: $(x_i - \bar{x})^2$, the squared deviation

This column is used in the process of calculating the standard deviation (see column 7). $(x_i - \bar{x})$ represents the amount that the class midpoint differs from the mean of the sample. If we want to measure how spread out around the mean the data are, then this would seem like a useful number. However, adding up $(x_i - \bar{x})$ for the whole sample will just give zero, as the positive and negative values will cancel each other. Hence, we take the square, so that all the numbers are positive. For the sixth class interval the class midpoint is 1450; subtracting the mean, 1682.7, gives −232.7, which when squared is 54149.2. This value is the square of the deviation from the mean, or just the squared deviation.

Column 7: $f_i(x_i - \bar{x})^2$, the variance and the standard deviation

For each class we multiply the frequency by the squared deviation, calculated in column 6. This gives an approximation to the total squared deviation for that class. For the sixth class we multiply the squared deviation of 54 149 by the frequency 103 to get 5 577 347. The sum of this column gives the total squared deviation from the mean for the whole sample. Dividing by the number of sample points gives an idea of the average squared deviation. This is called the variance. It is found by summing column 7 and dividing by 1000, the number in the sample, giving a variance of 39 120. A better measure of the spread of the data is given by the square root of this number, called the standard deviation and usually represented by σ. Here $\sigma = \sqrt{39\,120} \approx 198$.

Column 8: the relative frequency

If instead of 1000 data values in the sample we had 2000, 10 000, or 500, we might expect that the proportion falling into each class interval would remain roughly the same. This proportion of the total number is called the relative frequency and is found by dividing the frequency by the total number in the sample. Hence, for the third class, 1100–1200, we find $5/1000 = 0.005$.

Column 9: the cumulative relative frequency

By the same argument as above we would expect the proportion with a lifetime of less than 1900 h to be roughly the same whatever the sample size. This cumulative relative frequency can be found by dividing the cumulative frequency by the number in the sample. Notice that the cumulative relative frequency for the final class interval is 1. That is, the whole of the sample has a lifetime of less than 2000 h.

The data can be represented in a histogram as in Figure 21.3, which gives a simple pictorial representation of the data. The right-hand side has a scale equal to the left-hand scale divided by the number in the sample. These readings, therefore, give the relative frequencies and the cumulative relative frequency as given in Table 21.1.

We may want to sum up our findings with a few simple statistics. To do this we use a measure of the central tendency and the dispersion of the data, which are calculated as already described above. These are summarized below.

Central tendency – the mean

The most commonly used measure of the central tendency is the mean, $\bar{x}$

$$\bar{x} = \frac{1}{n}\sum_i f_i x_i = \sum_i \frac{f_i x_i}{n}$$

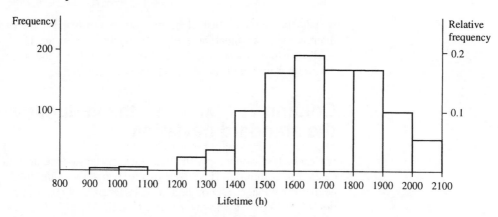

Figure 21.3 *Histogram of the lifetimes of a sample of light bulbs*

where x_i is a representative value for the class and f_i is the class frequency. The summation is over all classes.

If the data have not been grouped into classes then the mean is found by summing all the individual data values and dividing by the number in the sample:

$$\bar{x} = \frac{1}{n} \sum_i x_i$$

where the summation is now over all sample values

Dispersion – the standard deviation

The variance is the mean square deviation given by

$$v = \sigma^2 = \frac{1}{n} \sum_i f_i (x_i - \bar{x})^2 = \sum_i \frac{f_i}{n} (x_i - \bar{x})^2$$

where $\bar{x}$ is the sample mean, x_i is a representative value for the class, f_i is the class frequency, and the summation is over all classes. This value is the same as

$$v = \sigma^2 = \frac{1}{n} \sum_i f_i x_i^2 - \bar{x}^2$$

which can some times be quicker to calculate.

For data not grouped into classes we have

$$v = \sigma^2 = \frac{1}{n} \sum_i (x_i - \bar{x})^2$$

or

$$v = \sigma^2 = \frac{1}{n} \sum_i x_i^2 - \bar{x}^2$$

where the summation is now over all sample values.

The standard deviation is given by the square root of the variance:

$$\sigma = \sqrt{\frac{1}{n} \sum_i f_i (x_i - \bar{x})^2}.$$

We have already calculated the average lifetime and the standard deviation of the lifetimes of the sample of light bulbs. We repeat the calculation below. The mean is given by

class midpoint
$\downarrow$

$(2 \times 950 + 0 \times 1050 + 5 \times 1150 + 23 \times 1250 + 47 \times 1350 + 103 \times 1450$
$\uparrow$

class frequency

$\quad + 160 \times 1550 + 190 \times 1650 + 165 \times 1750 + 164 \times 1850 + 92 \times 1950$

$\quad + 49 \times 2050)/1000 = 1682.7.$

The variance is given by

class midpoint   sample mean
$\downarrow$ $\downarrow$

$(2 \times (950 - 1682.7)^2 + 0 \times (1050 - 1682.7)^2 + 5 \times (1150 - 1682.7)^2$
$\uparrow$

class frequency

$\quad + 23 \times (1250 - 1682.7)^2 + 47 \times (1350 - 1682.7)^2$

$\quad + 103 \times (1450 - 1682.7)^2 + 160 \times (1550 - 1682.7)^2$

$\quad + 190 \times (1650 - 1682.7)^2 + 165 \times (1750 - 1682.7)^2$

$\quad + 164 \times (1850 - 1682.7)^2 + 92 \times (1950 - 1682.7)^2$

$\quad + 49 \times (2050 - 1682.7)^2)/1000 = 39\,120\,420/1000 \approx 39120.4.$

Hence, the standard deviation is

$\sqrt{39120.4} \approx 198.$

Can we agree with the manufacturer's claim that the light bulbs have a lifetime of 1500 h? The number with a lifetime of less than 1500 h is 180 (the cumulative frequency up to 1500 h). This represents only 0.18, less than 20% of the sample. However, this is probably an unreasonable number to justify the manufacturers claim as nearly 20% of the customers will get light bulbs that are not as good as advertised. We might accept a small number, say 5%, failing to live up to a manufacturer's promise. Hence, it would be better to claim a lifetime of around 1350 h with a relative cumulative frequency between 0.03 and 0.077. Approximately 0.053 or just over 5% would fail to live up to the manufacturer's amended claim.

Example 21.1 A sample of 2000-Ω resistors were tested and their resistances were found as below:

1997 1998 2004 2002 1999 2000 2001 2002 1999 1998 1997 1999 2001
2003 2005 1996 2000 2000 1998 1999 1998 2001 2003 2002 1996 2002
1995 2000 2000 1999 1997 2004 2001 1999 1999 1998 2002 2003 2002
2001 1999 1998 1997 2002 2001 2000 1999 2001 1999 1998 2000 1999
1998 2000 2002 1999 2001 2000 2001 2002 2004 1996 2000 2002 2000
1998 2000 2005 2000 1999 2000 2000 2001 1999 2001 1997 2002 2001
2004 2000 2000 1999 1998 2001 1998 2000 2002 1998 1998 1999 2002
1999 2001 2002 2006 2000 2001 2001 2002 2003

Group the data in class intervals and represent it using a table and a histogram. Calculate the mean resistance and the standard deviation.

Solution   We follow the method of building up the table as described previously.

Column 1: class intervals. To decide on class intervals look at the range of values presented. The minimum value is 1995 and the maximum is 2006. A reasonable number of class intervals would be around 10, and in this case as the data are presented to the nearest whole number, the smallest possible class range we could choose would be 1 Ω. A 1-Ω class range would give 12 classes, which seems a reasonable choice. If we choose the class midpoints to be the integer values then we get class intervals of 1994.5–1995.5, 1995.5–1996.5, etc. We can now produce Table 21.2.

Table 21.2 *Frequency distribution of resistances of a sample of resistors. N represents the number of classes, n the number in the sample*

(1) Resistance (Ω)	(2) Class mid-point x_i	(3) Class frequency f_i	(4) $f_i(x)$	(5) Cumulative frequency F_i	(6) $(x_i - \bar{x})^2$	(7) $f_i(x_i - \bar{x})^2$	(8) Relative frequency f_i/n	(9) Relative cumulative frequency F_i/n
1994.5–1995.5	1995	1	1995	1	26.52	26.52	0.01	0.01
1995.5–1996.5	1996	3	5988	4	17.22	51.66	0.03	0.04
1996.5–1997.5	1997	5	9985	9	9.92	49.6	0.05	0.09
1997.5–1998.5	1998	13	25974	22	4.62	60.06	0.13	0.22
1998.5–1999.5	1999	17	33983	39	1.32	22.44	0.17	0.39
1999.5–2000.5	2000	19	38000	58	0.02	0.38	0.19	0.58
2000.5–2001.5	2001	16	32016	74	0.72	11.52	0.16	0.74
2001.5–2002.5	2002	15	30030	89	3.42	51.3	0.15	0.89
2002.5–2003.5	2003	4	8012	93	8.12	32.48	0.04	0.93
2003.5–2004.5	2004	4	8016	97	14.82	59.28	0.04	0.97
2004.5–2005.5	2005	2	4010	99	24.52	47.04	0.02	0.99
2005.5–2006.5	2006	1	2006	100	34.22	34.22	0.01	1
		100	200015	$F_N = n$		446.5	1	
		$\sum_i f_i = n$	$\bar{x} = \dfrac{200015}{100}$			$\sigma^2 = \dfrac{1}{n}\sum_i f_i(x_i - \bar{x})^2$	$\sum_i \dfrac{f_i}{n} = 1$	$\dfrac{F_N}{n} = 1$
			$\bar{x} = \dfrac{1}{n}\sum_i f_i x_i$			$\sigma^2 = \dfrac{446.5}{100}$		
			$\bar{x} = 2000.15$			$\sigma = 2.11$		

The other columns are filled in as follows:

Column 2. To find the frequency, count the number of data values in each class interval. The sum of the frequencies gives the total number in the sample; in this case 100.

Column 3. The cumulative frequency is given by adding up the frequencies so far. The number of resistances up to 1995.5 is 1. As there are three in the interval 1995.5–1996.5, add $1 + 3 = 4$ to get the number up to a resistance of 1996.5. The number up to 1997.6 is given by $4 + 5 = 9$, etc.

Column's 4, 6, and 7 help us calculate the mean and standard deviation. To find column 4 multiply the frequencies (column 3) by the class midpoints (column 2) to get $f_i x_i$.

The sum of this column is an estimate of the total if we added all the values in the sample together. Therefore, dividing by the number of items in the sample gives the mean:

$$\text{Mean} = (1 \times 1995 + 3 \times 1996 + 5 \times 1997 + 13 \times 1998$$
$$+ 17 \times 1999 + 19 \times 2000 + 16 \times 2001 + 15 \times 2002$$
$$+ 4 \times 2003 + 4 \times 2004 + 2 \times 2005 + 1 \times 2006)/100$$
$$= 200\,015/100 = 2000.15.$$

To find the standard deviation we calculate the variance first. The variance is the mean squared deviation. Find the difference between the mean and each of the class intervals and square it. This gives column 6. Multiply by the class frequency to get column 7. Finally, add them all up and divide by the total number. This gives the variance:

$$\sigma^2 = \frac{446.5}{100} = 4.465$$

and the standard deviation

$$\sigma = \sqrt{446.5} \approx 2.11.$$

The relative frequency (given in column 8) is the fraction in each class interval; that is, the frequency divided by the total number. For example, the relative frequency in the class interval 2000.5–2001.5 is $16/100 = 0.16$

The cumulative relative frequency is the cumulative frequency divided by the total number n. The cumulative frequency up to 2000.5 is 58. Divided by the total number gives 0.58 for the relative cumulative frequency.

We can sum up this sample by saying that the mean resistance is 2000.15 with a standard deviation of 2.11. These simple figures can be used to sum up how closely the claim that they are resistors of 2000Ω can be justified. We can also picture the frequency distribution using a histogram as in Figure 21.4

Returning to the lifetimes of the light bulbs we might want to ask what average lifetime and standard deviation might lead to less than 5% of the light bulbs having a lifetime less than 1500 h?

To answer this sort of problem we need to build up a theory of statistical models and use probability theory.

21.3 Random systems and probability

We are dealing with complicated systems with a number of random factors affecting its behaviour; for instance, those that we have already seen, production of resistors, lifetimes of bulbs. We cannot determine the exact behaviour of the system but using its frequency distribution we can estimate the probabilities of certain events.

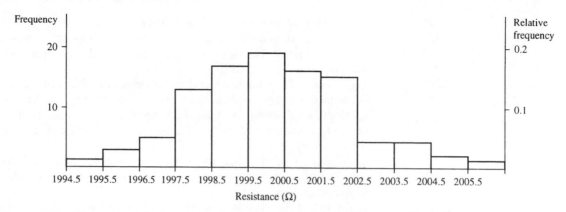

Figure 21.4 *Histogram of the frequency distribution of the resistances given in Table 21.2.*

Relative frequency and probability

The probability of an event is related to its relative frequency. If I chose a light bulb from the sample in Section 21.2, then the probability of its lifetime being between 1400 and 1500 h is 0.103. This is the same as the relative frequency for that class interval, that is, the number in the class interval divided by the total number of light bulbs in the sample. Here, we have assumed that we are no more likely to pick any one light bulb than any other, that is, each outcome is equally likely. The histogram of the relative frequencies (Figure 21.3 using the right-hand scale) gives the probability distribution function (or simply the probability function) for the lifetimes of the sample of light bulbs.

As an introduction to probability, examples are often quoted involving throwing dice or dealing cards from a pack of playing cards. These are used because the probabilities of events are easy to justify and not because of any particular predilection on the part of mathematicians to a gambling vocation!

Example 21.2 A die has the numbers one to six marked on its sides. Draw a graph of the probability distribution function for the outcome from one throw of the die, assuming it is fair.

Solution If we throw the die 10 000 times we would expect the number of times each number appeared face up to be roughly the same. Here, we have assumed that the die is fair; that is, any one number is as likely to be thrown as any other. The relative frequencies would be approximately 1/6. The probability distribution function, therefore, is a flat function with a value of 1/6 for each of the possible outcomes of 1, 2, 3, 4, 5, 6. This is shown in Figure 21.5

Notice two important things about the probabilities in the probability distribution for the die:

1. each probability is less than 1;
2. the sum of all the probabilities is 1:

$$1/6 + 1/6 + 1/6 + 1/6 + 1/6 + 1/6 = 1.$$

Example 21.3 A pack of cards consists of four suits, hearts, diamonds, spades, and clubs. Each suit has 13 cards; that is, cards for the numbers 1 (the ace) to 10 and a Jack, Queen, and King. Draw a graph of the probability distribution function for the outcome when dealing one card

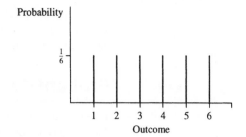

Figure 21.5 *Probability distribution for a fair die*

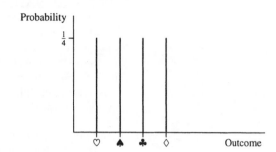

Figure 21.6 *The probability distribution for suits from a pack of cards.*

from the pack, where only the suit is recorded. Assume the card is replaced each time and the pack is perfectly shuffled.

Solution   If a card is selected, the suit recorded hearts, spades, clubs, or diamonds, the card is placed back in the pack and the pack is shuffled. If this is repeated (say 10 000 times) then we might expect that each suit will occur as often as any other, that is, 1/4 of the time. The probability distribution function again is a flat distribution, and has the value of 1/4 for each of the possible four outcomes. The probability distribution for suits is given in Figure 21.6.

Notice again that in Example 21.3 each probability is less than 1 and that the sum of all the probabilities is 1:

$$\tfrac{1}{4} + \tfrac{1}{4} + \tfrac{1}{4} + \tfrac{1}{4} = 1.$$

We have seen that a probability distribution can be represented using a graph. Each item along the x-axis has an associated probability. Probability is a function defined on some set. The set is called the sample space, S, and contains all possible *outcomes* of the random system. The probability distribution function is often abbreviated to p.d.f.

Some definitions

A *trial* is a single observation on the random system, for example, one throw of a die, one measurement of a resistance in the example in Section 21.2.

The *sample space* is the set of all possible outcomes, for example, for the die it is the set {1, 2, 3, 4, 5, 6}, and for the resistance problem it is the set of all possible measured resistances. This set may be discrete or continuous. An *event* is a set of outcomes. For instance, A is the event of throwing less than 4 and B is the event of throwing a number greater than or equal to 5.

An event is a subset of the sample space S.

Notice that in the case of a continuous sample space an event is also a continuous set, represented by an interval of values. For example, C is the event that the resistance lies in the interval 2000 ± 1.5.

Probability

The way that probability is defined is slightly different for the case of a discrete sample space or a continuous sample space.

Discrete sample space

The outcome of any trial is uncertain; however, in a large number of trials the proportion showing a particular outcome approaches a certain number. We call this the probability of that outcome. The probability distribution function, or simply probability function, gives the value of the probability that is associated with each outcome. The probability function obeys two important conditions:

1. all probabilities are less than, or equal to 1, that is, $0 \leqslant p(x) \leqslant 1$, where x is any outcome in the sample space;
2. the probabilities of the individual outcomes sum to 1, that is, $\sum p(x) = 1$.

Continuous sample space

Considering a continuous sample space. We assign probabilities to intervals. We find the probability of, for instance, the resistance being in the interval 2000 ± 0.5. Hence, we assign probabilities to events and not to individual outcomes. In this case, the function that gives the probabilities is called the probability density function and we can find the probability of some event by integrating the probability density function over the interval. For instance, if we have a probability density function $f(x)$, where x can take values from a continuous sample space, then the probability of x being in the interval a to b is given by, for example,

$$P(a < x < b) = \int_a^b f(x)\,\mathrm{d}x.$$

The probability density function obeys the condition:

$$\int_{-\infty}^{\infty} f(x)\,\mathrm{d}x = 1$$

that is, the total area under the graph of the probability density function must be 1.

Equally likely events

If all outcomes are equally likely then the probability of an event E from a discrete sample set is given by

$$\frac{\text{The number of outcomes in E}}{\text{The total number of outcomes in S}}.$$

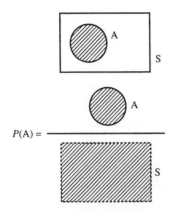

$$P(A) = \frac{\text{[hatched circle A]}}{\text{[hatched rectangle S]}}$$

Figure 21.7 $A \subseteq S$ (A is a subset of S) where all the outcomes in S are equally likely. The probability of A can be pictured as the proportion of S than is in A: the ratios of the areas.

21.4 Addition law of probability

That is, the probability of E is equal to the proportion of the whole sample space that is in E, when each of the outcomes are equally likely.

Example 21.4   What is the probability on one throw of a die getting a number less than 3?

Solution   Give this event the name A. Then A $= \{1, 2\}$ and S $= \{1, 2, 3, 4, 5, 6\}$. The number in A is 2 and the number in S is 6. Therefore, as each outcomes is equally likely, $p(A)$, the probability of A is $2/6 = 1/3$.

This particular result leads to another way of representing probability – by using area.

If we use a rectangle to represent the set of all possible outcomes, S, then an event is a subset of S, that is, one section of the rectangle and if all outcomes are equally likely then its probability can be pictured by the proportion of S that is in A. We will put a dotted line around the picture representing the set A to indicate the number in A or the area of A (see Figure 21.7).

This way of picturing the probability of an event can help in remembering some of the probabilities of combined events.

Disjoint events

Disjoint events are events with no outcomes in common. They cannot happen simultaneously.

Example 21.5   A is the event that a card chosen from a playing pack is under 6 (counting ace as low, i.e. $= 1$) and B is the event of choosing a picture card (Jack, Queen, or King). Find the probability that a card chosen from the pack is under 6 or is a picture card.

Solution

A $= \{1\heartsuit, 2\heartsuit, 3\heartsuit, 4\heartsuit, 5\heartsuit, 1\diamondsuit, 2\diamondsuit, 3\diamondsuit, 4\diamondsuit, 5\diamondsuit, 1\spadesuit, 2\spadesuit, 3\spadesuit, 4\spadesuit, 5\spadesuit,$
$1\clubsuit, 2\clubsuit, 3\clubsuit, 4\clubsuit, 5\clubsuit\}$

B $= \{J\heartsuit, Q\heartsuit, K\heartsuit, J\diamondsuit, Q\diamondsuit, K\diamondsuit, J\spadesuit, Q\spadesuit, K\spadesuit, J\clubsuit, Q\clubsuit, K\clubsuit\}$.

A and B are disjoint. As each outcome is considered to be equally likely the probabilities are easy to find:

$$p(A) = \frac{\text{Number in A}}{\text{Number in S}}.$$

There are 52 cards; therefore, 52 possible outcomes in S, so

$$p(A) = \tfrac{20}{52} = \tfrac{5}{13}.$$

Also

$$p(B) = \frac{\text{Number in B}}{\text{Number in S}} = \frac{12}{52} = \frac{3}{13}.$$

We want to find the probability of A or B happening; that is, the probability that the card is either under 6 or is a picture card:

$A \cup B = \{1\heartsuit, 2\heartsuit, 3\heartsuit, 4\heartsuit, 5\heartsuit, 1\diamondsuit, 2\diamondsuit, 3\diamondsuit, 4\diamondsuit, 5\diamondsuit,$
$1\spadesuit, 2\spadesuit, 3\spadesuit, 4\spadesuit, 5\spadesuit, 1\clubsuit, 2\clubsuit, 3\clubsuit, 4\clubsuit, 5\clubsuit, J\heartsuit, Q\heartsuit, K\heartsuit,$
$J\diamondsuit, Q\diamondsuit, K\diamondsuit, J\spadesuit, Q\spadesuit, K\spadesuit, J\clubsuit, Q\clubsuit, K\clubsuit\}$

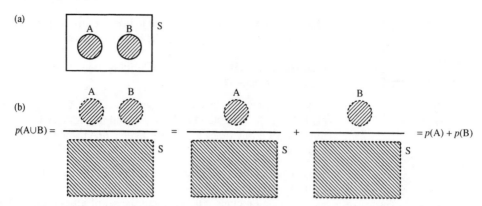

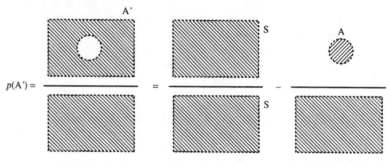

Figure 21.8 *The two events A and B are disjoint; that is, they have no outcomes in common. On a Venn diagram, as in (a), they do not intersect. The probability of either A or B can be found by $P(A \cup B) = P(A) + P(B)$, as in (b).*

Figure 21.9
$p(A') = 1 - p(A)$.

We can see that:

$$p(A \cup B) = \frac{\text{Number in } A \cup B}{\text{Number in S}} = \frac{32}{52} = \frac{8}{13}$$

$$\frac{\text{Number in A} + \text{number in B}}{\text{Number in S}} = \frac{\text{Number in A}}{\text{Number in S}} + \frac{\text{Number in B}}{\text{Number in S}}$$

$$p(A \cup B) = p(A) + p(B)$$

$$\frac{8}{13} = \frac{5}{13} + \frac{3}{13}.$$

We can also see this by using the idea of area to picture it, as in Figure 21.8.

We can also consider the probability of A not happening; that is, the probability of the complement of A, which we represent by A′. As A and A′ are disjoint.

$$p(A') + p(A) = 1$$
$$p(A') = 1 - p(A).$$

This is shown in Figure 21.9.

Non-disjoint events

Example 21.6 Consider one throw of a die

$A = \{a \mid a$ is an even number$\}$

$B = \{b \mid b$ is a multiple of 3$\}$.

Find the probability that the result of one throw of the die is either an even number or a multiple of 3.

Solution

$A = \{a \mid a$ is an even number$\}$

$B = \{b \mid b$ is a multiple of 3$\}$

then

$$A = \{2, 4, 6\}$$
$$B = \{3, 6\}$$
$$A \cup B = \{2, 3, 4, 6\}$$

and

$$p(A) = \tfrac{3}{6} = \tfrac{1}{2}$$
$$p(B) = \tfrac{2}{6} = \tfrac{1}{3}$$
$$p(A \cup B) = \tfrac{4}{6} = \tfrac{2}{3}$$

and

$$p(A \cup B) \neq p(A) + p(B).$$

Looking at the problem and using the idea of areas in the set, we can see from Figure 21.10 that the rule becomes

$$p(A \cup B) = p(A) + p(B) - p(A \cap B)$$

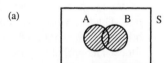

Figure 21.10   *(a) $A \cup B$.*
(b) $p(A \cup B) =$
$p(A) + p(B) - p(A \cap B)$.

21.5 Repeated trials, outcomes, and probabilities

Supposing we start to consider more complicated situations, like throwing a die twice. Then the outcomes can be found by considering all the possible outcomes of throwing the die the first time combined with all the possible outcomes of throwing the die the second time. As there are six possible outcomes for the first throw and six possible outcomes for the second throw, then there are $6 \times 6 = 36$ outcomes of throwing the die twice.

If we would like to find the probability that the first throw is a 5 and the second throw is a 5 or 6 then we can do this by listing all the 36 outcomes and finding the proportion that fall into our event.

The set S of all possible outcomes has 36 elements:

$$S = \{(1,1),(1,2),(1,3),(1,4),(1,5),(1,6),(2,1),(2,2),(2,3),(2,4),$$
$$(2,5),(2,6),(3,1),(3,2),(3,3),(3,4),(3,5),(3,6),(4,1),(4,2),$$
$$(4,3),(4,4),(4,5),(4,6),(5,1),(5,2),(5,3),(5,4),(5,5),(5,6),$$
$$(6,1),(6,2),(6,3),(6,4),(6,5),(6,6)\}$$

and each has an equally likely outcome. E = the first throw is a 5 and the second is a 5 or 6. Hence

$$E = \{(5,5),(5,6)\}$$

$$p(E) = \frac{\text{Number in A}}{\text{Number in S}} = \frac{2}{36} = \frac{1}{18}.$$

This sort of problem can be pictured more easily using a probability tree.

21.6 Repeated trials and probability trees

Repeatedly tossing a coin

The simplest sort of trial to consider is one with only two outcomes; for instance, tossing a coin. Each trial has the outcome of head or tail and each is equally likely.

We can picture repeatedly tossing the coin by drawing a probability tree. The tree works by drawing all the outcomes and writing the probabilities for the trial on the branches. Let us consider tossing a coin three times. The probability tree for this is shown in Figure 21.11.

There are various rules and properties we can notice.

Rules of probability trees and repeated trials

1. The probabilities of outcomes associated with each of the vertices can be found by multiplying all the probabilities along the branches leading to it from the top of the tree. HTH has the probability

$$\tfrac{1}{2} \times \tfrac{1}{2} \times \tfrac{1}{2} = \tfrac{1}{8}$$

2. At each level of the tree, the sum of all the probabilities on the vertices must be 1.
3. The probabilities along branches out of a single vertex must sum to 1. For example, after getting a head on the first trial we have a probability of 1/2 of getting a head on the second trial and a probability of 1/2 of getting a tail. Together they sum to 1.

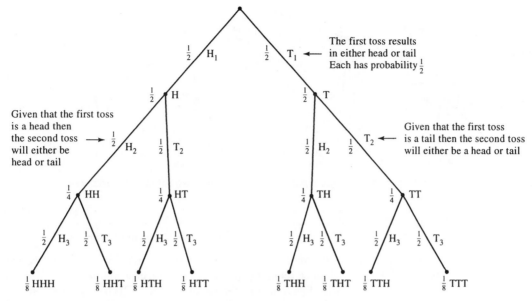

Figure 21.11 *A probability tree for three tosses of a coin.*

The fourth property is one that is only true when the various repeated trials are independent, that is, the result of the first trial has no effect on the possible result of the second trial, etc.

4. For independent trials, the tree keeps repeating the same structure with the same probabilities associated with the branches. Here, the event of getting a head on the third toss is independent of the event of getting a head on the first or second toss.

Using the tree we can find various probabilities, as we see in Example 21.7.

Example 21.7 What is the probability on three tosses of the coin that exactly two of them will be heads?

Solution Count up all the ways that we could have two heads and one tail, looking at the foot of the probability tree in Figure 21.11. We find three possibilities:

HHT, HTH, THH

each of these has probability of $\frac{1}{8}$. Therefore, the probability of exactly two heads is $\frac{3}{8}$.

Picking balls from a bag without replacement

We have 20 balls in a bag. Ten are red and ten are black. A ball is picked out of the bag, its colour recorded and then it is not replaced into the bag. There are two possible outcomes of each trial, red (R) or black (B). Let us consider three trials and their associated outcomes in a probability tree, as shown in Figure 21.12.

To find the probabilities we consider how many balls remain in each case. If the first ball chosen is red then there are only 19 balls left of which 9 are red and 10 are black; therefore, the probability of picking a red ball on the second trial is only 9/19 and the probability of picking black is 10/19.

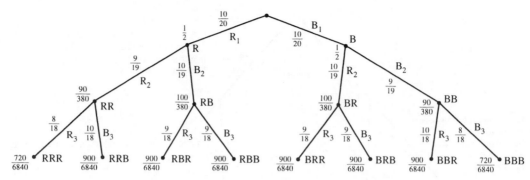

Figure 21.12 *Probability tree for three trials of picking a ball out of a bag without replacement.*

The first three rules given in the last example apply.

1. The probabilities of outcomes associated with each of the vertices can be found by multiplying all the probabilities along the branches leading to it from the top of the tree. BRB has the probability:

$$\frac{10}{20} \times \frac{10}{19} \times \frac{9}{18} = \frac{900}{6840} = \frac{1}{76}.$$

2. At each level of the tree the sum of all the probabilities on the vertices must be 1; for example, at the second level we have

$$\frac{90}{380} + \frac{100}{380} + \frac{100}{380} + \frac{90}{380} = \frac{9 + 10 + 10 + 9}{38} = \frac{38}{38} = 1.$$

3. The probabilities along branches out of a single vertex must sum to 1; for example, after picking red on the first trial we have a probability of 9/19 of picking red and 10/19 of picking black. Together they sum to 1.

The fourth property is no longer true as the various repeated trials are not independent. The result of the first trial has an effect on the possible result of the second trial, etc.

Using the tree we can find the probability of various events as in Example 21.8.

Example 21.8 Of 20 balls in a bag 10 are red and 10 are black. A ball is picked out of the bag, its colour recorded and then it is not replaced into the bag. What is the probability that of the first three balls chosen exactly two will be red?

Solution We can use the probability tree in Figure 21.12 to solve this problem. We look at the foot of the tree, which gives all the possible outcomes after three balls have been selected.

The ways of getting two red are RRB, RBR, BRR and the associated probabilities are

$$\frac{900}{6840}, \frac{900}{6840}, \frac{900}{6840}$$

or

$$\frac{10}{76}, \frac{10}{76}, \frac{10}{76}.$$

Summing these gives the probability of exactly two reds being chosen out of the three as 30/76.

21.7 Conditional probability and probability trees

The probabilities we have been writing along the branches of the probability tree are called conditional probabilities.

Example 21.9 What is the probability that the first throw of a die will be a 5 and the second throw will be a 5 or 6?

Solution A is the set of those outcomes with the first throw a 5 and B is the set of those outcomes with the second throw a 5 or 6. We can use a probability tree in the following way. After the first throw of the die either A is true or not, that is, we have only two possibilities, A or A′. After that, we are interested in whether B happens or not. Again we either get B or B′. We get the probability tree as in Figure 21.13.

Here, $p(B|A)$ means the probability of B given A, similarly $p(B|A')$ means the probability of B given (not A). We can fill in the probabilities using our knowledge of the fair die. The probability of A is 1/6. The second throw of the die is unaffected by the throw first throw of the die; therefore

$$p(B) = p(\text{throwing a 5 or a 6 on one throw of the die}) = \tfrac{2}{6} = \tfrac{1}{3}.$$

Working out the other probabilities gives the probability tree in Figure 21.14.

The probability that the first throw of a die will be a 5 and the second throw will be a 5 or 6 is $p(B \cap A)$, given from the tree in Figure 21.14 as

$$\tfrac{2}{6} \times \tfrac{1}{3} = \tfrac{1}{18}.$$

Notice that the probability we have calculated is the intersection of the two events A and B, that is, we calculated the probability that both occurred. When multiplying the probabilities on the branches of the probability tree we are using the following:

$$p(A \cap B) = p(A)p(B|A).$$

Furthermore, because of independence we have used the fact that $p(B|A) = p(B)$. That is, the probability of B does no depend on whether A has happened or not: B is independent of A. We, therefore, have the following important results.

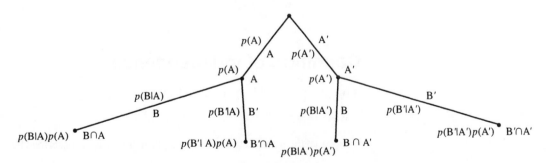

Figure 21.13 *A probability tree showing conditional probabilities.*

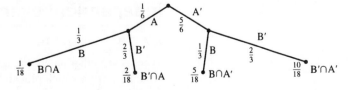

Figure 21.14 *The probability for Example 21.9.*

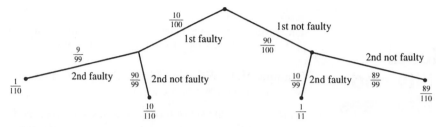

Figure 21.15 *Probability tree for Example 21.10.*

Multiplication law of probability

$p(A \cap B) = p(A)p(B|A)$

This law applies for any two events A and B. It is the law used in finding the probabilities of the vertices of the probability trees.

Example 21.10 It is known that 10% of a selection of 100 electrical components are faulty. What is the probability that the first two components selected are faulty if the selection is made without replacement?

Solution The probability we are looking for is:

$p(\text{first faulty} \cap \text{second faulty})$

$= p(\text{first faulty})p(\text{second faulty} \mid \text{first faulty})$

Here there are only two possibilities at each stage, faulty or not faulty. The probability tree is as shown in Figure 21.15.

Notice that their are only 99 components left after the first trial and whether the first was faulty or not changed the probability that the second is faulty or otherwise. Each branch of the tree, after the first layer, represents a conditional probability.

The answer to our problem is therefore

$p(\text{first faulty} \cap \text{second faulty})$

$= p(\text{first faulty})p(\text{second faulty} \mid \text{first faulty})$

$= \dfrac{10}{100} \times \dfrac{9}{99} = \dfrac{1}{110}.$

Condition of independence

If two events A and B are independent then

$p(B|A) = p(B).$

Notice that the multiplication law changes in this case, as described below.

Multiplication law of probability for independent events

$p(A \cap B) = p(A)p(B).$

There is one other much-quoted law of probability that completes all the basic laws from which probabilities can be worked out. That is Bayes's theorem and it comes from the multiplication law.

Bayes's theorem

As $p(A \cap B) = p(A)p(B|A)$ then as $A \cap B = B \cap A$ we can also write $p(B \cap A) = p(B)p(A|B)$ and putting the two together gives $p(A)p(B|A) = p(B)p(A|B)$ or

$$p(B|A) = \frac{p(A|B)p(B)}{p(A)}.$$

Bayes's theorem is important because it gives a way of swapping conditional probabilities that may be useful in diagnostic situations where not all of the conditional probabilities can be found directly.

Example 21.11   In a certain town there are only two brands of hamburgers available, Brand A and Brand B. It is known that people who eat Brand A hamburger have a 30% probability of suffering stomach pain and those who eat Brand B hamburger have a 25% probability of suffering stomach pain. Twice as many people eat Brand B compared to Brand A hamburgers. However, no one eats both varieties. Supposing one day you meet someone suffering from stomach pain who has just eaten a hamburger, what is the probability that they have eaten Brand A and what is the probability that they have eaten, Brand B?

Solution   First we define the sample space S, and the other simple events.

S = people who have just eaten a hamburger

A = people who have eaten a Brand A hamburger

B = people who have eaten a Brand B hamburger

C = people who are suffering stomach pains

We are given that:

$p(A) = \frac{1}{3}$

$p(B) = \frac{2}{3}$

$p(C|A) = 0.3$

$p(C|B) = 0.25.$

Note also that $S = A \cup B$.

As those who have stomach pain have either eaten Brand A or B, then $A \cap B = \emptyset$

$$p(C) = p(C \cap S) = p(C \cap A) + p(C \cap B)$$
$$= p(C|A)p(A) + p(C|B)p(B)$$
$$= 0.3 \times \frac{1}{3} + 0.25 \times \frac{2}{3} = \frac{8}{30}.$$

Then

$$p(A|C) = \frac{p(C|A)p(A)}{p(C)} = \frac{0.3 \times (1/3)}{8/30} = \frac{3}{8}$$

and

$$p(B|C) = \frac{p(C|B)p(B)}{p(C)} = \frac{0.25 \times (2/3)}{8/30} = \frac{5}{8}.$$

Hence, if they have stomach pain the probability that they have eaten Brand A is 3/8 and the probability that they have eaten Brand B is 5/8.

21.8 Application of the probability laws to the probability of failure of an electrical circuit

Components in series

We denote 'The probability that A does not fail' as $p(A)$. Then the probability that S fails is $p(S') = 1 - p(S)$.

As they are in series (see Figure 21.16), S will function if both A and B function: $S = A \cap B$. Therefore

$$p(S) = p(A \cap B)$$

as A and B are independent

$$p(S) = p(A)p(B)$$
$$p(S') = 1 - p(A)p(B).$$

Example 21.12   An electrical circuit has three components in series (see Figure 21.17). One has a probability of failure within the time of operation of the system of 1/5; the other has been found to function on 99% of occasions and the third component has proved to be very unreliable with a failure once for every three successful runs. What is the probability that the system will fail on a single operation run.

Solution   Using the method of reasoning above we call the components A, B, and C. The information we have is

$$p(A') = \tfrac{1}{5}$$
$$p(B) = 0.99$$
$$p(C'):p(C) = 1 : 3.$$

The probability we would like to find is $p(S')$, the probability of failure of the system and we know that the system is in series so $S = A \cap B \cap C$; that is, all the components must function for the system to function. We can find the probability that the system will function and subtract it from 1 to find the probability of failure:

$$p(S') = 1 - p(S).$$

As all the components are independent,

$$p(A \cap B \cap C) = p(A)p(B)p(C).$$

We can find $p(A)$, $p(B)$, and $p(C)$ from the information we are given: $p(A) = 1 - p(A') = 1 - 1/5 = 0.8$ and $p(B)$ is given as 0.99.

Given that $p(C') : p(C) = 1 : 3$, C fails once in every $1 + 3$ occasions, so $p(C') = 1/4 = 0.25$ and $p(C) = 1 - p(C') = 0.75$.

We can now find $p(A)p(B)p(C) = 0.8 \times 0.99 \times 0.75 = 0.594$. This is the probability that the system will function, so the probability it will fail is given by $1 - 0.594 = 0.406$.

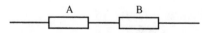

Figure 21.16   *Components in series. S fails if either A or B fails.*

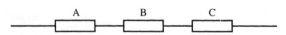

Figure 21.17   *Three components in series.*

Components in parallel

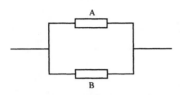

Figure 21.18 *Two components in parallel. S fails if both A and B fail.*

If a system S consists of two components A and B in parallel, as in Figure 21.18, then S fails if both A and B fail.

Again denote 'The probability that A does not fail' as $p(A)$. Then the probability that S fails is

$$p(S') = 1 - p(S).$$

As they are in parallel, S will function if either A or B functions: $S = A \cup B$. Therefore

$$p(S) = p(A \cup B).$$

A and B are independent, but not disjoint. They are not disjoint because both A and B can function simultaneously. From the addition law of probabilities given in Section 21.4

$$p(A \cup B) = p(A) + p(B) - p(A \cap B).$$

As A and B are independent,

$$p(A \cap B) = p(A)p(B).$$

Therefore

$$p(S) = p(A \cup B) = p(A) + p(B) - p(A)p(B)$$
$$p(S') = 1 - (p(A) + p(B) - p(A)p(B))$$
$$= 1 - p(A) - p(B) + p(A)p(B).$$

As this is a long-winded expression it may be more useful to look at the problem the other way round. S fails only if both A and B fail: $S' = A' \cap B'$. So

$$p(S') = p(A' \cap B') = p(A')p(B')$$

as A and B are independent. Finally, $p(S') = p(A')p(B')$.

This is a simpler form that we may use in preference to the previous expression we derived. It must be equivalent to our previous result, so we should just check that by substituting $p(A') = 1 - p(A)$ and $p(B') = 1 - p(B)$. Hence

$$p(S') = p(A')p(B')$$
$$p(S') = (1 - p(A))(1 - p(B))$$
$$p(S') = 1 - p(A) - p(B) + p(A)p(B)$$

which is the same as we had before.

Let us try a mixed example with some components in series and others in parallel.

Example 21.13 An electrical circuit has three components, two in parallel, components A and B, and one in series, component C. They are arranged as in Figure 21.19. Component's A and B are identical components with a 2/3 probability of functioning and C has a probability of failure of 0.1%. Find the probability that the system fails.

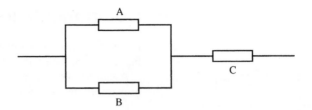

Figure 21.19   *Two components in parallel and one in series (Example 21.12).*

Solution   We denote 'The probability that A functions' as $p(A)$. The information we have is

$$p(A) = \tfrac{2}{3}$$

$$p(B) = \tfrac{2}{3}$$

$$p(C') = 0.001 \quad \Rightarrow \quad p(C) = 0.999.$$

The probability we would like to find is $p(S')$, the probability of failure of the system and we know that the system will function if (either A or B function) and C functions. So

$$S = (A \cup B) \cap C.$$

From the multiplication law for independent events we have:

$$p((A \cup B) \cap C) = p(A \cup B)p(C).$$

From the addition law for non-disjoint events (A and B are not disjoint because both A and B can occur), we have:

$$p(A \cup B) = p(A) + p(B) - p(A)p(B)$$

$$p(A \cup B) = \frac{2}{3} + \frac{2}{3} - \frac{2}{3} \times \frac{2}{3} = \frac{8}{9} \approx 0.8889.$$

So $p(S) = p(A \cup B)p(C) = 0.8889 \times 0.999 = 0.888$. Hence, the probability that the system functions is 0.888.

21.9 Statistical modelling

Suppose, as in Example 21.2, we have tested 5000 resistors and recorded the resistance of each one to an accuracy of 0.01Ω. We may wish to quickly decide whether the manufacturers claim of producing $2000\text{-}\Omega$ resistors is correct. One way of doing this is to divide our resistors into class intervals and draw up a table and a histogram. We can then count the percentage of resistors that are outside of acceptable limits and assume that the population behaviour is the same as the sample behaviour. Hence, if 98% of the resistors lie between $2000 \pm 0.1\%$ we may be quite happy.

A quicker way of doing this is to use a statistical model. That is, we can guess what the histogram would look like based on our past experience. To use such a model we probably only need to know the population mean, μ, and the population standard deviation, σ.

In the rest of this chapter, we will look at four possible ways of modelling data, the normal distribution, the exponential distribution, both continuous models, and the binomial distribution and the Poisson distribution which are discrete models.

21.10 The normal distribution

The normal distribution is symmetric about its mean. It is bell-shaped and the fatness of the bell depends on its standard deviation. Examples are given in Figure 21.20 and 21.21.

The normal distribution is very important because of the following points:

1. *Many practical distributions approximate to the normal distribution.* Look at the histograms of lifetimes given in Figure 21.3 and of resistances given in Figure 21.4 and you will see that they resemble the normal distribution. Another common example is the distribution of errors. If you were to get a large group of students to measure the diameter of a washer to the nearest 0.1 mm, then a histogram of the results would give an approximately normal distribution. This is because the errors in the measurement are normally distributed.
2. *The central limit theorem.* If we take a large number of samples from a population and calculate the sample means then the distribution of the sample means will behave like the normal distribution for all populations (even those populations which are not distributed normally). This is as a result of what is called the central limit theorem. There is a project exploring the behaviour of sample means given in the Projects and Investigations available on the companion website for this book.
3. Many other common distributions become like the normal distribution in special cases. For instance, the binomial distribution, which we shall look at in Section 21.12, can be approximated by the normal when the number of trials is very large.

Finding probabilities from a continuous graph

Before we look at the normal distribution in more detail we need to find out how to relate the graph of a continuous function to our previous idea of probability. In Section 21.3, we identified the probability of a class with its relative frequency in a frequency distribution. That was all right when we had already divided the various sample values into classes. The problem with the normal distribution is that is has no such divisions along the x-axis and no individual class heights, just a nice smooth curve.

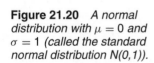

Figure 21.20   *A normal distribution with $\mu = 0$ and $\sigma = 1$ (called the standard normal distribution N(0,1)).*

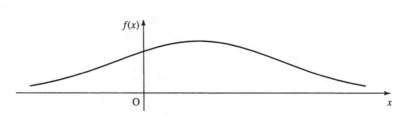

Figure 21.21   *Normal distribution with $\mu = 2$ and $\sigma = 3$.*

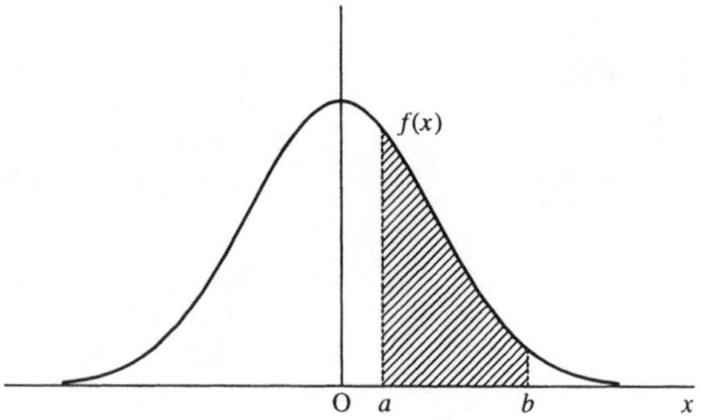

Figure 21.22   *A continuous probability density function and the probability that the outcome lies in the interval a ⩽ x ⩽ b.*

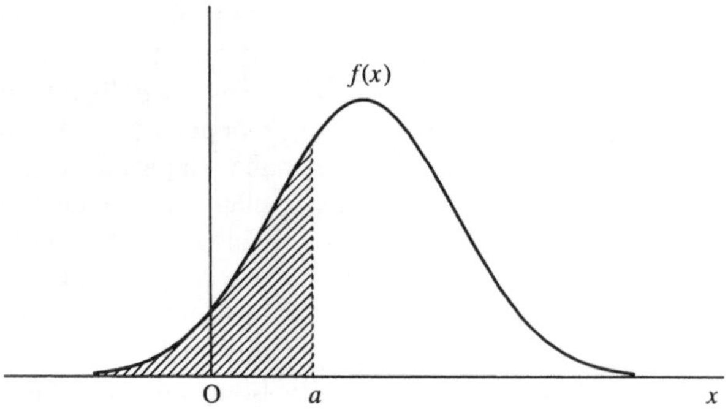

Figure 21.23   *The value of the cumulative distribution F(a) marked as an area on the graph of the probability density function, f(x), of a continuous distribution.*

To overcome this problem we define the probability of the outcome lying in some interval of values, as the area under the graph of the probability function between those two values as shown in Figure 21.22.

As we found in Chapter 7, the area under a curve is given by the integral; therefore, for a continuous probability distribution, $f(x)$, we define

$$p(x \text{ lies between } a \text{ and } b) = \int_a^b f(x)\, \mathrm{d}x.$$

The cumulative distribution function gives us the probability of this value or any previous value (it is like the cumulative relative frequency). A continuous distribution thus becomes the 'the area so far' function and therefore becomes the integral from the lowest possible value that can occur in the distribution up to the current value.

The cumulative distribution up to a value a is represented by

$$F(a) = \int_{-\infty}^a f(x)\, \mathrm{d}x$$

and it is the total area under to graph of the probability function up to a; this is shown in Figure 21.23.

We can also use the cumulative distribution function to represent probabilities of a certain interval. The area between two values can be found by subtracting two values of the cumulative distribution function as in Figure 21.24.

However, there is a problem with the normal distribution function in that is not easy to integrate! The probability density function for x, where x is $N(\mu, \sigma^2)$ is given by

$$f(x) = \frac{1}{\sqrt{2\pi}\sigma} \mathrm{e}^{-(x-\mu)^2/2\sigma^2}.$$

Figure 21.24 *The
probability p(a < x < b) can
be found by the difference
between two values of the
cumulative distribution
function F(b) − F(a).
Compare the difference in
areas with Figure 21.22.*

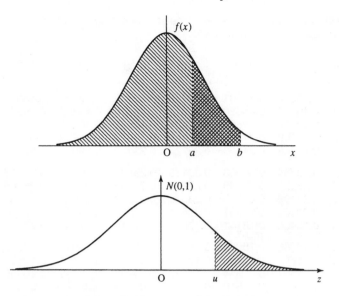

Figure 21.25 *The area in
tail of the standardized normal
curve $\int_u^\infty f(z)\,dz$ tabulated in
Table 21.3.*

It is only integrated by using numerical methods. Hence, the values of
the integrals can only be tabulated. The values that we have tabulated are
the areas in the tail of the standardized normal distribution; that is

$$\int_u^\infty f(z)\,dz$$

where $f(z)$ is the probability distribution with 0 mean ($\mu = 0$) and
standard deviation of 1 ($\sigma = 1$). This is shown in Figure 21.25 and
tabulated in Table 21.3. In order to use these values we need to use ideas
of transformation of graphs from Chapter 2 to transform any normal
distribution into its standardized form.

The standardized normal curve

The standardized normal curve is obtained from the normal curve by the
substitution $z = (x - \mu)/\sigma$ and it converts the original distribution into
one with zero mean and standard deviation 1. This is useful because we
can use a table of values for z given in Table 21.3 to perform calculations.

Finding the probability that *x* lies between a given range of values

Supposing we have decided that a sample of resistors have a mean of
10.02 and a standard deviation of 0.06, then what percentage lie inside
an acceptable tolerance of 10 ± 0.1?

We want to find the area under the normal curve $N(10.02, 0.06^2)$
between $x = 9.9$ and $x = 10.1$, that is, the shaded area in Figure 21.26.

First, convert the x values to z values, by using $z = (x - \mu)/\sigma$:

$$x = 9.9 \;\Rightarrow\; z = \frac{9.9 - 10.02}{0.06} = -2$$

$$x = 10.1 \;\Rightarrow\; z = \frac{10.1 - 10.02}{0.06} = 1.3333$$

So we now want to find the shaded area for z values (which will be the
same area as above), shown in Figure 23.27.

Table 21.3 *Areas in the tail of the standardized normal distribution. $P(z > u)$ values are given where z is a variable with distribution $N(0, 1)$*

u	0.00	0.01	0.02	0.03	0.04	0.05	0.06	0.07	0.08	0.09
0.0	0.50000	0.49601	0.49202	0.48803	0.48405	0.48006	0.47608	0.47210	0.46812	0.46414
0.1	0.46017	0.45620	0.45224	0.44828	0.44433	0.44038	0.43644	0.43251	0.42858	0.42465
0.2	0.42074	0.41683	0.41294	0.40905	0.40517	0.40129	0.39743	0.39358	0.38974	0.38591
0.3	0.38209	0.37828	0.37448	0.37070	0.36693	0.36317	0.35942	0.35569	0.35197	0.34827
0.4	0.34458	0.34090	0.33724	0.33360	0.32997	0.32636	0.32276	0.31918	0.31561	0.31207
0.5	0.30854	0.30503	0.30153	0.29806	0.29460	0.29116	0.28774	0.28434	0.28096	0.27760
0.6	0.27425	0.27093	0.26763	0.26435	0.26109	0.25785	0.25463	0.25143	0.24825	0.24510
0.7	0.24196	0.23885	0.23576	0.23270	0.22965	0.22663	0.22363	0.22065	0.21770	0.21476
0.8	0.21186	0.20897	0.20611	0.20327	0.20045	0.19766	0.19489	0.19215	0.18943	0.18673
0.9	0.18406	0.18141	0.17879	0.17619	0.17361	0.17106	0.16853	0.16602	0.16354	0.16109
1.0	0.15866	0.15625	0.15386	0.15151	0.14917	0.14686	0.14457	0.14231	0.14007	0.13786
1.1	0.13567	0.13350	0.13136	0.12924	0.12714	0.12507	0.12302	0.12100	0.11900	0.11702
1.2	0.11507	0.11314	0.11123	0.10935	0.10749	0.10565	0.10383	0.10204	0.10027	0.09853
1.3	0.09680	0.09510	0.09342	0.09176	0.09012	0.08851	0.08691	0.08534	0.08379	0.08226
1.4	0.08076	0.07927	0.07780	0.07636	0.07493	0.07353	0.07215	0.07078	0.06944	0.06811
1.5	0.06681	0.06552	0.06426	0.06301	0.06178	0.06057	0.05938	0.05821	0.05705	0.05592
1.6	0.05480	0.05370	0.05262	0.05155	0.05050	0.04947	0.04846	0.04746	0.04648	0.04551
1.7	0.04457	0.04363	0.04272	0.04182	0.04093	0.04006	0.03920	0.03836	0.03754	0.03673
1.8	0.03593	0.03515	0.03438	0.03362	0.03288	0.03216	0.03144	0.03074	0.03005	0.02938
1.9	0.02872	0.02807	0.02743	0.02680	0.02619	0.02559	0.02500	0.02442	0.02385	0.02330
2.0	0.02275	0.02222	0.02169	0.02118	0.02068	0.02018	0.01970	0.01923	0.01876	0.01831
2.1	0.01786	0.01743	0.01700	0.01659	0.01618	0.01578	0.01539	0.01500	0.01463	0.01426
2.2	0.01390	0.01355	0.01321	0.01287	0.01255	0.01222	0.01191	0.01160	0.01130	0.01101
2.3	0.01072	0.01044	0.01017	0.00990	0.00964	0.00939	0.00914	0.00889	0.00866	0.00842
2.4	0.00820	0.00798	0.00776	0.00755	0.00734	0.00714	0.00695	0.00676	0.00657	0.00639
2.5	0.00621	0.00604	0.00587	0.00570	0.00554	0.00539	0.00523	0.00508	0.00494	0.00480
2.6	0.00466	0.00453	0.00440	0.00427	0.00415	0.00402	0.00391	0.00379	0.00368	0.00357
2.7	0.00347	0.00336	0.00326	0.00317	0.00307	0.00298	0.00289	0.00280	0.00272	0.00264
2.8	0.00256	0.00248	0.00240	0.00233	0.00226	0.00219	0.00212	0.00205	0.00199	0.00193
2.9	0.00187	0.00181	0.00175	0.00169	0.00164	0.00159	0.00154	0.00149	0.00144	0.00139
3.0	0.00135	0.00131	0.00126	0.00122	0.00118	0.00114	0.00111	0.00107	0.00104	0.00100
3.1	0.00097	0.00094	0.00090	0.00087	0.00084	0.00082	0.00079	0.00076	0.00074	0.00071
3.2	0.00069	0.00066	0.00064	0.00062	0.00060	0.00058	0.00056	0.00054	0.00052	0.00050
3.3	0.00048	0.00047	0.00045	0.00043	0.00042	0.00040	0.00039	0.00038	0.00036	0.00035
3.4	0.00034	0.00032	0.00031	0.00030	0.00029	0.00028	0.00027	0.00026	0.00025	0.00024
3.5	0.00023	0.00022	0.00022	0.00021	0.00020	0.00019	0.00019	0.00018	0.00017	0.00017
3.6	0.00016	0.00015	0.00015	0.00014	0.00014	0.00013	0.00013	0.00012	0.00012	0.00011
3.7	0.00011	0.00010	0.00010	0.00010	0.00009	0.00009	0.00008	0.00008	0.00008	0.00008
3.8	0.00007	0.00007	0.00007	0.00006	0.00006	0.00006	0.00006	0.00005	0.00005	0.00005
3.9	0.00005	0.00005	0.00004	0.00004	0.00004	0.00004	0.00004	0.00004	0.00003	0.00003

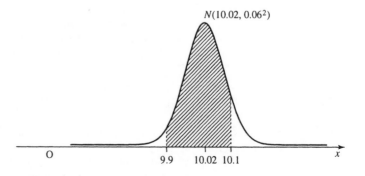

Figure 21.26 *The area under the normal curve of mean 10.02 and standard deviation 0.06 between 9.9 and 10.1.*

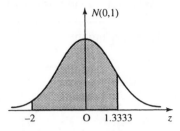

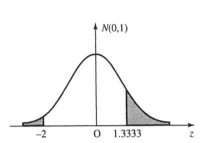

Figure 21.27 *The area under the standardized normal curve of mean 0 and standard deviation 1 between $z = -2$ and $z = 1.333$. The area is equivalent to that shown in Figure 21.26.*

Figure 21.28 *The area outside of the tolerance limits as given in Figure 21.27.*

In order to use Table 21.3, we need to express the problem in terms of the proportion that lies outside of the tolerance limits, as in Figure 21.28.

We use the table of the standardized normal distribution to find the proportion less than $z = -2$. As the curve is symmetric this will be the same as the proportion greater than $z = 2$. From the table this gives 0.02275.

The proportion greater than $z = 1.33$ from the table is 0.09176. Hence, the proportion that lies outside of our limits is

$$0.02275 + 0.09176 = 0.11451.$$

As the total area is 1, the proportion within the limits is $1 - 0.11451 = 0.88549$.

21.11 The exponential distribution

The exponential distribution is also named as the failure rate function, as it can be used to model the rate of failure of components.

Consider a set of 1000 light bulbs, a similar make to those tested in Section 21.1. However, now consider a batch of bulbs at random that have already been in use for some, unknown time. They are, therefore, of mixed ages. On measuring the time of failure we get Table 21.4.

These data are represented in a histogram given in Figure 21.29, giving the frequencies and relative frequencies, and Figure 21.30, giving the cumulative frequencies and relative cumulative frequencies.

Notice that Figure 21.29 looks like a dying exponential. This is not unreasonable as we might expect failure rates to be something like the problem of radioactive decay of Chapter 8, that is, a dying exponential.

We could think of it in a similar way to a population problem. The proportion that have failed after time t is given by the cumulative distribution function F. The proportion that are still functioning is therefore $1 - F$. The increase in the total proportion of failures is given by the failure rate multiplied by the number still functioning, if λ is the failure rate this gives

$$\frac{\mathrm{d}F}{\mathrm{d}t} = \lambda(1 - F).$$

This differential equation can be solved to give:

$$F = 1 - A\,\mathrm{e}^{-\lambda t}.$$

Using the fact that at time 0 there are no failures then we find $A = 1$. This gives the cumulative distribution of the exponential distribution as

$$F = 1 - \mathrm{e}^{-\lambda t}$$

Table 21.4   *Time to failure of a sample of light bulbs*

Time of failure (h)	Class mid-point	Frequency	Cumulative frequency	$f_i x_i$
0–200	100	260	260	26 000
200–400	300	194	454	58 200
400–600	500	154	608	77 000
600–800	700	100	708	70 000
800–1000	900	80	788	72 000
1000–1200	1100	60	848	66 000
1200–1400	1300	38	886	49 400
1400–1600	1500	33	919	49 500
1600–1800	1700	23	942	39 100
1800–2000	1900	14	956	26 600
2000–2200	2100	12	968	25 200
2200–2400	2300	10	978	23 000
2400–2600	2500	9	987	22 500
2600–2800	2700	13	1000	33 800
		1000		638 300

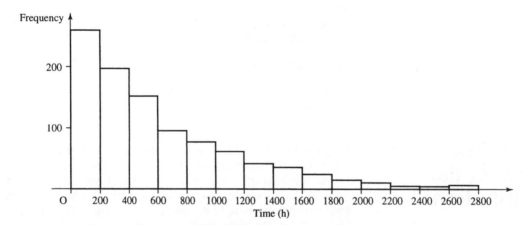

Figure 21.29   *Histogram of frequencies given in Table 21.4.*

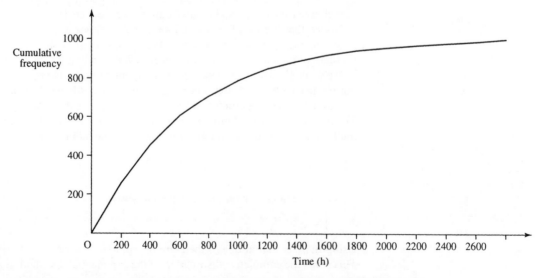

Figure 21.30   *Cumulative frequency of time to failure of a sample of 1000 light bulbs.*

where λ is the failure rate, that is, the proportion that will fail in unit time. The probability distribution can be found from the cumulative distribution by differentiating, giving

$$f = \frac{\mathrm{d}F}{\mathrm{d}t} = \lambda \, \mathrm{e}^{-\lambda t}$$

Mean and standard deviation of a continuous distribution

We can find the mean and standard deviation of a continuous distribution by using integration to replace the summation over all values. The mean is given by

$$\mu = \int x f(x) \, \mathrm{d}x$$

where the integration is over all values in the sample space for x. For the exponential distribution this gives

$$\mu = \int x \lambda \, \mathrm{e}^{-\lambda x} \, \mathrm{d}x.$$

which can be found by using integration by parts (Chapter 7) to be $1/\lambda$.

The standard deviation is given by

$$\sigma = \sqrt{\int (x - \mu)^2 f(x) \, \mathrm{d}x}$$

where the integration is over all values of x. For the exponential distribution this gives

$$\sigma = \sqrt{\int_0^\infty (x - 1/\lambda)^2 \lambda \, \mathrm{e}^{-\lambda x} \, \mathrm{d}x}.$$

Again using integration by parts, we obtain $\sigma = 1/\lambda$.

So we see that the mean is $1/\lambda$, as is the standard deviation for the exponential distribution.

Comparison of the data with the model

We can now compare a statistical model with the data given in Table 21.4. To do this we calculate the cumulative frequencies for the maximum value in each of the class intervals. The mean of the sample is 638.3. We calculated that the mean of the exponential distribution is given by $1/\lambda$ the inverse of the failure rate.

$$638.3 = \frac{1}{\lambda} \quad \Rightarrow \quad \lambda = 1.567 \times 10^{-3}.$$

Using $F(t) = 1 - \mathrm{e}^{-\lambda t}$

$$F(200) = 1 - \mathrm{e}^{-1.567 \times 10^{-3} \times 200} \approx 0.269$$

$$P(0 < x < 200) = F(200) - F(0) = 0.269$$

$$F(400) = 1 - \mathrm{e}^{-1.567 \times 10^{-3} \times 400} \approx 0.466$$

$$P(200 < x < 400) = F(400) - F(200) = 0.197$$

and so on, giving the values as in Table 21.5. The model's predictions agree quite well with the data. To find the model predicted frequencies and cumulative frequencies we multiply by the number in the sample, 1000.

Table 21.5   *Time to failure of a sample of light bulbs compared with values obtained by modelling with the exponential distribution*

Data						Model predictions		
Time of failure (h)	Class mid-point	Frequency	Cumulative frequency	$f_i x_i$	$F(x)$	Probabilities	Cumulative frequency	Frequency
0–200	100	260	260	26 000	0.269	0.269	269	269
200–400	300	194	454	58 200	0.466	0.197	466	197
400–600	500	154	608	77 000	0.609	0.143	609	143
600–800	700	100	708	70 000	0.714	0.105	714	105
800–1000	900	80	788	72 000	0.791	0.077	719	77
1000–1200	1100	60	848	66 000	0.847	0.056	847	56
1200–1400	1300	38	886	49 400	0.888	0.041	888	41
1400–1600	1500	33	919	49 500	0.918	0.03	918	30
1600–1800	1700	23	942	39 100	0.940	0.022	940	22
1800–2000	1900	14	956	26 600	0.956	0.016	956	16
2000–2200	2100	12	968	25 200	0.968	0.012	968	12
2200–2400	2300	10	978	23 000	0.977	0.009	977	9
2400–2600	2500	9	987	22 500	0.983	0.006	983	6
2600–2800	2700	13	1000	33 800	0.988	0.005	988	5

638 300

21.12 The binomial distribution

Consider a random system with a sequence of trials, the trials being such that:

1. Each trial has two possible outcomes (e.g. non-defective, defective), which we assign the outcomes of 1 (success) and 0 (failure). This type of trial is called a Bernoulli trial.
2. On each trial $p(1) = \theta$ and $p(0) = 1 - \theta$ and θ is the same on all trials.
3. The outcome of the n trails are mutually independent.

$p_n(r)$ is the outcome of r successes in n trials and

$$p_n(r) = \binom{n}{r} \theta^r (1 - \theta)^{n-r}$$

where

$$\binom{n}{r} = {}^n C_r = \frac{n!}{(n-r)!r!} = \frac{n(n-1)\ldots(n-r+1)}{r!}.$$

Setting $\alpha = 1 - \theta$, the probability of r successes in n trials is given by the rth term in the binomial expansion:

$$(\theta + \alpha)^n = \alpha^n + n\theta\alpha^{n-1} + \frac{n(n-1).}{2!}\theta^2\alpha^{n-2}$$
$$+ \frac{n(n-1)\ldots(n-r+1)}{r!}\theta^r\alpha^{n-r}\ldots + \theta^n.$$

Hence, the term binomial distribution.

Example 21.14   In five tosses of a coin find the probability of obtaining three heads.

Solution   Assign the outcome of obtaining a head to 1 and tail to 0. Assume that the coin is fair and therefore $\theta = \frac{1}{2}$, $1 - \theta = \frac{1}{2}$. The probability of obtaining three heads in five tosses of a coin is given by the binomial probability:

$$p_5(3) = \frac{5!}{3!2!}\theta^3(1-\theta)^2 = \frac{5 \times 4}{2!}\left(\frac{1}{2}\right)^3\left(\frac{1}{2}\right)^2 = 0.3125.$$

Mean and variance of a single trial

The mean of a discrete distribution can be found by using $\mu = \sum xp(x)$ and the variance is

$$\sigma^2 = \sum(x-\mu)^2 p(x)$$

where the summation is over the sample space.

We can use these to find the mean and variance of a single trial with only two outcomes, success or failure. The outcome of success has the value 1 and occurs with probability θ and the outcome of failure has the value 0 with probability $1-\theta$. Then, the mean is given by $1 \times \theta + (1-\theta) \times 0 = \theta$.

The variance of a single trial is given by

$$(1-\theta)^2\theta + (0-\theta)^2(1-\theta) = \theta - 2\theta^2 + \theta^3 + \theta^2 - \theta^3 = \theta(1-\theta).$$

$$\uparrow \quad \uparrow \quad \uparrow \quad \uparrow \quad \uparrow \quad \uparrow$$
$$x \quad \mu \quad p(1) \quad x \quad \mu \quad p(0)$$

The standard deviation is the square root of the variance:

$$\sigma = \sqrt{\theta(1-\theta)}.$$

The mean and standard deviation of the binomial distribution

The expressions involving a summation over the entire sample space can be used to find the mean and standard deviation of the binomial distribution but they take a bit of manipulation to find. Instead, we can take a short cut and use the fact that each trial is independent. The mean of the union of n trials is given by the sum of the means of the n trials. Similarly (for independent trials only), the variance of the union of the n trials is given by the sum of the variances.

Therefore, the mean of the binomial distribution for n trials is given by the number of trials × mean for a single trial $= n\theta$. The variance is given by $n\theta(1-\theta)$ and therefore the standard deviation is

$$\sigma = \sqrt{n\theta(1-\theta)}.$$

Example 21.15   A file of data is stored on a magnetic tape with a parity bit stored with each byte (8 bits) making 9 bits in all. The parity bit is set so that the 9 bits add up to an even number. The parity bit allows errors to be detected, but not corrected. However, if there are two errors in the 9 bits then the errors will go undetected, three errors will be detected, four errors undetected, etc. A very poor magnetic tape was tested for the reproduction of 1024 bits and 16 errors were found. If on one record on the tape there are 4000 groups of 9 bits, estimate how many bytes will have undetected errors.

Solution   Call 1 the outcome of a bit being in error and 0 that it is correct. We are given that in 1024 (n) trials there were 16 errors. Taking 16 as the mean over 1024 trials and using

$$1024\theta = 16$$

$$\theta = \frac{16}{1024} = \frac{1}{64}.$$

Errors go undetected if there are 2, 4, 6, etc. The probability of two errors in 9 bits is given by

$$P_9(2) = \binom{9}{2}\theta^2(1-\theta)^7 = \frac{9!}{7!2!}\left(\frac{1}{64}\right)^2\left(\frac{63}{64}\right)^7 = 0.0078717.$$

Multiplying by the number of data bytes of 4000 gives approximately 31 undetected errors.

The probability of four errors will obviously be much less.

$$P_9(4) = \binom{9}{4}\theta^4(1-\theta)^5 = \frac{9!}{4!5!}\left(\frac{1}{64}\right)^4\left(\frac{63}{64}\right)^5 = 0.0000008.$$

This probability is too small to show up only 4000 bytes. As the probability of six or eight errors is even smaller then they can safely be ignored.

The probable number of undetected errors is 31.

21.13 The Poisson distribution

The Poisson distribution is used to model processes where the distribution of the number of incidents occurring in any interval depends only on the length of that interval. Examples of such systems are:

1. incoming telephone calls to an exchange during a busy period;
2. customers arriving at a checkout counter in a supermarket from 4 to 6 p.m.;
3. accidents on a busy stretch of the M1;
4. number of misprints in a book.

When modelling situations in a Poisson process we use four assumptions:

1. If A is the event of n incidents in an interval and B the event of m incidents in another non-overlapping interval then A and B are independent, that is, $p(A \cap B) = p(A)p(B)$.
2. If A is the event of n incidents in an interval then $P(A)$ depends only on the length of the interval – not on the starting point of the interval.
3. The probability of exactly one incident in a small interval is approximately proportional to the length of that interval, that is, $P_1(t) \approx \lambda t$ for small t.
4. The probability of more than one incident in a small interval is negligible. Thus, for small t, $P_2(t) \approx 0$ and we can also say that

$$\lim_{t \to 0} \frac{P_n(t)}{t} = 0 \quad \text{for } n > 1.$$

It follows that $P_0(t) + P_1(t) \approx 1$ and as by assumption (3), $P_1(t) \approx \lambda t$, we get $P_0(t) \approx 1 - \lambda t$. We now think about the number of incidents in an interval of time of any given length, $(0, t)$, where t is no longer small. We can divide the interval into pieces of length h, where h is small, and use the assumptions above. We can see that in each small interval of length h there is either no event or a single event. Therefore, each small interval is approximately behaving like a Bernoulli trial. This means that we can approximate the events in the interval $(0, t)$ by using the Binomial distribution for the number of successes r in n trials. The probability of r incidents in n intervals, where the probability of an incident in any one interval is λh, is given by

$$P_r(t) = \lim_{h \to 0} \binom{n}{r} (\lambda h)^r (1 - \lambda h)^{n-r}.$$

Substituting $h = t/n$ gives

$$P_r(t) = \lim_{n \to \infty} \frac{n!}{(n-r)!r!} \frac{\lambda^r t^r}{n^r} \left(1 - \frac{\lambda t}{n}\right)^{n-r}.$$

We can reorganize this expression, by taking out of the limit terms not involving n

$$P_r(t) = \frac{(\lambda t)^r}{r!} \lim_{n \to \infty} \frac{n!}{(n-r)!n^r} \left(1 - \frac{\lambda t}{n}\right)^n \left(1 - \frac{\lambda t}{n}\right)^{-r}.$$

We can rewrite the first term inside the limit to give

$$P_r(t) = \frac{(\lambda t)^r}{r!} \lim_{n \to \infty} \frac{n}{n} \frac{n-1}{n} \cdots \frac{n-r}{n} \left(1 - \frac{\lambda t}{n}\right)^n \left(1 - \frac{\lambda t}{n}\right)^{-r}.$$

Now we notice that the first term inside the limit is made up of the product of r fractional expressions, which each have a term in n on the top and bottom lines. These will all tend to 1 as n tends to ∞. The last term is similar to the limit that we saw in Chapter 7 when calculating the value of e. There, we showed that

$$\lim_{n \to \infty} \left(1 + \frac{1}{n}\right)^n = e$$

and by a similar argument we could show that

$$\lim_{n \to \infty} \left(1 + \frac{x}{n}\right)^n = e^x.$$

It, therefore follows that

$$\lim_{n \to \infty} \left(1 - \frac{\lambda t}{n}\right)^n = e^{-\lambda t}.$$

The last expression involves a negative power of $(1 - \lambda t/n)$, which will tend to 1 as n tends to ∞.

This gives the Poisson distribution as

$$P_r(t) = \frac{(\lambda t)^r}{r!} e^{-\lambda t}.$$

This is an expression in both r and t where r is the number of events and t is the length of the time interval being considered. We usually consider the probability of r events in an interval of unit time, which gives the Poisson distribution as

$$P_r = \frac{\lambda^r}{r!} e^{-\lambda}$$

where λ is the expected number of incidents in unit time.

The mean and variance of the Poisson distribution

The Poisson distribution was introduced by considering the probability of a single event in a small interval of length h as (λh). We then used the binomial distribution, with $\theta = \lambda h$ and $h = t/n$ and n tending to ∞, to derive the expression for the Poisson distribution. As the mean of the binomial distribution is $n\theta$, it would make sense that the mean of the Poisson distribution is $n\lambda h$. Using $n = t/h$ we get the mean as λt over an interval of length t and therefore the mean is λ over an interval of unit length.

By a similar argument we know that the variance of the Binomial distribution is $n\theta(1 - \theta)$. Substituting $\theta = \lambda h$ we get the variance as $n\lambda h(1 - \lambda h)$. As n tends to infinity and h to 0 we get the limit λt. Therefore, the variance in unit time is λ.

Example 21.16   The average number of 'page not found' errors on a web server is 36 in a 24-h period. Find the probability in a 60-min period that:

(a)   there are no errors;
(b)   there is exactly one error;
(c)   There are at most two errors;
(d)   there are more than three errors.

Solution   Assuming that the above process is a Poisson process, then we have that the average number of errors in 24 h is 36 and therefore the average is 1.5 in 1 h. As the mean is λ, we can now assume that the number of errors in 1 h follows a Poisson distribution with $\lambda = 1.5$, giving

$$P_r = \frac{(1.5)^r}{r!} e^{-1.5}.$$

(a)   We want to find $P(\text{no errors}) = P_0 = ((1.5)^0/0!)e^{-1.5} = e^{-1.5} \approx 0.2231$.
(b)   We want to find $P(\text{exactly one error}) = P_1 = (1.5)^1/1!\, e^{-1.5} = 1.5e^{-1.5} \approx 0.3347$.
(c) $P(\text{at most two errors}) = P_0 + P_1 + P_2 = 0.2231 + 0.3347 + (1.5)^2/2!\, e^{-1.5} = 0.8088$.
(d) $P(\text{more than three errors}) = 1 - P\,(\text{at most three errors}) = 1 - (P_0 + P_1 + P_2 + P_3)$.

Using the result from Part (c) we get

$$P(\text{more than three errors}) = 1 - 0.8808 - (1.5)^3/3!\, e^{-1.5}$$

$$\approx 1 - 0.8808 - 0.1255 = 0.0657.$$

21.14 Summary

1.   The mean of a sample of data can be found by using

$$\bar{x} = \frac{1}{n} \sum_i x_i$$

where the summation is over all sample values and n is the number of values in the sample. If the sample is divided into class intervals then

$$\bar{x} = \frac{1}{n} \sum_i f_i x_i$$

where x_i is a representative value for the class, f_i is the class frequency, and the summation is over the all classes.

2. The standard deviation of a sample of data can be found by using

$$\sigma = \sqrt{\frac{1}{n} \sum_i (x_i - \bar{x})^2}$$

where the summation is over all sample values, n is the number of values in the sample, and $\bar{x}$ is the sample mean. If the sample is divided into class intervals then

$$\sigma = \sqrt{\frac{1}{n} \sum_i f_i (x_i - \bar{x})^2}$$

where x_i is a representative value for the class, f_i is the class frequency, $\bar{x}$ is the sample mean, and the summation is over all classes. The square of the standard deviation is called the variance.

3. The cumulative frequency is found by summing the values of the current class and all previous classes. It is the 'number so far'.

4. The relative frequency of a class is found by dividing the frequency by the number of values in the data sample – this gives the proportion that fall into that class. The cumulative relative frequency is found by dividing the relative frequency by the number in the sample.

5. In probability theory the set of all possible outcomes of a random experiment is called the sample space. The probability distribution function, for a discrete sample space, is a function of the outcomes that obeys the conditions:

$$0 \leqslant p(x_i) \leqslant 1$$

where x_i is any outcome in the sample space and

$$\sum_i p(x_i) = 1$$

where the summation is over all outcomes in the sample space.

6. An event is a subset of the sample space. The probability of an event (if all outcomes are equally likely) is

$$p(\mathrm{E}) = \frac{\text{The number of outcomes in the event}}{\text{The number of outcomes in the sample space}}.$$

7. The addition law of probability is given by $P(A \cup B) = P(A) + P(B) - (A \cap B)$ for non-disjoint events.

   If $A \cap B = \emptyset$ this becomes $P(A \cup B) = P(A) + P(B)$ for disjoint events.

8. Multiplication law of probabilities: $p(A \cap B) = p(A)p(B|A)$ if events A and B are not independent.

9. The definition of independence is that B is independent of A if the probability of B does not depend on A: $p(B|A) = p(B)$ if B

is independent of A. In this case the multiplication law becomes $p(A \cap B) = p(A)p(B)$, where A and B independent events.

10. Bayes's theorem is

$$p(B|A) = \frac{p(A|B)p(B)}{p(A)}.$$

11. The normal, or Gaussian, distribution is a bell-shaped distribution. Many things, particularly involving error distributions, have a probability distribution that is approximately normal.

12. To calculate probabilities using a normal distribution we use areas of the standard normal distribution in table form, so the variable must be standardized by using the transformation

$$z = \frac{x - \mu}{\sigma}$$

where μ is the mean and σ the standard deviation of the distribution.

13. The exponential distribution is used to model times to failure.

14. The probability density function of the exponential distribution is $f(t) = \lambda e^{-\lambda t}$ and the cumulative density function is given by $F = 1 - e^{-\lambda t}$.

15. The mean and standard deviation of a continuous distribution can be found by

$$\mu = \int xf(x)\,dx \quad \text{and} \quad \sigma = \sqrt{\int_0^\infty (x - \mu)^2 f(x)\,dx}$$

where the integration is over the sample space. For the exponential distribution these give $\mu = 1/\lambda$, that is, the mean time to failure is the reciprocal of the failure rate. Also $\sigma = 1/\lambda$.

16. The binomial distribution is a discrete distribution that models repeated trials where the outcome of each trial is either success or failure and each trial is independent of the others. Its probability function is

$$p_n(r) = \binom{n}{r}\theta^r(1 - \theta)^{n-r}$$

where

$$\binom{n}{r} = {}^nC_r = \frac{n!}{(n-r)!r!} = \frac{n(n-1)\ldots(n-r+1)}{r!}$$

and r is the number of successes in n trials.

17. The mean of a discrete distribution is given by $\mu = \sum xp(x)$ and the variance is $\sigma^2 = \sum (x - \mu)^2 p(x)$. The mean of the binomial distribution, for n trials, is $n\theta$, the variance is $n\theta(1 - \theta)$, and the standard deviation $\sigma = \sqrt{n\theta(1 - \theta)}$.

18. For the Poisson distribution the number of incidents occurring in any interval depends only on the length of that interval. Its probability function is

$$P_r = \frac{\lambda^r}{r!}e^{-\lambda}$$

where λ is the expected number of incidents in unit time. The mean and the variance are λ.

21.15 Exercises

21.1. An integrated circuit design includes a capacitor of 100 pF (picofarads). After manufacture, 80 samples are tested and the following capacitances found for the nominal 100-pF capacitor (the data are expressed in pF)

```
90 100 115  80 113 114   99 105   90   99 106 103
95 105  95 101  87  91 101 102 103   90  96  97
99  86 105 107  93 118  94 113   92 110 104 104
95  93  95  85  96  99  98  83  97  96  98  84
102 109  98 111 119 110 108 102 100 101 104 105
93 104  97  83  98  91  85  92 100  91 101 103
101  86 120  96 101 102 112 119
```

Express the data in table form and draw a histogram. Find the mean and standard deviation.

21.2. What is the probability of throwing a number over 4 on one throw of a die?

21.3. What is the probability of throwing a number less than 4 or a 6 on one throw of a die?

21.4. What is the probability of throwing an odd number or a number over 4 on one throw of a die?

21.5. What is the probability of drawing any Heart or a King from a well shuffled pack of cards?

21.6. On two throws of a die, what is the probability of a 6 on the first throw followed by an even number on the second throw?

21.7. Throwing two dice, what is the probability that the sum of the dice is 7?

21.8. What is the probability that the first two cards dealt from a pack will be clubs.

21.9. A component in a communication network has a 1% probability of failure over a 24-h period. To guard against failure an identical component is fitted in parallel with an automatic switching device should the original component fail. If that also has a 1% probability of failure what is the probability that despite this precaution the communication will fail?

21.10. Find the reliability of the system, S, in Figure 21.31. Each component has its reliability marked in the figure. Assume that each of the components is independent of the others.

21.11. A ball is chosen at random out of a bag containing two black balls and three red balls and then a selection is made from the remaining four balls. Assuming all

outcomes are equally likely, find the probability that a red ball will be selected:
(a) at the first time;
(b) the second time;
(c) both times.

21.12. A certain brand of compact disc (CD) player has an unreliable integrated circuit (IC), which fails to function on 1% of the models as soon as the player is connected. On 20% of these occasions the light displays fail and the buttons fail to respond, so that it appears exactly the same as if the power connection is faulty. No other component failure causes that symptom. However, 2% of people who buy the CD player fail to fit the plug correctly, in such a way that they also experience a complete loss of power. A customer rings the supplier of the CD player saying that the light displays and buttons are not functioning on the CD. What is the probability that the fault is due to the IC failing as opposed to the poorly fitted plug?

21.13. If a population, which is normally distributed, has mean 6 and standard deviation 2, then find the proportion of values greater than the following:

(a) 9   (b) 10   (c) 12   (d) 7

21.14. If a population, which is normally distributed, has mean 3 and standard deviation 4, find the proportion of values less than the following:

(a) 1   (b) -5 (c) -1 (d) 0

21.15. If a population, which is normally distributed, has mean 10 and standard deviation 3, find the proportion of values that satisfy the following:

(a) $x > 3$ (b) $x < 12$ (c) $x > 9$

(e) $x < 11$ (f) $9 < x < 11$ (g) $3 < x < 12$

21.16. A car battery has a mean life of 4.2 years and a standard deviation of 1.3 years. It is guaranteed for 3 years. Estimate the percentage of batteries that will need replacing under the guarantee.

21.17. A certain component has a failure rate of 0.3 per hour. Assuming an exponential distribution calculate the following:
(a) the probability of failure in a 4-h period;
(b) the probability of failure in a 30-min period;
(c) the probability that a component functions for 1 h and then fails to function in the second hour;

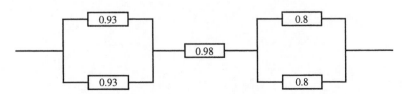

Figure 21.31   *A system for Exercise 21.10.*

Table 21.6 *Data for Exercise 21.20*

Number of defects in 74 min of recording time	Number of CDs with r defects	Chosen model probabilities	Chosen model frequencies
0	23		
1	32		
2	26		
3	12		
4	5		
5	2		
6+	0		

(d) of a group of five components the probability that exactly two fail in an hour.

21.18. A certain town has 50 cash dispenser machines but due to inaccessibility is only visited for repairs once a week after which all the machines are working. The failure rate of the machines is approximately 0.05 per 24 h. A town councillor makes a public complaint in a newspaper that on average at least one in 10 of the machines does not function. Assume an exponential model and calculate the number that are not functioning 1 day, 2 days, ..., 7 days after the day of the visit. Take the mean of these results to assess whether the councillor is correct.

21.19. A bag contains two red balls and eight green balls. A ball is repeatedly chosen at random from the bag, its colour recorded and then replaced. Find the following probabilities:

(a) the first three picked were green;
(b) in a selection of five there were exactly two red balls;
(c) there were no more than three red balls out of the first 10.

21.20. One hundred CDs, each containing 74 min of recording time, were tested for defects. The frequency of defects is given in Table 21.6. Calculate the mean number of defects per 74 min of recording time and choose an appropriate probability model. Using your model copy, complete the two empty columns of the table. Comment on the agreement between the number of incidents and the chosen model frequencies.

21.21. Telephone calls are received at a call centre at a rate of 0.2 per second on average. Calculate the probability that more than 11 calls are received in 1 min.

21.22. Tankers arrive at a dock at a rate of four per day. Assume that the arrivals are a Poisson process and find the following:

(a) the probability that less than five tankers arrive during 1 day;
(b) the probability that there are over five arrivals;
(c) the probability that there is exactly five arrivals;
(d) The probability that there are between two and five inclusive, arrivals.

Answers to exercises

Chapter 1

1.1. (a) {b}, (b) {a, b, c, d, e, f}, (c) {a, b}, (d) {c, d, e},
(e) {c, d, e}, (f) {b, c, d, e}, (g) {b, c, d, e},
(h) {a, b, c, d, f, g}, (i) {a, b, c, d, f, g}.

1.3. (a) {0, 1, 2, 3, 4}, (b) {3, 4, 5, 6, 7, 8, 9},
(c) {0, 1, 2, 3, 4, 5, 6, 7, 8, 9}, (d) {5, 6, 7, 8, 9},
(e) {5, 6, 7, 8, 9}.

1.4. (a) False, (b) False, (c) True, (d) False.

1.5. (a) A, (b) $A \cup B$, (c) $A \cap B$.

1.6. (a) $\{(x, y)|0 \leqslant x \leqslant 79 \text{ and } 0 \leqslant y \leqslant 24\}$,
(b) (i) $\{(x, y)|0 \leqslant x \leqslant 79 \text{ and } 13 \leqslant y \leqslant 24\}$
(ii) $\{(x, y)|0 \leqslant x \leqslant 79 \text{ and } 0 \leqslant y \leqslant 24 \text{ and } y \geqslant (79/24)x\}$

1.7. 1

1.9. (a) (i) 3, (ii) 3, (iii) 3/5 (iv) $2\frac{5}{6}$, (v) 9/125, (vi) 2,
(vii) $2\frac{1}{4}$, (viii) 3.
(b) (i) $f \circ g : x \mapsto \frac{2}{3}x^2 - 1$,
(ii) $g \circ f : x \mapsto \frac{1}{3}(2x - 1)^2$,
(iii) $f \circ g : x \mapsto 9/x^2$,
(iv) $f^{-1} : x \mapsto (x + 1)/2$,
(v) $h^{-1} : x \mapsto 3/x$.
(d) (i) 1, (ii) 9/25, (iii) $16\frac{1}{3}$.

Chapter 2

2.1. (a) 3, (b) 2, (c) -5, (d) 1/2.

2.2. (a) 3, (b) -5, (c) 4/7, (d) 0.

2.3. (a) $y = -5x + 11$.

2.4. (a) $y = -3.1x + 2$, (b) $y = 2x$, (c) $y = -x + 0.7$.

2.5. (a) $y = -3x + 1$, (b) $3y = 5x - 2$, (c) $5y = 2x + 3$,
(d) $y = -4$.

2.6. (a) 2, -2 (b) 1/2, $-l$ (c) 3 (d) 4, -4 (e) 2, -3.
(f) -3, -4, (g) 4, -3.

2.8. (a) Even, (b) Neither, (c) Odd.

2.9. (a) Even, (b) Even, (c) Neither, (d) Odd, (e) Even,
(f) Even.

2.10. (a) Yes, (b) No, (c) Yes.

2.11. (a) $t \leqslant 3.3$, (b) $x > -2/7$, (c) $y > -8/5$,
(d) $t < -3$.

2.12. (a) $x > -11$, (b) $t \geqslant 2$, (c) $u > 2$ or $u < -5$.

2.13. (a) $-3 \leqslant x \leqslant 3$, (b) $-1 < x < 3/2$ or $x > 5$,
(c) $-7 \leqslant t \leqslant 3$. (d) $-3\frac{1}{2} \leqslant w \leqslant 2\frac{1}{2}$.

2.14. (a) $A = 119, k = -0.2$, (b) $A = l, k = 3.1$.

2.15. $y = 1.52, c = 1.38$, volume $= 1.23 \, \text{m}^3$.

Chapter 3

3.1. (a) F, (b) F, (c) T, (d) T, (e) F, (f) F, (g) T, (h) T.
(i) T for all t, (j) T for $t = 3$, F for all other values of
t, (k) F.

3.2. (a) 2, (b) 0, 5, (c) $-2\frac{1}{5}, \frac{1}{5}$, (d) $3\frac{1}{4}, \frac{1}{2}$, (e) $t \leqslant 1/10$,
(f) $x < -1/2$.

3.3. (a) $x \geqslant 2 \lor x < -1$, (b) $t > l$, (c) $x > 3$.

3.4. (a) $\Leftarrow$, (b) $\Leftrightarrow$, (c) No implication, (d) $\Leftrightarrow$, (e) $\Rightarrow$,
(f) $\Leftrightarrow$, (g) $\Leftarrow$, (h) $\Leftrightarrow$, (i) No implication, (j) $\Rightarrow$, (k) $\Leftrightarrow$.

3.5. (a) F, (b) F, (c) T, (d) T, (e) F.

3.6. (a) $x/3 \notin \mathbb{Z}$, (b) $3 \leqslant y \leqslant 60$, (c) $w/2 \in \mathbb{Z} \land w > 20$,
(d) $|t - t_{n-l}| < 0.001$.

3.7. (a) (0,6), (1,4), (2,3), (3,1), (4,0), (b) 5 m.

3.8. $13.15 \, \text{m s}^{-1}$.

Chapter 4

4.2. (b) (i) F, (ii) T, (iii) F, (iv) T, (v) T, (vi) T.

4.4. (a) a, (b) abc, (c) a, (d) $a\bar{b}$,

4.5. (b) (i) 1, (ii) 1, (iii) 0, (iv) 0.

4.6. (a) $a(\bar{b}, +c)$, (b) $a + b + c$, (c) $a(b + \bar{c})$, (d) $\bar{a}\bar{b}\bar{d} + ab$.

4.7. (a) $\overline{(ab + c)} \cdot c$, (b) $ab\bar{c} + \bar{c}d + c\bar{d}$.

4.8. $r = \bar{c} + cd + \bar{a}b$.

Chapter 5

5.1. (a) 120°, (b) 720°, (c) 108°, (d) 360°, (e) 90°.

5.2. (a) $\pi/4$, (b) $3\pi/4$, (c) $\pi/18$, (d) $5\pi/6$.

5.3. (a) $\sqrt{3}/2$, (b) $\sqrt{3}$, (c) $-1/2$, (d) $\sqrt{3}/2$. (e) $\sqrt{3}$.

5.5. (a) $-\pi/4$, 2,2, 1/2, (b) $-\pi/8$, 1/2, 4, 1/4.

5.6. (a) 3, $\pi/2$, 4, $\pi/2$, (b) 1, $2\pi/377$, 377, $0.4 - (\pi/2)$,
(c) 40, $\pi/1500$, 3000, -0.8.

5.7. (a) $0.5, \pi, 1/\pi, 2$, (b) $2, \pi/36, 36/\pi, 2$,
(c) $52, \pi/40, 40/\pi, 80$.

5.8. (c) $0.4\,\mathrm{m\,s^{-1}}$.

5.9. 110 dB.

5.10. -5.74 dB.

5.12. $0.025/\pi$ Hz.

5.14. (a) 1.5 m, 0.75 m, 0.5 m, (b) 442 Hz.

5.15. $C = 2\cos(\pi/3)$, $d = -2\sin(\pi/3)$.

5.16. (a) $\sin(104°)$, (b) $\sin(4°)$, (c) $\cos(52°)$,
(d) $-\cos(63°)$, (e) $\sin(3x)$, (f) $\cos(2x)$.

5.17. $\cos(x)\cos(y)\cos(z) - \sin(x)\sin(y)\cos(z)$
$- \sin(x)\cos(y)\sin(z) - \cos(x)\sin(y)\sin(z)$.

5.18. (a) $\pi + 0.5236$, $2\pi - 0.5236$, $3\pi + 0.5236$,
$4\pi - 0.5236$, $5\pi + 0.5236$, $6\pi - 0.5236$,
(b) 0.3218, $\pi + 0.3218$, $2\pi + 0.3218$, $3\pi + 0.3218$,
$4\pi + 0.3218$, $5\pi + 0.3218$,
(c) 2.4981, $2\pi - 2.4981$, $2\pi + 2.4981$,
$4\pi - 2.4981$, $4\pi + 2.4981$, $6\pi - 2.4981$,
(d) 0.5236, $\pi - 0.5236$, $\pi + 0.5236$, $2\pi - 0.5236$,
$2\pi + 0.5236$, $3\pi - 0.5236$, $3\pi + 0.5236$,
$4\pi - 0.5236$, $4\pi + 0.5236$, $5\pi - 0.5236$,
$5\pi + 0.5236$, $6\pi - 0.5236$,
(e) $\pi/2$, $3\pi/2$, 2.0944, $2\pi - 2.0944$, $5\pi/2$,
$7\pi/2$, $2\pi + 2.0944$, $4\pi - 2.0944$, $9\pi/2$,
$11\pi/2$, $4\pi + 2.0944$, $\quad 6\pi - 2.0944$,
(f) No solutions.

Chapter 6

6.1. (a) $8.667\,\mathrm{m\,s^{-1}}$.
(b)
$$v = \begin{cases} \frac{1}{5}t^2 + 2 & \text{for } 0 \leqslant t \leqslant 10 \\ 22 & \text{for } t \geqslant 10. \end{cases}$$

(c) (i) $7\,\mathrm{m\,s^{-1}}$, (ii) $22\,\mathrm{m\,s^{-1}}$, (iii) $22\,\mathrm{m\,s^{-1}}$,
(d) $2\,\mathrm{m\,s^{-2}}$, (e) $0\,\mathrm{m\,s^{-2}}$,
(f)
$$a = \begin{cases} \frac{2}{5}t & \text{for } 0 \leqslant t \leqslant 10 \\ 0 & \text{for } t \geqslant 10 \end{cases}$$

(g) (i) $2\,\mathrm{m\,s^{-2}}$, (ii) $0\,\mathrm{m\,s^{-2}}$, (iii) $0\,\mathrm{m\,s^{-2}}$.

6.2. (1) $6x + 6$, (2) $\frac{1}{2}x^{-1/2} + \frac{1}{2}x^{-3/2}$, (3) $\frac{3}{2}\sqrt{2x} + \frac{5}{3x^3}$,

(4) $(9x^2 + 1)\cos(3x^3 + x)$, (5) $-12\sin(6x - 2)$,

(6) $2x\sec^2(x^2)$, (7) $-2/(2x - 3)^2$, (8) $24(4x - 5)^5$,

(9) $-x/\sqrt{(x^2 - 1)^3}$, (10) $-2/\sqrt{1 - (5 - 2x)^2}$,

(11) $-\sec^2(1/x)/x^2$,

(12) $x/\sqrt{x^2 + 2}$, (13) $-\frac{3}{2}(x + 4)^{5/2}$,

(14) $2\sin(x)\cos(x)$,

(15) $-15\cos^2(x)\sin(x)$, (16) $-3\cos(x)/\sin^4(x)$,

(17) $-10\cos(5x)\sin(5x)$,

(18) $3x^2\sqrt{x + 1} + \frac{1}{2}x^3/\sqrt{x + 1}$,

(19) $5\cos(x) - 5x\sin(x)$,

(20) $12x\sin(x) + 6x^2\cos(x)$,

(21) $3\tan(5x) + (15x + 5)\sec^2(5x)$,

(22) $3x\cos^{-1}(x) - (x^3/\sqrt{1 - x^3})$,

(23) $(3x^2\cos(x) + x^3\sin(x))/\cos^2(x)$,

(24) $-2\cos(x)/\sin^3(x)$,

(25) $((2x + 10)\cos(x) - 2\sin(x))/(2x + 10)^2$,

(26) $(2x\tan(x) - x^2\sec^2(x))/\tan^2(x)$,

(27) $(6x/\sqrt{x - 1}) - (3x^2/2\sqrt{(x - 1)^3})$,

(28) $20x/(5x^2 + 1)^2$,

(29) $(-\cos(x)/(x + 1)^2) - (\sin(x)/(x + 1))$,

(30) $2x/\sqrt{1 - x^4}$,

(31) $[(\frac{5}{2}x - 2)x\sin(x) + x^2\cos(x)]/\sqrt{x - 1}$,

(32) $-4x\cos(x^2)\sin(x^2)$,

(33) $[5\tan(\sqrt{5x - 1})\sec^2(\sqrt{5x - 1})]/\sqrt{5x - 1}$

6.3. $-(9.475 \times 10^5\cos(20\pi t) + 3.553\cos(30\pi t))$

Chapter 7

7.1. (a) $(x^2/4) + (x^3/3) + c$, (b) $-2\cos(x) + \tan(x) + c$,
(c) $\frac{-1}{2x^2} + c$, (d) $x + (x^3/3) + \frac{3x^4}{4} + c$,
(e) $x - (5x^2/2) + c$, (f) $-\frac{1}{4}\sin(2 - 4x) + c$,
(g) $(\sqrt{(2x - 1)^3}/3) + c$, (h) $2\sqrt{x + 2} + c$,
(i) $[(x^2 - 4)^4/8] + c$, (j) $(\sqrt{(1 + x^2)^3}/3) + c$,
(k) $[-1/(1 + \sin(x))] + c$ (l) $(\frac{1}{2})(x^2 + x - 6)^2 + c$,
(m) $\dfrac{-4}{3(x^3 - 7)} + c$, (n) 0.0207,
(o) $x^2\sin(x) + 2x\cos(x) - 2\sin(x) + c$, (p) 8.633,
(q) $(2x - 3)^5(10x + 3)/120 + c$,
(r) $-\cos(x) + (2\cos^3(x)/3) - (\cos^5(x)/5) + c$,
(s) $(\sin(4x)/32) + \frac{1}{4}\sin(2x) + (3x/8) + c$,
(t) $-(\cos(8x)/16) + \frac{1}{2}\cos(2x) + c$.

7.2. $s = 3t - \frac{1}{2}t^2 + 5$; when $t = 2\,\mathrm{s}$, $s = 9\,\mathrm{m}$.

7.3. $y = -5t$

7.4. $y = x - (2x^3/3) + 1$.

7.5. $0.0187\,\mathrm{A}$.

7.6. $F = (1 - \cos(3\pi t))/\pi$.

7.8. $28\frac{1}{2}$

7.9. 9

7.10. 2

7.11. 4.76

7.12. 1.382

7.13. (a) 0.9450788, 0.94583209, (b) 0.9461459, 0.94608693

7.14. (a) 1.106746, (b) 1.098942

7.15. 0.1667

Chapter 8

8.2. $P_0 = 1, k = 1/1200$.

8.3. $N_0 = 5 \times 10^{-6}, k = -4.3 \times 10^{-4}$.

8.4. $A = 100, k = -0.1$

8.7. (a) 4.14431, (b) 0.99505, (c) 0.56882, (d) Not defined, (e) Not defined.

8.8. (a) $2t\,e^{t^2-2}$, (b) $e^{-t}(2\sinh(2t) - \cosh(2t))$,
(c) $(2x \sinh(x) - (x^2 - 1)\cosh(x))/(\sinh^2(x))$,
(d) $(3x^3 - 3)/(x^3 - 3x)$,
(e) $1/(\ln(2)x)$, (f) $4\ln(a)a^{4t}$, (g) $2^t(\ln(2)t^2 + 2t)$,
(h) $-2/(e^{t-1})^2$.

8.9. (a) $\frac{1}{4}e^{4t-3} + c$, (b) 0.113, (c) $\frac{1}{4}\cosh(2x^2) + C$,
(d) $-\frac{1}{2}x^2\ln(x) - (x^2/4) + c$, (e) 0.7182824,
(f) $\ln(\cosh(t)) + c$, (g) $\ln(x^2 - 2x - 4) + c$,
(h) $2\ln(t - 3) - \ln(t - 1) + c$, (i) -0.40547.

8.10. $I = -0.01\,e^{-10t}, 0.0693\,\text{s}$

8.11. (a) $-0.0025\,\text{A}$, (b) $-1.684 \times 10^{-5}\,\text{A}$,
(c) $-1.135 \times 10^{-7}\,\text{A}$.

Chapter 9

9.1. (a) $\mathbf{b} - \mathbf{a}$, (b) $\mathbf{a} - \mathbf{b}$, (c) $\mathbf{a} + \mathbf{b}$, (d) $-\mathbf{b}$,
(e) $-\mathbf{a}$, (f) $\mathbf{b} - \mathbf{a}$, (g) $\mathbf{a} - \mathbf{b}$, (h) $2\mathbf{a}$, (i) $-2\mathbf{b}$,
(j) $\mathbf{a} + \mathbf{b}$.

9.2. (a) (0,5), (b) (2,1), (c) (−2,−1),
(d) (2,1), (e) (−2,4), (f) (−1,7),
(g) (4,7), (h) (−3,2,3), (i) (30,60,20),
(j) (39,30,−4), (k) (12,32,19).

9.3. (a) $3.162\angle 1.249$, (b) $3.162\angle -0.322$,
(c) $3.162\angle -1.893$, (d) $7.810\angle -0.876$.

9.4. (a) (−5,0), (b) (−1,0), (c) (0.354,−0.354),
(d) (1.5,2.598).

9.5. (a) $-1.248\cos(3t) + 2.728\sin(3t)$,
(b) $2.837\cos(20t) + 9.589\sin(20t)$.

9.6. (a) $5\cos(10t + 0.644)$, (b) $10.20\cos(157t - 1.768)$.

9.7. (a) $13.040\cos(2t + 2.457)$, (b) 0,
(c) $6.325\cos(628t - 2.820)$.

9.8. (a) (0.6,0.8), (b) (5/13,12/13), (c) (5/12,−12/13),
(d) (0.707,0.707), (e) (0.832,0.555), (f) (1,0),
(g) (0,−1), (h) (1/3, 2/3,2/3),
(i) (0.408,−0.408,0.816),
(j) (0.707,0,−0.707).

9.9. (a) $5\mathbf{i} + 2\mathbf{j}$, (b) $-\mathbf{i} - 2\mathbf{j}$, (c) $-6\mathbf{i} + 2\mathbf{j}$,
(d) $-\mathbf{i} + 2\mathbf{j} - 3\mathbf{k}$, (e) $0.2\mathbf{i} - 1.6\mathbf{j} + 3.3\mathbf{k}$.

9.10. (a) −3, (b) 3, (c) −3.

9.11. (a) 1.305, (b) $\pi/2$, (c) 1.616.

9.13. (a) 1.107 radians to x-axis and 0.4636 radians to
y-axis;
(b) 1.726 radians to x-axis, 2.236 radians to y-axis,
and 2.452 radians to z-axis.

9.14. (a) (i) 2.828, (ii) −4.243, (iii) 1.961, (iv) −5.099,
(v) 0.2425;
(b) (i) 4.619, (ii) 1.265,

9.16. (a) 7, (b) 10, (c) 11.

9.17. (a) $(-\lambda, 1 + 3\lambda), \lambda \in \mathbb{R}$, (b) $(1 - 3\lambda, 1 - 5\lambda)$,
$\lambda \in \mathbb{R}$, (c) $(1 + 5\lambda, 1 + 2\lambda), \lambda \in \mathbb{R}$,
(d) $(-1 - 2\lambda, -4), \lambda \in \mathbb{R}$.

Chapter 10

10.1. (b) (i) $4 + j$, (ii) $-2 + j6$, (iii) $1 + j2$;
(c) (i) $3 + j5$, (ii) $5 - j3$, (iii) 5, (iv) $-\frac{1}{6} - \frac{j}{2}$,
(v) $7 + j6$.

10.2. (a) 1, (b) $-j$, (c) 1.

10.3. (a) $34 - j2$, (b) $-3 - j4$, (c) $\frac{23}{26} - j\frac{11}{26}$,
(d) $\frac{57}{97} + j5\frac{95}{97}$.

10.4. $j3/2$.

10.5. $x = 3, y = 5$.

10.6. (a) 1,2, (b) $1 + j\sqrt{5}, 1 - j\sqrt{5}$,
(c) $(1/6) + j\sqrt{11}/6, (1/6) - j\sqrt{11}/6$,
(d) $-(1/4), 2$ (e) $j\sqrt{\frac{3}{2}}, -j\sqrt{\frac{3}{2}}$;

10.7. (a) $-1 - j3$, (b) $b = 2, c = 10$.

10.8. (a) $4.899\angle 1.030$, (b) $6.708\angle 2.678$,
(c) $6.403\angle -2.246$, (d) $4.899\angle -2.601$.

10.9. (a) $-3.536 - j3.536$, (b) $3.464 - j2$,
(c) $-1.827 + j0.813$, (d) $3.992 - j3.010$.

10.10. $x = 5\frac{1}{4}, y = 4\frac{1}{4}$.

10.11. (a) $36\angle 23\pi/20$, (b) $4\angle 7\pi/20$, (c) $13.626\angle 2.159$,
(d) $10.968\angle -0.539$, (e) $12\angle -3\pi/4$, (f) $9\angle 4\pi/5$.

10.12. $16\angle 3.083$.

10.13. $Z = 409174\angle -0.212, V = 2.046 \times 10^6\,\text{V}$,
relative phase $= -0.212$.

10.14. $Y = 3.916 \times 10^{-4}\angle -1.561, I = 3.916 \times 10^{-3}\,\text{A}$,
relative phase $= -1.561$.

10.15. (a) $313\angle 32.4°$, (b) $9.7347 \times 10^{-4}\angle 106.5°$.

10.16. (a) $4.899\,e^{1.030j}$, (b) $6.708\,e^{2.678j}$, (c) $6.403\,e^{-2.246j}$,
(d) $8\,e^{0.384j}$, (e) $3\,e^{2.13j}$, (f) $6\,e^{1.9j}$.

10.17. (a) $4\angle 2, -1.665 + 3.637j$, (b) $1\angle -\pi/2, -j$
(c) $2\angle \pi, -2$, (d) $6\angle 0.858, 3.922 + j4.541$,
(e) $\frac{1}{2}\angle 1.283, 0.142 + j0.479$, (f) $3\angle 11\pi/12$,
$-2.898 + j0.776$,

(g) $2.906\angle2.017, -1.255 + j2.621$.

10.18. (a) $-2, 0$, (b) $-2.633, -1.438$, (c) $0.183, 0.285$,
(d) $-1.821, 0.260$, (e) $-3.539, -12.133$.

10.19. (a) $36\,e^{j23\pi/20}$, (b) $4\,e^{j\pi7/20}$, (c) $13.627\,e^{j2.159}$,
(d) $10.969\,e^{-j0.539}$, (e) $12\,e^{-j3\pi/4}$, (f) $4\,e^{-j23\pi/20}$,
(g) $144\,e^{j0}$.

10.20. $27\angle1.38, 27\,e^{j1.38}$.

10.21. (a) $1, j, -1, -j$,
(b) $0.866 + j0.5, j, -0.866 + j0.5$,
$-0.866 - j0.5, -j, 0.866 - j0.5$,
(c) $1.618 + j1.176, -0.618 + j1.902, -2$,
$-0.618 - j1.902, 1.618 - j1.176$,
(d) $0.437 + j0.757, -0.874, 0.437 - j0.757$,

10.22. $\cos^3(\theta) - 3\cos(\theta)\sin^2(\theta)$.

10.23. $\frac{3}{4}\sin(\theta) - \frac{1}{4}\sin(3\theta)$.

10.24. $1.174 + j0.540, -0.151 + j1.284, -1.267 + j0.252$,
$-0.631 - j1.128, 0.877 - j0.949$.

10.25. (a) $1 - j, -1 - j$,
(b) $0.327 + j3.035, -0.327 - j0.035$.

10.26. (a) 4, (b) 0.5

Chapter 11

11.1. (a) $(5/2, -17/4)$ is a local minimum,
(b) $(2/3, 4/3)$ is a local maximum,
(c) $(1/3, -2/9)$ is a local minimum and
$(-1/3, 2/9)$ is a local maximum,
(d) $(10, 40)$ is a local minimum and $(-10, -40)$ is
a local maximum,
(e) $(-4, -126)$ is a local minimum, $(0, 2)$ is a local
maximum and $(1, -1)$ is a local minimum.

11.2. $1/(2\sqrt{2}), -1/(2\sqrt{2})$.

11.5. $8, -27/256$.

11.6. 1.193 m, 2.384 m.

11.8. $0.21\omega^2, -0.106875\omega^2$.

11.9. $1/2, \frac{1}{2}(1 + \cos(\theta))$.

11.10. $\sqrt{h/3c}, 66\frac{2}{3}\%$.

Chapter 12

12.1. (a) $21, 25, 29, a_{n+1} = a_n + 4, a_1 = -3$,
(b) $0.25, 0.125, 0.0625, a_{n+1} = \frac{1}{2}a_n, a_1 = 8$,
(c) $0, -3, -6, a_{n+1} = a_n - 3, a_1 = 18$,
(d) $6, -6, 6, a_{n+1} = -a_n, a_1 = 6$,
(e) $2, 0, -2, a_{n+1} = a_n - 2, a_1 = 10$,
(f) $29, 37, 46, a_{n+1} = a_n + n, a_1 = 1$,
(g) $21, 28, 36, a_{n+1} = a_n + n + 1, a_1 = 1$.

12.2. (a) $2, 5, 8, 11, 14$, (b) $720, 360, 240, 180, 144$,
(c) $0, -3, -8, -15, -24$, (d) $6, 8, 10, 12, 14$,
(e) $2, 6, 18, 54, 162$,
(f) $-1, 2, -4, 8, -16$, (g) $\frac{1}{2}, 1, 1\frac{1}{2}, 2, 2\frac{1}{2}$,
(h) $2, 5, 8, 11, 14$, (i) $1, 3, 9, 27, 81$.

12.3. (a) $\sum_{n=0}^{n=10} x^n$, (b) $\sum_{n=1}^{n=8}(-2)^n$, (c) $\sum_{n=1}^{n=6} n^3$,

(d) $\sum_{n=1}^{n=8}\left(-\frac{1}{3}\right)^n$,

(e) $\sum_{n=2}^{n=10}\frac{1}{n^2}$, (f) $\sum_{n=1}^{n=8}(-4)\left(\frac{1}{4}\right)^{n-1}$.

12.4. (a) $0, 0.1987, 0.3894, 0.5646, 0.7174, 0.8415$,
$0.9320, 0.9854, 0.9996, 0.9738$,
(b) $1, 0.9553, 0.8253, 0.6216, 0.3624, 0.0707$,
$-0.2272, -0.5048, -0.7374, -0.9041$,
(c) $0, 2, 4, 6, 8, 6, 4, 2, 0, -2$,
(d) $1, 1, 1, 1, -1, -1, -1, -1, 1, 1, 1$.

12.5. (a) $22, 42, 6 + (n - 1)4$,
(b) $1, -1.5, 3 + (n - 1)(-0.5)$,
(c) $17, 47, -7 + (n - 1)6$.

12.6. (a) $-10 + (n - 1)4, 560$, (b) $-5 + (n - 1)0.5, -5$,
(c) $25 + (n - 1)(-3), -70$.

12.7. $2/3, 2 + (n - 1)\frac{2}{3}, 2, 2\frac{2}{3}, 3\frac{1}{3}, 4, 4\frac{2}{3}, 5\frac{1}{3}$.

12.8. 32

12.9. (a) $8, 128, 2^{n-1}$, (b) $1/192, 1/49152, \frac{1}{3}(\frac{1}{4})^{n-1}$,
(c) $1/3, 1/243, -9(-\frac{1}{3})^{n-1}$, (d) $29.296\,875$,
$71.525574, 15(3/4)^{n-1}$.

12.10. (a) $(8/25)5^{n-1}, 31249.92$,
(b) $(-1.3867)(-2.0801)^{n-1}, 157.3377$,
(c) $64(-1/2)^{n-1}, 42.5$.

12.11. 8

12.12. 17

12.13. (a) 30, (b) 1.99609375, (c) -2.2499619.

12.14. (a) $\dfrac{1 - z^n}{1 - z}$, (b) $\dfrac{1 - (-1)^n y^{2n}}{1 + y^2}$, (c) $\dfrac{2(1 - (2/x)^n)}{1 - 2/x}$

12.15. (a) Convergent, 4 (b) Not convergent,
(c) Convergent, 20.25, (d) Convergent, 1/3.

12.16. (a) 4/9, (b)1/6, (c) 1/45.

12.17. (a) $1 + 3x/2 + 3x^2/4 + (x^3/8)$,
(b) $1 - 4x + 6x^2 - 4x^3 + x^4$,
(c) $x^3 - 3x^2 + 3x - 1$,
(d) $1 - 8y + 24y^2 - 32y^3 + 16y^4$,
(e) $1 + 8x + 28x^2 + 56x^3 + 70x^4 + 56x^5$
$+28x^6 + 8x^7 + x^8$,
(f) $8x^3 + 12x^2 + 6x + 1$,
(g) $8a^3 + 12a^2b + 6ab^2 + b^3$,
(h) $x^7 + 7x^5 + 21x^3 + 35x + (35/x) + (21/x^3)$
$+(7/x^5) + (1/x^7)$,
(i) $a^4 - 8a^3b + 24a^2b^2 - 32ab^3 + 16b^4$.

12.18. (a) 1.331, (b) 0.6561, (c) 8.120601,

12.19. (a) $1 + 10x + 40x^2 + 80x^3 + \cdots$,
(b) $1 - 24x + 252x^2 - 1512x^3 + \cdots$
(c) $64 - 192z + 240z^2 + 160z^3 + \cdots$,
(d) $1 + 8x + 30x^2 + 70x^3 + \cdots$

(e) $1 - 6x + 15x^2 - 20x^3 + \cdots$

(f) $(1/32) - (5x/8) + 5x^2 - 20x^3 + \cdots$.

12.20. $5\cos^4(\theta)\sin(\theta) - 10\cos^2(\theta)\sin^3(\theta) + \sin^5(\theta)$.

12.21. (a) 0, 8, (b) −7, −24, (c) −12, 316.

12.22. (a) 0.9227, (b) 1.0721, (c) 74.2204.

12.23. (a) $1 - x^2/2! + (x^4/4!) - (x^6/6!) + \cdots$, all x
(b) $1 + x^2/2! + (x^4/4!) + (x^6/6!) + \cdots$, all x
(c) $x - (x^2/2) + (x^3/3) - (x^4/4) + \cdots$, $|x| < 1$
(d) $1 + 1.5x + (0.75x^2/2!) - (0.375x^3/3!) + \cdots$, $|x| < 1$,
(e) $1 - 2x + 3x^2 - 4x^3 + \cdots$, $|x| < 1$.

12.24. (a) $1 - x^2 + (x^4/3) - (17x^6/6!) + \cdots$,
(b) $1 - (x^3/3) + (x^5/5) - (x^7/7) + \cdots$,
(c) $x + x^2 + (x^3/3) - (4x^5/5!) + \cdots$,
(d) $1 - 2.5x + 2.875x^2 - 2.8125x^3 + \cdots$.

12.25. (a) 1.025, (b) 0.09967, (c) 0.03, (d) 0.9713.

12.26. (a) 0.1951, (b) 2.005.

12.27. (a) 3/25, (b) 6, (c) 0, (d) 1, (e) 1, (f) −0.2273.

12.28. (a) −0.618034, (b) 1.49535, (c) 0.450 184.

12.29. (a) $|x_n - x_{n-1}| < 0.0000005$,
(b) $|x_n - x_{n-1}| < 0.0000005|x_n|$.

Chapter 13

13.1. (a)

$$\begin{pmatrix} 2 & 0 & 0 \\ 3 & -4 & 2 \end{pmatrix}$$

(b)

$$\begin{pmatrix} 4 & -2 & 0 \\ 2 & -21 & 4 \end{pmatrix}$$

(c) Not possible
(d)

$$\begin{pmatrix} 4 & 10 \\ 2 & -19 \\ 0 & -4 \end{pmatrix}$$

(e)

$$\begin{pmatrix} 4 \\ 0 \\ -1 \end{pmatrix}$$

(f) Not possible, (g) (8 10), (h) (8 − 7 − 4),
(i) (24 30)
(j)

$$\begin{pmatrix} 3 & 5 \\ -2 & 3\frac{1}{3} \end{pmatrix}$$

(k)

$$\begin{pmatrix} -3/2 & 8 \\ -5/2 & 37/3 \end{pmatrix}$$

(l) Not possible,

(m)

$$\begin{pmatrix} 3 & -8 & -1 \\ 8 & 38 & 10 \\ 3 & 30 & 19 \end{pmatrix}$$

(n)

$$\begin{pmatrix} 16\frac{1}{3} & 10\frac{8}{9} \\ -5\frac{4}{9} & -3\frac{17}{27} \end{pmatrix}$$

(o) Not possible
(p)

$$\begin{pmatrix} 16 \\ 4 \end{pmatrix}$$

13.3.

$$\begin{pmatrix} 0 & 0 \\ 0 & 0 \end{pmatrix}$$

13.4. (a)

$$\begin{pmatrix} -1 & 1 & 0 & 0 & 0 & 0 & 0 \\ 1 & 0 & -1 & 0 & 0 & 1 & 0 \\ 0 & 0 & 0 & 0 & 0 & -1 & 1 \\ 0 & -1 & 1 & 1 & -1 & 0 & -1 \\ 0 & 0 & 0 & -1 & 1 & 0 & 0 \end{pmatrix}$$

$$\begin{pmatrix} -1 & 1 & 0 & 0 & 0 & -1 & 0 & 0 \\ 0 & -1 & 1 & 0 & 0 & 0 & -1 & 0 \\ 0 & 0 & -1 & -1 & 0 & 0 & 0 & -1 \\ 1 & 0 & 0 & 1 & -1 & 0 & 0 & 0 \\ 0 & 0 & 0 & 0 & 1 & 1 & 1 & 1 \end{pmatrix}$$

13.5. (a) (0,0), (−0.5, 0.866), (−1.366, 0.366), (−0.866, −0.5),
(b) (−4, 1), (−3, 1), (−3, 2), (−4, 2),
(c) (0,0), (1,0), (1, −1), (0, −1),
(d) (0, 0), (1, 0), (1, 5), (0, 5),
(e) (−2, 3), (−2.5, 3.866), (−3.366, 3.366), (−2.866, 2.5),
(f) (−2.098, −2.366), (−2.598, −1.5), (−3.464, −2), (−2.964, −2.866),
(g) (0, 0), (−0.5, 4.33), (−1.366, 1.83), (−0.866, −2.5),
(h) (0, 0), (0, 1), (1, 1), (1, 0),
(i) (0, 0), (3.25, 1.299), (4.549, 3.049), (1.299, 1.75),
(j) (a) Rotation through −120° about the origin,
(b) Translation by (4, −1),
(c) Reflection in the x-axis,
(e) Translation by (2, −3) followed by rotation through −120° about the origin,
(g) Scaling by 1/5 in the y-direction followed by rotation through −120° about the origin,
(i) Scaling along a line at an angle of 30° to the x-axis by a factor of 1/4.

13.6. (a) Determined, $x = 0, y = 3$,
(b) Determined, $x = 5, y = -2$,
(c) Inconsistent, (d) Indeterminate,
(e) Determined, $x = 1, y = -10$,
(f) Inconsistent,
(g) Inconsistent, (h) Determined, $x = 6, y = 4$.

13.7. (a) Determined, $x = -2, y = -3, z = 1$,
(b) Determined $x = 400/21, y = 430/7$,
$z = -265/51$,
(c) Indeterminate,
(d) Inconsistent.

13.8. (a)

$$\begin{pmatrix} 1 & 2 \\ 0 & 1 \end{pmatrix}$$

(b)

$$\begin{pmatrix} 2/5 & -1/5 \\ 1/15 & 2/15 \end{pmatrix}$$

(c) No inverse,
(d)

$$\begin{pmatrix} 3/4 & 1/4 & -1/4 \\ -1/4 & 1/4 & -1/4 \\ 5/4 & 1/4 & 1/4 \end{pmatrix}$$

(e)

$$\begin{pmatrix} 1/3 & -1/3 & 1/3 \\ 4/15 & -1/15 & -1/30 \\ -1/3 & 1/3 & 1/6 \end{pmatrix}$$

(f) No inverse.

13.9. (a) -9, (b) 16, (c) 0, (d) 50.

13.10. (a) $-6\mathbf{i} - 8\mathbf{j} + 5\mathbf{k}$, (b) $\frac{3}{2}\mathbf{i} - \frac{3}{2}\mathbf{j}$.

13.11. (a) (i) 10, (ii) 0

13.12. (a)

$$\begin{pmatrix} \sigma_x \\ \sigma_y \\ \sigma_z \end{pmatrix} = \frac{E}{1 - 3v^2 - 2v^3} \begin{pmatrix} 1 - v^2 & v + v^2 & v + v^2 \\ v + v^2 & 1 - v^2 & v + v^2 \\ v + v^2 & v + v^2 & 1 - v^2 \end{pmatrix}$$

13.13. (a) 6, $\begin{pmatrix} 1 \\ 0 \end{pmatrix}$, 2, $\begin{pmatrix} 1 \\ 4 \end{pmatrix}$,

(b) 2, $\begin{pmatrix} 5 \\ 1 \end{pmatrix}$, -4, $\begin{pmatrix} 1 \\ -1 \end{pmatrix}$,

(c) 1, $\begin{pmatrix} 1 \\ -1 \end{pmatrix}$, 4, $\begin{pmatrix} 2 \\ 1 \end{pmatrix}$,

13.14. $v_R = 4.905I + 1.147$.

13.15. (a) Length $= 9.882 + 1.905 \times$ load,
(b) (i) 14.6 cm, (ii) 19.4 cm.

13.16. $P = 11.327 - 0.0556T$.

13.17. (a) $y = (1/70)(-82 + 123x + 215x^2)$,
(b) $y = 1.29 - 2.22333x + 2.03333x^2$.

Chapter 14

14.1. (a) Linear, 1, (b) Linear, 2, (c) Linear, 2,
(d) Not linear, 1, (e) Not linear, 2.

14.2. (a) $1.6\cos(3t) + 1.2\sin(3t) + t^2 - \frac{1}{2}t + \frac{1}{2}$
(b) $0.8\cos(3(t - 10)) + 0.6\sin(3(t - 10))$

14.3. (a) $(t^2 - 2\sin(t))/\cos(t)$

14.4. (a) $y = e^{3t}$ (b) $y = 2e^{t/4} - 4$,
(c) $y = (12/169)e^{-3t} - (12/169)\cos(12t)$
$+(5/169)\sin(12t)$,
(d) $y = 4e^{-2t/3} - e^{-t}$, (e) $y = 3t^2 + 3$,
(f) $y = (18/169)e^{-12t} + (18/169)\cos(5t)$
$+(92/169)\sin(5t)$.

14.5. (a) $x = -0.064e^{-5t/2} - 0.56e^{-t}$
$+0.624 - 0.56t + 0.2t^2$
(b) $x = (-0.2t + 0.08)e^{-2t} - 0.08\cos(6t)$
$+0.06\sin(6t)$,
(c) $x = e^{-2t}(-0.5\cos(t) - 0.5\sin(t)) + 0.5e^{-t}$.

14.6. (a) $q = -\frac{5}{13} \times 10^{-3}\cos(1000t) +$
$\frac{12}{13} \times 10^{-3}\sin(1000t)$
$V_C = -\frac{50}{13}\cos(1000t) + \frac{120}{13}\sin(1000t)$
$V_R = \frac{600}{13}\sin(1000t) + \frac{1440}{13}\cos(1000t)$
$V_L = \frac{300}{13}\cos(1000t) - \frac{720}{13}\sin(1000t)$
(b) $q = -10^{-9} + 10^{-6}t$, $V_c = t - 10^{-3}$,
$V_R = 10^{-3}$, $V_L = 0$.

14.7. (a) $q = \frac{cv_i}{Rcj\omega + 1}e^{j\omega t}$, (b) (i) $v_c = \frac{v_i}{Rcj\omega + 1}e^{j\omega t}$,
(ii) $V_R = \frac{Rcv_ij\omega}{Rcj\omega + 1}e^{j\omega t}$,
(c) 0.99995, 0.995, 0.707, 0.0995, 0.0099995,
0.001, $f_{hc} = 10^3/(2\pi)$.

14.8. (a) $x_1 = -(e^{-t}/10) - (e^{-11t}/90) - (e^{-2t}/9)$,
(b) $x_1 = (8e^{-5t/2}/15) - (5e^{-4t}/24)$
$-(13/40) + (1/2)t$.

14.9. (a) $y_n = 4(1/2)^n - 3$
(b) $y_n = 49(-1/4)^n/25 + (1/25) + (n/5)$
(c) $y_n = (-22 - 49n/3)(0.6)^n + 25$
(d) $y_n = (767/390)(0.2)^n - (209/130)(-0.8)^n$
$+(25/39)(0.5)^n$.
(e) $y_n = (3/5)(-(1/2))^n + (2/5)\cos(\pi n/2)$
$-(1/5)\sin(\pi n/2)$.

Chapter 15

15.1. (a) $\mathrm{Re}(s) > 4$, (b) $\mathrm{Re}(s) > -2$, (c) $\mathrm{Re}(s) > 0$,
(d) $\mathrm{Re}(s) > 0$.

15.2. (a) $15/(s^2 + 9)$, (b) $s/(s^2 + (1/4))$, (c) $8/s^5$,
(d) $1/(2(s + 5))$.

15.3. (a) e^{4t}, (b) $3e^{-t}$, (c) $4u(t)$, (d) $2\sin(\sqrt{3}t)/\sqrt{3}$,
(e) $(4/9)\cos((2/3)t)$,
(f) $t^2/2$, (g) $5t^3/6$.

15.4. (a) $1/(s+3)^2$, (b) $\dfrac{4}{(s+2)^2+1}$, (c) $\dfrac{8s}{(s^2-16)^2}$,

(d) $\dfrac{e^{-s}}{s}$, (e) $\dfrac{e^{-\pi s}}{s^2+1}$.

15.5. (a) $(1/3)e^{2t}\sin(3t)$, (b) $e^{-4t}\cos(t)$,
(c) $e^{-(t-2)}u(t-2)$, (d) $e^t\cos(3t)$.

15.6. (a) $u(t)-e^{-t}$, (b) $(1/2)u(t)-(1/2)\cos(\sqrt{2}t)$,
(c) $-(1/3)\cosh(2t)+(2/3)\sinh(2t)+(1/3)e^{-t}$,
(d) $-(2/5)\cos(t)-(6\sin(t)/5)+2e^{3t}/5$.

15.7. (a) $3e^{-5t}$, (b) $-1+t+e^{-t}$,
(c) $19\sin(2t)/8+(9t/4)$, (d) $4e^{2t}+4te^{2t}-7e^t$,
(e) $\cos(t)-3\sin(t)+t$,

(f) $2-2e^{-t/2}\cos\left(\dfrac{\sqrt{7}}{2}t\right)-\dfrac{2}{\sqrt{7}}e^{t/2}\sin\left(\dfrac{\sqrt{7}}{2}t\right)$.

15.8. $V_0 c\,e^{-t/RC}$.

15.9. $\frac{1}{2}e^{-t}-e^{-2t}+\frac{1}{2}e^{-3t}$.

15.10. (a) $(3u(t)/4)-(3e^{-4t}/4)$,
(c) (i) $-(4/3)-(s/(s+2))-[2s/(3(s+3))]$,
(ii) $-2e^{-2t}-e^{-3t}$.

15.11. $0.08\delta(t)+t$

15.12. (a) $(-17-20j)e^{j5t}/689$,
(b) $(-17\cos(5t)+20\sin(5t))/689$,
$(-20\cos(5t)-17\sin(5t))/689$.

15.13. (a) All z, (b) $|z|>1$, (c) $|z|>3$.

21.14. (a) $\dfrac{2z}{z-1}-\dfrac{z}{2(z-1)^2}$

(b) $\dfrac{z^2-z\cos(3)+2z\sin(3)}{z^2-2z\cos(3)+1}$

(c) $\dfrac{4z}{z-0.2}-\dfrac{6z}{z-2}$, (d) $\dfrac{2z}{z-e^{j4}}$.

15.15. (a) $2u_n+2\delta_n$, (b) $\frac{1}{2}n+(-3)^n$,
(c) $(3/2)^n+2(-3/2)^n$,
(d) $\cos(n)+\sin(n)$.

15.16. (a) $\dfrac{z^3}{z-3}-(z^2+3z+9)$, (b) $\dfrac{z}{2(z-(\frac{1}{2}))^2}$,

(c) $\dfrac{ze^j}{(z-e^j)^2}$, (d) $\dfrac{z^3+4z^2+z}{(z-1)^4}$.

15.17. (a) $n4^{n-1}$, (b) $2^n e^{j4n}$, (c) $n(0.4)^{n-1}$, (d) $2^{n+2}-2$.

15.18. (a) $u_n 2+(-1)^n/2$,
(b) $-u_n+(-\sqrt{2})^n/(2+\sqrt{2})+(\sqrt{2})^n/(2-\sqrt{2})$,
(c) $-(0.1)^n/15+(0.4)^n/15$,
(d) $2u_n/9+[n/3]-2(-2)^n/9$,
(e) $(j)^n/2+(-j)^n/2$.

15.19. (a) $3(-5)^{n+1}$ (b) $(1/4)u_n+(1/2)n-(1/4)(-1)^n$
(c) $(9u_n/5)+16(-4)^n/5$,
(d) $17u_n-32(2)^n+8n(2)^n$,
(e) $(1/2)u_n+(1/2)n-(1/2)\cos(\pi n/2)$,
(f) $(1/3)u_n+(1/2)^n/14-11(-1/5)^n/105$.

15.20. $3\frac{3}{4}(0.3)^n-(0.1)^n+5(-0.5)^n/4$.

15.21. (a) $5u_n-4(0.8)^n$ (c) $\dfrac{z-1}{24(z-0.2)}-\dfrac{5}{24}$

(ii) $(5(0.5)^n/6)+2(0.2)^n/3$

15.22. $\delta_n+0.5n$

15.23. (a) $\dfrac{e^{j5n(n+1)}}{10e^{j5}-3}$,

(b) $\dfrac{\cos(5(n+1))(10\cos(5)-3)+\sin(5(n+1))\sin(5)}{(10\cos(5)-3)^2+\sin^2(5)}$,

$\dfrac{\sin(5(n+1))(10\cos(5)-3)-\cos(5(n+1))\sin(5)}{(10\cos(5)-3)^2+\sin^2(5)}$.

Chapter 16

16.1. $\dfrac{1}{4}+\displaystyle\sum_{n=1}^{\infty}\left(\dfrac{\sin(n\pi/2)\cos(n\pi t/2)}{n\pi}\right.$

$\left.+\dfrac{(1-\cos(n\pi/2))\sin(n\pi t/2)}{n\pi}\right)$

$\dfrac{1}{4}+\displaystyle\sum_{\substack{n=-\infty\\n\neq 0}}^{\infty}\left(\dfrac{\sin(n\pi/2)}{2n\pi}+j\dfrac{\cos(n\pi/2)-1}{n\pi}\right)e^{jn\pi t/2}$.

16.2. $\dfrac{1}{6}+\displaystyle\sum_{n=1}^{\infty}\left(\dfrac{2}{n^2\pi^2}(-1)^n\cos(n\pi t)\right.$

$\left.+\left(\dfrac{-(-1)^n}{n\pi}+\dfrac{2((-1)^n-1)}{n^3\pi^3}\right)\sin(n\pi t)\right)$

$\dfrac{1}{6}+\displaystyle\sum_{\substack{n=-\infty\\n\neq 0}}^{\infty}\left(\dfrac{(-1)^n}{n^2\pi^2}\right.$

$\left.+j\left(\dfrac{(-1)^n}{2n\pi}-\dfrac{((-1)^n-1)}{n^3\pi^3}\right)\right)e^{jn\pi t_n}$.

16.3. $\frac{1}{2}+\frac{1}{2}\cos(2t)$ $\frac{1}{2}+\frac{1}{4}e^{j\pi t}$

16.4. $\dfrac{1}{4}+\displaystyle\sum_{n=1}^{\infty}\left(\left(\dfrac{(-1)^n-1}{n^2\pi^2}\right)\cos(n\pi t)\right.$

$\left.-\dfrac{(-1)^n}{n\pi}\sin(n\pi t)\right)$

$\dfrac{1}{4}+\displaystyle\sum_{\substack{n=-\infty\\n\neq 0}}^{\infty}\left(\dfrac{(-1)^n-1}{2n^2\pi^2}+\dfrac{j(-1)^n}{2n\pi}\right)e^{jn\pi t}$

16.5. $\dfrac{1}{2}+\displaystyle\sum_{n=1}^{\infty}\left(\left(\dfrac{2((-1)^n-1)}{n^2\pi^2}\right)\cos(n\pi t)\right.$

$\left.-\dfrac{4((-1)^n-1)}{n^3\pi^3}\sin(n\pi t)\right)$

$\dfrac{1}{2}+\displaystyle\sum_{\substack{n=-\infty\\n\neq 0}}^{\infty}\left(\dfrac{(-1)^n-1}{n^2\pi^2}+j\left(\dfrac{2(1-(-1)^n)}{n^3\pi^3}\right)\right)e^{jn\pi t}$

16.6. $0 + \frac{1}{2}\cos(2\pi t) + \sum_{\substack{n=1 \\ n \neq 2}}^{\infty} \frac{n(1-(-1)^n)}{\pi(n^2-4)}\sin(n\pi t)$

16.7. $0 + \frac{1}{4}(e^{j2\pi t} + e^{-j2\pi t})$

$+ \sum_{\substack{n=-\infty \\ n \neq 0,2,-2}}^{\infty} \frac{nj((-1)^n - 1)}{2\pi(n^2-4)}e^{jn\pi t}$

$0 + \sum_{n=1}^{\infty} 2\left(\frac{((-1)^n - 1)}{n^2\pi^2}\cos(n\pi t)\right)$

$0 + \sum_{\substack{n=-\infty \\ n \neq 0}}^{\infty} \frac{((-1)^n - 1)}{n^2\pi^2}e^{jn\pi t}$

16.8. (a) Odd, (b) Even, (c) Odd.

16.9. (a) $2\pi, c_1 = 1, c_3 = 1, c_n = 0$ for all other n,
$\phi_1 = \pi/2, \phi_3 = -\pi/2, \phi_n = 0$ for all other n,
(b) $\pi, c_1 = 1/\sqrt{2}, c_n = 0$ for all other n,
$\phi_1 = \pi/4, \phi_n = 0$ for all other n,
(c) $2\pi, c_n = 2/n^2, n \neq 0, c_0 = 0, \phi_n = 0$ for all n.

16.10. (16.1)

$c_n = \begin{cases} 0 & \text{when } n = 4m \text{ for some } m \\ \sqrt{2}/n\pi & \text{when } n = 4m + 1 \text{ or} \\ & n = 4m + 3 \text{ for some } m \\ 2/\pi n & \text{when } n = 4m + 2 \text{ for some } m \end{cases}$

$\phi_n = \begin{cases} 0 & \text{when } n = 4m \text{ for some } m \\ -\pi/4 & \text{when } n = 4m + 1 \text{ for some } m \\ -\pi/2 & \text{when } n = 4m + 2 \text{ for some } m \\ -3\pi/4 & \text{when } n = 4m + 3 \text{ for some } m \end{cases}$

(16.3) $c_1 = 1/2, c_n = 0$ for all other $n, \phi_n = 0$ for all n.
(16.7)

$c_n = \begin{cases} 0 & \text{when } n \text{ is even} \\ 4/(\pi n)^2 & \text{when } n \text{ is odd} \end{cases}$

$\phi_n = \begin{cases} 0 & \text{when } n \text{ is even} \\ \pi & \text{when } n \text{ is odd} \end{cases}$

16.12. $(1/4) + (1/8)(\cos(2t) + \sin(2t))$.

Chapter 17

17.1. $\partial f/\partial x = \ln(y^2), \partial f/\delta y = 2xy/y^2$.

17.2. (a) $\partial u/\delta x = -2yx/(x^2 + y^2)^2$,
$\partial u/\partial y = -x^2 - y^2/(x^2 + y^2)^2$;
(b) $\partial u/\delta x = 2x - y/x^2, \partial u/\partial y = 2y + (1/x)$

17.3. (a) 3, (b) 0, (c) 0, (d) 1, (e) 3/8.

17.4. $\partial z/\partial r = 2t, \partial z/\partial \theta = 0$.

17.5. $6 + 24t$

17.6. $\partial u/\partial t = -8\,e^{-8t}(A\cos(2x) + B\sin(2x))$,
$\partial^2 u/\partial x^2 = e^{-8t}(-4A\cos(2x) - 4B\sin(2x))$.

17.7. 10.048.

17.8. 2.315×10^{-4}.

17.9. 3%.

Chapter 18

18.1. (a) $(2x, -z, -y + 15z^2)$, (b) $(0,0,0.26)$,
(c) $\nabla\cdot\mathbf{F} = y + 3xz - 1$, (d) $(0,3,4)$,
(e) $\mathbf{k}(2xy^2z - z)$, (f) $6xy^2z^2 + 8x^3yz + 2x^2y^2$,
(g) maximum slope is 4 in direction $(0,0,1)$.

18.3. (a) -2, (b) $13\frac{1}{3}$.

18.4. $31\frac{1}{2}$.

18.5. (a) 0, (b) 0, (c) 2.644.

18.6. (a) $(a^4/2) - (a^3/2)$, (b) $-8/3$, (c) 1/12.

Chapter 19

19.1. (a), (e), and (g) are isomorphic; (d) and (i) are isomorphic.

19.3. K_n is planar for $n \leqslant 4$.
K_{mn} is planar for $m \leqslant 2$ or $n \leqslant 2$.

19.4. (a)

Incidence matrix   Adjacency matrix

$\begin{pmatrix} 1 & 0 & 0 & 1 & 0 & 1 \\ 1 & 1 & 1 & 0 & 0 & 0 \\ 0 & 1 & 0 & 0 & 0 & 0 \\ 0 & 0 & 1 & 1 & 1 & 0 \\ 0 & 0 & 0 & 0 & 1 & 1 \end{pmatrix}$ $\begin{pmatrix} 0 & 1 & 0 & 1 & 1 \\ 1 & 0 & 1 & 1 & 0 \\ 0 & 1 & 0 & 0 & 0 \\ 1 & 1 & 0 & 0 & 1 \\ 1 & 0 & 0 & 1 & 0 \end{pmatrix}$

$\begin{pmatrix} 1 & 1 & 1 & 0 & 0 & 0 \\ 1 & 0 & 0 & 1 & 0 & 1 \\ 0 & 1 & 0 & 0 & 1 & 1 \\ 0 & 0 & 1 & 1 & 1 & 0 \end{pmatrix}$ $\begin{pmatrix} 0 & 1 & 1 & 1 \\ 1 & 0 & 1 & 1 \\ 1 & 1 & 0 & 1 \\ 1 & 1 & 1 & 0 \end{pmatrix}$

$\begin{pmatrix} 1 & 1 & 0 & 0 & 0 & 0 \\ 0 & 0 & 1 & 1 & 0 & 0 \\ 0 & 0 & 0 & 0 & 1 & 1 \\ 1 & 0 & 1 & 0 & 1 & 0 \\ 0 & 1 & 0 & 1 & 0 & 1 \end{pmatrix}$ $\begin{pmatrix} 0 & 0 & 0 & 1 & 1 \\ 0 & 0 & 0 & 1 & 1 \\ 0 & 0 & 0 & 1 & 1 \\ 1 & 1 & 1 & 0 & 0 \\ 1 & 1 & 1 & 0 & 0 \end{pmatrix}$

(c) (3 3 1 3 2), (3 3 3 3), (2 2 2 3 3)
These represent the degrees of the vertices.

19.5. A minimum spanning tree has total weight 21. There are more than one.

19.6. The shortest path is sbdt. It has length 15.

19.7. Maximum flow is 18.

19.8. Maximum flow is 6.

19.10. (a) Path SB, BB, BB, BH accepted; (b) Path begins SA, then no more legal moves. Not accepted;
(c) Path begins with arcs SA, AH. No more legal

moves and string is not exhausted. Not accepted.

Chapter 20

20.1. $L = \{a^n bc | n \geqslant 2\}$

20.3.   (a) S $\Rightarrow$ aSa $\Rightarrow$ aaa
(b) S $\Rightarrow$ aSa $\Rightarrow$ abSba $\Rightarrow$ abbSbba $\Rightarrow$ abbcbba

20.4.   S ::= A"b"   A ::= aC*   C ::= "c"|"a"

20.6.
```
<marks>
    <subject name="Electronics 1">
        <student id="121">
            <semester1>40</semester1>
            <semester2>58</semester2>
        </student>
        <student id="122">
            <semester1>35</semester1>
            <semester2>38</semester2>
        </student>
        <student id="123">
            <semester1>75</semester1>
            <semester2>65</semester2>
        </student>
    </subject>
    <subject name="Engineering Maths">
        <student id="121">
            <semester1>66</semester1>
            <semester2>54</semester2>
        </student>
        <student id="122">
            <semester1>54</semester1>
            <semester2>45</semester2>
        </student>
        <student id="123">
            <semester1>72</semester1>
            <semester2>22</semester2>
        </student>
    </subject>
    <subject name="Communications">
        <student id="121">
            <semester1>58</semester1>
            <semester2>60</semester2>
        </student>
        <student id="122">
            <semester1>44</semester1>
            <semester2>55</semester2>
        </student>
        <student id="123">
            <semester1>70</semester1>
            <semester2>68</semester2>
        </student>
    </subject>
</marks>
```

Chapter 21

21.1.   99.5625, 8.9496

21.2.   1/3

21.3.   2/3

21.4.   2/3

21.5.   4/13

21.6.   1/12

21.7.   1/6

21.8.   3/51

21.9.   1/(10000)

21.10.   0.9362

21.11.   (a) 3/5, (b)3/5, (c) 3/10

21.12.   0.091

21.13.   (a) 0.06681, (b) 0.02275, (c) 0.00135, (d) 0.30854

21.14.   (a) 0.30854, (b) 0.02275, (c) 0.15866, (d) 0.22663

21.15.   (a) 0.99018, (b) 0.74751, (c) 0.63056, (d) 0.63056, (e) 0.26112, (f) 0.73769

21.16.   17.8%

21.17.   (a) 0.699, (b) 0.139, (c) 0.192, (d) 0.273

21.18.   8.858, the councillor is correct.

21.19.   (a) 0.512, (b) 0.2048, (c) 0.879

21.20.   Model probabilities: 0.2231, 0.3347, 0.2510, 0.1255, 0.0471, 0.0141, 0.0045;
Model frequencies: 22, 33, 25, 13, 5, 1, 1
There is good agreement between the model and the actual number of incidents.

21.21.   0.5402

21.22.   (a) 0.4823, (b) 0.3614, (c) 0.1563, (d) 0.6936

Index